Mechanical Tests for Bituminous Mixes

Characterization, Design and Quality Control

Other RILEM Proceedings available from Chapman and Hall

Publisher's Note
This book has been produced from camera ready copy provided by the individual contributors. This method of production has allowed us to supply finish copies to the delegates at the Symposium.

Mechanical Tests for Bituminous Mixes

Characterization, Design and Quality Control

Proceedings of the Fourth International Symposium held by RILEM (The International Union of Testing and Research Laboratories for Materials and Structures) and organized by RILEM Technical Committee 101-BAT (Bitumen and Asphalt Testing) in cooperation with the Hungarian Chemical Society, the Scientific Society for Transport, the Chair for Road Construction of the Technical University of Budapest, Hungary.

Budapest
23–25 October, 1990

EDITED BY

H.W. Fritz

and

E. Eustacchio

CRC Press
Taylor & Francis Group
Boca Raton London New York

CRC Press is an imprint of the
Taylor & Francis Group, an **informa** business

A CHAPMAN & HALL BOOK

CRC Press
Taylor & Francis Group
6000 Broken Sound Parkway NW, Suite 300
Boca Raton, FL 33487-2742

First issued in paperback 2019

ISBN-13: 978-0-412-39260-3 (hbk)
ISBN-13: 978-0-367-86346-3 (pbk)

British Library Cataloguing in Publication Data
Mechanical tests for bituminous mixes.
   1. Hydrocarbons
   I. Fritz, H. W.   II. Eustacchio, E.
   547.01

Library of Congress Cataloging-in-Publication Data
Available

Visit the Taylor & Francis Web site at
http://www.taylorandfrancis.com

and the CRC Press Web site at
http://www.crcpress.com

Scientific Committee

H.W. Fritz*	*Swiss Federal Laboratories for Materials Testing and Research (EMPA), Dübendorf, Switzerland (Chairman)*
C.D.M. Beverwijk	*Koninklyke Shell Laboratorium, Amsterdam, Netherlands*
K.E. Cooper	*Department of Civil Engineering, University of Nottingham, UK*
M. Csicsely*	*General Directorate of Public Roads, Budapest, Hungary*
E. Eustacchio*	*Technical University of Graz, Austria*
U. Isaacson	*Division of Urban Road and Street Engineering, Lulea University of Technology, Sweden*
I. Ishai	*Faculty of Civil Engineering, Technion Israel Institute of Technology, Haifa, Israel*
M. Luminari*	*Autostrade Spa, Rome, Italy*
L. Nagy*	*Central Research and Design Institute for Silicate Industry, Budapest, Hungary*
S. Pallotta	*Autostrade Spa, Rome, Italy*
G. Peroni	*Autostrade Spa, Rome, Italy*
G. Ramond	*Laboratoire Central des Ponts et Chaussées (LCPC), Paris, France (Secretary)*
B. Rubio	*Centro de Estudios de Carreteras, El Goloso, Madrid, Spain*
J. Sanchez-Caba	*Repsol Petroleo SA, Madrid, Spain*
A. Šolc*	*Gradevinski Institut, Zagreb, Yugoslavia*
D. Svetel	*Institut za Puteve, Belgrade, Yugoslavia*
R. Urban	*Arbeitsgemeinschaft der Bitumen, Industrie e.V., Hamburg, West Germany*
J. Verstaeten*	*Centre de Recherches Routières, Brussels, Belgium*
E. Nemesdy	*Technical University of Budapest, Hungary (Consultant)*

* Members of the Organizing Committee
I. Pallos, Technical University of Budapest, was also a member of the Organizing Committee.

Contents

Preface

The Fourth International Symposium is being organized by RILEM Technical Committee TC 101–BAT 'Bitumen and Asphalt Testing' and is entitled 'The Role of Mechanical Tests for the Characterization, Design and Quality Control of Bituminous Mixes'. Previous RILEM Technical Committees have organized the following international symposia and seminars dealing with bitumen and asphalt materials:

   1968 in Dresden and 1978 in Darmstadt on bitumen and its testing.
   1975 in Budapest and 1983 in Belgrade on asphalt materials and its testing.
   1986 Scientific Seminar held in Olivet, France on Mix Design and Quality Control for Bituminous Mixes.
   1988 Residential Seminar held in Dubrovnik, Yugoslavia on the theme Formulation, Control and Behaviour of Polymer-Bitumens for Waterproofing or for Road Construction.

As a follow up to these meetings, the Fourth International Symposium has the purpose of emphasizing the role of mechanical tests in the characterization, design and quality control of bituminous mixes. Its basic objectives are:

● to evaluate all information enabling the assessment of methods, their advantages and drawbacks and also the manner in which problems can be solved,
● to establish a better understanding and a documentation of results of research and practical engineering in order to establish a common ground between research, road authorities and contractors.

Thus a new way of thinking, that is, a new philosophy, is being advocated concerning the role of mechanical testing of asphalt materials.

   Traditional test methods help avoid large errors, but the significance of the results is very restricted. Fundamental methods are more complex; they yield better, more significant parameters but the application of these methods is limited to experts and special laboratories. Consequently, every effort should be made to develop tests having the advantages of both the traditional as well as modern methods and avoiding the restrictions of both.

   The particular aim of this symposium is to promote tests for the characterization, design and control of bituminous mixes, considering the needs of practice and science and the connection between them.

   Part One concerns specimen preparation. The production of representative specimens is one of the most important aspects of materials testing. In this part, experience relating to form and dimensions, production conditions and technological and morphological properties of test specimens is presented and discussed.

Part Two refers to tests with one-time loading. This subject concerns the traditional testing methods by one-time loading such as the Marshall test, uniaxial tension test, diametrical compression test, uniaxial creep test and other methods. Their purpose, for example, mix design or rapid control of mechanical properties, is discussed in these papers.

Part Three relates to tests with repeated loading. This subject concerns modern fundamental tests and other tests by repeated loading with the purpose of providing significant information on fatigue, permanent deformation and moduli especially for mix design. In these papers the newest research results are discussed.

Forty three papers from nineteen countries were selected for the Fourth International Symposium and appear in this volume of the Proceedings. It is intended that all prepared discussions will be printed in a further volume of Proceedings which will also include various addresses, presentations and the final conclusions of each session from the entire symposium.

H.W.FRITZ
President of RILEM TC 101–BAT
and the Fourth International Symposium

PART ONE
SPECIMEN PREPARATION

1 DUBAI EXPERIENCE IN DEVELOPMENT OF LARGE STONE ASPHALT MIXTURES

A.A. ELIAN
Dubai Municipality, Dubai, United Arab Emirates

Abstract
This report describes a development of new asphalt mix design procedure which is better indicator of road performance in Dubai. The Marshall method has served us well over the years but as wheel loads and tire pressures increase, and as temperature is high, there was a need for an update of mix design procedures. This new method allowed the incorporation of aggregate size larger than 25mm, and the use of vibratory compaction apparatus with 6" mold size diameter.
Results from the mix design study indicated that when compared to the conventional mix characteristics, the introduction of larger stone to the asphalt mixture improves its properties considerably.
Keywords: Marshall Mix Design, Large Stone, Vibration, Indirect Tensile Strength, Fatigue Prediction.

1 INTRODUCTION

During the last few years, there has been considerable increase in volume of heavy trucks with large axle loads and high tire pressure. The actual maximum single axle load can reach 25 to 30 tons, and the total predicted number of standard single axle load applications for trucks is about 150 to 160 psi. On the other hand, high temperature weather prevail in Dubai for most time of year. The ambient temperature is about 40 to 45 degree centigrade for few months of year, and measured temperature of pavement surface is about 80 - 85 degree centigrade. Although the structural design of pavement is quite satisfactory, after some of the highways were opened to traffic, they suffered from different types of distress such as fatigue cracking, bleeding and rutting. The method of asphalt mix design, which has been used for long time in Dubai, is Marshall method, which is currently used by many agencies all over the world. This method of mix design has many shortcomings such as:

3

1) The maximum size of aggregate is limited to 1 inch.
2) The method of compacting Marshall specimens is by impact, which is different from the one actually taking place in field by rollers.
3) The thickness of Marshall sample is almost constant, and it does not take into consideration the different layers thickness in field.
4) The aggregate orientation in Marshall mould is different from that actually occurring in field
5) Marshall mix design method does not take into consideration resistance of mix to many types of distress such as fatigue and rutting. Stability and flow are used which are measured at 60 degree centigrade. Therefore, the method ignores the effect of temperature.

In current roads contracts specification in Dubai, one of the following maximum aggregate size and method of mix design, is normally specified:

1) 25.4mm aggregate maximum size, with Marshall method of mix design.
2) 37.5mm aggregate maximum size, with Marshall method of mix design, but with substituting aggregate larger than 25.4mm with an equivalent weight of 19 - 25.4mm.
3) 37.5mm aggregate maximum size, with Marshall method of mix design, without substitution.

A lot of problems were faced during construction, one of which is the large difference between in-situ density and Marshall density and high variation in daily Marshall densities due to presence of large size aggregate. It is believed that using this large aggregate is the correct step in the right direction, but a proper method of mix design shall be used.

In this study, a new method of mix design was developed, which would suit available materials and local conditions. In this proposed method, most disadvantages of conventional methods were avoided and the new mix has better performance under heavy traffic and high temperature.

2   MATERIALS CHARACTERIZATION

2.1 Asphalt Cement Selection

One asphalt cement was selected for use in the study: AC 60/70 penetration grade supplied by Dubai Bitumen Supply Company.

The asphalt cement was subjected to a series of standard laboratory tests to determine its physical properties. These tests, with applicable standard test methods, were as follows:

```
Penetration              -   ASTM D5
Kinematic Viscosity      -   ASTM D-2170
Ring and Ball Softening  -   ASTM D-36
Point
Specific Gravity         -   ASTM D-70
```

Results of these tests are shown in Table 1

TABLE 1   PROPERTIES OF ASPHALT CEMENT

TEST	RESULT
Penetration at 25 degree centigrade	62
Viscosity at 135 degree centigrade (Cst)	460
Ring and Ball Softening Point (C°)	51
Specific Gravity	1.03

2.2 Aggregate Selection

2.2.1 The following aggregate sizes and sources were used in this work:

AGGREGATE SIZE (mm)		SOURCE	CRUSHER
37.5	-	28	HATTA	AL SHAALI
22	-	12	WADI FILLI	AL THANI
12	-	5	WADI FILLI	AL THANI
5	-	0	WADI FILLI	AL THANI

2.2.2 Aggregate Samples were tested for the following:

AGGREGATE TESTING	STANDARD TEST METHOD
Sand Equivalent	ASTM D-2419
Specific Gravity	BS 812 Part 2
Water Absorption	BS 812 Part 2
Abrasion	ASTM C 131
Crushed Faces	DM 300
Aggregate Crushing Value	BS 812
Flakiness Index	BS 812
Elongation Index	BS 812
Aggregate Gradation	BS 812
Soundness	ASTM C-88

Results of all these tests are shown in Table 2

TABLE 2   AGGREGATE PROPERTIES

	AGGREGATE SIZE (mm)			
	37.5 - 28	22 - 12	12 - 5	5 - 0
SAND EQUIVALENT	-	-	-	78%
APPARENT SPECIFIC GRAVITY	2.957	2.976	2.977	2.970*
CROSS SPECIFIC GRAVITY	2.926	2.952	2.951	2.921
SATURATED SURFACE	2.936	2.960	2.960	2.938
DRY SPECIFIC GRAVITY				
WATER ABSORPTION	0.35%	0.27	0.30%	0.56%
ABRASION	13%	14%	14%	
CRUSHED FACES	41%	65%	79%	
AGGREGATE CRUSHING VALUE	22%	15%	15%	
FLAKINESS INDEX	7%	15%	25%	
ELONGATION INDEX	16%	25%	29%	
SOUNDNESS	5%	3%	3%	7%
AGGREGATE GRADING				
PASSING PERCENT OF				
THE FOLLOWING SIZES				
50mm				
37.5mm	100			
28mm	41.1	100		
14mm		44.4	100	
6.3mm		0.1	33.1	100
3.35mm			0.8	93
0.300mm				28
0.075mm				9.5

* Specific Gravity values were determined on fine aggregate without filler   Relative Density of Filler = 2.918

3   EXPERIMENTATION

The main objective of this research was to develop new mix design procedure which is better indicator of road performance than existing method. In order to achieve this objective, the following laboratory work was conducted:

3.1 Marshall Method

Two mixtures were prepared using Marshall method of mix design as detailed in Asphalt Institute Manual MS-2. The aggregate grading of these two mixes are as follows:

SIEVE SIZE (mm)	GRADING LSA-3	GRADING LSA-8	SPECIFICATION LIMITS TABLE 17 BS 4987		
50	100	100		100	
37.5	100	100	95	-	100
28	83	78	70	-	94
14	65	65	56	-	76
6.3	49	49	44	-	60
3.35	41	41	32	-	46
0.300	13	13	7	-	21
0.075	3	3	2	-	8

Another two mixture designs were prepared using same grading as (LSA-3 & LSA-8) but with substituting aggregate larger than 25.4mm with the equivalent weight of 19 - 25.4mm (LSA-4 and LSA-12) respectively. Figure 1 shows grading of all these mixes together with specification limits.

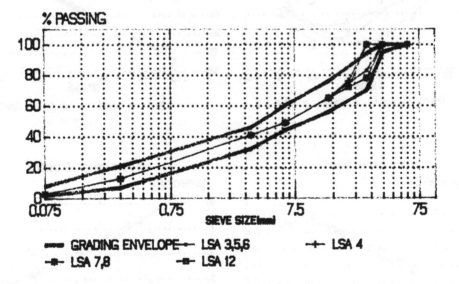

figure 1   LSA MIX DESIGNS

The following six Marshall design curves were plotted for each mix in order to determine the optimum binder content:

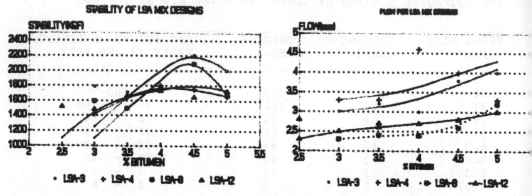

Fig. 2

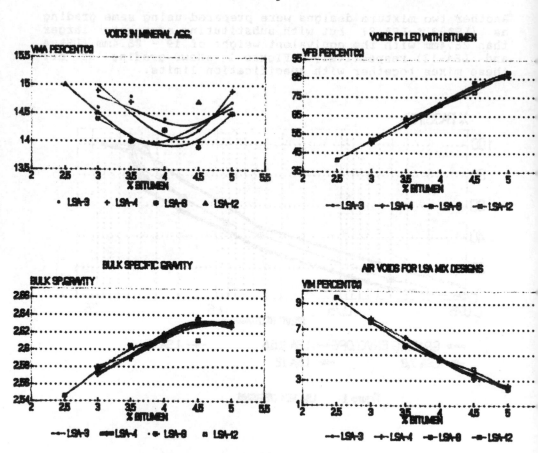

Fig. 3

The optimum binder content was determined for all these
mixtures, and the corresponding properties are shown in
Table 3.

TABLE 3 MARSHALL MIX DESIGN

	LSA-3	LSA-4	LSA-8	LSA-12
OPTIMUM BINDER CONTENT	4.4	4.25	4.5	4.2
BULK SPECIFIC GRAVITY AT OPTIMUM BINDER CONTENT	2.628	2.623	2.630	2.612
V.I.M. %	3.6	4.0	3	4.4
V.M.A. %	14.2	14.3	14	14.4
V.F.B. %	79	75	83	74
STABILITY (kgf)	2000	1810	2000	1966
FLOW (mm)	3.5	2.6	2.5	2.7

Marshall specimens were prepared at optimum binder content
for above mixes and then tested for indirect tensile
strength and fatigue prediction.

3.2 Development of New Mix Design Procedure

3.2.1 Background

Aggregate with maximum size larger than 1" cannot be used in
the 4" mold. Many agencies simply specify aggregate
quality, aggregate grading, and a design asphalt cement
content based on experience. These recipe - type methods
are difficult to adjust according to changes in traffic and
environment.
There is a modification to the Marshall procedure for large
stone sized mixtures by the U.S. Corps of Engineers. The
complete aggregate sample is mixed with the asphalt cement
and then screened through a 1" sieve. The +1" portion is
discarded and the balance is used to manufacture samples for
the Marshall test in the conventional manner. In
consideration of the design criteria, the stability and flow
values are left unadjusted; whereas, adjustments are made to
the measured voids according to the percentage and specific
gravity of the +1" material.

9

Pennsylvania DOT developed procedures for fabricating and testing 6" diameter samples. They have found that the Marshall stability of a 6" diameter was at least twice that of a 4" specimen and the corresponding flow was at least 1.5 times. Several agencies have developed procedure for 2" maximum stone size mixture where 11 lbs. of loose mix is compacted by a vibrating hammer in a 6" diameter mold. The sample is allowed to cool, extruded and then tested for air voids and voids in mineral aggregate.

In the Refusal test developed by the Transport and Road Research Laboratory in the United Kingdom, a compaction standard has been developed. The density of the oven - dried core is determined. It is then placed in a 6" mold and heated to 284 degree farenheit. The sample is then compacted to refusal using a 750 watt vibrating hammer for two minutes at each face.

A summary of all compaction methods for large stone mixtures is given in Table 4.

TABLE 4   COMPACTION PROCEDURES FOR SPECIMEN FABRICATION FOR HEAVY DUTY MIXTURES (1)

TEST PROCEDURE	COMPACTIVE EFFORT	MAXIMUM AGGREGATE - SIZE	MOLD SIZE DIAMETER
MARSHALL	75 BLOWS EACH END OF SPECIMEN, 10 lb. HAMMER, 18" FALL	1"	4"
MODIFIED MARSHALL PENNSYLVANIA DOT	75 BLOWS EACH END OF OF SPECIMEN 22.5 lb HAMMER, 18" FALL	1.5" - 2"	6"
GYRATORY TESTING MACHINE	CAN BE ADJUSTED ACCORDING TO ANTICIPATED TRAFFIC LEVEL	1" 1.5"	4" 6"
TRRL REFUSAL TEST	COMPACTED TO REFUSAL 2-4 MINS. USING 750 WAATT, 50Hz VIBRATING HAMMER, 4" TAMPING FOOT	1.5" - 2"	6"
MINNESOTA DOT	VIBRATING HAMMER-30 SECS. EACH FACE 5 7/8" TAMPING FOOT	1.5" - 2"	6"
CALIFORNIA KNEADING COMPACTOR	COMPACTED BY 3.1" IN COMPACTION RAM, 20 TAMPING BLOWS AT 250 psi FOLLOWED BY A STATIC LEVELING LOAD	1"	4"

3.2.2 Proposed New Method

A 750 watt 50 Hz vibrating hammer with 100mm and 145mm tamping feet were used in compacting 80mm thick asphalt specimen in 6" diameter mold. This assembly is similar to the one used in the Refusal test developed by the Transport and Road Research Laboratory in United Kingdom. The mixing and compacting temperatures were determined as per ASTM D1559. The mixture is then placed in the preheated mold. The 100mm tamping foot is moved around the mold giving a few seconds compaction at each of eight equispaced, diametrically opposite positions. The tamping foot is moved from one position to the next before material pushes above its edge. After required time of continuous compaction, a 145mm tamping foot is used to smooth irregularities on the surface of the specimen.

The mold is then turned over, the specimen is pushed down using the 145mm tamping foot and compacted for same time as before. After 24 hours, the specimen is removed from the mold. The specimen is tested for specific gravity as per ASTM D-2726. A 100mm diameter core sample is taken from the compacted specimen, which is then tested for stability, flow, indirect strength, and fatigue prediction.

Figures 4 & 5 show equipment and compaction of specimens.

COMPACTING THE SAMPLE BY VIBRATING HAMMER

DIFFERENT TAMPING FEET USED IN COMPACTION OF SAMPLES
(FIGURE 4)

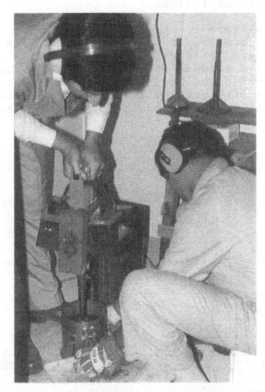

COMPACTING THE SAMPLE BY VIBRATING HAMMER

6" DIAMETER MOULD USED IN MAKING THE SAMPLES
(FIGURE 5)

13

Two asphalt mixtures (LSA-6 & LSA-7) were prepared using vibration method, as follows :

3.2.2.1 LSA-6 Mixture

The aggregate grading of this mix is the same as that of LSA-3 (figure 1) Different time of vibration ranging from 30 seconds to 240 seconds at different binder contents were used in preparation of specimens. These samples were then tested for specific gravity as per ASTM D-2726. Figure 6 shows relation between vibration time and bulk specific gravity.

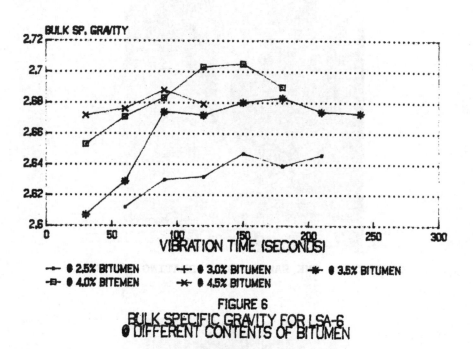

FIGURE 6
BULK SPECIFIC GRAVITY FOR LSA-6
@ DIFFERENT CONTENTS OF BITUMEN

Each point represents average of two samples. Air voids and voids in mineral aggregate were then determined at refusal (i.e. at maximum bulk specific gravity of each binder content). Figures 7 & 8 show relation between VIM and VMA at refusal with binder content.

14

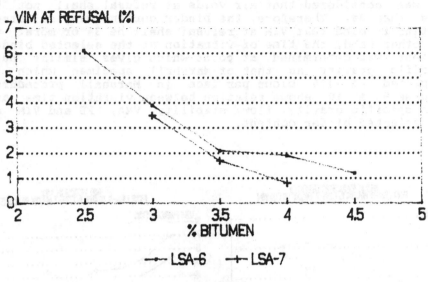

FIGURE 7
AIR VOIDS AT REFUSAL FOR
LSA-6 & LSA-7

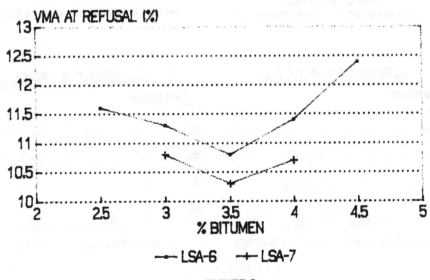

FIGURE 8
VOIDS IN MINERAL AGGREGATES AT REFUSAL
FOR LSA-6 & LSA-7

It was considered that air voids at refusal shall not be less than 3%. Therefore, the binder content was selected, putting in mind that VIM at refusal shall be 3% or more. On the other hand, the time of vibration at the selected binder content was determined, at point which gives similar bulk specific gravity as that of Marshall specimen which is compacted at 100 blows per face in Marshall procedure. Figures 9 & 10 shows relation between vibration time and bulk specific gravity, flow, stability, VMA, VFB and VIM at the selected binder content.

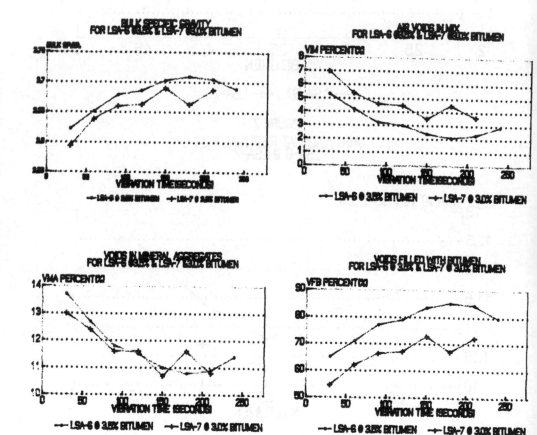

FIGURE 9

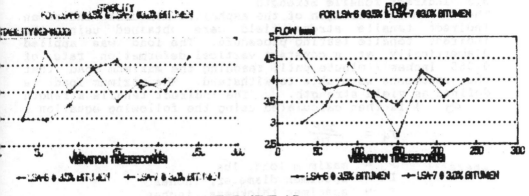

FIGURE 10

3.2.2.2 LSA-7 Mixture

The aggregate grading of this mix is the same as that of LSA-8 (figure 1) It should be mentioned that the difference between aggregate grading of LSA-6 and LSA-7 is in the percentage retained on sieve 25.4mm. In grading of LSA-6, the percent retained on sieve 25.4mm is 19%, while it is 23% in grading of LSA-7 mix. This is in order to examine the effect of increasing the retained on sieve 1" in both Marshall method and vibration method. A similar procedure was followed in LSA-7 mix as that of LSA-6 in determining binder content and time of vibration. Figures 7, 8, 9, 10 & 11.

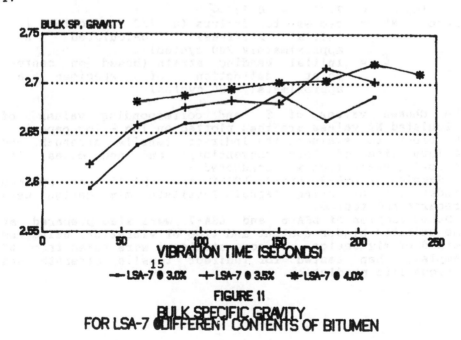

FIGURE 11
BULK SPECIFIC GRAVITY
FOR LSA-7 @DIFFERENT CONTENTS OF BITUMEN

3.3 Indirect Tensile Strength

The ultimate strength of the asphaltic mixtures under an indirect tensile stress field were obtained using the indirect tensile testing procedure. The load was applied diametrically at a constant vertical deformation rate of 0.065 inches / minute until reaching the maximum load that the specimens were able to withstand. The maximum load is defined as yield strength. The indirect tensile strength (σ_y) is then calculated using the following equation :

$$\sigma_y = \frac{2P}{\pi Dt}$$

where
P = maximum load, lbs
D = specimen diameter, inches
t = specimen thickness, inches

3.4 Fatigue Curve Prediction

Constant strain fatigue curve
The fatigue relation

$$N_f = K_2 \left(\frac{1}{\varepsilon}\right)^{n_2}$$

can be calculated from the following equations (According to Virginia Highway Transportation Research Council)

$$n_2 = 0.0374\,\sigma_y - 0.744$$

$$\log K_2 = 7.92 - 0.122\,\sigma_y$$

where
Nf = cycles to failure (a 1/3 reduction in initial stiffness calculated at approximately 200 cycles)
ε = initial bending strain (based on center point deflection of specimen at approximately 200 cycles)

The chosen values of ε and corresponding values of calculated Nf values are then plotted on log - log paper.
In order to evaluate the indirect tensile strength and fatigue life of both conventional and new mixes, the following evaluation was conducted :-
1 Samples of LSA-3, LSA-4, LSA-8 and LSA-12 at optimum binder content using Asphalt Institute mix design were prepared and tested.
2 Other samples of LSA-6 and LSA-7 were also prepared at the selected binder content and time of vibration, using new method of mix design. 4" core specimens were taken from 6" samples, then tested for indirect tensile strength and fatigue life prediction.

4   ANALYSIS OF RESULTS

4.1 Conventional Mix Design

Marshall method of mix design as per Asphalt Institute manual MS-2 was used in preparation of LSA-3 and LSA-8. The difference in grading between these two mixes is in the percentage retained on sieve 25.4mm. In grading of LSA-3 mix, the percent retained on sieve 25.4mm is 19 %, while it is 23% in grading of LSA-8. Table 3 shows optimum binder contents, and corresponding properties at these contents. By examining the mix design data of these two mixes, the following comments can be noticed :

1   The variation in bulk specific gravity at same binder content for LSA-8 was higher than that of LSA-3. This was reflected on air voids and VMA values.
2 It is expected that even higher variation in bulk specific gravity, and voidage contents, will be noticed in plant samples, due to the fact that unequal distribution of aggregate will occur in the 4 inches mold.
3   The Insitu specific gravity of 6 inches diameter core samples will normally show higher values than the 4 inches laboratory Marshall samples. Therefor, very high degree of compaction will be noticed.
Two other mixes were prepared (LSA-4 & LSA-12) in which aggregate larger than 25.4mm were substituted with the equivalent weight of 19-25.4mm. The optimum binder contents, and corresponding properties at these contents are shown in table 3. The following comments can be said on these mixes.
1 The variation in bulk specific gravity values at same binder content is much less than that of LSA-3 and LSA-8.
2 It is very difficult to do this substitution when making Marshall specimens from plant mixes. Therefore, Marshall specimens using laboratory mixes will be different from those made from plant. This will give different values of specific gravity and voidage contents even at same binder content.
3 The Insitu bulk specific gravity of core specimens, will vary from those of Marshall specimen, because they are of different grading.
4   The mix design does not actually represent the Insitu mix.

4.2 Mix Design by Vibration Method
Two mixes were prepared (LSA-6, LSA-7) using new method. These two mixes have aggregate with same grading as LSA-3 and LSA-8. Table 5 shows comparison between characteristics of large stone mixes by vibration method and by conventional method. The following comments can be mentioned on the large stone mixed, prepared by vibration:

1   The bulk specific gravity values are almost similar for two methods.
2   The variation in bulk specific gravity at   same binder   content   is   low   for   mixes   prepared   by vibration. This is due to large molds used.
3   The optimum binder contents of mixes prepared by vibration are lower than those of conventional methods
4   The VMA and VFB values are lower for mixes prepared by vibration than those prepared by conventional method.
5   The   air   voids   values   for   mixes   prepared   by vibration are higher than those prepared by conventional methods.
6   Stability and flow values are close for two method.
7   Stiffness:
According   to   shell   recommendation,   getting   high Marshall stability should not be our aim.   Also obtaining the flow and stability requirements separately is not the target.   The ratio of stability to flow is important and not their individual values.
This ratio gives a measure of what is termed the stiffness of the mix which can be related to tire pressure. In order to prevent permanent deformation for the mix under high stress the Marshall stability / flow ratio should not be less than 1.2 times the tire pressure. Shell procedure, for calculating stiffness is shown in table 5.
If we consider, the tire pressure to be 150 psi, then stiffness should not be less than.
$$1.2 \ x \ 150 \ = \ 180$$
All mixes, which are shown in Table 5   have stiffness higher than 180.   The mix LSA-6 in Table 5 gave the highest stiffness value.

4.3 Procedure of New Mix Design by Vibration
The following steps, which are proposed to be followed in designing large stone asphalt mixture.
1   The aggregate fractions are combined by different percentage to be within required envelope of grading.
2   By use of vibrating hammer and 6 inches mold, specimens are prepared at different time of vibration and different binder contents.
3 Draw curves, which shows relation between bulk specific gravity and time of vibration for different binder contents.
4   Air voids and voids in mineral aggregate at refusal (maximum bulk specific gravity) are determined for each binder content.
5   Draw two graphs showing relation between air voids and VMA at refusal versus binder content.
6 Choose   binder content, which gives minimum 3% of air voids at refusal. This binder content is the optimum.
7   If the Asphalt mixture does not look workable at the selected binder content, then change grading and repeat steps from 1 to 7.

TABLE 5   COMPARISON   BETWEEN   CHARACTERISTICS   OF   LARGE   STONE   MIXES
BY VIBRATION METHOD AND CONVENTIONAL METHOD

	LARGE STONE MIXES			
	VIBRATION	CONVENTIONAL	VIBRATION	CONVENTIONAL
	LSA 6	LSA 3	LSA 7	LSA 8
Gradation retained on 25.4mm	19%	19%	23%	23%
Optimum binder content %	3.5	4.4	3.0%	4.5%
Bulk specific gravity at optimum binder content	2.641	2.628	2.620	2.630
VIM %	4.5	3.6	6%	3%
VFB %	69	79	58	83
VMA %	13	14.2	13%	14
* Stability (Kgf)	2400	2000	1267	2000
Flow (mm)	3.7	3.5	3.4	2.5
Stiffness Kgf/mm 648	571	372	800	

*

In case of 6 inches samples, core specimens of 4 inches diameter were taken and tested
for stability and flow.

8   The time of vibration shall be determined, by determining
the bulk specific gravity, which correspond to the average
bulk density of three Marshall specimen compacted at 100
blows per face.

4.4   Indirect Tensile Strength and Fatigue Predicton
The samples, which were prepared as per clause 3.4 were
tested for indirect tensile strength, then fatigue life was
predicted according to procedure developed by Virginia
Highway Transportation Research Council. Table 6 and
figures 12, 13 & 14 show results of indirect tensile
strength and predicted fatigue curves for all these mixes.
LSA-6, which was prepared by vibration method, gave highest
value of indirect strength.

TABLE 6   INDIRECT TENSILE STRENGTH AND FATIGUE PARAMETERS

TYPE OF MIX	RETAINED ON SIEVE 25.4mm	METHOD OF MIX DESIGN	BY psi	N2	log K2	K2	
LSA 3	19%	Conventional	133	4.23	-8.306	4.94	E-09
*LSA 6	19%	Vibration	206	6.96	-17.212	6.14	E-18
LSA 8	23%	Conventional	144	4.64	-9.648	2.25	E-10
*LSA 7	23%	Vibration	181	6.03	-14.162	6.89	E-15
LSA 4	Zero	Conventional	174	5.76	-13.308	4.92	E-14
LSA 12	Zero	Conventional	188	6.29	-15.016	9.64	E-16

4 inches Core samples were taken from 6 inches specimens, then tested for indirect
tensile strength.

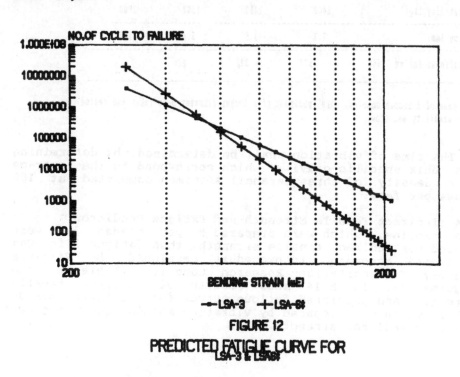

FIGURE 12
PREDICTED FATIGUE CURVE FOR
LSA-3 & LSA-6

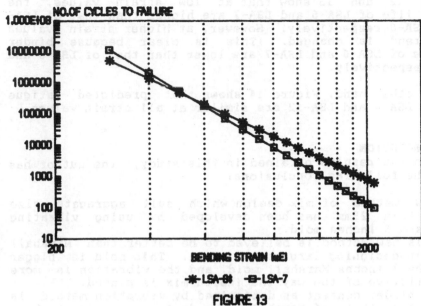

FIGURE 13

PREDICTED FATIGUE CURVE FOR
LSA-6# & LSA-7

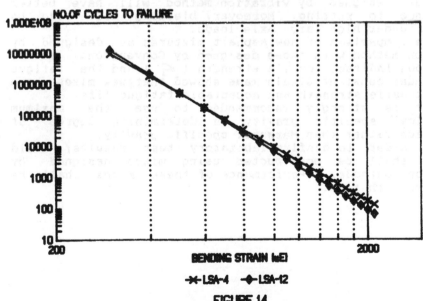

FIGURE 14

PREDICTED FATIGUE CURVE FOR
LSA-4 & LSA-12

Figures 12 and 13 show that at low strain values, the fatigue life of LSA-6 and LSA-7 are higher than that of LSA-3 and LSA-8 respectively. However, at higher strain values this trend is changed. This is clear because binder contents of LSA-6 and LSA-7 are lower than that of LSA-3 and LSA-8 respectively.

On the other hand, Figure 14 shows that predicted fatigue life of LSA-4 and LSA-12 are similar at all strain values.

5   CONCLUSION

Based on the results obtained in this study, the author has drawn the following conclusions:

5.1 New method of mix design which suit aggregate size larger than 25mm has been developed by using vibrating hammer and 6 inches mold.

5.2 This new method is believed to be better than Marshall method in designing large stone mixes. This mold is bigger than the 4 inches Marshall mold, and the vibration is more representative of the way the insitu mix is placed.

5.3 The binder content as determined by vibration method is lower than that of Marshall mix design as per Asphalt Institute MS-2

5.4 Due to low binder content, and high stiffness values, mixes as designed by vibration method will have better resistance to rutting. Moreover, bleeding will also be avoided, under high large axle loads.

5.5 The comparison of new asphalt mixtures as designed by vibration method with those designed by Conventional method based upon indirect tensile strength (σ_t) and the fatigue life under constant strain made showed that new mixes have higher tensile strength and acceptable fatigue life.

5.6 It is strongly recommended to use the maximum theoretical specific gravity in determining degree of compaction rather than Marshall specific gravity.

5.7 In order to confirm laboratory test results, field trials shall be constructed using mixes designed by vibration methods. Performance of these mixes shall be evaluated with time.

6 REFERENCES

1 Acott, M. "The Design of Hot Mix Asphalt for Heavy Duty Pavements." National Pavement Association, Riverdale, Maryland.
2 Acott, M. "Today's Traffic Calls for Heavy Duty Asphalt Mixes." National Asphalt Pavement Association, Riverdale, Maryland.
3 Ramsey, S. "The Percentage Refusal Density Test". The Journal of The Institution of Highways and Transportation, February 1985.
4 McNical, A., & Bell, G., "Air Voids Versus PRD". The Journal of the Institution of Highways and Transportation." February 1986.
5 "Design and Performance Study of A Heavy Duty Large Stone Hot Mix Asphalt Under Concentrated Punching Shear Conditions." National Asphalt Pavement Association, Riverdale, Maryland.
6 Powell, W.D., & Leech, D., "Standards For Compaction of Dense Roadbase Macadam." Transport and Road Research Laboratory, Crowthorne, Berkshire, 1982.

ACKNOWLEDGMENTS

The author would like to express his appreciation to technicians of Asphalt Sub unit for their help in conducting the experiments.

Thanks are also due to Mr. Khalil Hamdi for drawing the graphs, and for Mr. D. C. Fernandes for typing the manuscript.

2 COMPACTION OF ASPHALT FOR LABORATORY TESTING

R. LETSCH and M. SCHMALZ
Prüfamt für bituminöse Baustoffe und Kunststoffe,
Technical University Munich, Germany

Abstract

Asphalt concretes on roads and bridges are compacted by rollers with and without vibration. The compaction of the asphalt mixes in the laboratory is in general done by impact. The Marshall test is up to now the main criterion for the design of asphalt mixes in many countries (e.g.'Germany DIN 1996). Using a Proctor hammer discs of 15 cm diameter of compacted asphalt can be produced which are suitable for permeability tests or for obtaining prisms for bending tests. Depending on the composition of the asphalt mixes, the way of compaction has a decisive influence on the properties of the mix. For the compaction of asphalt mixes a steel roller with vibrator was constructed. The weight, the speed and the vibrating force of the roller can be varied in a wide range to achieve the required compaction. By exchanging the roller with two rubber-lined wheels rutting tests can be performed.

Keywords: Asphalt Mix, Compaction, Vibrating Roller, Laboratory, Rutting

1 Introduction

The compaction of asphalt concretes on roads and bridges is in general done by rollers with and without vibration. The work required for the complete compaction depends on the properties of the mix, the temperature and the thickness of the layer. For laboratory testing the compaction is in general performed by impact e.g. by the Marshall hammer as it is described in DIN 1996. The Gyrator test is another method for determining the compactibility of an asphalt concrete mix /1,2/. The measure for the compaction is the void content, e.g. determined according to DIN 1996. Depending on the composition of the asphalt mix, the way of compaction can have a decisive influence on the properties of the mix.

Figure 1 shows the principles of the compaction by the Marshall hammer in comparison to a vibrating roller. By the second method not only a vertical but also a rotating movement of the aggregate is possible with it's effect on

26

the compaction and the density of the mix.

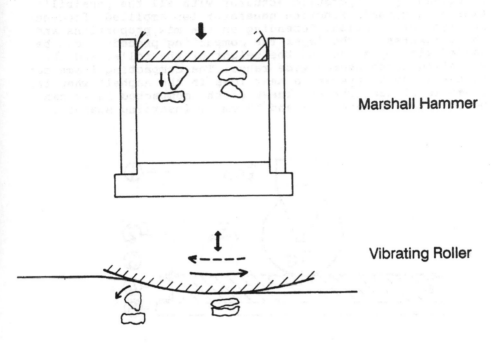

Marshall Hammer

Vibrating Roller

Fig. 1: Compaction by Marshall hammer and by vibrating
roller

In the laboratory very often specimens have to be
prepared for determining the tensile strength, the water
impermeability or the rutting depth. Such specimens of
compacted asphalt cannot be obtained from Marshall or
Gyrator specimens. Using a Proctor hammer (DIN 18127) discs
of 15 cm diameter are produced which can be suitable for
permeability tests or for obtaining prisms for bending
tests. Rutting tests have to be performed on slabs of at
least 25 * 35 cm² up to full scale pavements /3/.
In the past two years the problem has arisen how to
compact asphalts on waterproofing layers which were applied
to specially prepared concrete slabs. A special type of a
waterproofing layer is the so-called "sand asphalt". This
sand asphalt has to be roller compacted to give a layer
with a thickness of about 12 mm.

2 Roller Compaction

To meet these problems a roller compacting machine as it
is shown in figure 2 was constructed. It consists of a
steel roller which can be heated up to about 80°C and of a

vibrator which can be turned on independantly. The roller
is driven by an hydraulic actuator with all the possibilities of a dynamic function generator for amplitude,freqency
and type of function. Depending on the mix proportions and
the thickness of the layer the compacting procedure can be
varied also by adjusting the weight of the roller and the
vibrating force over a wide range. The compacting frame can
be heated up to prevent a heat loss in the asphalt when the
frame is filled. The thickness of the compacted layer can
be adjusted between 1 cm and 12 cm by inserting spacers.

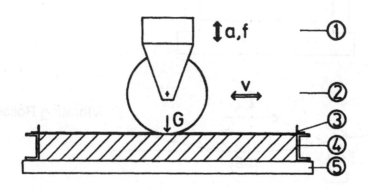

```
1   vibrator:  f = 50Hz,  a = 0.01 to 0.2 mm
2   roller:    G = 2 to 6 kg/cm,  v = 0.01 to 2 m/s
3   filling frame
4   compacting frame: 36*60 cm²
5   base plate
```

Fig. 2: Vibrating roller for compaction of asphalt

Tests conducted with different asphalt concretes
show that an optimum compaction may be reached at a certain
thickness of the compacted layer (see figure 3). The
movement of the roller corresponded to a sine-wave, this
means that the speed was highest in the middle of the 60 cm
long slab with the slowing down towards the ends. The
compaction of the asphalt concrete was therefore not
constant over the length of the slabs as it is shown in
figure 4. The greatest compaction was not at the ends of
the slabs where the time of loading by the roller was
longest but about 5 to 10 cm to each side of the middle of
the slabs, with the least compaction at the ends.

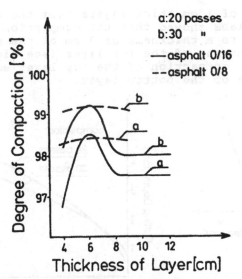

Fig. 3: Relation between degree of compaction according to
Marshall (DIN 1996) and the thickness of compacted
asphalt layers for 20 and 30 passes of a vibrating roller

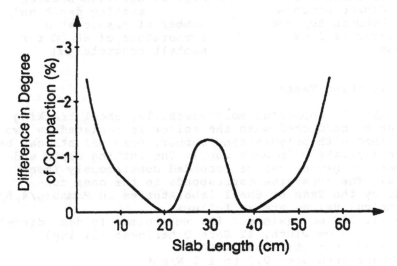

Fig. 4: Difference in degree of compaction over the length
of 6 cm thick slabs of asphalt concrete

The flexural strength of an aspalt concrete is greatly
influenced by the strength of the outermost layers and
therefore by the compaction of these layers. Measurements

of the void content of 1 cm thick layers from the top and
the bottom of the slabs showed that the compaction of the
bottom layer was up to a thickness of 8 cm of the compacted
slab greater than the one of the top layer (see figure 5).
At thicker slabs the compaction of the top layer was
greater than the one of the bottom layer.

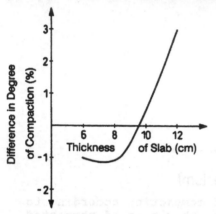

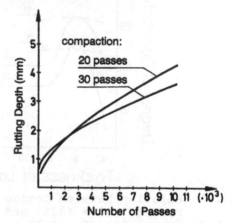

Fig. 5: Difference in
degree of com-
paction between top and
bottom layer of 1 cm
thickness

Fig. 6: Relation between
rutting depth and
number of passes at a
temperature of 40 °C for
asphalt concrete 0/8

3 Wheel Tracking Tests

To make the apparatus more versatile, wheel tracking
tests can be conducted when the roller is replaced by two
wheels lined with polyurethane-rubber. Dead weights can be
attached separately to each wheel. The rutting depth and
the number of passes can be recorded continuously (see
figure 6). The apparatus corresponds to the ones constructed by the Esso or Shell laboratories in Hamburg/4,5/.
The technical values are as follows:
    Wheel: 200 mm maximum / 194 mm minimum (sides) diameter
           50 mm width, 42 Sore D hardness (lining)
    Area of contact: about 1000 mm²
    Contact pressure: 0.2 to 1.2 N/mm²
    Length of rutting: maximum 26 cm for 30 cm slabs
    Speed: maximum 0.5 m/s (0.75 Hz)

4 Waterproofing layers

Tests with sand asphalt, which is used beside other
materials as waterproofing layer on concrete bridges in
Germany, show big differences related to the watertightness

depending on the kind of compaction. Specimens with 12 mm thickness compacted with the Proctor hammer were water permeable at void contents greater than 3 %, while specimens compacted with the vibrating roller were water impermeable at void contents below 4.5 %. The water pressure was 1 bar, the test duration 24 hours. In practice sand asphalt with a void content of 5 % or less has proven to be water impermeable.

Roller compacted asphalts on waterproofing layers of bitumen sheetings exhibit a very good contact to the sheetings with about 100 % compaction in the contact zone. This was also demonstrated by using coloured asphalt concrete to give a distinct contact line.

These tests were conducted in relation to the "ZTV-BEL-B" in which requirements and testing procedures for waterproofing layers on concrete bridges in Germany are defined /6/.

References

/1/ Motz, H.: Laboratory Investigation to Assess the Compactibility and the Resistance to Permanent Deformation of Asphaltic Concretes by Static and Dynamic Test Methods. Third International Eurobitumen Symposium, Den Haag 1985.
/2/ Gauer, P.: Eine Analyse der Verdichtungswilligkeit und des Verformungswiderstandes von bituminösen Mischgut bei der Verdichtung im Gyrator. Dissertation, Darmstadt 1975.
/3/ Kast, O.; von der Weppen, W.: Beurteilung der Standfestigkeit von Asphaltbeton-Deckschichten mit Hilfe des Spurbildungstestes. Bitumen 40(1978) pp 180 - 183
/4/ Mitteilung der Shell AG: "Caribit", Deutsche Shell AG Hamburg, 1985
/5/ Hilmer, A.: Einfluß der Radlasten und Reifeninnendrücke auf die Spurrinnenbildung an Asphaltstraßen. Mitteilungen des Prüfamtes für Bau von Landverkehrswegen der Technischen Universität München, Heft 43, 1984
/6/ Zusätzliche Technische Vorschriften und Richtlinien für die Herstellung von Brückenbelägen auf Beton (ZTV-BEL-B). Der Bundesminister für Verkehr, Verkehrsblatt-Verlag, Dortmund, 1987

3 COMPACTION IN THE LABORATORY AND IN PRACTICE

P. RENKEN
Institut für Straßenwesen, Technische Universität Braunschweig,
Germany

1 Introduction

There is a great number of material tests in existance
showing a high degree of technical achievement and precision. Such tests are nevertheless not helpfull in improving the mechanical properties of coated material, as
compaction under laboratory conditions differs greatly
from compaction outside. The specimens for these tests
must have the same properties as those of the ready compacted layer in the field. Unfortunately formerly coated
materials being compacted in the marshall compactor often
have been characterized stiffer than in the roller compacted layer in the road. Moreover, the dimensions of the
small marshall specimens are really not suitable for most
of the important material tests. To avoid this defect it
is necessary to get knowledge about the difference between
the compaction in the laboratory and in the field such as
the resulting mechanical properties of the asphalt concrete. Therefore an extensive program of investigation
was started comparing systematically the mechanical properties of cores taken from the road and specimens which
have been prepared under laboratory conditions.

2 Method of the investigation

In this program of investigation there have been examined
three different kinds of coated materials for wearing
course, four different kinds of compaction in the field
and three different kinds of compaction in the laboratory.
For each of these 21 steps of varation finally specimens
were prepared. These specimens were taken to test some
important mechanical material properties, namely wheel-
tracking-tests, dynamic bending tests and tests to examine the characteristics of the coated material under
conditions of low temperatures, cooling down tests and
tension tests.

2.1 Variants of the coated material

Looking for suitable variants of the coated material it was necessary to confine to asphalt concrete mixture to save money. Finally in the program of investigations there were included an asphalt concrete with a maximum grain size of 11 mm for roads with heavy traffic loadings, then another asphalt concrete with a maximum grain size of 11 mm for roads with normal traffic loadings and moreover an asphalt concrete with a maximum grain size of 8 mm for roads with easy traffic loadings. In the northern part of Germany we have found some suitable building sites, one building site on highway with heavy traffic, one building site on a national road with normal traffic and one for a local road construction with slight traffic to build in the asphalt concrete with a maximum grain size of 8 mm. The asphalt mixture has the following composition: The asphalt concrete for the heavy traffic was mixed with 5.7 % bitumen with a penetration of 6.5 mm, 9 % filler and 55.1 % chippings up to 11 mm and has been endowed with a compaction resistance D of 23.1 [21 Nm]. The asphalt concrete for normal traffic was mixed with a 6.1 % bitumen with a penetration of 8 mm, 7.7 % filler and 54 % chippings up to 11 mm and has been endowed with a compaction resistance D of 23.3 [21 Nm]. The asphalt concrete for the slight traffic was mixed with 6.6 % bitumen with a penetration of 8 mm, 10.3 % filler and 50.6 % chippings up to 8 mm and has been endowed with a compaction resistance D of 15.5 [21 Nm]. You can remove from these dates that it was not possible to obtain an extremely heavy compactable asphalt with an exceptional resistance against deformations. These facts show the difference between optitude test in the laboratory and actual conditions during the asphalt production in the coating plant.

2.2 Compaction in the field

The coated materials were transported from the coating plant to the building site and then built in the road surface with a paver under normal circumstances. There were given no restrictions or instructions with regard to the speed of the paver, frequency of the vibrating plank and the lift of the tamper. Therefore there was no control about the precompaction of the asphalt concrete directly after paving. During the following required rolling compaction however, there was consequently paid attention that in each of these fields of investigation of one building site were used only one type of roller. The sort of roller has been in the field 1 always the vibrating roller, in the field 2 always the three-wheel-tandem roller and in the field 3 only the rubber tyred roller. The number of roller passages has to be decided by the crew of the roller in a way, that they achieve a degree of compaction of 98 % with regard to the density of a compacted marshall-specimen. Then there has been

taken a number of cores from each of the three fields
with a diameter of 30 cm. From these cores the specimens
to test the mechanical properties of the asphalt concrete
were to be sawed. During treating of the coated material,
a number of samples has been taken from the pail of the
paver. Specimens have been prepared by different types of
compaction in the laboratory.

2.3 Methods of compaction in the laboratory

The results of an extensive study of literature has been
that actually four typical groups of types of compaction
were used in the laboratories all over the world, namely
the blow compaction, the vibrating compaction, the pressure and knead compaction (or gyratorical compaction) and
the rolling compaction. To compact the coated material in
the laboratory with these kinds of compaction there have
been constructed suitable compactors. All samples should
be produced in same thickness, like the cores from the
street, namely 4 cm. Another point to consider have been
the dimensions of the specimens. The small dimensions of
the marshall-specimen are not suitable for most of the pre-
tentious material tests. Therefore we had to develop basi-
cal equipments to prepare round or rectangular specimens.

2.3.1 Blow compaction

Blow compacted plates have been included in this investigation to check the procedure similar to the procedure of
marshall compactor. This method is spread world-wide to
prepare specimens in the laboratory. In the progress of
this procedure a certain number of successive vertical
acting blows has to be applied on coated materials filled
in a cylindrical form. This principle of the marshall-
compactor has been retained in a modified way. There was
constructed a cylindrical form with a diameter of 25 cm
to produce great round plates. For compacting, a plate of
steel was put on the coated material to distribute contant the acting load. The compactor has been conceived in
a way that a drop hammer strikes the plate excentricly .
By rotating the ground plate in the same moment it is
guaranteed that the specimens actually are compacted
equally over the whole specimen. The temperature during
the compaction progress must be constantly 135 °C.

2.3.2 Vibrating compaction

The vibrating compaction in the laboratory has to imitate
the vibration of the roller in the field. There were
taken two electric motors and fixed on a plate and then
put on a rectangular form with a dimension of 19,5 x 29
cm. The motor works conter-rotating. Motors and plate
have a weight of 20 kg, the motors make 3000 vibrations
in the minute. Extensive investigations have shown, that
to obtain a degree of compaction of 98 % the time of

compaction must be 30 seconds. The temperature during the compaction progress must be constantly 135 °C.

2.3.3 Pressure and knead compaction
To prepare specimens with a pressure and knead compaction there was developped an engine which is able to increase the vertical load stepwise. Each step of the load can hold for any time. Controlled horizontal changing movements can work simultaneously because the side-boards of the compactor are hinged. The engine of the compactor is regulated with a hydraulic aggregate. There can be prepared specimens with the dimension of 20 x 30 cm. The temperature during the compaction must be consequently 135 °C.

2.3.4 Rolling compaction
The procedure to do rolling compaction is the result of many trials. The coated material must be heated up to 180 °C and is then filled into a form. It takes 24 minutes to compact the asphalt alternatively with a rubber wheel and with a smooth wheel with 22 roller passages. During the compaction the temperature decreases from 180 °C to 110 °C. The size of the compacted plate is 28 x 36 cm.

2.4 Material tests
With the described methods of compaction there was produced a great number of different compacted specimens and then examined with regard to the different asphalt concrete properties.

2.4.1 Wheel tracking tests
The wheel tracking tests were done with a load of a 0,71 N/mm² for 3 hours and a speed of rolling about 0,3 m/sec. The way of the wheel for each passage was 22,8 cm. The specimens were tested with a temperature of 40 °C and 50 °C. The rutting in the wheel-tracking-testing machine was measured continously.

2.4.2 Dynamic bending tests
For the dynamic bending tests there were prepared quadratic prisms with a section area of 4 x 4 cm and a length of 16 cm. These specimens were sticked into the testing machine with the quadratic sectional area, on the other side a thin metal plate was sticked. On this metal plate acts the power of a vibrating power drive, which makes the specimens vibrate. The amplitude of these vibrations was measured on the head plate. The whole testing machine is put into a climatic cabinet which can be regulated for temperatures between - 30 °C and + 50 °C. With aid of this testing machine, temperatures between - 20 °C and + 40 °C and for a changing power with frequencies from 2 cps up to 120 cps the phase angle between power and

swing out was measured and then the elastic modulus was calculated.

2.4.3 Tests with low temperatures

To investigate the properties of the specimens compacted with different methods in the presence of low temperatures, were done cooling down tests and tensile tests. For these tests prismatic samples with dimensions of 4 x 4 x 16 cm were applied. The tests have been made with a process controlled tension testing machine equipped with a thermal indifferent measuring base. With aid of the tensile tests the specimens were drawed asunder with a speed of 1 mm/min and with a temperature of - 25 °C and - 10 °C until the specimens rupture. Out of the relation between stress and extensions, registrated during the test tensile strength and fracture strain were determined for all investigated variants. For the cooling down tests prismatic specimens with the dimension of 4 x 4 x 16 cm were applied, too. These specimens were cooled down with 10 °C/h beginning with a temperature of + 20 °C. The length of the specimen, measured in the beginning, was kept constant during this test. The tensile stress, occured on account of the wanted handicap of the thermic contraction, was measured until the fracture of the specimen in dependance of temperature and time. Then the maximum kryogenic tensile stress and the appertaining fracture temperature were determined.

3 Results

Comparing the results for the mechanical material properties, determined as mentioned before, a realistic compaction under laboratory conditions is to be found with the aid of multivariate statistical analysis which should be improved in further tests. In any case it will be possible to prepare asphalt concrete specimens with aid of this improved kind of compaction under laboratory conditions.

   Aim of this improvement is, that the mechanical properties of the laboratorycally prepared specimens will be close to the mechanical properties of asphalt concrete compacted by rollers on the road. The analysis of all test results is completed, so that the improvement of a realistic compactor can go on. This analysis turned out to be very difficult and it is part of a contract of investigation, conducted for the ministry of traffic. In agreement with the ministry, partial results are not allowed to be published now. However, it can be expected that it will be necessary to apply a compacting effort similar to a roller for gaining realistic properties of the coated material. Probably the last test results will be found until autumn 1990. Then a final recommendation for the construction of a realistic compactor might be given.

4 FIELD AND LABORATORY EVALUATION OF A NEW COMPACTION TECHNIQUE

A.O. ABD EL HALIM
Civil Engineering Department,
Carleton Univeristy, Ottawa, Canada
O.J. SVEC
Institute for Research in Construction, National Research Council,
Ottawa, Canada

Abstract
In most countries of the world asphalt pavement of road networks represents the largest single investment in the transportation sector. In Canada, asphalt. pavements built over the past 50 years are worth more than 70 billion dollars (in 1984 Dollars). In order to maintain the current road network at the present level of service and prevent it from further deterioration, over 8 billion dollars is needed annually.

Traditionally, asphalt layers are compacted using the conventional steel drum rollers. These rollers may have vibratory abilities as well as rubber coated drums. In North America as well as in many parts of the world, the asphalt mats are compacted using one of these heavy rollers followed by a multi wheeled rubber roller. It is believed that the rubber roller will seal the surface cracks induced during the passes made by the steel roller. Recent research work based on the concept of relative rigidity has indicated that the rapid deterioration of new asphalt pavements is directly related to current compaction equipment. It has been shown analytically and experimentally that steel rollers used in compacting the asphalt layer will result in surface cracks during construction and will remain even after the multi wheeled rubber rollers are used. Traffic and environmental influences will accelerate the failure of the new layer.

This paper presents a brief summary of the new theoretical approach and results of the first field tests carried out in Canada using a new compactor termed Asphalt Multi-Integrated Roller (AMIR). The results showed that the concept of relative rigidity is valid and the AMIR compactor provides crack free asphalt pavement with higher and more uniform densities. Also, the paper gives the results of laboratory tests performed on beams, cores and slabs of asphalt pavement compacted in the field.
Keywords: Asphalt pavement, Compaction, AMIR, Field trials, Laboratory Testing

1 INTRODUCTION

The construction of asphalt pavements is traditionally carried out by laying the hot asphalt concrete mix over a base course or an existing road surface, and then compaction is done with three different types of compactors. The first operation in the compaction procedure is carried out by using heavy steel rollers to obtain the desired density. Smoothing out the surface is accomplished with a multi wheeled rubber roller followed by a light steel roller. The finished product is assumed to be structurally sound and free of defects. In many cases it has been observed that new pavements start to crack prematurely and therefore the maintenance work has to begin shortly after construction. This problem of premature failure of asphalt overlays is commonly recognized and reported by Sherman 1982, Goetzee. and Monismith 1980 and Kennepohl and Lytton 1984.

Most of these studies are based on the erroneous assumption that the new pavement is structurally sound after construction and that the phenomenon of crack initiation and propagation is due to other factors such as traffic loading, inadequate base compaction, substandard asphalt mixes, thermal stresses and other environmental influences.

Abd El Halim 1984, Abd El Halim et all 1987 pointed out that in almost all cases, premature cracking of asphalt pavements is initiated at the time of construction due to the method and type of compaction equipment employed. Present compaction methods have not changed considerably for the last five decades. Compacting the relatively soft asphalt mix with rigid cylindrical steel rollers invariably leads to an initially cracked structure. The other factors such as traffic and environment, which are generally held responsible for pavement cracks, only aggravate and accelerate the failure process.

This paper presents a brief summary of the results of a theoretical analysis carried out to identify the main parameters for prevention of "initiation of cracks" in pavements during construction. This paper also reports results of field tests carried out in Canada using the first AMIR prototype in North America to verify the analytical investigation.

2 PREVENTION OF CONSTRUCTION INDUCED CRACKS

The phenomenon of initial cracking of new asphalt overlays during compaction has been well established through various field and laboratory investigations by Abd El Halim and Bauer 1986. In order to overcome this problem one must make the rigidities as well as the relative geometry of the rollers and of the soft asphalt mix more compatible. There is very little one can do to increase the stiffness of the asphalt mix during compaction to bring the modulus ratio of the steel and of the mix closer to unity. Therefore, the only alternative is to decrease the stiffness of the steel compactor. The second parameter is the relative geometry of the system. This will require increasing the diameter of the roller considerably to make the interface between the roller and the compacted asphalt surface as flat as possible.

A flat steel plate compactor would not meet these requirements because of its high stiffness in addition to the lack of mobility. The above requirements for a crack free pavement system led to the development of a revolutionary new compactor called AMIR (Asphalt Multi-Integrated Roller). Figure 1 shows the first Canadian AMIR prototype. It has been designed and built to satisfy the main requirements for crack free asphalt overlays which are :

1) The modulus ratio between the rubber belt and the asphalt mix is close to unity.

Fig. 1.   The AMIR Prototype.

2) The belt /drum arrangement resulted in a flat contact surface to keep the upper and lower interfaces of the compacted asphalt layer parallel during compaction.

The following sections briefly outline the experimental testing program, results and main conclusions.

3 FIELD TRIALS

Three field trials were carried out to test the performance of the new Asphalt Multi-Integrated Roller. The first field trial was done on an unpaved parking lot in Toronto in June 1989.
Two more field trials took place in Ottawa in August and in November of 1989. The details of the first and second field trials are discussed below. The testing is still underway on asphalt samples from the third trial at the time of preparing this paper.

3.1 Toronto
This field test was carried out on a gravel parking lot using HL-3 hot asphalt mix specified by the Ministry of Transportation of Ontario. It consisted of four test sections as described below:

a. 70 mm thick layer of HL-3 mix , 12 meter long by 1.8 meter wide, compacted by the AMIR prototype.

b. 70 mm thick layer, 12 meter long by 1.8 meter wide, compacted by a steel roller having the same total weight as AMIR (approximately 8000 kg.).

c. 130 mm thick layer, 26 meter long by 1.8 meter wide, compacted using AMIR.

d. 130 mm thick layer, 26 meter long by 1.8 meter wide, compacted using the steel drum roller.

Only the static steel roller was employed in the Toronto Trial. The main objective of the Toronto field trial was to evaluate the ability of the AMIR compaction method in comparison to the static steel compaction method under the same number of passes.

3.2 Ottawa I
The results of the Toronto field test led to the second field test in Ottawa. The test site was selected from one of the paved service roads on the campus of the National Research Council. The field trial consisted of

overlaying two- 150 meter long by 6.0 meter wide sections. Each section was subdivided into 3 subsections. The first two were overlaid by 50 mm thick layer of the Ministry of Transportation of Ontario HL-4 standard hot asphalt mix. While the two remaining subsections were overlaid with 75 mm thick layer of the same asphalt mix. One 150 meter-section was compacted using the AMIR roller while the other 150 meter-section was compacted using a standard vibratory roller followed by rubber wheeled roller. The main objective of the field trial was to achieve the highest possible densities with each compaction method. In total six (6) passes were made by the AMIR roller while over twenty (20) passes were made by the vibratory/rubber tire rollers. Three weeks after this compaction test, asphalt beams, cores and slabs from each section were recovered and transported to the laboratories of the National Research Council and Carleton University for further testing.

4 TESTING PROGRAM

In order to objectively evaluate the influence of two distinctly different compaction techniques on the properties of the asphalt material the following laboratory tests were employed:

1. Specific gravity and density
2. Indirect tensile strength
3. Direct tensile strength
4. Flexural strength

These four tests formed the basis of the experimental program that was designed and carried out to achieve the following objectives :

1. To evaluate the influence of compaction method on the physical and mechanical properties of the compacted asphalt mixes.
2. To determine the most critical mode of stress.
3. To identify relationships that may exist between direct, indirect and bending strength of asphalt samples recovered from the field.

4.1 Density Test
The densities and specific gravities of asphalt core samples 75 mm in diameter, recovered from the compacted field sections, were determined according to ASTM Standard Method D2726-79. Densities are recognized as the most important measure of the quality of the finished pavement. However, it must be born in mind that the results of the density tests do not reveal the actual mechanical properties of the pavement.

41

4.2 Indirect Tensile Strength Test

The indirect tensile strength test is commonly used for its simplicity. The test is performed by loading an asphalt core specimen diametrically until failure. The maximum load is used to calculate the indirect tensile strength as follows :

$$I.T.S. = \frac{2 P_{max}}{\pi d t} \qquad [1]$$

Where

 I.T.S. = Indirect Tensile Strength
 P_{max} = Maximum Load
 d = Diameter of the specimen
 t = Thickness of the specimen

While the indirect tensile strength test is a popular one, the test results are usually affected by the fact that asphalt material is not homogeneous. Furthermore, early work carried out by Abd El Halim 1986 has shown that the direction of the rollers during the compaction phase can significantly affect measured strength of the tested core samples. In addition, because the size of the core samples is relatively small the effect of the induced hairline cracks during compaction may be difficult to quantify.

4.3 Direct Tensile Strength Test

The problems with the indirect tensile strength test and the need to evaluate the effect of construction induced cracks on the strength of the compacted mix resulted in the development of a new testing equipment. This new testing equipment measures the direct tensile strength of relatively large asphalt slabs . It consists of two steel plates, one is fixed and the other moves horizontally on a frictionless surface. The movable plate is linked to a mechanical system consisting of a load cell, electric motor and an L.V.D.T. The plate is normally moved at a rate of displacement of 50 mm per minute. The movement of the plate simulates the response of an old pavement to a drop in the temperature. The size of the asphalt specimens is 500 mm by 250 mm without any restriction for thickness. The sample is attached to the top of the plates using 24-hour epoxy glue. The test can be performed at large range of temperatures. Figure 2 illustrates the main components of this test table. The use of this test equipment provides the opportunity to calculate the strain energy absorbed by the sample. Also,the test allows a researcher to observe the history of the cracking process during loading of the sample.

42

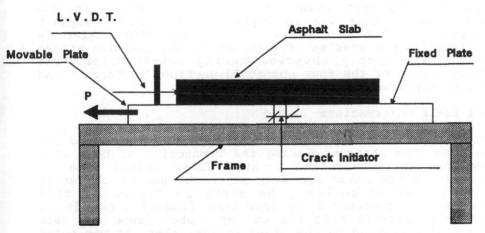

Fig. 2.   Direct tensile strength test table.

4.4 Flexural Strength Test

The indirect and direct tensile strength tests described above can provide invaluable information about the response of the pavement to the thermal and traffic conditions. However, most pavement design methods consider the strength of the asphalt pavement under flexural loading as an important design parameter . Also, the maximum bending stresses can be determined more accurately than the tensile strength by the indirect and direct test methods. The main problems with the flexural tests are associated with the effect of the weight of the asphalt sample (in the case of large size beams) and the effect of creep. This is particularly serious for the tests performed under room temperature. A large scale bending test apparatus was designed and built for this experimental program. A two-point bending is used so that the central section of the beam specimen is loaded by pure bending only. The size of the test specimens were 600 mm long by 150 mm wide.

It is obvious from the above discussion that evaluation of asphalt pavements, which would simulate the actual field conditions, can not be achieved by performing one or two tests. It is the opinion of the authors that the four tests described above should be employed to fully understand the mechanical behavior of asphalt pavements.

5 TEST RESULTS

As was mentioned earlier the main objective of this investigation was to compare and evaluate the influence of compaction method on the properties of the field compacted asphalt mixes. The first criterion in the comparison process was to evaluate the finished asphalt surfaces of the two field trials. The AMIR roller is designed on the premise of eliminating the hairline cracks that are commonly observed during construction. The observations of the four compacted sections in Toronto and Ottawa confirmed the following results.

5.1 Field Observations
I. Conventional Compaction Equipment :
Numerous hairline surface cracks ranging from 2 mm to 100 mm long were observed during the compaction of both test sections in Toronto and Ottawa. These cracks remained visible to the naked eye even after the use of the multi rubber wheeled roller , as shown in Figure 3. This observation represents an important finding since it is commonly believed that the use of rubber drums can cure the cracks induced by the steel roller. Also, it was noted that the Vibratory roller generated more surface cracks than a standard Static steel roller. It is also interesting to note that the cracks were observed on two different mixes and two different asphalt structures. This observation confirms that the main contributor to the phenomenon of "hairchecking" is the drum of the rollers. The advantage of the Ottawa test site was that the paved access road was closed to traffic after compaction. Thus, an opportunity existed to monitor the pavement surface condition during and after the 1990 winter. Recent observations confirmed that the construction induced cracks still exist after 8 months and in some locations these cracks extended through the full width of the paved lane.

II. AMIR Compactor:
In contrast to the observations reported above, the asphalt surfaces compacted by the AMIR prototype showed no evidence of any surface cracking. This observation holds true for both asphalt mixes compacted in Toronto and Ottawa. Furthermore, the AMIR compacted surfaces had a much tighter texture when compared to the conventionally compacted pavements. Also, it was observed that longitudinal joints compacted by the AMIR roller were much smoother and crack free. Figure 4 gives a photo of the Ottawa test section compacted by AMIR.

The observations of the field tests verified the results of early analytical and laboratory investigations carried out to explain how construction cracks occurred. Clearly, a crack free surface would be expected to be more resistant to thermal stresses, intrusion of surface water to the underlying layers and the effects of freeze/thaw cycles. These benefits are achieved in addition to significant improvement in the mechanical properties of AMIR compacted pavement as discussed below.

Fig. 3. Asphalt surface compacted by vibratory and rubber tire rollers.

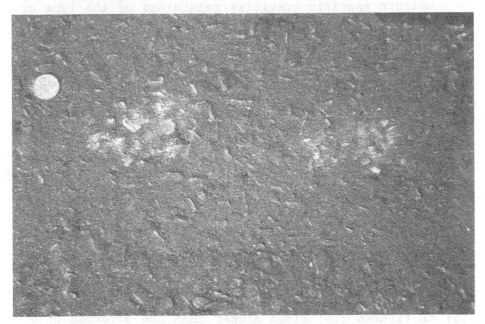

Fig. 4. Asphalt surface compacted by AMIR.

45

5.2 Results Of Laboratory Testing Program
I. Density Tests
The results of performing the specific gravity and density tests on asphalt cores recovered from the Toronto and Ottawa field trials are given in Tables 1 and 2. The results obtained from both field trials showed that the new AMIR compaction method is capable of achieving higher densities as indicated by the overall mean values of the measured specific gravities. Also, the results indicated that the density distributions across and along the AMIR compacted sections are more uniformly distributed as indicated by the lower values of the standard deviation and coefficient of variation. It should be noted that each value given in the tables, represents the average of three core specimens.

The data is organized according to the distance between a core-set and the outside edge of the paved lane. For example, the first row in Table 1 corresponds to a set of cores 300 mm off the edge of the pavement, while the last row in the table provides the data of the closest core-set to the longitudinal joint between the AMIR and Conventional compacted sections. Analysis of the results given in Tables 1 and 2 support the following :

1. The AMIR compaction method achieved higher and more uniform specific gravities regardless of the type of the mix, the thickness of the asphalt lift and the underlaying layer.

2. The new compaction method can compact thicker asphalt lifts to higher densities than currently used rollers. As can be seen in Tables 1 and 2, the AMIR method achieved specific gravities of 2.352 and 2.421 on the thick lifts while the current compaction methods gave specific gravities of 2.328 and and 2.413 on the thinner lifts.

3. The quality of the compacted pavements at the edges of the paved lanes was noted to be poor in the case of the conventional compaction method. This result is confirmed by comparing the results of the first three rows in Table 1 and the first four in Table 2 to the overall mean of the compacted sections.

The results discussed above require further explanation. The total weight of the AMIR roller was equal to the weight of the static steel roller used in Toronto and smaller than the vibratory one used in Ottawa. The stresses under these conventional rollers are therfore ten (10) to fifteen (15) times higher than the stresses under the larger area of the AMIR roller. One can therefore ask, how can the AMIR roller achieve higher densities, especially when the number of passes was the same in

46

TABLE 1: Results of Specific Gravity Tests
Toronto Field Trial

| | Average of 3 asphalt cores | | | |
| | AMIR Method Specific Gravity | | Conventional Method Specific Gravity | |
Thickness	70 mm	130 mm	70 mm	130 mm
	2.348	2.293	2.255	2.238
	2.356	2.323	2.303	2.283
	2.361	2.344	2.333	2.352
	2.350	2.357	2.352	2.357
	2.354	2.365	2.318	2.352
	2.348	2.366	2.339	2.359
	2.339	2.370	2.331	2.349
	2.323	2.372	2.340	2.362
	2.336	2.377	2.344	2.378
Mean	2.346	2.352	2.328	2.337
Standard Deviation	0.012	0.026	0.033	0.043
C.O.V.	0.50 %	1.10%	1.42%	1.84%

TABLE 2 : Results of Specific Gravity Tests
Ottawa Field Trial

| | Average of 3 asphalt cores | | | |
| | AMIR Method Specific Gravity | | Conventional Method Specific Gravity | |
Thickness	50 mm	75 mm	50 mm	75 mm
	2.406	2.412	2.400	2.397
	2.418	2.411	2.403	2.384
	2.423	2.402	2.401	2.384
	2.421	2.409	2.401	2.387
	2.442	2.424	2.397	2.410
	2.446	2.414	2.442	2.409
	2.438	2.416	2.427	2.417
	2.441	2.417	2.431	2.414
	2.432	2.443	2.427	2.423
	2.437	2.441	2.430	2.424
	2.445	2.427	2.430	2.426
	2.442	2.434	2.425	2.429
Mean	2.433	2.421	2.413	2.409
St. Dv.	0.013	0.016	0.022	0.019
C.O.V.	0.57%	0.60%	0.91%	0.80%

Toronto and much lower in Ottawa in comparison to the number of passes made by conventional rollers ? The improvement achieved by the AMIR compaction method can be explained as follows:

1. The AMIR compactor provides a longer time of contact between the compacted asphalt mix and the loading device. While, the current cylindrical shape of the rollers will provide a contact length of 200 mm at the most, the flat belt of the AMIR roller will provide a contact length of at least 1500 mm. Thus, for the same speed the AMIR roller will be in contact with the mix at least seven (7) times longer than a current roller.

2. Due to the geometry of the AMIR roller which provides a rectangular contact area of 3000 mm long by 1600 mm wide, the asphalt mat under the AMIR roller is much more confined than the mat under a cylindrical drum. Furthermore, the effect of the larger confining length in the direction of rolling, as it is in the case of the AMIR, minimizes the rebounding of the asphalt mat.

3. While the total compaction effort of both rollers may be the same for a certain period of time, the useful energy induced by each roller is significantly different. Clearly, the construction induced cracks in the case of current rollers would result in wasting portion of the total compaction effort. The compaction by AMIR was shown to be crack free and therefore, one expects no waste of energy of compaction.

II. Indirect Tensile Strength Tests
Results of performing the indirect tensile strength tests on core samples recovered from Toronto and Ottawa are given in Tables 3 and 4. The results indicated the following :

1. Asphalt cores compacted by the AMIR roller gave 22.8 and 19.5 % higher strength than cores compacted by the static steel roller in the Toronto field Trial. The strength of the AMIR compacted pavement was more uniform as indicated by the lower values of the calculated Coefficient of Variation, i.e. 6.3 and 9.2 % in the case of AMIR and 22.0 and 15.6 % in the case of the static steel roller.

2. Similar results were obtained when the tests were carried out on asphalt cores taken from the Ottawa trial. As shown in Table 4, the AMIR compaction resulted in 8.5 and 5.8 % increase in the indirect tensile strength. Also, the values of the Coefficient of Variation, i.e. 2.0 and 5.2 % in the case of AMIR, and, 10.2 and 7.2 % in the case of the

TABLE 3: Test Results of Indirect Tensile Strength Tests Toronto Field Trial

| | Strength (in MPa) | | | |
| | AMIR Method | | Conventional Method | |
Thickness	70 mm	130 mm	70 mm	130 mm
	0.656	0.780	0.302	0.588
	0.782	0.647	0.538	0.478
	0.691	0.636	0.673	0.626
	0.664	0.763	0.582	0.566
	0.674	0.787	0.623	0.796
	0.660	0.799	0.643	0.636
Mean	0.688	0.735	0.560	0.615
Ratio:				
AMIR/Steel	122.9 %	119.5 %		
St.Dv.	0.044	0.067	0.123	0.096
C.O.V.	6.3 %	9.2 %	22.0 %	15.6 %

TABLE 4 : Test Results of Indirect Tensile Strength Ottawa Field Trial

| | Strength (in MPa) | | | |
| | AMIR Method | | Conventional Method | |
Thickness	70 mm	130 mm	70 mm	130 mm
	0.760	0.731	0.802	0.671
	0.776	0.665	0.825	0.631
	0.785	0.654	0.679	0.691
	0.802	0.665	0.658	0.592
	0.774	0.637	0.684	0.611
	0.755	0.621	0.638	0.558
Mean	0.775	0.662	0.714	0.626
Ratio				
AMIR/ Steel	108.5%	105.8%		
St.Dv.	0.016	0.035	0.072	0.045
C.O.V.	2.0 %	5.2 %	10.2 %	7.2 %

vibratory/rubber tire combination suggest that characteristics of the AMIR compacted mixes are unique and can not be obtained by any other currently used compaction methods.

3. All the tested cores where loaded perpendicular to the rolling direction. Thus, the higher values of the standard deviation of the steel compacted samples can be explained as a result of the variation in the stress distribution under the more rigid steel drum.

The test results of the indirect tensile strength showed consistently that the AMIR compacted asphalt concrete mixes will have higher strength. However, the influence of the hairline cracks on the indirect tensile strength may not be best measured by this test. The authors adopted the direct tensile strength for the purpose of quantifying the effect of construction cracks on the structural integrity of the compacted mix.

III. Direct Tensile Strength Tests

The results of performing the direct tensile strength test on 9 asphalt slabs taken from each of the compacted sections of the Ottawa field trial are given in Table 5 and Figures 5 and 6. As can be seen in Figure 5 and Table 5 the direct tensile strength of the crack free AMIR compacted samples are 11 % higher on the average than the ones compacted by the vibratory/rubber tire combination. The effect of the construction induced cracks is more evident when the strain energy results are compared. It can be seen in Figure 6 that the AMIR compacted sample required more energy in order to reach the same failure criterion of the conventional compacted sample. This result would suggest that asphalt pavements compacted with current equipment can not meet their expected design life.

TABLE 5 : Results of Direct Tensile Strength Tests Ottawa Field Trial

	AMIR Method	Average of 3 Tests Strength (in MPa) Conventional Method	Ratio
	0.412	0.333	124 %
	0.405	0.356	114 %
	0.453	0.452	100 %
Mean	0.423	0.380	111 %
St.Dv.	0.033	0.061	
C.O.V.	7.8 %	16.0 %	

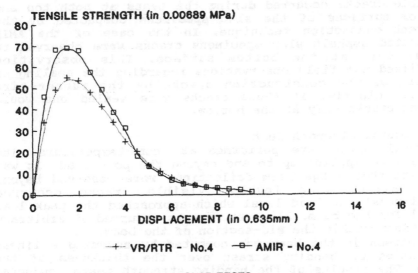

SLAB TESTS

FIGURE 5: RESULTS OF DIRECT STRENGTH
NRC TEST SECTION - MIDDLE OF LANE
BLOCK - 3

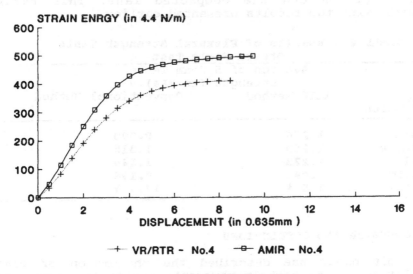

SLAB TESTS

FIGURE 6: RESULTS OF STRAIN ENERGY
NRC TEST SECTION - MIDDLE OF LANE
BLOCK - 3

Visible cracks occurred during the tests at both top and bottom surfaces of the slab specimens compacted by the current compaction technique. In the case of the AMIR compacted asphalt slab specimens cracks were observed to occur only at the bottom surface. This observation confirmed the field observations regarding the sealing or curing of the construction cracks by the rubber tire roller. Clearly, if these cracks were sealed one would expect cracks only at the bottom.

IV Flexural Strength Test

Flexural tests were performed at room temperature. The loads were applied up to and beyond the peak load. It was evident that large beam deflections were observed beyond the maximum load levels. Visible cracks occurred continuously at load level which approached the peak load level for the beam. Also, the cracks occurred at arbitrary locations within the mid-section of the beam.

The stress in the beam was computed by assuming a linear variation of bending stress over the thickness of the beam. The results of the bending strength tests conducted on asphalt concrete compacted by the AMIR and a standard vibratory roller are presented in Table 6. It is evident that the AMIR compaction produces asphalt concrete with higher flexural strength. It was also found that the current compaction procedure introduces larger variation of strength across the compacted lane. This result supports with the results presented earlier.

TABLE 6 : Results of Flexural Strength Tests Ottawa Field Test

| Location | Average of 5 Beam Tests Strength (in MPa) | |
	AMIR Method	Conventional Method
Edge	1.276	0.998
Centre	1.269	1.315
All	1.273	1.140
St.Dv.	0.096	0.198
C.O.V.	7.5 %	17.3 %

6 DISCUSSION AND CONCLUSIONS

This paper has described the phenomenon of crack initiation in new asphalt pavements during the compaction phase. The main factors responsible for initial crack formation are the large difference in stiffness between the steel drums and the asphalt mix and secondly, the curvature imposed on the surface of the asphalt overlay by the circular shape of the roller.

Consideration of these aspects has led to the development of a model and proto-type compactors which have, as much as practically possible, overcome the problem of crack initiation during construction. The soundness of this new concept has been proven by the field verification and the laboratory tests. In the experimental study a direct comparison of results was made between AMIR and steel compacted asphalt specimens. The AMIR compacted specimens showed a considerable improvement in mechanical properties. In addition to the improvement in the measured density and strength achieved by the AMIR method, their consistent uniformity was evident from the laboratory test results. On the other hand the conventionally compacted asphalt pavements showed numerous surface cracks which were predominantly aligned perpendicular to the direction of rolling. These cracks contributed to the lower values of the measured asphalt concrete strength and the higher levels of variation in density and strength across the compacted lane. The benefits of using the AMIR compaction are expected to be achieved with the use of the a smaller number of compaction machines and without changing the currently employed asphalt mixes.

7 ACKNOWLEDGMENT
The authors would like to acknowledge the financial assistance for this study through the National Research Council Canada, the Ministry of Transportation and of Ontario and through operating grants provided by the Natural Science and Engineering Research Council of Canada.

8 REFERENCES

Sherman , G. (1982). Minimizing Reflection Cracking of Pavement Overlays, National Cooperative Highway Research Program Synthesis of Highway Practice No. 92.

Goetzee , N .F. and C.L.Monismith (1980). Reflection Cracking: Analysis, Laboratory Studies and Design Considerations, Proceedings of the Association of Asphalt Paving Technologists, Vol. 49.

Kennepohl , G. J. A. and B.L.Lytton (1984). Pavement Reinforcement With Tensar Geogrid For Reflection Cracking Reduction. Proceedings of Paving in Cold Areas. Mini- Workshop Canada/Japan Science and Technology Consultations. Tsukuba Science City, Japan.

Abd El Halim, A.O. (1984). The Myth of Reflection Cracking, Canadian Tech. Asphalt Association, Proceedings Vol. XXIX.

Abd El Halim, A.O. and G.E. Bauer (1986). Premature Failure Of Asphalt Overlay At Time Of Construction. Journal of Transportation Forum, Road and Transportation Association of Canada, Vol. 3.2, Sept.

Abd El Halim, A.O. , W.Phang and R.Haas (1987). Realizing Structural Design Objectives Through Minimizing Of Construction Induced Cracking. Sixth International Conference on Structural Design Of Asphalt Pavements, Ann Arbor ,U.S.A. July 13-16, Vol.I.

PART TWO
TESTS WITH UNIQUE LOADING

5 INFLUENCE OF POLYMER MODIFICATIONS ON THE VISCOELASTIC BEHAVIOUR OF BITUMENS AND ON MECHANICAL PROPERTIES OF BITUMINOUS MIXTURES: EVALUATION TESTS

A. APICELLA
Dip. di Ingegneria dei Materiali e della Produzione, Università
Federico II, Napoli, Italy
C. CALIENDO, B. FESTA and P. GIANNATASIO
Dip. di Ingegneria dei Trasporti, Università Federico II,
Napoli, Italy

Abstract:
A commercial type bitumen and its modifications with different amounts of an aliphatic copolymer were characterized by means of some ordinary methods such as Fraass, penetration, softening and viscosity tests and with more sophisticated calorimetric (DSC) and dynamic-mechanical (DMA) analysis.
The viscoelastic behaviour of the bitumen was significantly influenced by the polymer addition. The modified material presented higher shear viscosities and tensile moduli which influenced the results of the mechanical tests (Marshall and Indirect tensile tests) done on bituminous mixtures.
Among the ordinary characterization methods for bitumens, the penetration test only has been found to be sensitive to the polymer modification.
As concerns the mixtures, the indirect tensile test has been found to be more sensitive and adequate to quantify the influence of the polymer addition.
The need of more sophisticated dynamic-mechanical characterization even for the mixtures is discussed.
Critical evaluation of the results of the ordinary tests made on both bitumen and mixtures, however, can still give useful informations on the behavioural response in use of these materials once modified.

Introduction

The structural design of asphalt pavements is generally achieved by elastic methods based on multilayer elastic theory. However, the application of these methods contains an intrinsic approximation related to the fact that the bitumen based layers are not completely elastic but behave viscoelastically. This means that the pavement behaviour in different loading conditions may change as a function of temperature and rate of application of the load. Moreover, even the optimization of the asphalt paving mix composition is achieved by means of some empirical tests such as the *indirect tensile test* or the *Marshall test* which completely neglect the viscoelastic nature of the bituminous binder. The same mechanical characteristics of bituminous binders are also determined by using empirical tests such as the measure of the softening point, the Fraass breaking point or penetration. These tests, as currently structured, are often analyzing only part of the complex behaviour of these viscoelastic systems; tests made at relatively high temperatures and at slow rates may only disclose the viscous nature of the material while those run at lower temperatures and high rates are testing the material elasticity. It is then difficult to critically analyze the results of these tests without the use of a complementary investigation which distinguishes the elastic and viscous contributions in different conditions of temperature and loading rates.

Several investigations evidenziate as long loading times results are accessible through transient experiments and short loading time results through dynamic experiments such as the oscillatory tests made at variable frequencies. Moreover, it has been also shown through the theory of linear viscoelasticity that transient and dynamic modes are equivalent in nature and that interrelations between transient and dynamic results exist.

The empirical tests currently used are generally transient in nature (time dependent) and only when run on viscoelastic materials having the same ratio between the elastic and viscous components or in temperature ranges in which one of the two components is predominant can adequately compare the behavioural response of a more complex systems containing these materials, such a bituminous mix is, and only in conditions similar to those of tested.

In some road applications, in order to increase either the material resistance to the viscoplastic deformations (rutting) and to fatigue failure (cracking) it has been suggested the addition of polymeric additives to the bituminous binder. The function of the addition of a polymer is related to the increase of the systems elasticity and, to some extent, to the increase of the system tenacity and rigidity. The polymeric phase is generally not compatible and uniformly dispersed in the bitumen phase. It is then evident from the previous discussion that appropriate tests accounting for the system viscoelasticity are needed in order to correctly identify the function of the polymeric additive and hence the optimum amount to add to the bituminous phase.

The scope of this work is to critically analyze the type of tests that are currently used for the characterization of a bitumen and of a bituminous mixture. In particular, we focused our attention on a commercial type bitumen and its modification with different amounts of an aliphatic copolymer characterized either with the ordinary tests generally required for the binder and with a calorimetric analysis (Differential Scanning Calorimetry) and oscillatory measurements (Dynamic Mechanical Analysis). The latter is a test particularly sensitive to the difference in the viscous and elastic behaviour. Moreover, the results of the tests used for the optimization of the amount of bituminous phase of a conglomerate (*indirect tensile test* or the *Marshall test*) are discussed for systems made with the unmodified and polymer additivate bitumen.

Experimental

Apparatus and Procedures
The materials were thermally characterized by means of Differential Scanning Calorimetry (DSC, Mettler) operating between -100 and 200 °C at the heating rate of 10° C/min and mechanically by using a Dynamic-Mechanical-Analyzer (DMA, Du Pont) operating at the same heating rate

between -50 and 60°C. The 1 cm x 5 cm samples for the DMA tests were 1 mm thick strips obtained by pouring the fluid bitumen between two thin sheets of paper. The softening point, penetration and Fraas breaking point were determined according to the standard test methods (ASTM D 36, D 5, CNR 43/1974). The Marshall and Indirect tensile (ASTM D 1559, D4123) tests were carried out on an Universal Testing Machine. Viscosities were determined by means of a coaxial cylinders rotational rheometer between 60 and 160°C; measures were taken when stationary conditions were achieved.

Material Characteristics
A commercial grade bitumen 80/100 has been used as binder. Its composition was:

Asphalthenes	% by weight	14
Saturated solvents	% by weight	15
Naphtenic aromatics	% by weight	45
Polar aromatics	% by weight	26

Chips of an ethylene-propylene-buthylene copolymer (Betaplast) has been added to the bitumen in 4.0, 5.5 and 7.0% by weight. The polymer chips and the bitumen were pre-heated and melted at 160°C, added to the hot and fluid bitumen and then manually mixed for 2 min.

A type A1a (according to the HRB classification) limestone and basalt mineral aggregate with maximum stone size of about 15 mm has been used. The grading distribution is reported in figure 1.

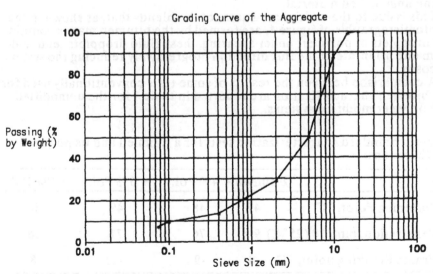

Figure 1: Mineral aggregate grading

The bituminous mixtures were obtained by first mixing at 170 °C the polymer chips with the pre-heated aggregate and then pouring the fluid bitumen heated at 160 °C. The overall mixing time in a compounder working at 150 revolutions per minute was of about 100 seconds.

The thermograms obtained from DSC analysis for the polymer, bitumen and blends are reported in figure 2.

The copolymer clearly shows the glass transitions associated with the Poly-ethylene, Poly-buthylene and Poly-propylene blocks, while the bitumen shows a complex behaviour reflecting the multicomponent nature of this material. A glass transition, however, is located around -10° C. The bitumen-copolymer blend tends to preserve the characteristics of the single components even if the fractions of the copolymer seem to be slightly plasticized by the presence of some of the bitumen components (probably low molecular weight solvents).

The steady state shear viscosities for the bitumen and the three blends measured at different temperatures are reported in figure 3 as a function of the inverse temperature (Arrhenius plot). In this type of representation the local slope of the curve is proportional to the process activation energy which is higher for the modified bitumen.

All the measurements were taken at temperatures above 60° C where the materials were fluid and essentially viscous (absence of elastic components). However, some differences in the rheological behaviour were observed; the polymer addictivated bitumen was, in fact, highly thixotropic. While steady state constant values of the viscosity were rapidly achieved by the bitumen at all the temperature investigated, the polymer modified material, especially at low temperatures, needed longer times to reach a stationary value. The initial viscosities of the polymer blends were significantly higher than those of the unmodified bitumen. Nevertheless, the steady state viscosities of the modified systems were still higher than those of the unmodified material.

This is due to the multiphase nature of the blends that, as shown in the photomicrograph's of figure 4, are formed by the bitumen matrix containing dropplets of polymer; under shearing, these fluid dropplets tend to deform and orientate in the flux direction progressively reducing the system viscosity.

A comparison between the results of some tests conventionally used for the bitumen characterization are reported in table 1 for the unmodified and polymer modified systems.

Table 1: Standard characterization tests for a bitumen and its polymer modifications:

	Bitumen	+ 4% Pol.	+ 5% Pol.	+ 7% Pol.
Softening Point, °C	47	48	49	50
Penetration, mm/10 (25°C)	90	76	71	66
Fraass Breaking point, °C	-8	-9	-11	-8

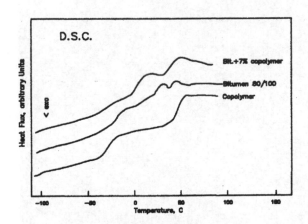

Figure 2: DSC thermograms of the polymer, bitumen and 7% blend

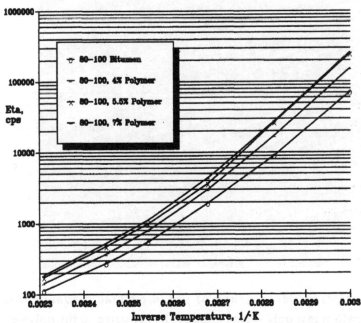

Figure 3: Arrhenius plot of the steady state shear viscosity of 80/100 bitumen and polymer additivate blends.

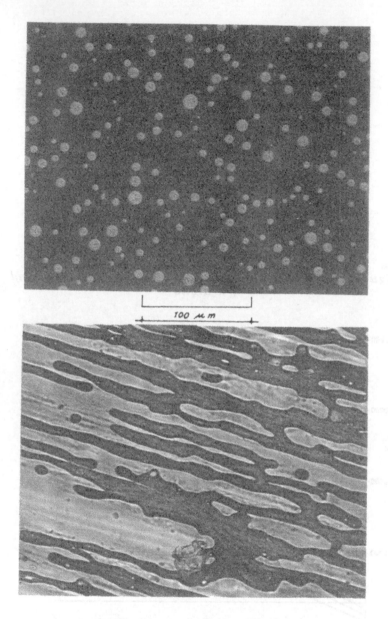

Figure 4: Optical Micrograph of pol. modified bitumen (7%)

The penetration test only seems to be really sensitive to the polymer modification while the changes observed in the other two tests are very small. This is to be ascribed to a lack of sensitivity of these two methods rather than to the additive ineffectiveness.

At temperature between -10°C and 60°C it is relevant to ignore the viscoelastic nature of the system. A dynamic-mechanical characterization of the systems, in fact, indicate that in this range of temperature significant differences exist between the modified and unmodified systems. The diagram of figure 5 reports the *elastic modulus* and the *tangent Delta* as a function of the temperature for a flexural test run at a frequency of 5 Hz on the unmodified bitumen and its modification with 7% of copolymer. In an oscillatory test made on a viscoelastic material a delay between the application of the load and the material deformation exists which increases as the material viscous nature prevails; the *Tan Delta* is the ratio between the viscous and elastic components of the mechanical properties. This parameter is near to zero for a completely elastic materials and approaches infinity for viscous systems (its value is unity when the elastic and viscous components are equal).

It is evident from figure 5 that the elastic component of the tensile modulus of the modified material (left side of the figure) is significantly higher than that of the unmodified bitumen, moreover, at lower temperatures, it tends to become still higher. It is only near 50°C that the modified material become completely fluid and viscous losing the elastic modulus components. This is confirmed by the analysis of the *Tan Delta* values of the modified system (right side of the figure) which are always lower (higher elasticity) than those of the unmodified material. The modified material preserves some elastic characteristics even at temperatures near 50°C.

From this additional test we can confirm that the penetration test is effective since run at a temperature at which the differences in the elastic behaviour were significant.

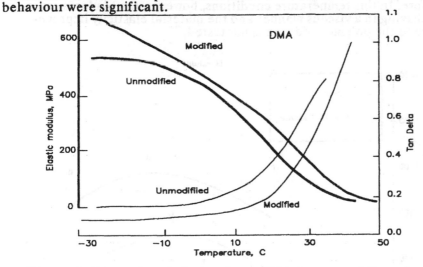

Figure 5: Viscoelastic behaviour of bitumen and its polymer modification

Mixture Characteristics
The inert aggregates were pre-heated at 170°C and mixed with the polymer chips at 160°C until a homogeneous coating was achieved around the particles, then the bitumen was added. Different amounts of binder were used to prepare samples for the mechanical tests. This procedure gave better results compared to those obtained from previous mixing of the

polymer and bitumen and then addition of the aggregates.

This can be probably due to the fact that part of the polymeric coating of the inert material tends to remain on the aggregate surface even after the addition of the bitumen. Acting as an elastic bridge between the bitumen and the rigid aggregates, the polymer improves the distribution of load and deformations over the multiphase material. It could be expected, then, that the actual concentration of the polymer in the bituminous phase is lower than calculated one due to the fact that polymer richer zones are present around the aggegates.

Samples for the indirect tensile tests and Marshall tests were obtained from the different composition mixtures. According to the ASTM standards, the Marshall tests were run at 60°C and, as clearly indicated in figure 6, significant differences between the systems obtained from the modified and unmodified bituminous binder exist. The values of the failure loads were always higher than those of the unmodified systems and the optimum composition for the modified system, even if not clearly located, was found at a higher binder content. According to the viscosity data, these changes can be attributed to the increase of binder viscosity that, due to the highly thixotropic nature of the modified bitumen is still higher than that reported in figure 3. In the initial stages of the deformation, in fact, the bituminous phase containing the polymer dropplets tends to preserve a high viscosity until they become highly oriented in the deformed matrix. This condition is not usually achieved since the mixture sample generally fails before. In this temperature conditions, however, the binder is essentially behaving in a viscous manner and the material elasticity improvement due to the polymer addition is not tested.

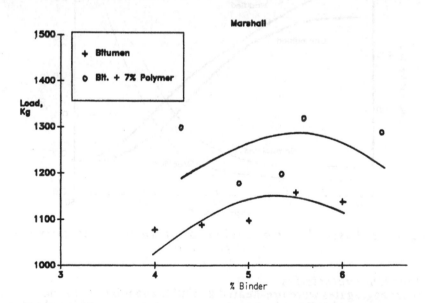

Figure 6: Marshall tensile tests for unmodified and polymer modified bituminous mixtures at different binder contents

More significant informations can be obtained from the indirect tensile test which is run at room temperature where the binder behaves viscoelastically (see figure 5). The improvements due to the presence of the polymer modifier, which is reported in figure 7, is surely more evident.

The lack of this test, however, can be find out in the testing rate that can strongly influence the material response. In slower tests, in fact, the viscoelastic material relaxes the stress and plastically deforms, even if the material is characterized by a predominant elastic behaviour when loaded at higher rates. A more effective characterization should imply the use of dynamic tests made at different frequencies in order to fully identify the behavioural response of the system in different loading and temperature conditions.

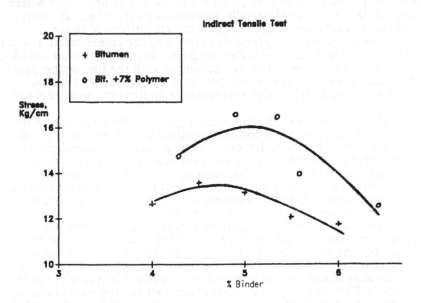

Figure 7: Indirect tensile tests for unmodified and polymer modified bituminous mixtures of different binder content

Conclusions

Ordinary testing methods and more sophisticated calorimetric and dynamic-mechanical characterizations have been used to analyze the influence of a polymer modification on the properties of a commercial type bitumen blended with different amounts of an ethylene-propylene-buthylene copolymer.

The polymer is not dissolved in the bitumen but forms a second phase present as dropplets of 5 to 10 microns. Under shearing these particles highly orient in the flux modifying the viscosity and, hence, improving the system thixotropicity.

The mixtures made with the polymer compounded with the inert aggregate before the addition of the bitumen gave the best mechanical results due to the fact that the polymer forms an elastic coating on the particles improving the stress distribution in the multiphase system.

The polymer modification, usually, increases the bitumen elasticity, as indeed observed from the dynamic-mechanical analysis (DMA). Some of the ordinary characterizations performed were testing, depending on the temperature range at which they were done, only one of the elastic or viscous components of the material viscoelastic response. However, from the analysis of the available literature and from our experimental data, it can be evinced that some of the ordinary characterization tests, conventionally used for bitumens and their mixtures with aggregates, if appropriately correlated between them can still give a correct recognition of the influence of a polymer modification on the viscoelastic response of these complex materials.

The use of the penetration test, which is made at a temperature where the material behaves viscoelastically, can evidentiate this effect. Moreover, the low and relatively high temperature transitions of these multicomponent systems are precisely defined by the calorimetric analysis (DSC), however, the Fraas and softening points evaluations can even identify the limits of an extreme rigidity, associated with the low temperature glassyfication of the system components, or of an excessively fluid behaviour, associated with the high temperature melting of the system components.

The comparison between the indications given by the penetration test and by the softening point of bitumens modified with different types of polymer, can disclose if the polymer is finally acting as a second phase, as in our case, or is dissolved in the bituminous matrix acting as a plasticizer (in this case a reduction of the softening point should be observed).

Finally, further evaluation of the mechanical properties of bituminous mixture modified with different amounts of polymer can be associated with the preliminary observations of the binder characteristics.

In our case, the Marshall test made on samples made from mixture containing different amounts of a polymer modified bitumen, was not able to clearly identify the presence and the degree of improvement of the related mechanical properties. In fact, the test was run at a temperature where only the viscosity change of the bitumen were tested. Conversely, the Indirect Tensile Test, made at room temperature where the system was fully viscoelastic, clearly indicate the enhancements due to the presence of the polymer phase. An optimization of the mixture composition was also obtained from these data: the maximum resistance of the modified and unmodified systems was found at different binder compositions (higher for the modified bitumen).

Even for the complete analysis of the mixture viscoelastic behaviour, however, the use of the dynamic-mechanical properties made by using dynamic-tensile tester could result useful to clarify the characteristics of these complex materials in actual service conditions.

REFERENCES

APICELLA, A. (1985) Effect of Chemorheology on the Epoxy resin Properties, **Development in Reinforced Plastics-5**, Elsevir Applied Science Pub. Ltd.

CASTAGNETTA V. (1988) I leganti ed i comglomerati bituminosi modificati. **Le esperienze internazionali.** Giornata di Studio. Napoli.

COSTANTINIDES G., LOMI C., SHROMEK N. (1987) Mescole di bitumi con polimeri. Conoscenze attuali e ricerche future. **Rassegna del bitume,** No. 2.

DE LA TAILLE G., BOYER G. (1980) Correspondence between transient and dynamic viscoelastic functions in the case of asphalt. **International Congress on Rheology, 8th,** Naples.

DINNEN A. (1985) Bitumen-thermoplastic rubber bends in road applications. **Eurobitume Symposium,** The Hague.

DUMONT A.G. (1986) Methodes basees sur les essais mecaniques modernes. **Rilem,** Olivet, sept.

FRITZ H.W. (1986) Methodes de formulation pour la fabrication des enrobes basees sur les essais mecaniques traditionnels. **Rilem,** Olivet, sept.

GIANNATTASIO P. I leganti ed i comglomerati bituminosi modificati . Relazione introduttiva. Giornata di Studio. Napoli, 1988.

KOLB K. H. (1985) Laboratory evaluation of polymer-modified asphalt **Eurobitume Symposium,** The Hague.

NIEVELT G. (1988) Les liantes et les enrobes bitumineux modifies. **Les expriences autrichiennes.** Giornata di Studio. Napoli.

PALLOTTA S. (1988) Significant methods based on modern phisical and phisico-chemical test. **Rilem,** Dubrovnik , sept.

PERONI G. (1988) I leganti ed i conglomerati bituminosi modificati. **Esperienze italiane.** Giornata di Studio. Napoli.

VALERO L. (1986) L'emploi de la Gyratory Testing Machine dans l'etude des melanges bitumineux. **Rilem,** Olivet, sept.

6 BEHAVIOUR OF ASPHALT AGGREGATE MIXES AT LOW TEMPERATURES

W. ARAND
Institut für Straßenwesen, Technische Universität Braunschweig,
Germany

Abstract
In the end of the seventieth the more frequent occurance
of cracks on the roads of the Federal Republic of Germany
was the motive to develop a process controlled testing
machine for the investigation of the behaviour of asphalt
pavements at low temperatures. By tensile tests and cooling down tests the influence of the composition of asphalt concrete specimens on their behaviour at low temperatures has been investigated. On the way of mathematical analysis by multiple linear regressions the results
of the investigations are demonstrated in charts.
Keywords: Low temperature behaviour, Process controlled
testing machine, Asphalt concrete, Influence of composition, Influence of binder viscosity.

1 Introduction

Solid bodies expand being warmed up. They shrink being
cooled down. Asphalt aggregate mixes in compacted condition react correspondingly.

In asphalt concrete pavements a cooling induced shrinkage cannot occur. As a result of the restrained shrinkage tensile stresses arise. At high temperatures the
tensile stresses are compensated by relaxation. At low
temperatures the asphalt concrete pavements becomes more
elastic and the capability for relaxation decreases. That
is the reason for increasing tensile stresses. If the
tensile stress reaches the tensile strength cracking happens.

In the end of the seventies progressive cracking was
observed on the roads of the Federal Republic of Germany.
The reason for cracking was the application of stiffer
asphalt aggregate mixes. These stiffer materials were
used to avoid rutting.

The more frequent occurance of cracks was the motive
for the Institute for Highway Engineering of the Technical
University of Braunschweig to develop a process controlled testing machine for the investigation of the beha-

viour of asphalt aggregate mixes in compacted condition at low temperatures.

2 Process controlled testing machine

The main elements of the testing machine are a very stiff frame, strain transducers for the measurement of the length of the specimens and a step motor with a gear drive to hold constant respectively to correct the length of the specimens during the test. The strain transducers for the measurement of the length of the specimens are working independently from changes of the temperature with a limit of resolution of 5×10^{-5} mm. They are fixed on thermal indifferent rods which do not change their length when temperature is changing. The step motor works in connection with a gear drive with the same outstanding

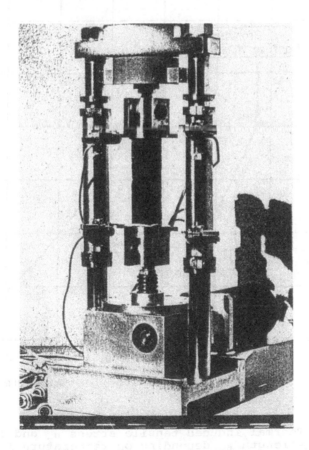

Picture 1: Process controlled testing machine

accuracy. The testing machine is shown in picture 1. The process control is done by a personal computer with accompanying electronical instruments.

The process controlled testing machine allows the realization of test procedures like
- tensile tests at different constant temperatures in the range of + 30 until - 35 °C,
- cooling down tests with constant length of the specimens in the same range of temperatures,
- relaxation tests and retardation tests at temperatures whatever you choose.

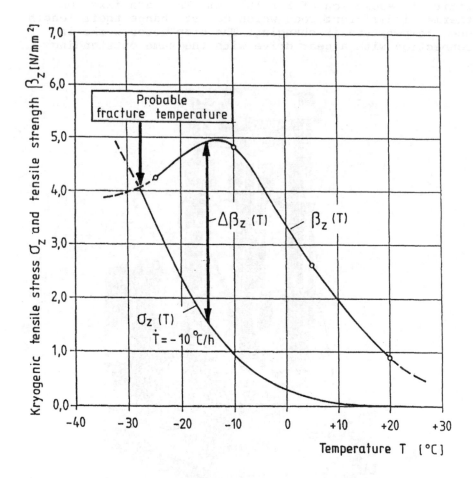

Figure 1: Thermal induced tensile stress σ_z and tensile strength β_z depending on temperature T

The tensile tests at stepwise constant temperatures
deliver the tensile strength β_z and the fracture strain
ϵ_{fr} as a function of the chosen temperature T. The results
of the cooling down tests are the thermal induced (kryogenic) stresses σ_z including the fracture stress σ_{fr} at
the moment of the failure of the specimen and the fracture
temperature T_{fr}. The relations between the different features are shown in figure 1. Attention: The difference
between the tensile strength β_z and the thermal induces
stress σ_z is the so-called reserve capacity of tensile
strength $\Delta\beta_z$ which allows the asphalte concrete pavement
to bear load induced tensile stresses, too.

The relaxation tests and the retardation tests ar more
suitable for fundamental scientific research work.

3 Test results obtained on specimens of the road

With support of the authorities of some countries in the
Federal Republic of Germany samples of the wearing courses of roads in the northern, middle and southern part
of Germany have been taken. These samples showed different degrees of deterioration, that means different
lengths of cracks in cm/m². Specific crack lengths between 0 and more than 300 cm/m² were observed.

From the road samples there were cut prismatic specimens with sectional areas of 40 x 40 mm² and lengths of
160 mm. These specimens were used for cooling down tests
to determine the fracture temperature T_{fr}.

Other portions of the samples were used to extract the
bituminous binder and to find out the softening point
ring and ball.

The results of the investigations led to the following
equations found by simple linear regression analysis:

$$T_{fr} = 0,64 \cdot SP - 64,7 \quad [°C] \qquad (1)$$

$$L_{fr} = 5,8 \cdot SP - 314 \quad [cm/m^2] \qquad (2)$$

In the formulas (1) and (2) means:

T_{fr} = fracture temperature in °C,

L_{fr} = specific crack length in cm/m²,

SP = softening point ring and ball in °C.

The degree of accuracy is

- 88 % for formula (1) and
- 68 % for formula (2).

The coefficient of regression in formula (1) with the value 0,64 indicates that the fracture temperature increases about 6,4 K when the softening point ring and ball arises 10 K. Mathematical transforming of equation (2) leads to the statement that the risk of cracking starts at a softening point ring and ball of about 55 °C. It is true, the last statement is uncertain because of the weak degree of accuracy of 68 %.

4 Influence of composition of asphalt aggregate mixes

To investigate the influence of composition of asphalt aggregate mixes in compacted condition on their behaviour at low temperatures the composition of the mixes was varied systematically. In detail the gradation of the aggregate mix 0/11 mm was changed in 4 steps using the Talbot function (see table 1). The small exponent m = 0,43 means an aggregate mixture with a high content of filler and a small content of chips. The exponent m = 0,55 indicates a low amount of filler and a greater portion of chips.

Table 1: Gradations of the aggregate mixes, described by the Talbot funktion

Exponent m	Gradation of Aggregate Mix in % by Weight							Sum
	< 0,09	0,09/0,25 mm	0,25/0,71mm	0,71/ 2,0 mm	2,0/5,0 mm	5,0/8,0 mm	8,0/11,2 mm	
-	%	%	%	%	%	%	%	%
0,43	12,6	6,9	11,0	17,2	23,0	15,8	13,5	100,0
0,47	10,4	6,3	10,7	17,1	24,0	16,9	14,6	100,0
0,51	8,5	5,9	10,1	17,0	24,8	17,9	15,8	100,0
0,55	7,0	5,4	9,5	16,9	25,4	18,9	16,9	100,0
Talbot Funktion $D = 100 \cdot \left(\frac{d}{11,2}\right)^{m}$ [%]								

The binder content was changed in 3 steps, namely 5,0; 5,6 and 6,2 % by weight.

Additionally 4 kinds of asphalt cement were chosen, namely B 200, B 80, B 65 and B 45, with the features of viscosity shown in table 2.

Additionally 4 kinds of asphalt cement were chosen, namely B 200, B 80, B 65 and B 45, with the features of viscosity shown in table 2.

Table 2 : Characteristic values of the bitumens

Feature	Dimension	Kind of Bitumen			
		B 200	B 80	B 65	B 45
Penetration at T = + 25 °C	1/10 mm	189	89	50	36
Softening Point Ring and Ball	°C	39,0	45,5	51,0	55,5
Penetration Index	-	- 0,73	- 0,98	- 0,95	- 0,65
Break Point Fraaß	°C	-25,0	-17,5	-12,0	-10,5
Duktility	cm	⟩100	⟩100	8,1	5,0
Density at T = + 25 °C	g/cm³	1,010	1,020	1,013	1,019

The asphalt aggregate mixes were prepared in the laboratory by aid of a laboratory mixer. The above mentioned prismatic specimens were manufactured by blow compaction using a Marshall compactor. The specimens were used for simple tensile tests at different, stepwise constant temperatures with a constant deformation rate of 1 mm/min and cooling down tests with constant length of the specimens and a constant temperature rate of $\dot{T}$ = - 10 K/h.

4.1 Results of tensile tests
The results of the tensile tests are given in table 3 and figure 2. Table 3 contents the tensile strengths β_z and the fracture strains ε_{fr} of asphalt concrete specimens 0/11 mm depending on aggregate gradation, binder content and temperature. At temperatures above the freezing point the tensile strength increases with decreasing content of bitumen. The influence of the aggregate gradation is more or less indifferent. At low temperatures a higher binder content benefits the tensile strength as well as a higher content of filler. The highest values for the tensile strengths are obtained in a range of temperatures about T = - 10 °C.

For a high fracture strain high temperatures, high binder contents and high contents of filler are favourable. The last mentioned influence disappears at very low temperatures.

Table 3 : Tensile strengths β_z and fracture strains ε_{fr} of asphalt concrete specimens 0/11 mm depending on aggregate gradation (exponent m), binder content and temperature

Temperature [°C]	Binder content % by weight	Tensile strength β_z [N/mm²]				Fracture strain ε_{fr} [‰]			
		Exponent m =				Exponent m =			
		0,43	0,47	0,51	0,55	0,43	0,47	0,51	0,55
+ 20	5,0	0,65	0,51	0,58	0,44	5,83	5,15	4,18	4,20
	5,6	0,55	0,43	0,52	0,43	6,26	6,15	5,22	4,61
	6,2	0,43	0,41	0,44	0,38	8,58	6,53	5,98	5,10
+ 5	5,0	2,95	2,84	3,19	2,77	2,63	2,71	2,60	2,57
	5,6	2,76	2,65	3,13	2,88	3,25	3,16	2,99	2,73
	6,2	2,45	2,62	2,98	2,71	4,02	3,60	3,25	3,04
- 10	5,0	5,51	5,86	5,57	4,36	0,44	0,54	0,45	0,39
	5,6	6,07	6,10	5,81	5,17	0,58	0,58	0,48	0,48
	6,2	6,66	6,31	5,71	5,02	0,76	0,67	0,54	0,49
- 25	5,0	4,53	4,59	4,46	4,01	0,19	0,20	0,18	0,18
	5,6	4,68	4,64	4,51	4,27	0,19	0,19	0,18	0,19
	6,2	5,33	5,36	4,35	3,85	0,22	0,25	0,19	0,16

With respect to the influence of the kind of binder on the tensile strength and the fracture strain figure 2 shows: A harder bitumen is better for the tensile strength at higher temperatures, a softer one at lower temperatures. Independently of the temperature softer bitumens are giving the greater fracture strain to the asphalt aggregate mix in compacted condition.

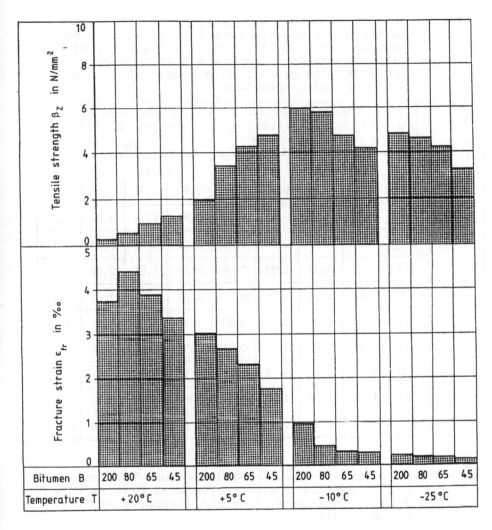

Figure 2: Tensile strength β_z and fracture strain ε_{fr} depending on kind of bitumen B and tempe- rature T

4.2 Results of cooling down tests

The results of the cooling down tests with a constant rate of cooling $\dot{T} = -10$ K/h are given in tables 4 and 5 as well as in figure 3. As table 4 shows on the left side the thermal induced - the so-called kryogenic - stresses increase with decreasing temperatures. A systematic in- fluence of the composition of the asphalt aggregate mix is not detectable. Opposite to this the fracture stresses

Table 4 : Thermal induced (kryogenic) stresses σ_z and
differences between tensile strengths β_z and
thermal induced (kryogenic) stresses σ_z of
asphalt concrete specimens 0/11 mm depending
on aggregate gradation (exponent m), binder
content and temperature

Temperature [°C]	Binder content % by weight	Thermal induced (kryogenic) stress σ_z [N/mm²]				Difference $\Delta\beta_z$ between β_z and σ_z [N/mm²]			
		Exponent m =				Exponent m =			
		0,43	0,47	0,51	0,55	0,43	0,47	0,51	0,55
+ 5	5,0	0,10	0,10	0,11	0,09	2,85	2,74	3,08	2,68
	5,6	0,09	0,09	0,09	0,10	2,68	2,56	3,04	2,79
	6,2	0,07	0,09	0,09	0,09	2,39	2,54	2,89	2,62
- 10	5,0	0,83	0,84	0,90	0,77	4,69	5,02	4,67	3,59
	5,6	0,75	0,74	0,80	0,80	5,34	5,36	5,01	4,37
	6,2	0,63	0,73	0,73	0,75	6,00	5,61	4,98	4,29
- 25	5,0	4,17	4,13	4,15	3,83	0,36	0,46	0,31	0,18
	5,6	4,18	4,08	4,08	4,08	0,66	0,65	0,43	0,19
	6,2	3,93	4,05	4,01	3,94	1,39	1,32	0,34	-0,08

Table 5 : Fracture stresses σ_{fr} and fracture temperatures
T_{fr} of asphalt concrete specimens 0/11 mm depending on aggregate gradation (exponent m) and
binder content

	Binder content % by weight	Fracture stress σ_{fr} [N/mm²]				Fracture temperature T_{fr} [°C]			
		Exponent m =				Exponent m =			
		0,43	0,47	0,51	0,55	0,43	0,47	0,51	0,55
	5,0	4,67	4,61	4,47	3,97	27,0	27,4	26,2	26,0
	5,6	5,41	5,14	4,93	4,28	29,0	28,6	28,0	26,2
	6,2	5,33	4,91	4,78	4,33	29,4	28,2	27,8	27,0

and the fracture temperatures are a little better if the content of binder and the content of filler are a bit higher (see table 5).

The strongest influence to the fracture temperature comes from the kind of binder (see figure 3).

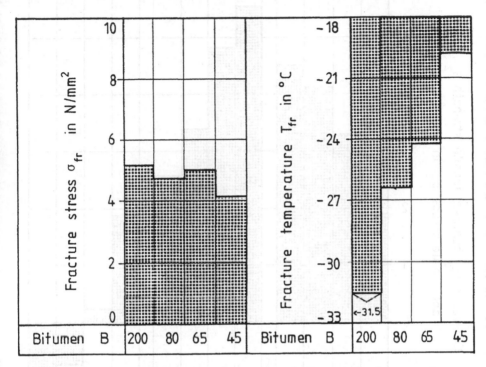

Figure 3: Fracture stress σ_{fr} and fracture temperature T_{fr} depending on kind of bitumen

4.3 Results of tensile tests and cooling down tests in comparison

As shown in figure 1 with falling temperatures the tensile strength increases, reaches a maximum value about T = - 10 °C and falls down. Opposite to this the tensile stress rises monotonously with falling temperatures. If the tensile stress meets the tensile strength cracking occurs.

The difference between the tensile strength and the kryogenic stress at the same temperature is the so-called tensile strength reserve capacity $\Delta\beta_z$. The values for this features are registered in table 4 depending on aggregate gradation, binder content and temperature. At temperatures above the freezing point no significant influence of the composition of the asphalt aggregate mix

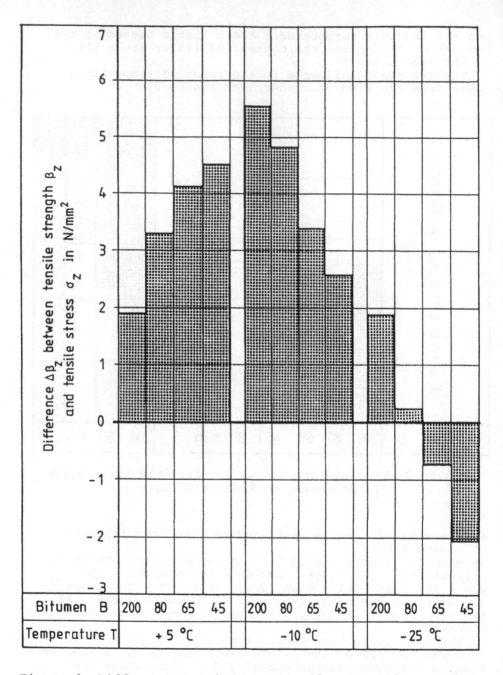

Figure 4: Difference $\Delta\beta_z$ between tensile strength β_z and kryogenic tensile stress σ_z depending on the kind of bitumen B and the temperature T

can be observed. At lower temperatures a highter content
of asphalt cement and a higher content of filler seem to
be advantageously.

The most significant influence owns to the kind of
binder (see figure 4). At higher temperatures - here T =
+ 5 °C - harder bitumens and at lower temperatures softer
bitumens promote the tensile strength reserve capacity.
At very low temperatures - here T = - 25 °C - bitumens
with high viscosity lead immediately to thermal cracking,
recognizable on the negative values for the differences
between the tensile strength and the thermal induced
stress.

5 General view

To obtain a general view multiple linear regression ana-
lysises have been conducted. The results of these analy-
sises are given in table 6. Of importance are the amounts
of the coefficients of the multiple regression and the
positive or negative signs as well as the degrees of accuracy.

Table 6 : Coefficients of multiple linear regression
analysis

Feature	Temper- rature	Coefficients of multiple Regression					Degree of Accuracy
		Constant	Content of Binder	Exponent m	Broken Sand	SP RaB	
-	°C	-	% by weight	-	%	°C	%
Fracture strain ε_{fr} [‰]	+ 20	10,270	1,5286	- 19,357	- 0,0210	- 0,0475	79,6
	+ 5	4,978	0,7049	- 4,402	- 0,0059	- 0,0720	82,1
	- 10	2,231	0,1019	- 1,187	- 0,0012	- 0,0348	63,7
	- 25	0,426	0,0161	- 0,260	- 0,00004	- 0,0042	32,3
Difference $\Delta\beta_z$ between β_z and σ_z [N/mm²]	+ 20	- 1,527	- 0,1227	- 0,710	0,0015	0,0645	85,7
	+ 5	- 4,516	- 0,2103	1,502	0,0002	0,1696	75,8
	- 10	15,635	0,6029	- 10,543	- 0,0118	- 0,1768	75,8
	- 25	12,237	0,3122	- 6,299	- 0,0022	- 0,2242	67,5
σ_z [N/mm²]		8,369	0,2637	- 7,126	- 0,0041	- 0,0281	61,2
T_{fr} [°C]		- 61,315	- 1,0368	16,758	0,0055	0,6783	81,3

To make obviously the relations between the most important features graphical representations are given. In

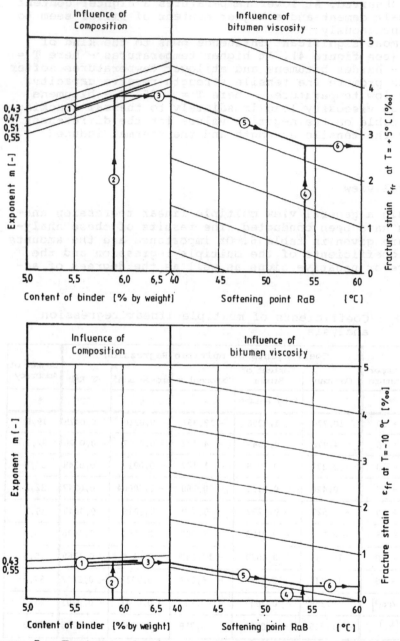

Figure 5: Frature strain ε_{fr} at temperatures of T = + 5 and T = – 10 °C depending on the composition of the mix and the viscosity of the binder

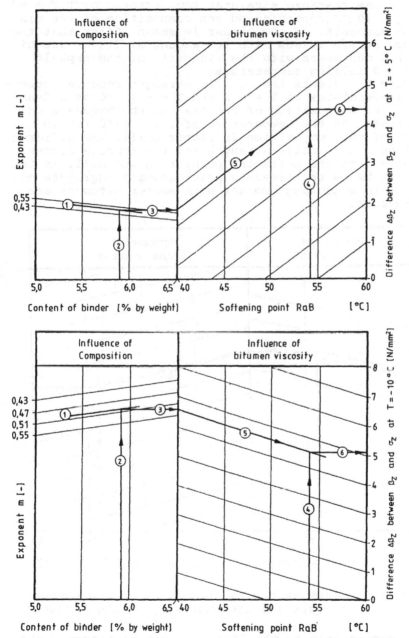

Figure 6: Differences $\Delta\beta_z$ between β_z and σ_z at temperatures of T = + 5 and T = - 10 °C depending on the composition of the mix and the viscosity of the bitumen

figure 5 the fracture strain at temperatures of T = + 5
and T = - 10 °C depending on the composition of the mix
and the viscosity of the binder is shown. After that the
fracture strain arises with the amount of filler and bitumen and decreases with the viscosity of the asphalt
cement ·and falling temperatures.

Figure 6 contents the tensile strenght reserve capacity at temperatures of T = + 5 and T = - 10 °C as a function of the composition of the mix and the hardness of
the bitumen. At the temperature of T = + 5 °C the influence of the composition is rather small. The difference between the tensile strength and the tensile stress
increases with the softening point ring and ball. At temperatures below the freezing point asphalt aggregate mixes rich in bituminous mortar with softer bitumens are
advantageous.

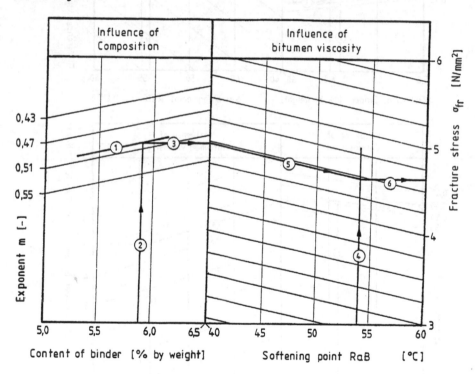

Figure 7: Fracture stress σ_{fr} depending on the composition
of the mix and the viscosity of the bitumen

Correspondingly to the lower diagram of figure 6 the
fracture stress obtained by a cooling down test at the
moment of failure of the specimen goes up with the amount
of bituminous mortar and drops with the viscosity of the
asphalt cement (see figure 7).

Remarkable is the zoom up of the fracture temperature
with the softening point ring and ball as demonstrated by
figure 8. The slope of the curve is represented by a coefficient of regression with a value of 0,6783. This value is very simular to that found on specimens of the
road (see formula 1: 0,64). This allows the assumption
that the fracture temperature generally changes with 2/3
of the softening point ring and ball.

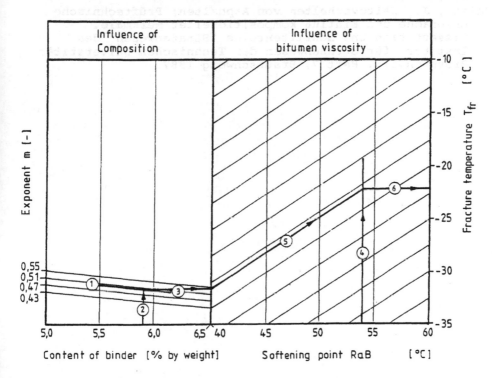

Figure 8: Fracture temperature T_{fr} depending on the composition of the mix and the viscosity of the bitu-
men

Literature:

Arand, W.: Zum Relaxationsverhalten von Asphalten. Das
    stationäre Mischwerk - Der bituminöse Straßenbau 15
    (1981) 4, 14 - 22
Arand, W.: Verhalten von Aspahlten bei tiefen Tempera-
    turen. Die Asphaltstraße - Das stationäre Mischwerk
    17 (1983) 3, 101 - 105

Arand, W.; Steinhoff, G.; Eulitz, J.; Milbradt, H.: Verhalten von Asphalten bei tiefen Temperaturen. Entwicklung und Erprobung eines Prüfverfahrens. Schriftenreihe "Forschung Straßenbau und Straßenverkehrstechnik" des Bundesministers für Verkehr, Abteilung Straßenbau, Heft 407, Bonn-Bad Godesberg 1984

Arand, W.: Kälteverhalten von Asphalt. Die Asphaltstraße - Das stationäre Mischwerk 21 (1987) 3, 5 - 16 sowie 4, 7 - 19

Eulitz, J.: Kälteverhalten von Asphalten; Prüftechnische Ansprache und Einfluß kompositioneller Merkmale. Dissertation und Schriftenreihe "Straßenwesen" des Instituts für Straßenwesen der Technischen Universität Braunschweig, Heft 7, Braunschweig 1987

7 EVALUATION OF THE COHESION OF BITUMEN MIXTURES BY MEANS OF THE INDIRECT TRACTION RESISTANCE TEST

U. BONVINO
Road and Transport Department, Bari University, Italy

Abstract

As it as now become common road design practice to use criteria based on the theory of elasticity or viscoelasticity, it is increasingly important to have simple, quick and, if possible, not very expensive laboratory tests for characterizing the materials used in pavements. As the single load laboratory test for bitumen mixtures, the indirect traction resistance one was chosen both because, despite its simplicity, this approach to analysing the materials used in paving better corresponds to current trends in flexible pavement structural analyses. Furthermore, if one considers particular strenght criteria, this test can quickly and accurately determine the probable asphalt mix cohesion value. This means that this kind of experimental approach can also probably be used in quality control during the production and laying of the seal coats. Thus, for example, the quadratic parabola – assumed to be an intrinsic curve whose vertex is near the tangent to the traction fracture circle – gives extremely simple formulae which, for calculating the mechanical characteristic cohesion use at most just **one** indirect traction test.

<u>Keywords:</u> Indirect Traction, Procedure, Experimental Programme, Numerical Results, Marshall Test, References.

1 Introduction

Extra-urban road avings are subjected to frequent and remarkable stresses, mainly caused by the traffic of heavy vehicles. Therefore the wearing course needs to have particular technological requisites and strength properties.

Among the first nes we can mention the kind and amount of bitumen, and among the second ones the flexible pavement resistance to tensile stress. This last aspect is particularly relevant, as it makes it possible to:

a) define, as we shall explain later, a parameter capable of characterising bitumen mixtures;

b) optimize, in function of the percentage of bitumen, the cohesion of the wearing course and, therefore, its resistance to shearing stresses, which are among the most important stresses to which the pavement is subjected;

c) prepare a quick and efficient test, by means of a simple laboratory equipment;

d) determine the probable values of the mechanical properties of flexible pavements, considering particular fracture criteria;

e) control the quality of bitumen concretes during the phases of production and distribution.

Besides, the choice of the resistance to tensile stress is also determined by other reasons. We considered the fact that the value of this quantity influences the response of bitumen mixtures in the presence of cracks.

2 Indirect traction test resistance

The indirect traction test consists, as it is known, in subjecting a cylindrical test piece to a diametral compression load distributed along two opposed generating lines (figure 1). In this way, an almost homogeneous tensile stress is determined, the direction of which is perpendicular to the diametral plane containing the two loaded generating lines.

In general, taking into account the limits of test temperature and bitumen percentage, bitumen concrete test pieces break as a consequence of cracks along the above quoted plane. If A is the generic point placed in a plane transversal to the test piece longitudinal axis, the stresses caused in it during the indirect traction test are derived from the theory of plane elasticity. Using as a system of coordinates the one shown in figure 2, and indicating with:

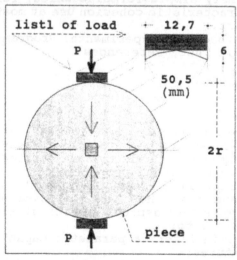

Fig.1. The indirect traction. test.

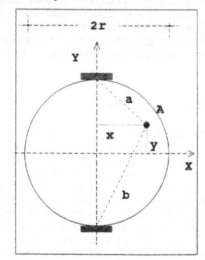

Fig.2. System of coordinates.

- P the applied load;
- 2r and H respectively diameter and height of the test piece;
- x,y the rectangular coordinates (in the adopted X-Y frame of reference) of the considered point;
- a,b its polar coordinates;

we obtain, by means of easy calculations, which are too long to be described here, the following final expressions, referring to the stress components:

$$\sigma_x = -\frac{2\,P}{\pi\,h}\left[\frac{(r-y)\,x^2}{a^4} + \frac{(r+y)\,x^2}{b^4} - \frac{1}{2\,r}\right] \tag{1}$$

$$\sigma_y = -\frac{2\,P}{\pi\,h}\left[\frac{(r-y)^3}{a^4} - \frac{(r+y)^3}{b^4} - \frac{1}{2\,r}\right] \tag{2}$$

$$\tau_{xy} = +\frac{2\,P}{\pi\,h}\left[\frac{(r-y)^2\,x}{a^4} - \frac{(r+y)^2\,x}{b^4}\right] \tag{3}$$

Considering the particular case in which point A is placed along the axis of the abscissas, that is:

$$y = 0 \tag{4}$$

$$a = b = \sqrt{r^2 + x^2} \tag{5}$$

The previous equations become:

$$\sigma_x = -\frac{P}{\pi\,h\,r}\left[1 - \frac{64\,r^2\,x^2}{(4r^2 + 4x^2)^2}\right] \tag{6}$$

$$\sigma_y = +\frac{P}{\pi\,h\,r}\left[1 - \frac{64\,r^4}{(4r^2 + 4x^2)^2}\right] \tag{7}$$

$$\tau_{xy} = 0 \tag{8}$$

On the contrary, should point A be placed on the axis of the ordinates, that is:

$$x = 0 \tag{9}$$

$$a = r - y \tag{10}$$

$$b = r + y \tag{11}$$

The expressions (1), (2),and (3) would take the following form:

$$\sigma_x = \frac{P}{\pi h r} \qquad (12)$$

$$\sigma_y = -\frac{P}{\pi h}\left[\frac{2}{r-y} + \frac{2}{r+y} - \frac{1}{r}\right] \qquad (13)$$

$$\tau_{xy} = 0 \qquad (14)$$

Finally, these equations are furtherly simplified when point A is placed on extreme positions.

Tables 1 and 2 show the variations experienced in these cases by the stress components; in particular when the considered point is placed, respectively, along each orthogonal X-Y diameter of the cylindrical test piece.

TABLE 1. $\qquad 0 \leqslant X \leqslant r$

STRESS	X=0	X = r	PATH
traction	$\sigma_x = \frac{P}{\pi h r}$	$\sigma_x = 0$	non linear
compression	$\sigma_y = -\frac{3P}{\pi h r}$	$\sigma_y = 0$	non linear
shear	$\tau_{xy} = 0$	$\tau_{xy} = 0$	

TABLE 2. $\qquad 0 \leqslant Y \leqslant r$

STRESS	Y=0	Y = r	PATH
traction	$\sigma_x = \frac{P}{\pi h r}$	$\sigma_x = 0$	constant
compression	$\sigma_y = -\frac{3P}{\pi h r}$	$\sigma_y = \infty$	non linear
shear	$\tau_{xy} = 0$	$\tau_{xy} = 0$	

Figures 3 and 4 show, even though from the point of view of quality, the moment of tensile and compression stresses along the two above quoted fiducial axes. In particular, if we examine simultaneously table 1 and figure 3, we find out that the test piece breaks in consequence of traction if its resistance to compression is at least three times higher than its resistance to tensile stress. On the contrary, we infer from figure 4 that the horizontal tensile stress σ_x has a constant value, equal to the one provided by the equation (12); whereas the vertical compression stress σ_y changes from a minimum, in the centre of the test piece, to an infinite value in correspondance of the points of application of the P load on the circumference.

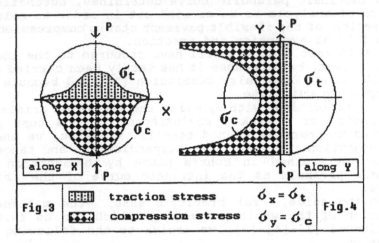

Fig.3	▦ traction stress	$\sigma_x = \sigma_t$	**Fig.4**
	▨ compression stress	$\sigma_y = \sigma_c$	

In these conditions, the test piece should tend to break in consequence of compression, near the loaded areas, rather than in consequence of traction in the central part of the vertical diameter.

The reduce the amount of compression stresses we introduced, near the points of load application, two steel listels, described in detail in figure 1.

3 Arrangement of the procedure

Former studies concerning bitumen concrete wearing courses have confirmed that the intrinsic curve is not linar and that the well known Mohr-Coulomb criterion must be adopted with great caution, fully respecting particular procedures, as it provides extremely dispersed values of cohesion and of the angle of internal friction [1].

The suggested procedure, though belonging to the more general Mohr's theory, allows us to study the $\sigma = F(\tau)$ law and to express it as follows:

$$\sigma = -m + p\,\tau^2 \qquad\qquad (15)$$

It represents, in the $\sigma - \tau$ plane, the equation of a parabola having the σ axis as its own geometrical axis. Finally, to simplify the representation, we considered compression stresses as ositive and tensile stresses as negative.

The adoption of any intrinsic curve implies the possibility of an experimental realization of any stress state. Thus, supposing that these stress states are variable inside the test pieces (but in short intervals and near, anyway, to the origin of the $\sigma - \tau$ frame of reference) and that the levels of crisis are only characterised by peak resistances, the acceptation of the intrinsic parabolic curve determines, automatically, the kind of analysis to be carried out in order to calculate the cohesion of the flexible pavement static compression with simple lateral expansion and traction.

For this last one we shall have recourse to the indirect traction test, both because it has lately been carried out by the CNR Road Materials Commission [2], and because it is of simple and quick use.

Thus, indicating with σ_c and σ_t the levels of crisis characteristic of ' **peak** ' resistances of the test pieces subjected to compression and tensile stresses, we analysed the connection between the perpendicular and tangential stresses determined in Mohr's plane by the adoption of the quadratic parabola as the intrinsic curve of the bitumen mixture.

Leaving aside, for brevity's sake, the mathematical expression of the whole procedure, which can be furtherly investigated [3]-[4]-[5], we arrive to the following final expression:

$$\sigma = -\frac{1}{8}\frac{(\sigma_c - \sigma_t)^2}{\sigma_c + \sigma_t} + \frac{2}{\sigma_c + \sigma_t}\tau^2 \qquad (16)$$

Reminding that $\sigma_t < 0$, because we considered tensile stresses as negative.

The equation (16) can be applied to bitumen mixtures because point T2 always admits the real ordinate (figure 5).

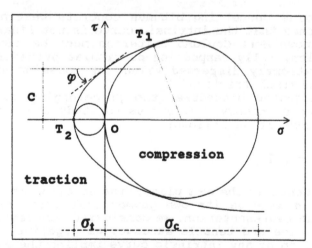

Fig.5. The intrinsic curve: quadratic parabola.

This condition leads to:

$$|\sigma'_t| \leqslant |\sigma_t| \qquad (17)$$

being:

- σ_t the abscissa regarding the traction breaking value,

- σ'_t the abscissa of the tangential point T2 of the cor-respondant circle of crisis provided by the following expression (3):

$$\sigma'_t = - \frac{\sigma_c - \sigma_t}{4} \qquad (18)$$

and for the:

$$- \frac{\sigma_c - \sigma_t}{4} \leqslant \sigma_t \qquad (19)$$

eliminating the negative sign, as a consequence of the geometrical formulation given to the problem, we obtain:

$$\frac{\sigma_c}{|\sigma_t|} \leqslant 3 \qquad (20)$$

Then the osculating circle at V is the circle of crisis in uniaxial traction regimen (figure 5).

The said inequality represents a crucial point in the whole procedure, provided that the experiments confirm that:

$$\frac{\sigma_c}{|\sigma_t|} \cong 3 \qquad (21)$$

In that case the equation (16) can be simplified, once the formal correctness of the signs ($\sigma_c = -3\sigma_t$) is restored, as follows:

$$\sigma = \sigma_t - \frac{1}{\sigma_t} \tau^2 \qquad (22)$$

or is rewritten in its equivalent:

$$\sigma = - \frac{\sigma_c}{3} + \frac{3}{\sigma_c} \tau^2 \qquad (22bis)$$

These last expressions can be applied to bitumen mixtures, provided that the previous hypotheses are respected (21), in order to determine quickly the required mechanical characteristics.

As a matter of fact, as the cohesion is geometrically defined by the intercept of the intrinsic curve on the axis of ordinates, cancelling the equations (16), (22), and (22 bis) we obtain, respectively, the following relations:

$$- \frac{1}{8} \frac{(\sigma_c - \sigma_t)^2}{\sigma_c + \sigma_t} + \frac{2}{\sigma_c + \sigma_t} \tau^2 = 0 \qquad (23)$$

$$\sigma_t - \frac{1}{\sigma_t} \tau^2 = 0 \qquad (24)$$

$$- \frac{\sigma_c}{3} + \frac{3}{\sigma_c} \tau^2 = 0 \qquad (25)$$

with $\sigma_t < 0$ and $\sigma_c > 0$.

Leaving aside, as usual, the intermediate stages and introducing, where it is necessary, the symbol of absolute value (in order to simplify the practical application of the whole procedure), we obtain the cohesion value of the bitumen mixture through quick and simple formulas, which utilize the two commonest laboratory tests:

$$C = \frac{\sigma_c + |\sigma_t|}{4} \qquad (26)$$

$$C = |\sigma_t| \qquad (27)$$

$$C = \frac{\sigma_c}{3} \qquad (28)$$

In conclusion, even though with the limit of the (21), the equation (26) determines the cohesion through the general formulation of the procedure; the (27), on the contrary, allows the calculation of the same quantity, but through a single test, the traction test. Finally, the equation (28) achieves the same result by means of the single static compression test with simple lateral expansion.

It is necessary to remind that, from a theoretical point of view, the two repeatedly quoted laboratory tests are not capable, if they are utilized separately, of a univocal characterisation of bitumen mixtures, because, as it is known, the pavements wearing courses are constantly subjected to composite stresses, which are much more complex than simple ones, such as compressive or tensile stresses.

However, a more careful examination of the three final equations allows us to affirm that, as far as the described procedure is concerned, the two repeatedly quoted laboratory tests, even though they are utilized separately, provide the 'variation limits' of the cohesion value of the bitumen concrete.

In particular, as far as the validity of the equation (21) is concerned, we can see that:

- the inferior limit is provided by the indirect traction breaking stress,

- the superior limit is provided by one third of the simple compression breaking test.

4 Experimental programme

Indirect traction tests were carried out acting on a single type of bitumen concrete; the granulometric curve of the granular mix (composed by crushed aggregates) turned out to be completely internal to the curves estimated by the ANAS specifications, normally used for expressways wearing courses.

The composition of the mix and sizes are shown in table 3, whereas the following table shows the numerical results derived by a few tests carried out on bitumen.

TABLE 3.

CRUSHED AGGREGATES				APPARENT SP.GR.
sand ...... 55 %				
shot ...... 45 %				2650
100 %				(daN/mc)

GRADING ANALYSIS (passing sieve)			
U.N.I.	Mix	Shot	Sand
15	100.0	100.0	100.0
10	95.6	90.2	100.0
5	64.4	22.7	98.5
2	33.9	5.9	56.7
0.40	15.2	3.2	25.1
0.18	9.4	2.6	15.0
0.075	7.4	2.3	11.5

TABLE 4.

PENETRATION at 25°C	(dmm)	94
SOFTENING POINT	(°C)	45.6
FRASS POINT	(°C)	-10
DUCTILITY at 25°C	(cm)	>100
VOLATILITY at 163°C	(%)	0.21
SOLUBILITY in CS_2	(%)	99.14

All indirect traction tests were carried out according to the rules shown in [2], among which we can mention:

- diameter of test pieces: d=4 inches;

- height of test pieces around 63,5 mm;

- aggregates and bitumen temperatures, besides mixing and compaction temperatures, according to CNR rules (8);

- utilization of an automatic Marshall compactor (75 blows per basis);

- test pieces time of seasoning 12h;

- test pieces climatic conditioning before the test: 25°C for 6 hours;

- utilization of a pair of duly shaped steel listels, having a transversal section as shown in figure 1;

- constant approach speed between the press plates, measuring 0,85±0,05 mm/sec.

For each variable bitumen percentage in the interval 3%÷7% (with regard to the aggregates weight), we prepared 9 groups of test pieces, each of which consisting of 4 cylinders, which makes 180 specimens on the whole. After the phase of seasoning, we measured the geometrical characteristics of the test pieces. On their cylindrical surface we drew a generating line and, on each basis, two 90° diameters; then two more generating lines which defined, together with the one already drawn, two mutually orthogonal axial planes.

For each test we could thus measure 4 heights, and we derived the corresponding arithmetical mean. Whenever we observed, between two single values, a deviation of more than 10% from the mean of the four measures, we proceeded to discard the test piece. Afterwards, we determined the average diameter for each specimen, carrying out two orthogonal measures along the intermediate section and the same number on each of the two bases; finally, we calculated the arithmetical mean of the six quantities thus obtained. In this case, we discarded those test pieces which showed a difference between the measures superior to the tolerance stated in one millimeter.

After the necessary climatic conditioning, we proceeded to the breaking of the test pieces belonging to each group, placing the two steel listels on the two diametrically opposed generating lines, for which the difference between the measured values turned out to be very slight.

After carrying out the test, we calculated the P breaking loads (not the strains, these quantities being irrelevant to the aims of this research) and lastly we calculated the corresponding stresses with the known expression:

$$\sigma_t = \frac{P}{\pi h r} \tag{29}$$

5 Numerical results

Later on , the numerical results obtained from the experiments-
were elaborated with the aid of statistical analysis, and
through different phases.

- Phase 1. For each percentage of bitumen the following
statistical indexes were determined: average value, mean
square value, variancy, standard deviation, coefficient of
variation, variancy degree.

- Phase 2. Analysis and comparison of the previous statistical indexes.

- Phase 3. Through the regression and correlation analysis,
we individuated the most significant connections existing
between the percentage of bitumen and the cohesion measure,
this last one being provided by the indirect traction test.

Thus, table 5 shows, for each percentage of bitumen, the
average strength values of the four test pieces of each
group. On the contrary, we sometimes only calculated the mean
of three specimens, having discarded the fourth value by
means of the fidelity criterion described below.

TABLE 5.

GROUPS	BITUMEN CONTENT				
n.	3%	4%	5%	6%	7%
1	6,11	8,01	9,25	7,86	6,91
2	7,05	6,89	9,81	8,28	8,11
3	5,68	8,64	10,36	8,85	7,01
4	6,79	8,81	9,39	9,54	8,28
5	6,85	9,38	10,44	9,41	6,87
6	5,91	8,39	10,92	9,77	8,26
7	7,15	9,08	10,32	8,32	7,85
8	6,58	8,84	9,60	9,01	8,19
9	7,37	8,99	9,19	9,60	8,60

The fidelity criterion consisted in verifying that, after di-
sposing the four values of the obtained strengths in increasing order:

$$R_1 < R_2 < R_3 < R_4$$

after calculating the $R_1 - R_m$ and $R_1 - R_m$ differences, the
greatest of them as absolute value turned out to be smaller
than, or at least equal to 1.46s, having indicated with R_m and
s, respectively, the mean and the standard deviation of the
four determinations; according to these rules we prepared
tables 5 and 6 from which we derive the following results.
a) Independently on the particular numerical values obtained, the increase of the percentage of bitumen determines

higher resistances to tensile stress. However, this phenomenon is observed below a given proportioning of bitumen,
because these resistances slowly decrease when it is
exceeded.

b) It is amazing to remark how the connection between the
two variables can be expressed, as we shall better explain in
the next paragraph, by means of a parabolic function (with
its concavity downwards), confirming that for a given bitumen
concrete there is a best value of bitumen which makes its
resistance to tensile stress become maximum.

c) The measure of the coefficient of variation, which reaches its minimum value (5,82%) in proximity to that best value
of percentage of bitumen, appears equally significant.

TABLE 6.

σ_t	BITUMEN CONTENT				
	3%	4%	5%	6%	7%
Mean value	6,61	8,56	9,92	8,96	7,72
Mean square value	43,99	73,74	98,74	80,69	59,93
Variancy	0,31	0,24	0,33	0,41	0,33
Standard deviation	0,55	0,49	0,58	0,64	0,57
$s/\sqrt{n}$	0,18	0,16	0,19	0,21	0,19
Coeff.of variation %	8,38	8,16	5,82	7,16	7,42
Variancy degree	debole	debole	debole	debole	debole

The investigation is completed by schedule 7, which sums up
the average values of the characteristics of bitumen concrete
test pieces subjected to the traction test.

TABLE 7.	BITUMEN CONTENT				
	3%	4%	5%	6%	7%
σ_t (daN/cmq)	6,61	8,56	9,92	8,96	7,72
Porosity (%)	15,2	12,0	9,5	7,5	4,5
Voids index	0,18	0,14	0,10	0,08	0,05
Density (daN/dmc)	2,14	2,19	2,23	2,25	2,29

6 The 'best value' of bitumen

The previous numerical results were furtherly elaborated with
the aid of the regression analysis. As a matter of fact, it
is known that any physical phenomenon can be studied
assembling the experimental data which express the relation

existing between two or more variables, of which the phenomenon is a function.

Besides, for several reasons, including the necessity of a synthetical representation of tests results with their successive elaboration, and also the opportunity of carrying out eventual operations of derivation, interpolation, etc., it seemed useful to express, by means of an empirical equation, the connection existing between the considered variables.

Consequently, after collecting all the numerical data, we tried to find the equation which would best represent the said connection.

As the considered variables are two:

- X: proportioning of bitumen
- Y: traction breaking stress,

the individuation of the mathematical connection allowed us to estimate the 'resistance to tensile stress', as a dependent variable of the percentage of bitumen.

Among the regression curves we employed the following ones:

REGRESSION	EQUATION
linear	$Y = aX + b$
parabolic	$Y = a + bX + cX^2$
geometric	$Y = aX^b$
exponential	$Y = ab^x$

In them X and Y represent, respectively, the percentage of bitumen (though expressed as a whole number) and the measure (in daN/cmq) of the resistance to tensile stress.

The analysis ended with the calculation of the a,b and c constants, carried out utilizing statistical procedures and determining, for each regression curve, the corresponding coefficient of correlation (r).

This last parameter promptly indicated the kind of connection existing between the two variables. Figure 6, in fact, shows the numerical results of the coefficients obtained by each regression curve, individuating at the same time in the parabolic curve the only one capable of expressing correctly (r=0,982), by means of a mathematical function, the required connection:

$$\sigma_t = -7,25 + 6,48\, b - 0,62\, b^2 \qquad (30)$$

Then the analysis just carried out allowed us to draw the curve of figure 7, which shows, in the %b $- \sigma_t$ plane, a characteristically bell shaped parabola (with its concavity downwards), to signify that the unitary values of the

97

resistance to tensile stress breaking increase as does the percentage of bitumen.

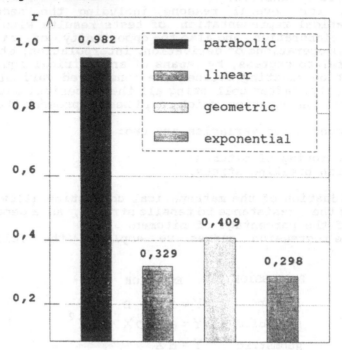

Fig.6. Coefficients of correlation.

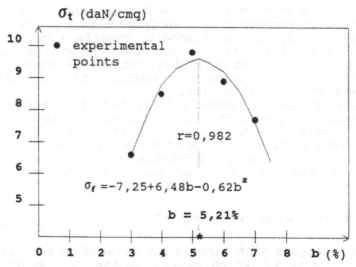

Fig.7. Relation between: % bitumen and
resistance to indirect traction test (25°C).

However, once a given value (best value) is exceeded, they start to decrease as quickly as they increase in the first part.

It follows that it is possible to individuate the %b which determines the maximum resistance to tensile stress and therefore, on the basis of what came out of the application of the previously described procedure, also the corresponding cohesion value.

This percentage, which in the case of the examined mixture turned out to be: b = 5,21%, can be considered as best value, with which the examined bitumen concrete, compatibly with the extent of strains, can be made.

7 The connection with the Marshall test

Once ascertained, then, that it is possible to individuate, by means of the indirect traction test, that bitumen content which gives the maximum cohesion to the bitumen mixture, the necessity of verifying the possible correspondance with the best value of bitumen provided by the classical Marshall test [8] followed as a natural consequence.

Therefore, we prepared with the same mixture five series of test pieces (of suitable number), each of which was characterised by a different bitumen percentage included, obviously, in the 3% ÷ 7% interval. The corresponding numerical results are summed up in table 8, which shows, besides the average stability values (compaction with n=75 blows), also the sliding and stiffness values and other significant characteristics of Marshall cylindrical test pieces.

TABLE 8.		BITUMEN CONTENT				
		3%	4%	5%	6%	7%
M	Stability (daN)	861	1196	1414	1300	1130
A	Flow (mm)	1,2	1,7	1,8	2,3	2,9
R S	Rigidity (daN/mm)	718	686	786	565	424
H A	Density (daN/dmc)	2,19	2,24	2,28	2,30	2,34
L	Porosity (%)	13,8	10,5	7,5	5,5	2,5
L	Voids index	0,16	0,12	0,08	0,06	0,02

Afterwards, using the usual statistical approach, we calculated the variation law of Marshall stability, in function of the percentage of bitumen, obtaining the following parabolic regression curve:

$$S = -1345 + 1023\,b - 96\,b^2 \qquad (31)$$

whose coefficient of correlation turned out to be, as it was expected, very near to the unit (r=0,988).

The previous equation represents, in the %b - S cartesian plane (figure 8), the classical and well-known Marshall parabola; from it we derived the best value of bitumen: b = 5,34% which appears as being very near to the one derived from the traction test.

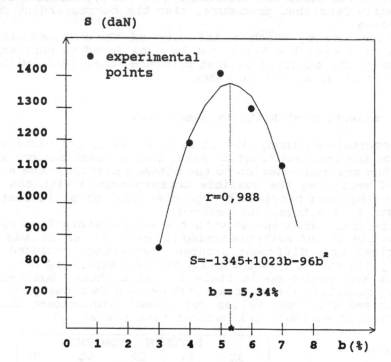

Fig.8. Relation between: % bitumen and Marshall stability test (60°C).

Finally, the investigation meant to ascertain a correspondance of the results provided by the indirect traction test and by the Marshall test, individuated a close linear relation between the two variables (r=0,984):

$$\sigma_t = + 1,351 + 0,593 \, S \tag{32}$$

This equation, graphically represented in figure 9, can be applied provided that the resistance to indirect traction (test temperature 25°C), is expressed in daN/cmq and the Marshall stability (test temperature 60°C) in KN.

It will be also interesting to determine the possible connection between the Marshall sliding values and the vertical diametral strain values of the test pieces subjected to the indirect traction test, but this will be studied later.

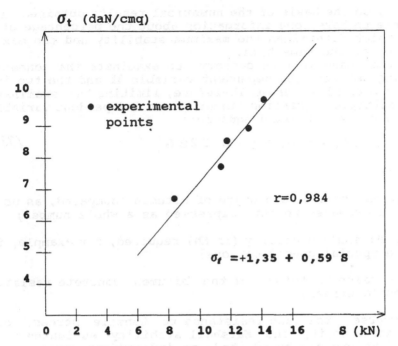

σ_t (daN/cmq)

experimental points.

$r=0,984$

$\sigma_t = +1,35 + 0,59 \ S$

S (kN)

Fig.9. Relation between: Marshall stability (60°C) and indirect traction test (25°C).

8 Conclusions

Independently on the numerical results obtained, the investigation carried out turned out to be particularly interesting. In fact, the verified possibility of determining bitumen concretes cohesion values by means of the adoption of the intrinsic parabolic curve, with the consequent utilization of the single indirect traction test, has the remarkable advantage of simplicity and quickness over better known procedures making use of more complex equipment.

The number of tests carried out, even though they were based on a single type of mixture, allows to think that the results obtained will always be qualitatively verified, provided that the simple compression average breaking stress is almost three times the indirect traction average breaking stress.

We believe therefore that the procedure used to determine the percentage of bitumen, which determines the maximum mixture cohesion, is more reliable than other ones (such as the Marshall Procedure), as it uses, as a parameter to optimize, the resistance to shearing stresses, which are among the most important stresses to which a pavement wearing course is subjected.

Besides, on the basis of the numerical results obtained, it is possible to have some information about the percentage of bitumen which determines the maximum stability and the maximum cohesion at the same time.

In that case it is necessary to examinate the connection existing between the dependent variable %b and the two independent variables S and σ_t. Therefore, limiting the examination to a multiple regression linear to two independent variables, we obtain the following equation:

$$b = 9,051 + 4,064 \ S - 6,228 \sigma_t \qquad (33)$$

in which:

- b represents the percentage of bitumen (compared, as usual, to the aggregates weight) expressed as a whole number;

- S the Marshall stability (in KN) required, for example, from the bid special specifications;

- σ_t the value in daN/cmq of the bitumen concrete resistance to tensile stress.

Of course, the same resistance to tensile stress, or in alternative to it, the Marshall stability estimated by bid specifications can be adopted as independent variable.

On the contrary, for what concerns the comparison between the results derived by the application of the Marshall test and of the indirect traction test, even though with the limits imposed by the utilization of a single bitumen mixture, we can affirm what follows.

1) The two said tests provide practically equivalent results, as the corresponding best values of bitumen turn out to be more or less coincident. In particular, their numerical difference only amounts to 0,13% of the weight of the aggregates mixture (figure 10).

2) It follows that the utilization of the best value of bitumen, derived from the traction test, allows, from an economiacal point of view, a slight reduction of the bitumen content; with regard to the examined mixture, it turned out to be inferior of almost 2,4% to the one which optimizes the stability.

3) The regression analysis showed the very good adaptability of the traction test results to the parabolic function; in fact, the corresponding coefficient of correlation, turned out to be very high (r=0,982) and aligned to the Marshall test coefficient (r=0,988).

4) Last but not least is the examination of the coefficient of variation on the traction test average breaking values (table

6); in fact, it is not by chance thath this important parameter reaches its minimum value (5,82%) in proximity to the best value of the percentage of bitumen.

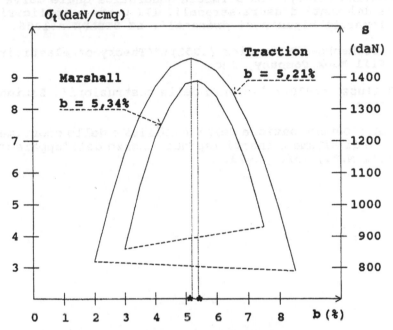

Fig.10. Best values of bitumen derived from the traction and Marshall test.

9 References

[1] U.Bonvino (1988): 'La definizione della coesione e dell'attrito dei conglomerati bituminosi e suoi limiti'. LE STRADE n.1248.

[2] C.N.R.: "Determinazione della resistenza a trazione indiretta e della deformazione a rottura di miscele di aggregati lapidei e bitume". (swelling)

[3] U.Bonvino (1989): 'La parabola quadratica quale curva intrinseca dei manti d'usura stradali. I parte -impostazione della procedura'. LE STRADE n.1254.

[4] U.Bonvino (1989): "La parabola quadratica quale curva in-

intrinseca dei manti d'usura stradali. II parte - applicazione ai conglomerati bituminosi impermeabili'. **LE STRADE n.1255.**

[5] U.Bonvino (1989): 'La parabola quadratica quale curva intrinseca dei manti d'usura stradali. III parte - applicazione ai conglomerati bituminosi drenanti'. **LE STRADE** n.1256.

[6] S.Timoshenko-J.N.Goodier (1951):"'Theory of elasticity'. **McGraw-Hill Book Company**, Inc.

[7] O.Belluzzi (1970): 'Scienza delle costruzioni'. **Zanichelli**, Bologna.

[8] C.N.R.:'Determinazione della stabilità e dello scorrimento di miscele di bitume e inerti lapidei a mezzo dell'apparecchio Marshall'. **N.30**, 15/03/1973.

8 STRESS DEVIATION IN BITUMEN MIXTURES

U. BONVINO
Road and Transport Department, Bari University, Italy

Abstract

It is know that triaxial tests give useful information on the constitutive law of asphalt mixtures even if it has been shown that this is only so in particular conditions of stress and strain. If the results of these tests are to be extended with even the minimum degree of reliability to real situations it is necessary to use parameters which are at least independent of the chosen system of coordinates. In other words, the parameters should be tied to the stress and strain constants. As numerical results are affected by different factors tied to the conditions the triaxial test is carried out in and the nature of the aggregate, some parameters were kept constant (test temperature, load application speed, type of bitumen, lithic elements, etc.) while others were varied (lateral pressure and the quantity and arrangement of spaces in the seal coat).
Keywords: Invariants, Laboratory tests, Experiments, Characteristics curves, The collapse line, Conclusions, References.

1 Introduction

This report is set against the background of the rheology of bitumen concretes, and aims at defining the response of the viscoelastic materials employed in the construction of flexible pavements.

In particular, this report aims at examining, on one side, the behaviour of bitumen mixtures when they experience state of stress changes; on the other side, it aims at individuating the main parameters which regulate their mechanical behaviour, in order to provide an experimental support to the observed phenomenons.

We paid special attention to the examination of test pieces strength properties. By means of a traditional triaxial equipment, test pieces were subjected to gradually increasing deviatoric stresses, until they reached cracking conditions. It is known that triaxial tests provide, in particular conditions, useful information on bitumen concrete resistance, and more generally, on their constitutive law [1]. It is also known that, by means of the construction of Mohr's circles, it is possible to represent, at any time, the stress-strain of any material element, whenever the body to which it belongs undergoes stress variations.

The behaviour of some ideal materials, such as perfectly elastic materials, only depends on their initial and final states, and it doesn't depend on the way they are connected. On the contrary, the behaviour of bitumen mixtures depends on the evolution of the stress-strain state.

Therefore, in order to represent as a whole the different stress states achieved during a generic loading, it would be necessary to construct all the different corresponding Mohr's circles in the same diagram. However, this kind of graphical representation would turn out to be complex, of little significance and perhaps of difficult practical use. This is the reason why the notion of 'stress and strain paths', applied in previous experimental researches on the same materials [2]-[3]-[4], has been introduced.

Using the same approach, and considering the bitumen concrete as comparable to a continuous solid, the stress state in each of its points is individuated, in general conditions, by the symmetric tensor having nine σ_{ij} components. If the frame of reference changes, the stress tensor changes too. However, it is obvious that the stress state at one point of the examined material cannot depend on the way it is represented. We can therefore infer that bitumen mixtures strength properties, as for any other solid material, will depend on some of its quantities which are **'invariable'** with regard to a change of the fiducial axis, rather than on the stress tensor as a whole. This research is in fact based on some of these quantities concerning 'closed' bitumen concretes, whereas 'draining' bitumen concretes will be considered later.

2 The choice of the invariants

If we make use of matrix notations, the stress tensor can be synthetically expressed in the following way:

$$\sigma_{ij} = \begin{bmatrix} \sigma_x & \tau_{xy} & \tau_{xz} \\ \tau_{yx} & \sigma_y & \tau_{yz} \\ \tau_{zx} & \tau_{zy} & \sigma_z \end{bmatrix} \tag{1}$$

Besides, as it is useful to divide the stress state into an isotropic and a deviatoric component, the same σ_{ij} tensor can be rewritten in the following way:

$$\sigma_{ij} = \begin{bmatrix} \sigma_m & 0 & 0 \\ 0 & \sigma_m & 0 \\ 0 & 0 & \sigma_m \end{bmatrix} + \begin{bmatrix} (\sigma_x - \sigma_m) & \tau_{xy} & \tau_{xz} \\ \tau_{yx} & (\sigma_y - \sigma_m) & \tau_{yz} \\ \tau_{zx} & \tau_{zy} & (\sigma_z - \sigma_m) \end{bmatrix} \quad (2)$$

where: $\quad \sigma_m = \dfrac{(\sigma_x + \sigma_y + \sigma_z)}{3}$ $\qquad\qquad$ (3)

represents the mean of normal stresses along the x, y and z axis (figure 1).

Therefore, the stress state around a point consists of a spherical or hydrostatic symmetry, individuated by the spherical tensor (responsible of volume changes without shape variations), and by another one, defined by the stress deviation tensor (responsible of shape variations).

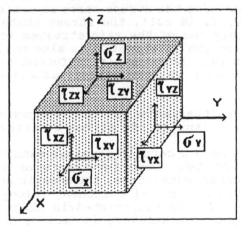

FIG.1 - Stress components

This last component measures the stress state deviation which, on the basis of what we have already said, turns out to be characterised by three invariants, that is by three quantities, the value of which doesn't change if there is a change in the orientation of the frame of reference, according to which the stress state components are measured.

The three state of stress invariants, called invariants of first, second and third rank are defined (in terms of main stresses) as follows, respectively:

$$\begin{cases} I_1 = \sigma_1 + \sigma_2 + \sigma_3 & (4) \\ I_2 = \sigma_1\sigma_2 + \sigma_1\sigma_3 + \sigma_2\sigma_3 & (5) \\ I_3 = \sigma_1\sigma_2\sigma_3 & (6) \end{cases}$$

The stress deviation invariants, expressed in function of the previous invariants, are:

$$\begin{cases} J_1 = 0 & (7) \\ J_2 = I_2 - I_1^2/3 & (8) \\ J_3 = I_3 - I_1 I_2/3 + 2\, I_1^3/27 & (9) \end{cases}$$

It is essential to notice that any other perfectly symmetrical combination of the stress tensor components is an invariant [5] (for example $\sigma_1^2 + \sigma_2^2 + \sigma_3^2$), which can be expressed as a combination of the above quoted ones. Besides, it is apparent that if I_3 is different from zero, so are the three main stresses, in that case the stress state is called triaxial. If, on the contrary, I_3 is null, the stress state can be biaxial or plane, and only one of the main stresses is null; finally, it is uniaxial if the invariant I_2 is also null. In this case, only one of the main stresses is different from zero.

According to what we have already asserted, the examined invariants are:

- the first rank invariants of the isotropic tensor I_1,
- the second invariant of the stress deviator J_3.

Besides, they have a certain physical meaning, which allows the definition of two other invariants on which the whole research will concentrate. Thus, the generic stress vector $\vec{\sigma}$ can be divided into a normal component (along n), and into a component tangential to the octahedric plane (figure 2).

In the theory of plasticity, in fact, a special attention is given to stresses acting on planes, the inclination of which is the same as the direction of the main stresses; such planes define, in fact, a octahedron, and therefore the stresses acting on them are called octahedric stresses. These stresses can be expressed, by means of procedures too long to be described here, with the following components, which are, generally speaking, expressed in the following way:

$$\sigma_{oct} = \frac{1}{3} \left(\sigma_x + \sigma_y + \sigma_z \right) \qquad (10)$$

$$\tau_{oct} = \frac{1}{3} \sqrt{(\sigma_x - \sigma_y)^2 + (\sigma_y - \sigma_z)^2 + (\sigma_z - \sigma_x)^2 + 6(\tau_{xy}^2 + \tau_{yz}^2 + \tau_{zx}^2)} \qquad (11)$$

invariable, in their turn, with regard to the three belonging to the frame of reference. We have in fact:

$$\sigma_{oct} = \frac{1}{3} I_1 \qquad (12)$$

$$\tau_{oct}^2 = -\frac{2}{3} J_2 \qquad (13)$$

In terms of main stresses, the aforesaid equations become:

$$\sigma_{oct} = \frac{1}{3} (\sigma_1 + \sigma_2 + \sigma_3) \qquad (14)$$

$$\tau_{oct} = \frac{1}{3} \sqrt{(\sigma_1 - \sigma_2)^2 + (\sigma_2 - \sigma_3)^2 + (\sigma_3 - \sigma_1)^2} \qquad (15)$$

and therefore, the stress state can also be represented in the space of the main stresses (figure 3).

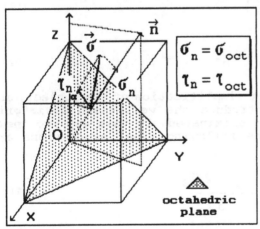

$\sigma_n = \sigma_{oct}$

$\tau_n = \tau_{oct}$

octahedric plane

FIG.2- State of stress in the physical space.

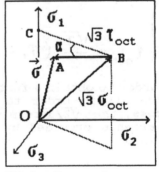

FIG.3 - State of stress in the space of principal stress.

From this figure we derive that, for a full description of the $\vec{\sigma}$ vector position, it is necessary to know a third invariant.

This can be usefully chosen as the α angle which the plane passing through OAB forms with the OCB plane.

To sum up, the stress state at point A can be described with the aid of the three following invariants:

1) the OB segment along the diagonal space (half-line along which $\sigma_1 = \sigma_2 = \sigma_3$):

$$OB = \sqrt{3}\ \sigma_{oct} \tag{16}$$

2) the BA segment in which lies a plane which is orthogonal to the diagonal space:

$$BA = \sqrt{3}\ \tau_{oct} \tag{17}$$

3) the α angle which defines the rotation of AB with regard to the vertical plane containing the diagonal space.

In the specific case of stress states determined by the triaxial test, for which $\sigma_2 = \sigma_3$ (stress state in conditions of radial symmetry), the α angle is null, and therefore, the two first invariants are sufficient to describe the stress state:

$$\begin{cases} \alpha = 0 & (18) \\[2mm] \sigma_{oct} = \dfrac{1}{3}\ (\sigma_1 + 2\sigma_3) & (19) \\[2mm] \tau_{oct} = \dfrac{\sqrt{2}}{3}\ (\sigma_1 - \sigma_3) & (20) \end{cases}$$

In order to obtain a further simplification of the subject, and above all in order to introduce the well-known **deviatoric stress**, the $\sqrt{2}/3$ term was eliminated. In this way two 'new' invariants provided by the following final expressions are defined:

$$\begin{cases} p = \sigma_{oct} = \dfrac{1}{3}\ I_1 = \dfrac{\sigma_1 + 2\sigma_3}{3} & (21) \\[3mm] q = \dfrac{3}{\sqrt{2}}\ \tau_{oct} = \sqrt{\dfrac{3}{2}\ J_2} = \sigma_1 - \sigma_3 & (22) \end{cases}$$

Later on, we'll refer to compression tests with simple and restrained lateral expansion ($\sigma_1 > \sigma_3$).

3 Arrangement of laboratory tests

A laboratory equipment can be considered as being the 'ideal' one, if it is able to modify the stress and strain state of the test piece which must be analysed, starting from initial conditions identical to those in situ, and following, in this way, a strain path similar to the one induced in the same element, placed in its natural seat, by a variation of conditions on the external surface.

However, all this is difficult to be obtained in practice, because sampling and preparation procedures determine an alteration of the stress state in test pieces, and also because complex variations of the strain state would require highly sophisticated (from the point of view of construction and use) experimental equipment. All this would make their utilization too difficult and would also make the interpretation of measures too complicated.

It follows that, considering the restrictions imposed by problems of construction and utilization, the traditional laboratory equipment can only partially reproduce stress state variations in situ; therefore, for the practical application of these experimental results, we tried to consider this last point, making use of a simple triaxial equipment owned by the Road and Trasports Department of the Faculty of Engineering in Bari.

The adjective triaxial is here improperly used, as the two main stresses (σ_2 and σ_3) are identical; therefore, it would be more correct to use the word 'biaxial' for these tests. However, we shall continue to use the word 'triaxial' because it is commonly used.

Besides, it is important to consider the fact that the standard triaxial test is characterized by the coincidence of the main stress and strain planes, and also by the absence of a rotation of the main directions during the whole carrying out of the test.

It is known that the triaxial cell consists of a cylindrical see-through container, which contains the test piece (test pieces 102 mm. large and 198÷201 mm. long were used, for obvious reasons of manufacture). The cell was filled with distillate water, used to convey a given hydrostatic pressure to the test piece which, in its turn, had been wrapped into a thin rubber sheath. A perfectly sealed piston, sliding along a lubricated slide, placed on the top base of the test piece by means of a rigid head, determined an axial load, which added itself to the one deriving from the pressure of the liquid.

It is also known that the stress state in the triaxial cell can be modified acting on the pressure of the liquid (cell pressure or isotropic pressure σ_3) and on the axial load which was increased imposing a shift to the top base of the test piece at the constant speed of 0,07 mm./sec. Measuring with a comparator the shifting of the piston, we derived the axial strain ε_1; the radial strain was derived from the axial strain and from the volume variation of the water in the cell. In particular, the stress and strain state of the asphalt concrete sample was formally represented in the following way.

Indicating with A_o, h_o and V_o respectively the area of the transversal section of the test piece, the initial height and volume, and indicating with A, h and V the corresponding quantities of the strained sample, the axial load P determines a homogeneous pressure on the bases the value of which is P/A and, then, an axial stress (which, to simplify, will be called σ_a) having the following expression:

$$\sigma_a = \sigma_r + \frac{P}{A_o} \qquad (23)$$

and in which $\sigma_r = \sigma_3$ represents cell pressure.

Axial strains are supposed to be homogeneous, because of the rigid surfaces of the machine in contact with the test piece bases; they were derived measuring the shifting of the top base. In fact, the axial strain is defined as:

$$\mathcal{E}_a = -\frac{\triangle h}{h_o} \qquad (24)$$

On the contrary, the radial strain (ε_r) was supposed to be constant along the height of the test piece, as we had accepted the hypothesis that the said test piece keeps the shape of a straight cylinder during the carrying out of the test.

In that case, because of the radial symmetry condition, we have:

$$\mathcal{E}_v = \mathcal{E}_a + 2\mathcal{E}_r \qquad (25)$$

from which:

$$\mathcal{E}_r = \frac{\mathcal{E}_v - \mathcal{E}_1}{2} \qquad (26)$$

being:

$$\mathcal{E}_v = -\frac{\triangle V}{V_o} \qquad (27)$$

the known volumetric strain.

Obviously, before applying the equation (23), we also considered the variation of area of the initial transversal section of the test piece, resulting from radial strain.

Thus if:
- A is the average value of the area of that transversal section (calculated along the axis of the test piece),
- the strains of the test piece are little,
the following equality must be verified:

$$V = A h = A_o h_o + \triangle V \tag{28}$$

from which we easily obtain:

$$A = \frac{1 + \dfrac{\triangle V}{V_o}}{1 + \dfrac{\triangle h}{h_o}} \; \frac{V_o}{h_o} \tag{29}$$

and, therefore:

$$A = A_o \; \frac{1 - \xi_v}{1 - \xi_a} \tag{30}$$

Finally, from the measure of the axial force P conveyed by the piston, the first invariant was easily derived.

In fact, from the equations (22) and (23), it follows that (being $\sigma_a = \sigma_1$ and $\sigma_r = \sigma_3$):

$$q = \sigma_1 - \sigma_3 = \frac{P}{A} \tag{31}$$

While the invariant p, proveided by the (21), easily takes, again by means of the (23), the following final expression:

$$p = \sigma_3 + \frac{P}{3 A} \tag{32}$$

4 Experiments

Some of the physical and mechanical characteristics of the impermeable bitumen concrete samples, prepared with the best local materials, were reported in Table 1; the percentage of bitumen used (4,5% of the weight of the aggregates) is the one which provided, statistically, the best Marshall stability, whereas the following table shows some tests carried out on bitumen.

TABLE 1.

GRADING ANALYSIS			CRUSHED AGGREGATES	
	U.N.I.	(%)		(%)
passing sieve	20	100,0		
"	15	97,6	thin stone	30
"	10	75,6	shot	25
"	5	44,2	sand	42
"	2	25,0	Portland cement	3
"	0,40	11,8		___
"	0,18	7,8		100
"	0,075	6,1		

TEMPERATURES (°C)			BITUMEN 80/100	
	aggregates	154		(%)
	cement	25		
	bitumen	140	by weight of	4,5
	compaction	144	aggregates	

CONGLOMERATE PROPERTIES

absolute specific weight	(daN/mc)	2575
Marshall density	(daN/mc)	2330
Marshall voids	(%)	8,11
number shots for compaction	(n)	75
Marshall stability	(daN)	1030
Marshall flow	(mm)	2,74
rigidity	(daN/mm)	376
permeability	(cm/sec)	$1*10^{-7}$

TABLE 2

PENETRATION at 25°C	(dmm)	84
SOFTENING POINT	(°C)	45,8
FRASS POINT	(°C)	-9
DUCTILITY at 25°C	(cm)	>100
VOLATILITY at 163°C	(%)	0,41
SOLUBILITY in C S	(%)	99,10

We paid special attention to the elimination of the inevitable leakage factors which always occur in experiments. We tried to divide them into two classes, each of which relates respectively to the kind of material examined, and to the modalities of the experiments. As far as first class factors are concerned, considering the inevitable heterogeneity of bitumen mixtures, we made sure that each test piece was prepared with the greatest care, and that it respected the planned basic mixture.

Finally, the main leakage factors associated to the second class, were limited as much as it was possible, as:

- during each test the piston progress speed was kept constant (0,07 mm./sec);
- the whole test was carried out at the same temperature (25°C);
- finally, the application of the research was carried out by the same laboratory technicians.

In the beginning, having remarked in some preliminary tests carried out with similar lateral expansion, that the deviator q constantly increased to a maximum (q max), beyond which it slightly decreased.

We thought it was important to individuate those quantities which influence bitumen mixture strength, and as a consequence the measure of the cracking deviator.

It was evident that the following variables had to be considered during the tests:

- the measure of cell pressure,
- the amount and disposition of spaces in the concrete.

As a matter of fact, the initial structural arrangement of the test pieces, which is later altered by the variation of the deviator q, depends on these variables.

In order to verify what is here affirmed, we prepared 50 cylindrical test pieces divided into 5 groups of equal number, each of which was characterised by different cell pressures (table 3).

TABLE 3.

GROUPS	I	II	III	IV	V
σ_3 (daN/cmq)	0	1	2	3	4

After applying the hydrostatic compression σ_3 to each test piece, we increased the vertical load until we exceeded the cracking stress σ_1. The results are shown in table 4 which shows, for each group and each test piece, the values of axial cracking stresses, expressed in daN/cmq, and also the correspondent axial strains.

On the following tables 5 and 6 show the average values and a few important statistical parameters, concerning the two said quantities.

From all this, we furtherly derived table 7, which sums up

the quantities already described in the previous chapters and some others we shall use in the future.

TABLE 4.

Groups :	I		II		III		IV		V	
N	σ_1	Σ_1%	σ_1	Σ_1%	σ_1	Σ_1%	σ_1	Σ_1%	σ_1	Σ_1%
1	17,63	1,13	22,59	1,50	24,06	1,24	26,19	1,74	28,76	1,75
2	17,10	1,03	22,88	1,25	24,12	1,53	26,74	1,74	28,88	1,66
3	16,89	0,89	23,50	1,25	24,60	1,58	27,02	1,87	28,96	1,88
4	17,41	1,01	22,09	1,25	24,86	1,49	26,35	1,74	28,50	1,86
5	17,62	1,03	22,63	1,21	24,32	1,35	25,98	1,68	28,76	1,88
6	17,26	0,91	22,89	1,32	24,58	1,51	26,35	1,71	29,01	1,92
7	17,41	0,93	22,74	1,30	23,99	1,28	26,72	1,76	28,51	1,89
8	17,60	0,94	22,43	1,28	24,19	1,39	27,09	1,78	29,12	1,97
9	16,91	0,91	23,09	1,29	24,51	1,42	26,31	1,68	29,28	1,98
10	17,83	1,02	22,81	1,22	24,86	1,46	27,00	1,75	27,87	1,90
σ_3	0		1		2		3		4	

TABLE 5.

σ_1	σ_3				
	0	1	2	3	4
Mean value	17,37	22,77	24,41	26,58	28,76
Mean square value	301,67	518,37	595,89	706,37	827,16
Variancy	0,092	0,129	0,092	0,137	0,143
Standard deviation	0,304	0,360	0,303	0,370	0,378
$s/\sqrt{n}$	0,096	0,114	0,096	0,117	0,120
Coeff.of variation %	1,75	1,58	1,24	1,39	1,31
Variancy degree	debole	debole	debole	debole	debole

TABLE 6.

Σ_1%	σ_3				
	0	1	2	3	4
Mean value	0,98	1,29	1,42	1,74	1,87
Mean square value	0,97	1,66	2,04	3,05	3,50
Variancy	0,005	0,006	0,011	0,003	0,008
Standard deviation	0,072	0,078	0,105	0,052	0,092
$s/\sqrt{n}$	0,023	0,025	0,033	0,016	0,029
Coeff.of variation %	7,36	6,08	7,36	2,97	4,92
Variancy degree	debole	debole	debole	debole	debole

TABLE 7.

σ_3	σ_1	q	p	$\Sigma_1\%$	$\Sigma_1\%/q$	q/p	$\Sigma_1\%\dfrac{q}{p}$
0	17,37	17,37	5,79	0,98	0,056	3,00	2,94
1	22,77	21,77	8,26	1,29	0,059	2,64	3,41
2	24,41	22,41	9,48	1,42	0,063	2,36	3,35
3	26,58	23,58	10,86	1,74	0,074	2,18	3,79
4	28,76	24,76	12,25	1,87	0,076	2,02	3,78

5 Characteristics curves obtained from the tests

The response of close concrete test pieces, function of cell pressure and amuont of spaces, is shown on the cartesian plane: deviatoric stress-vertical strain (q, ε,%); in the resulting curves corresponding to each of the five groups of examined test pieces are shown (figure 4).

We thus remarked that, generally speaking, the impermeable crust samples show a curvilinear movement which can be divided into four branches. Figure 5 shows a schematic representations of one of these curves, expressly accentuated to allow a correct graphical representation.

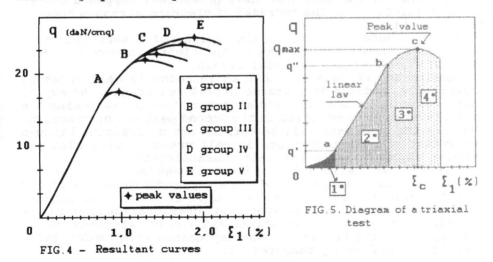

FIG.4 - Resultant curves

FIG.5. Diagram of a triaxial test

The first part (segment 0a), probably a part of settlement, of little significance and with a slightly curvilinear (certainly not linear) movement, is accounted for by the closing of part of the spaces (though they are modest in size and amount), and by the inevitable, though slight, discontinuities present, on a small scale, in the examined samples. We presume that it will be interesting to verify the existence and development of this first section for draining asphalt concretes.

In the second part (segment ab), a decidedly linear one, the test piece shows a pseudoelastic behaviour; in this part the first fractures begin: they develop evenly, if the load is kept constant, provided it is included in ab, they do not increase and only start to spread when the load is once again increased.

The third part (curvilinear segment bc) is connected to the mixture characteristics of viscosity; it ends where the compression strength of the test piece reaches its maximum (**peak stress**). The fractures spread unevenly, in that they do not grow even when the load is constant.

Finally, in the fourth and last part (curve beyond point c), we observed a plastic behaviour in which, therefore, as the strain increased, the ability of the sample to bear loads decreased. We believe it will be interesting to compare the width of this last section with the one corresponding to the draining bitumen mixture.

The tests carried out so far showed that bitumen concrete test pieces reveal a rather consistent peak stress, and their q value, because of which this happens, increases as does the isotropic pressure (figure 4). On the contrary, with any value of cell pressure, the first part always appears as almost imperceptible. The increase in the modulus of elasticity (in the interval of the examined cell pressures), appears almost insignificant, while the increase of cracking stresses appears rather consistent.

Besides, because of increasing values of isotropic pressure, the collapse points move to the right, in correspondance with higher values of axial strain.

Summing up, we remarked that an increase in cell pressure essentially produces an increase in peak resistance. We expect that, when the arrangement and amuont of the spaces inside the test pieces will be modified in the second part of the research, and when the same test will be carried out on draining bitumen concretes, the same increase in cell pressure will also determine higher elastic hardness characteristics.

Finally, we believe the evident linear movement (second field) to be clearly connected, in a way, with the value of the test temperature (25°C), because, as it is known, all bitumen mixtures are heterogeneous materials, characterised, as far as elasticity is concerned, by stress and strain relations, which are far from being linear, in correspondance with temperatures near to the softening temperature.

6 The collapse line

The place of the peaks shown in figure 4 is comparable to a line; in the plane defined by the two invariants p-q, it can be named 'collapse line', and it represents for a given p value, the maximum deviatoric stress which the test piece subjected to the triaxial test can bear.

In order to determine a possible relation between the invariants, we assembled all the experimental data which allowed (even if in a dishomogeneous way) to express the relation existing between the variable **p** and the deviatoric stress **q**.

Besides, for several reasons, such as the necessity of a synthetical representation of experimental information and their successive elaboration, and such as the opportunity of carrying out eventual operations of derivation, integration, etc., it is convenient to express analytically, by means of an empirical equation (which cannot be theorically derived by a deterministic physical law), the connection existing between the said variables.

In the following phase of this research, therefore, we determined the mathematical equation representing the said connection; apparently the best equation (and, therefore, the best regression curve among the possible ones) turned out to be the one which most approached itself to the experimental data, and at the same time, showed the smallest number of arbitrary constants.

In our case, with the aid of physics and mathematics, we tried to find in plane p-q a function which was able to express the test we had carried out. Among the regression curves, usually employed in these circumstances, we made use of the following ones: linear, parabolic, exponential, geometric, polynomial.

The mathematical analysis was concluded, by means of statistical procedures, with the determination of the constants and also of the corresponding correlation coefficient, in order to judge the opportunity of the adaptation of each curve to the experimental data. In the interval between the examined cell pressures, table 8 shows the results derived from the application of the described procedures, and the very good values of almost all the correlation coefficients, which confirms the connection existing between the two variables.

TABLE 8.

REGRESSION	EQUATION	CORRELATION COEFFICIENT
linear	$q = 11,659 + 1,106*p$	0,972
parabolic	$q = 3,197 + 3,116*p - 0,112*p^2$	0,993
exponential	$q = 13,311*1,054^p$	0,959
geometric	$q = 7,838*p^{0,465} \approx 7,838*\sqrt{p}$	0,983

No third rank curve was reported, as it showed an excessively winding movement, with a succession of concavity and convessity, and was therefore of little practical significance for this research.

Besides, in order to find out which curve was most adherent to our experiments, we analysed, with the aid of physics and mathematics, the limits of each of them from the point of view of coherence and validity.

The $q=f(p)$ function must necessarily pass through the origin of the frame of reference, as for $\sigma_1=0$ e $\sigma_3=0$, $p=0$ and $q=0$ must follow as a result. It follows that, in the initial section at least, that is in correspondance with low cell pressures, linear and exponential curves must be discarded.

The parabolic regression curve does not appear suitable either, as it provides $q<0$ for $p\leq0$ values, which would imply $\sigma_1<\sigma_3$ for (22). Besides, the parabolic curve turns the concavity downwards, to indicate that it intersects the axis of the abscissas in two different points; consequently, from a given q value, two different p values would derive.

These conditions already suffice to allow us to say that the curve which best defines the collapse line is, among the ones employed, the geometric regression curve (Fig.6).

7 Conclusions

The 'response' of bitumen concrete to a given system of external stresse depends not only on the kind and amount of loads, but also on the way they are applied. Therefore, it seemed important to visualize the way in which the stress state is modified as a consequence of the application of external loads; it was confirmed that this can be obtained by means of stress invariants.

Undoubtedly, the behaviour of bitumen concretes is very complex, and therefore it is difficult to understand their most relevant aspects with the only aid of these experimental results. However, even if these results do not consider the variables of temperature and duration of load application, they provide useful information on bitumen mixtures, as the behaviour of each element can be compared to a succession of physical conditions, characterised by the parameters q and p provided by the equations (5) and (6).

Once again, therefore, the employment of the triaxial test confirms its validity in allowing a deeper knowledge of concretes. Besides, even if it is not cyclically used, its utilization allows a characterization of concretes by means of the determination of parameters, which are useful in structural analyses of flexible parameters.

Even though, as it is known, the triaxial equipment aims at 'controlling' the three main stresses, in order to make possible the realization of general stress states, we used it in a simpler way in this research.

We subjected a sample of material (subjected to a lateral

pressure σ_3) to a compression test. Lateral pressure generates an isotropic stress and, therefore, to obtain the cracking of the test piece we acted on the axial load. The difference bet- ween axial pressure and cell pressure leads to the applied deviatoric stress, which was employed in this research.

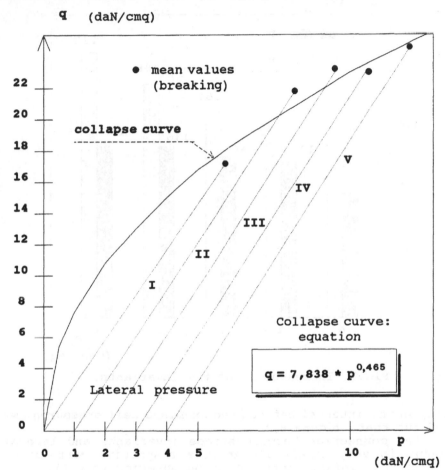

FIG.6. Results of tests and collapse curve

Finally, the variation of the isotropic component of the stress state, allowed us to examine the compressibility properties of the bitumen sheeting.

We thus observed that the employed invariants, though in various measure, as does cell pressure. As a matter of fact, considering for both of them, in correspondance with null lateral pressure, the basic value of 100, it followed that a few units of cell pressure are enough to double the invariant p, while the corresponding increase of the invariant q does not reach the 50%. The graphic below (figure 7) shows in detail the variations of the average values of p and q in correspondance with the examined low cell pressures.

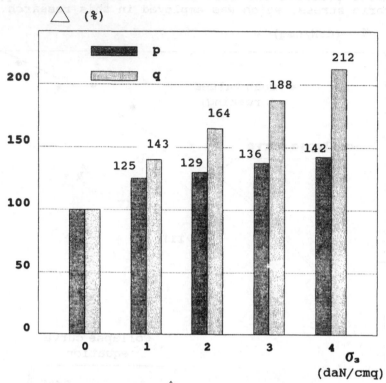

FIG.7. Increase △ of the invariants.

Again, in the interval between the examined cell pressures, we found out that (figure 8):

    - the connection between stress invariants and lateral pressure is very likely linear (the respective correlation coefficients, infact, turn out to be superior to 0,9);

    - the angular coefficients of the respective straight lines are more or less the same, thus confirming the constant difference between stress invariants when cell pressure is the same (table 7).

    As it was easily predictable, the average values of axial strains, corresponding to the moment of 'crisis' of the material, increase as lateral pressure increases.

    This strain also varies with a linear law when lateral pressure varies (figure 9); experimental results reveal, in fact, a very high correlation coefficient (r=0,989). It will be interesting to compare this movement with the one concerning permeable concrete.

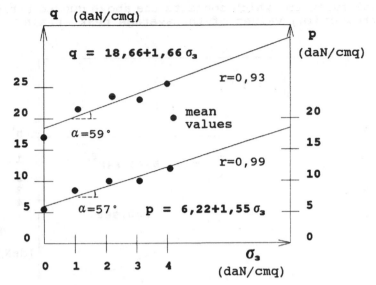

FIG.8. Connection with the
lateral pressure.

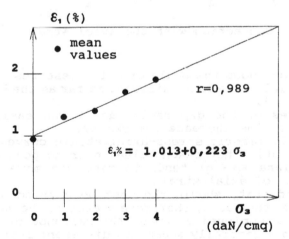

FIG.9. Connection between:
axial strain and lateral press

It is also very important to examine the values assumed by the
q/p ratio with the variation of cell pressure. In particular,
the highest limit in the ratio between the two invariants is
three; this result occurs in correspondance with null lateral
pressure. Figure 10 shows the graphic of the probable ex-

ponential function, which connects the above quoted ratio with the corresponding values of the average peak strain.

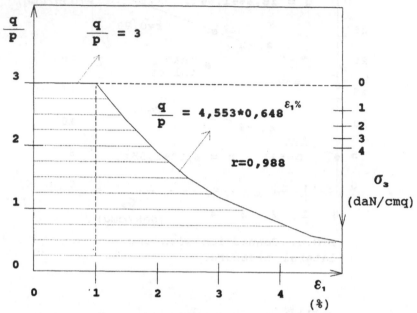

FIG.10. Connection with the axial strain.

The graphic also shows the scale of cell pressures. Obviously, the exponential curve is significant as far as the equation straight line q/p=3.
For low values of the q/p ratio, as it was easily predictable, peak strains increase remarkably.
Besides, all deviatoric stress-axial strain curves have a linear movement until they reach values near to peak stress, which (in the plane $\varepsilon_1\%- q$) tends to rise and move towards increasing values of axial strain.
It is believed that, should the test be carried out at a temperature higher than 25°C, therefore nearer to the softening temperature of the employed bitumen, the response of mixture will turn out to be, generally speaking, different from the one described in figure 4 and, in particular, devoid or almost so, of the linear movement section.
Finally, the tests which have been carried out, made it possible to correlate the two invariants p and q, and to individuate in the geometric function the only function able to represent correctly the 'line of the critical state'.
This last one, in its turn, can be simplified and compared to two segments of a straight line (figure 11), which define the field of the possible stress states.

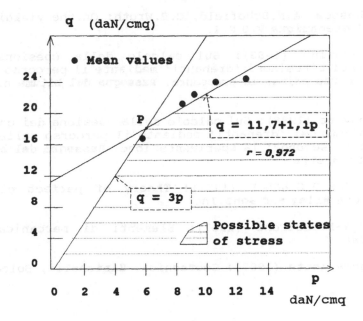

FIG.11 - Line of the critical condition.

In conclusion, the determination of two segments of a straight line in the plane of the invariants, represents a remarkable experimental result useful for the definition of a strength criterion for bitumen mixtures.

As a matter of fact, if bitumen concrete test pieces reach cracking, and the corresponding stress invariants are determined, the respective points of the p-q coordinates will turn out to be, very likely, to be lined on a half-line having the following equation (for the examined mixture): q=11,7+1,1p.

The said half-line has its origin at point P placed on the equation straight line: q=3p.

In fact for:

$$\sigma_3 = 0 \longrightarrow q = \sigma_1 \quad e \quad p = 1/3\ \sigma_1 \longrightarrow q/p = 3$$

As a consequence, on the same plane all points representing possible stress states must be placed on the right of the two segments of straight line (figure 11).

8 References

[1] U. Bonvino (1988): La definizione della coesione e dell'attrito dei conglomerati bituminosi e suoi limiti. **Le strade** n.1248.

[2] K.H.Roscoe, A.N.Schofield, C.P.Wroth: On the yielding of soils". **Geotechnique** V.8 N.1.

[3] U. Bonvino (1988): Sul calcolo della coesione dei conglomerati bituminosi 'drenanti' mediante il percorso delle tensioni. Parte prima: la procedura. **Rassegna del bitume** n.9 III trimestre.

[4] U. Bonvino (1988): Sul calcolo della coesione dei conglomerati bituminosi 'drenanti' mediante il percorso delle tensioni. Parte seconda: la sperimentazione. **Rassegna del bitume** n.10 IV trimestre.

[5] W.PRAGER-P.G.HODGE (1951): Theory of perfect plastic solids. **John Wiley and Sons**.Inc.

[6] G.Rega-F.Vestroni (1987): Elementi di meccanica dei solidi. **Esa**.

[7] R. Lancellotta (1987): Geotecnica. **Zanichelli**, Bologna.

9 EXPERIMENTAL TESTING ON BITUMINOUS DRAINING CONCRETE

A. BUCCHI and M. ARCANGELI
Ist Costruzioni Strade, Ferrovie e Aeroporti, Bologna, Italy

Abstract
The testing was aimed at ascertaining which experimental method is most appropriate for designing a draining concrete for the surface layer of a road pavement. The use of gap-grading on the aggregates and of binders modified with elastomers gives the mix mechanical characteristics which cannot be evaluated correctly by the usual testing systems.

1 Introduction

The continuous reserarch on higher comfort and safety levels for users on one side and a higher attention paid to environmental safeguard on the other, led to the formulation of a new kind of bituminous concrete in the field of road pavements, characterized by a discontinuous granulometry allowing remarkable characteristics of permeability and phonoabsorbency.

The granulometry formulated with the prevailing presence of big size elements allows for a mix with a high percentage of communicating empty spaces in order to quickly drain the surface water and minimize traffic noise. According to recent studies and experiences, it seems possible to minimize such noise by 4 — 6 decibels through the absorption of sound waves within the concrete and the reduction of the air compression-decompression phenomenon between tire and pavement (air pumping).

The non-stagnation of water on road surface and the remarkable roughness actually cancel the risk of aquaplaning thus granting the pavement good bond conditions under every meteorological situations. In this connection, frost formation on surface is not likely to occur, too.

In the end, the draining pavement remarkably acts on the reduction of the atomization effect of the water stagnating on the roadway when vehicles go along. A particular reference is made to heavy vehicles with further visibility improvement when raining for a better traffic security.

As already mentioned, the highly porous draining concretes are obtained through discontinous granulometry and need specially designed binders. In these binders, the resistance must be high since the faying surfaces among lithic elements are not many in the mix; a higher oxidation and ageing resistance must then be secured as the exposure to atmospheric agents is higher. The binder must be made up of bitumen modified through special elastomers in order to have such characteristics.

We can then assume that the draining concrete is a particular material having behavioural characteristics that differ from those of bituminous concretes. Therefore, the purpose of this report aims at assessing whether the methods employed for a survey and optimization of the mixes used for normal bituminous concretes are still valid for the draining concretes.

2 Experimentation

The experimentation has been carried out at the laboratory of the Road Railway and Airports Building Institute of the University of Bologna.

In order to compare the rheology of draining bituminous concretes with that of traditional bituminous concretes, four granulometric curves have been assumed as stated in Fig.1, where the granulometric zones of draining concretes and traditional wear concretes are reported.

In particular, MIX 1 represents the widest curve as it is near to the lower zone limit of draining concrete ; MIX 2 lies on the upper zone limit. MIX 3 and MIX 4 lie on the lower and upper zone limits of traditional wear bituminous concretes.

A bitumen modified with 5% in weight of SBS elastomer has been employed as a binder; we have therefore a penetration number of 60 dmm at 25°C, a P.A. softening point of 72°C, a +4 penetration

128

number index and a 1000 poises viscosity at 80°C. As to the bitumen percentage employed in concretes, a preventive survey has been carried out on the basis of granulometric curves by using Duriez formula and a formula especially developed at the laboratory of the Road Institute.

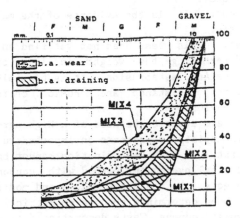

Fig.1

Sieves and riddles UNI	% in weight of undersize			
	MIX 1	MIX 2	MIX 3	MIX 4
15	100	1CO	100	-
10	65	80	70	1CC
5	20	4C	43	67
2	14	20	25	45
0,4	9	12	12	24
0,18	7	8	7	15
0,075	4,5	5	6	11

Optimum bitumen percentages ranging between 4.5 and 5.5% have been obtained by shifting from MIX 1 to MIX 4. In order to make the experimentation uniform, we operated in a bitumen percentage range between 3.5 and 5.5% in weight on inerts for all mixes.

The following tests have been carried out:
- Marshall test (CNR - B.U. No. 30/73). It is the test employed domestically and internationally in order to know the characteristics of bituminous concrete resistance; this test is usually employed for the optimization of mix constituents.
- Constant load deformation test (CNR - B.U. No. 106/85). This test studies the deformation of bituminous concretes by stressing the parameters of viscoelastic behaviour of these materials.
- Brazilian indirect traction test. This test has not yet been published in Italy but has already been drawn up by NRC. It

indirectly determines the traction resistance by stressing specific resistance characteristics of bituminous concretes.
- Impression test (DIN Rule 1996). Though it is not widely employed in Italy, it gives useful information on the deformation degree of bituminous concretes.
- Permeability test. It is the evidence of the main function of draining concretes.
- Cantabrian test. This test does not yet find an application in Italy. By determining the weight in loss of the concrete submitted to Los Angeles abrasion, it is possible to indirectly know the concrete resistance and cohesion characteristics and therefore its behaviour.

The concretes are draining because of a high percentage of empty spaces. This characteristic represents an important element that must be assessed accurately. Such determination cannot be carried out by means of the classical pycnometer method with coupons paraffinizing because paraffin goes into the empty spaces and modifies the results. The volume has been determined through geometrical method in the laboratory on regular coupons. Tests demonstrated that deviations of 7 −8 points may take place according to the various methods. In case of irregular coupons, the pycnometer can still be employed by waterproofing the surfaces with a thin plastic or tinfoil sheet.

3 Marshall test

Three coupons for each mix have been prepared as samples. The stability results as compared to bitumen percentage have been reported in Fig.2. The percentages of empty spaces have been reported in Fig.3.

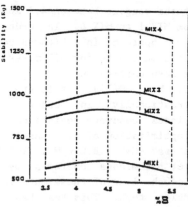

Fig. 2

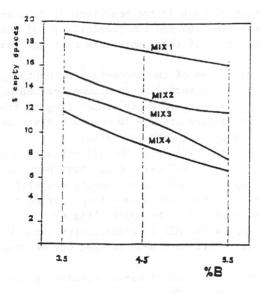

Fig.3

Tests stress that:
- stress/deformation graphs do not show a peak as it happens with
  traditional bitumen concretes but flatten out. It is therefore not
  possible to accurately determine the slippage and the rigidity as
  stability/slippage ratio is meaningless. Marshall test is
  actually not important for the knowledge of the characteristics of
  concrete deformation degree.
- Stability curves as to bitumen percentage for each mix type do not
  show a marked bell shape with an accurate detection of optimum
  bitumen percentage and for which we have greatest stability
  (fig.2); curves are fairly flat with modest stability differences
  at the various bitumen percentages. This is important because it is
  actually impossible to determine the optimum bitumen percentage
  with respect to the mix granulometric curve by means of Marshall
  test. Marshall test is therefore an evidence for the stability
  order according to the mix type, thus leaving any consideration
  about concrete resistance as an indefinite question.
- Fig.2 shows the great stability difference between the various
  mixes. In particular, we note that MIX 1, the widest curve, has a
  stability at about 600 Kg. This value is about the half the value
  usually present in a bituminous concrete for wear layer having a
  granulometric curve included between MIX 3 and MIX 4. A

131

consideration must be made on the behaviour of draining concretes and their length in time and the possibility of detecting the rheological behaviour of these materials must be looked for in other tests.

- Fig.3 shows the variation of the percentage of empty spaces for all mixes and for each mix according to bitumen percentage. We note that MIX 1 has percentages of empty spaces ranging from 16 to 19% and that such percentages are 9 – 10 points higher than those of MIX 3 and MIX 4 traditional wear concretes.

In order to know the behaviour of the bituminous concretes made up with modified bitumens, the following tests have been carried out:

- Concrete made up with MIX 1 granulometry and with normal non-modified bitumen. Marshall tests have been carried out on coupons at 60°C under NRC rule - B.U. No. 30/73 (fig.4).
- Concrete made up with MIX 1 granulometry and with modified bitumen. Marshall tests have been carried out on coupons at 50°C and 70°C.

Fig.4 shows that the use of normal bitumen gives the bitumen stability/percentage curve the characteristic bell shape, yet with modest resistances, independently from wide granulometry.

Marshall tests carried out on MIX 1 (discontinuous granulometry and modified bitumen) clearly showed average differences of stability results but have not yet detected a behaviour that shows the chance to solve the problem of characterization and optimization.

The Marshall test on mixes made up with modified bitumens and carried out according to present provisions and independently from granulometric curves, does not give an exhaustive picture of the characteristics of bituminous concretes.

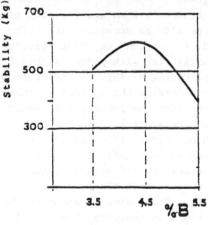

Fig.4

4 Constant load deformation tests

The constant load deformation test (creep test) defines the unit deformation trend affecting a bituminous concrete submitted to a constant normal stress under pre-established temperature conditions. Deformation is represented through the deformation function, ratio between the variation of deformation over a period of time and the applied stress. The test determines the following parameters:

j1 :   value of deformation function in 1 sec.; the most rigid mixes show values smaller than j1.

a :   slope value of deformation function according to time in a double logarithmic diagram; mixes with a more elastic behaviour show values smaller than a.

jp :   value of permanent unit deformation by loading time units; mixes with higher viscoplastic deformations show values higher than jp parameter.

Tests have been carried out under NRC rule - B.U. No. 106/85. The results are the mean of three determinations.

With the results of the tests carried out in the field of linear viscoelasticity it is possible to determine the absolute magnitude of E complex module at an f frequency, T temperature and the phase shift angle between the sinusoidal deformation and the applied sinusoidal stress.

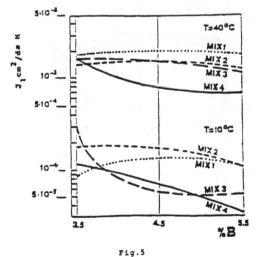

Fig.5

133

As to jl, Fig.5 shows the values according to bitumen percentage for the four mixes and at 10°C and 40°C. First of all, there are no substantial differences between the values of the mixes and the different bitumen percentages, thus stressing that the draining concretes have a rigidity that can be compared to the traditional ones. At low temperatures (10°) closed mixes show jl values lower than the open ones but this difference is notably reduced at high temperatures (40°C).

jp values are reported in Fig.6. This parameter, stating the resistance to permanent deformations, does not show big differences between the mixes even if the lowest values concern the closest mixes. This is important because it demonstrates that draining concretes have a good resistance to plastic deformations: it is a necessary condition to avoid treads in situ.

As to a parameter, as reported in Fig.7, draining concretes show elasticity characteristics similar to those of traditional concretes and this happens at both high and low temperatures.

Figs. 8 and 9 show the trends of E and φ. As to E, the differences are more marked in connection with high bitumen dosages and demonstrate that the absolute magnitude of the complex module increases as the mix compactness increases; this must be taken into account in the superstructure scantling. As to φ, an index of the concrete elastic behaviour, no remarkable differences among the mixes are noticed.

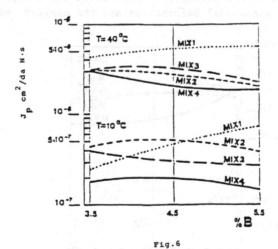

Fig.6

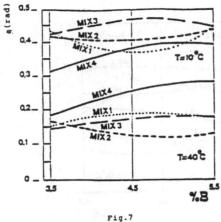

Fig.7

In the end, constant load deformation tests demonstrated that the behaviour of draining concretes is similar to that of traditional concretes; the observed inequalities have never been such as to believe that there are different magnitudes for the values of the single parameters. We also noted that there are no important behaviour differences of mixes when the bitumen percentage differs.

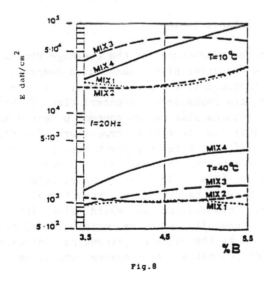

Fig.8

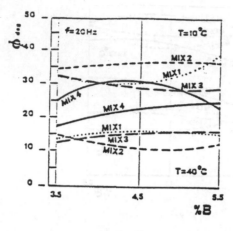

Fig.9

This confirms Marshall test and did not meet the expectations as to the chance of an optimization of the bitumen percentage by means of this kind of test. The responsibility of this situation in this case must be attributed to the presence of modified bitumen.

5 Indirect traction resistance tests

This test consists in the measurement of the breakage stress of the indirect traction on a diametral plane caused by a compression load applied on a plane orthogonal to the previous one; it gives useful information concerning the resistance characteristics of bituminous concretes. Tests have been carried out on coupons prepared according to Marshall terms and submitted to a 25°C temperature. The four mixes have been tested with different bitumen percentages. The mean results of three assessments are reported in Figs. No. 10 and No. 11. Note that traction resistances (Fig. 10) of draining bituminous concrete can be comparable to the resistances of the traditional bituminous concrete; the same observation is valid also for unitary deformations (Fig. 11). Generally, traction resistances of all mixes, except MIX 3, increase as the bitumen percentage increases; the unitary deformation of diametral compression increases too as

136

expected. Breakage stresses never go below 6.0 daN/sc and reach maximum values of 7.5 – 8.5 daN/sc. Unitary deformations of breakage compression range between 1.2 – 1.7%.

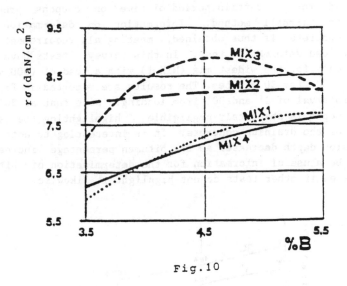

Fig.10

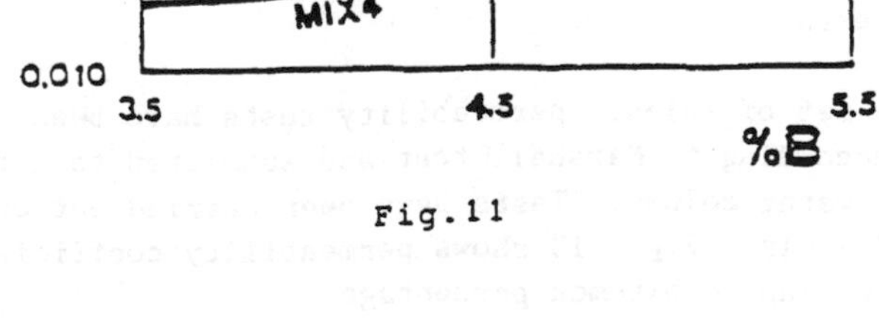

Fig.11

6 Impression tests

Impression tests, unofficial in Italy, have been carried out according the DIN 1996 rule updated to 1984. They measure the driving of a 22.2 mm metal punch at a pre-established temperature under a constant load for a certain period of time on coupons prepared according to Marshall method. Information on deformation of bituminous concrete is thus obtained, meeting all requirements of the constant load deformation test. In this survey, tests have been carried out only for the widest mix (MIX 1) with a load of (500 + 25) N applied at 40°C for 30 minutes. The results are reported in Fig. 12 for a time interval of 1' and 30' from loading. Note that at 30' the impression depth is absolutely negligible, highlighting the small deformation of the draining concrete. It is interesting to note that the impression depth decreases as the bitumen percentage increases.

This can be a useful information for the determination of bitumen percentage even if other tests do not highlight it likewise.

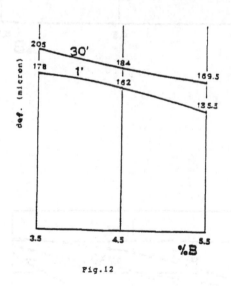

Fig.12

7 Permeability tests

In absence of a set of rules, permeability tests have been carried out on coupons according to Marshall test and submitted to a constant load of a 45 cm water column. Tests have been carried out only for the widest MIX 1 mix. Fig. 13 shows permeability coefficients of empty spaces according to bitumen percentage.

138

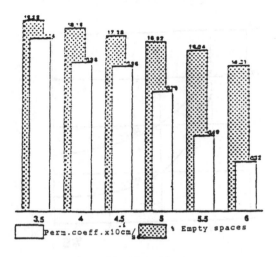

Fig.13

Note how permeability decreases as the percentage of empty spaces decreases; such relationship is not obvious but demonstrates that empty spaces are clearly communicating.

8 Cantabrian test

Cantabrian test is applied in order to know the behaviour in laboratory of bituminous mixes that show a degradation following a lack of cohesion and therefore have an inadequate resistance to the tangential actions caused by traffic. One test has turned out to be very important concerning the submission of concrete samples to an abrasion mechanism causing disaggregation with fall of lithic elements. Here are the formalities. The coupons are prepared according to what is provided in Marshall test. Coupons are put into the Los Angeles equipment without the abrasion mechanical charge and the cylinder turns at three hundred revs. The test takes place at 25°C. Coupons are weighed before and after the test and the results are expressed according to the percentage of the loss in weight during the test as compared to the initial weight.

The tests have been carried out for MIX 1 and MIX 4 to compare the behaviour of a draining porous bituminous concrete with that of a bituminous concrete with traditional continuous granulometry; both concretes have been made up with modified bitumen. Fig. 14 shows the results of the losses in weight, the mean of three determinations according to the bitumen percentage.

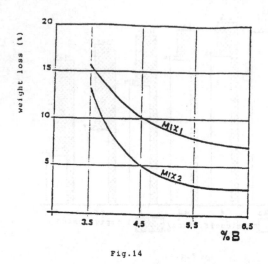

Fig.14

Note that for each mix the loss due to abrasion is high at low bitumen percentages and stabilizes over a bitumen percentage of 4.5 – 5.0%. This means that low bitumen percentages are to be avoided even if in other tests, such as the Marshall one, no remarkable resistance variations to the various bitumen percentages had not been stressed.

As to the detection of the optimum bitumen percentage concerning the loss due to abrasion, it is not possible to make an estimate on it since reliable data on an experimental basis about acceptable maximum values are not yet at disposal.

As to the comparison between MIX 1 (discontinuous granulometry mix) and MIX 4 (continuous granulometry mix), we note a remarkable difference in loss due to abrasion; actually, for bitumen percentages higher than 4.5% the loss due to abrasion of MIX 1 is about the double that of MIX 4. This factor must be carefully evaluated and considered when using draining concretes. We also think that a special experiment and a road control must be carried out in order to get reliable references and acceptability parameters to compare laboratory experiments with actual behaviour.

9 Final remarks

The survey tests gave useful information about the behaviour of draining bituminous concretes.

The survey has particularly relied on Marshall test that is generally employed to characterize the traditional bituminous concretes. As to draining concretes, we noted that this test gives little information. So as it is normalized, working on coupons at 60°C, the test does not let us know the deformation degree of concretes and does not show remarkable resistance differences when the bitumen percentage varies. It gives no importance to draining concretes showing remarkably lower stabilities than those of traditional concretes.

Constant load deformation tests actually show rigidity and elasticity characteristics that can be compared with those of traditional concretes and anticipate a good working behaviour of the materials under examination. Unfortunately also these tests do not show strong behaviour differences as the bitumen percentage varies even if it seems that, within certain limits, rigidity increases as the bitumen percentage increases.

Indirect traction resistance tests actually confirm the results of the constant load deformation tests, mostly stressing the mechanical characteristic having almost similar values both for draining concretes and for traditional ones made up with additioned bitumens.

The constant load deformation and indirect traction resistance tests therefore support the behaviour of draining concretes, unlike the results obtained by means of Marshall test.

The short survey with the impression test states that this test deserves consideration for its simple performance. The results have demonstrated that the deformation degree of the draining concretes decreases as the bitumen percentage increases, at least for the examined field.

Permeability tests state full conformity of draining concretes with the function they must perform.

Eventually, a particular attention is deserved to Cantabrian test for its simple performance through an equipment generally present in a road laboratory. It gives very useful information on the road behaviour of bituminous concretes; yet its acceptability limits have still to be predetermined.

We can now state that the surveys have demonstrated that the bitumen modified with elastometers is mainly responsible for the behaviour of bituminous mixes. By means of these bitumens, especially for draining concretes, the planning of the mix must take place on the basis of the consideration of the behaviour of these materials under different tests; the most important ones are: the constant load deformation test, the indirect traction test and the Cantabrian Test.

10 INDIVIDUATION OF LABORATORY TESTS TO QUALIFY POROUS ASPHALT FRICTION COURSES IN RELATION TO STRUCTURAL RESISTANCE

P. COLONNA
Road and Transport Department, Bari University, Italy

Abstract
In addition to the common tests of characterization on the aggregate (grading, imbibition, Los Angeles, etc.) and on the bitumen (penetration, Fraass, viscosity, etc.) it is suitable to carry out, on the porous asphalt, Cantabrian test and, on the whole asphalt concrete with all the layers, the compression test in common pipes of polyvinyl chloride (pvc) for pipelines of drain and ventilation in the buildings, defined by the Italian standars UNI 7443; from the examination of the stress - strain diagram relative to this one it can be drawn:

a) the breaking stress to compare with suitable limitvalues to establish for each one type of road, that express the attitude of the pavement to the structural resistance;
b) the interval of deformation Δ corresponding to the 95 % of the breaking stress to compare with minimum intervals, also them defined for each one type of road,that express the attitude of the pavement to the fatigue resistance.

The study just described in this work was carried out for porous pavements, but it's evident that the compression in pipe test, for its simplicity, can be usefully utilized also for ordinary pavements.
The same rather can give a good contribution to the knowledge of the real structural behaviour of the pavements.
Keywords: Porous asphalt, Compression in pipe test, Drainage courses.

1 Introduction

In a previous study the author has reported the results of a research set up with the intention to depeen the structural behaviours of the porous asphalt friction courses.
For doing this, it was realized a stretch of experimental road with six types of pavement, differing one from the other for thickness and materials. So, by means of Road and Transportation Department Laboratory of the Bari University,

current type tests (asphalt content, grading, Marshall, voids, bulk density, etcetera) and less traditional tests (compressive uniaxial strenght test, indirect tensile "Brazilian" test, creep test and bending test) have been done.

It is to emphasize that these last were done on the whole asphalt concrete composed of three layers, in order to study the real comportament of the pavement.

The conclusion of such work are briefly recalled in the next paragraph. The reader who already knows them can, so, omit examining it.

2 Results of the previous experiments

In conflict with the trend shown by the results of the unconfined compression tests, the bending tests put in experimental evidence that the use of the porous asphalt 5 cms thick improves the strength of the pavement by over 40 %, in comparison both with traditional layer as thick as the previous one, and the porous asphalt 2 cms thick.

The explanation given to the different behaviours of the porous asphalt concrete subjected to the different types of tests is the following.

The use of the porous asphalt in comparison with the traditional type improves the bending strength because, thanks to its discontinous sieving with a prevalent presence of thick elements, it has a stronger lithic structure. The bending action (fig. 1) tends to provoke indeed a "lateral" compression of the grains in the surface layer and, if there are points of contact among them, the presence of voids standing below and above them does not produce a reduction of the strength of the layer, because the factor that is involved at a structural level is the strength of grain to the compression.

In order that the strength difference exists, and that it is significant and concerning the whole pavement, some conditions should necessarily take place:

a) the porous asphalt should present a not negligible thickness (4 ÷ 5 cms);
b) the grains shouldn't have the possibility of shifting laterally.

The second condition can be actually verified because of the continuity of the pavement, whereas it is completely absent in laboratory check as in unconfined compression tests.

The fact that the structural strength of porous concretes is influenced by this circumstance is demonstrated by the aspect of the samples (cores) of pavement after the compressive collapse.

The samples with porous asphalt 5 cms thick show indeed,

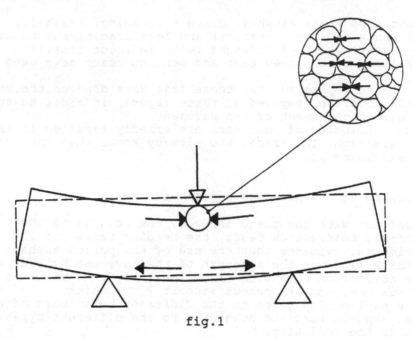

fig.1

if compared with these with traditional layer as thick as the previous one, that they collapsed because of a lack of lateral contrast (fig. 2).

It follows that in order to estimate the structural strength of a porous asphalt concrete, tests like the unconfined compression are completely unsuitable, whereas, in the light of what has been experimented till now, it can be assumed that the Marshall test can offer both partial and comparative results.

Interesting results can derive from the bending test performed on the whole ensemble of the pavement, although it involves some objective operative difficulties.

3 New trend of research

For simulating in laboratory, with efficacy and simplicity, the behaviour "in situ" of the pavement, lateral confined compresion tests were thought to effettuate.

But this condition of boundary was obtained not preventing quite the lateral expansion, just like an ideal core "in situ" of pavement is partially prevented to the lateral expansion for the presence of the surrounding pavement.

The condition of boundary was therefore obtained through a section of a not excessively rigid pipe: were chosen common pipes of polyvinyl chloride (pvc) for pipelines of drain and ventilation in the buildings, defined by the

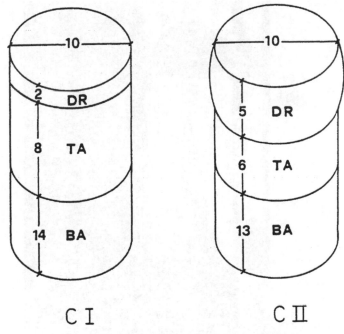

CI   CII

fig. 2

Italian standars UNI 7443. The characteristics of the chosen
pipes are shown in the next table:

Table 1. Pipe UNI 7443-302/100

outside diameter	100 mm
thickness	3,0 mm
bulk density	1,37 - 1,47 g/cmc
unit load of yelding	$\geq$ 48 MPa (480 kgf/cmq)
elongation at yelding	$\leq$ 10 %
modulus of elasticity	3000 MPa ($\approx$ 30000 kgf/cmq)

4 Results

Were utilized, obviously, pipes with inside diameter
coincident with the diameter of the cores of pavement.
Between the two materials was interposed a lubricant for
becaming the strains of the sample independents of the
superficial characteristics of the pipe. The compression was
exercised through an upper piston (set on the surface layer
of the pavement) with diameter just inferior to that of the
pipe, so that it could easy penetrate in this one.
    Was utilized a common Marshall press with forward speed
of the piston of 0,85 mm/s (photo n. 1).

photo n. 1

The temperature of the test was about 25C.

Were tested n. 4 samples of CI type and n. 4 of CII type. The characteristics of the materials utilized are indicated in the already cited previous work of the author.

In fig. 3 are shown the stress - strain average diagrams for the two types of samples.

5 Interpretation of the results

The results of the experimentation are so in all according with those of the bending tests, as they express that the use of the surface layer of porous asphalt 5 cms thick gives at all the pavement a better structural behaviour respect to the same situation with 2 cms of thickness.

The compression strengths of the compression tests in pipe are, on the average, about ten times upper to those verified in the unconfined compression test.

The trend of the stress - strain diagrams, besides, puts in evidence, owing to the different flatness near breaking zone, that the pavement with greater thickness of porous asphalt is able to support for longer time the plastic strains.

This is very important, because shows a probable better

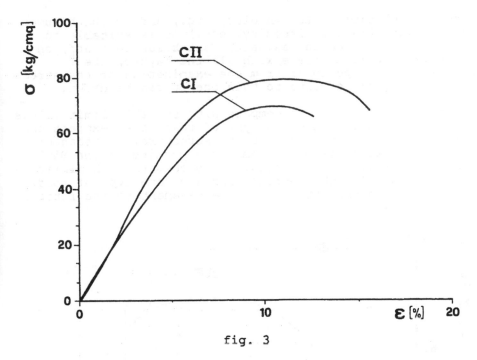

fig. 3

fatigue behaviour of the pavement, due to the fact that the accumulation of the deformations doesn't produce immediately the collapse of the structure.

Of course, in order that this one happens really, it's necessary that the next two conditions are respected:

a) the bitumen must be not brittle but, rather, must allow few displacements of the chippings particles without that the bonds are broken;
b) the aggregates must have high compression strength and small abrasion so that are not broken (as micro and macro level) owing to the action of the loads, producing so inadmissible further settlings and the partial filling of the voids with the reduction of the draining effect.

Therefore it's extremely useful the execution of the "Cantabrian" test consists in the determination of the loss (per cent) in weight of a Marshall sample subjected to 300 revolutions in the Los Angeles drum.

6 Conclusions

Therefore it is possible to affirme that in addition to the common tests of characterization on the aggregate

(grading, imbibition, Los Angeles, etc.) and on the bitumen
(penetration, Fraass, viscosity, etc.) it is suitable to
carry out, on the porous asphalt, Cantabrian test and, on
the whole asphalt concrete with all the layers, the
compression in pipe test; from the examination of the stress
- strain diagram relative to this one it can be drawn:

a) the breaking stress to compare with suitable limit values
   to establish for each one type of road, that express the
   attitude of the pavement to the structural resistance;
b) the interval of deformationΔ corresponding to the 95 %
   of the breaking stress (fig. 4) to compare with minimum
   intervals, also them defined for each one type of road,
   that express the attitude of the pavement to the fatigue
   resistance.

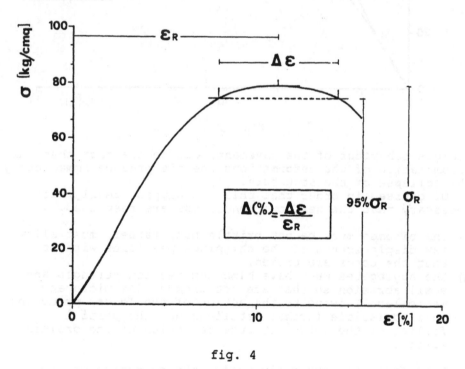

fig. 4

7. Extension of the methodology to non - porous pavements

   The study just described in this work was carried out for
porous pavements, but it's evident that the compression in
pipe test, for its simplicity, can be usefully utilized also
for ordinary pavements.
   The same rather can give a good contribution to the
knowledge of the real structural behaviour of the pavements.

8　References

Beaumont, J. Faure, B. Huet, M. and Verhee, F. (1989) The french experience about free draining bituminous mixes, in **Proceedings of 4th Eurobitume Symposium**.

Colonna, P. (1989) Study of the structural behaviour of the porous asphalt friction course with the purpose to recognize a criterion of design, in**Proceedings of 4th Eurobitume Symposium**.

Decoene, Y. (1989) Knowledge acquired after 10 years of research on porous asphalt in Belgium, in**Proceedings of 4th Eurobitume Symposium**.

Lefebvre, G. (1989) Belgian specifications for pervious coated macadams, in **Proceedings of 4th Eurobitume Symposium**.

Lefebvre, G. (1989) Prevention of clogging of pervious coated macadam, in **Proceedings of 4th Eurobitume Symposium**.

Legnani, G. (1989) Tappeti drenanti e fonoassorbenti, in **Rassegna del bitume** 12 - 13

Pallotta, S. Peroni, G. and Lucanara, R. (1989) Open grade wearing courses. Experiences by Societa Autostrade, in **Proceedings of 4th Eurobitume Symposium**.

Perez Jimenez, F. E. and Calcada Perez, M. A. (1989) Application of the Cantabrian test to open mix design, in **Proceedings of 4th Eurobitume Symposium**.

Weiringer, H. and Nievelt, G. (1989) Construction & evaluation of drainage courses. State of art - Austria, in **Proceedings of 4th Eurobitume Symposium**.

149

11 COEFFICIENTS OF RELATIVE PERFORMANCE FOR BITUMINOUS MIXES

L. DOMENICHINI, P. DI MASCIO and P. DI IORIO
University of L'Aquila, Rome, Italy

Abstract
The aim of the report is to determine the influence of variations outside specifications of the roadbase, binder and wearing courses bituminous mix composition characteristics on the behaviour of flexible pavements in service.

In case the properties do not meet specification requirements, it is necessary to judge to what extent the noticed deficiencies will affect the fatigue performance and rutting of the pavement under traffic. As a matter of fact, often a criterium based only on the amount of bituminous mix layed is adopted to value penalties to be applied instead of referring to performance criteria.

The report deals on defining criteria to judge the quality of layed bituminous mixes based on traditional acceptance tests and pavement performance under traffic. Such criteria allows to define coefficients of relative performance, with reference to both fatigue life and pavement rutting tendency, as function of variations outside specifications of the bitumen content and air voids of the bituminous mixes used in roadbase, binder and wearing courses.

Keywords: Bituminous Mix Design, Quality Controls, Flexible Pavement Fatigue Life, Flexible Pavement Rutting, Coefficients of Relative Performance.

1 Introduction

The operational performance of flexible pavements is largely conditioned by the mechanical properties and strength of the bituminous mixes which constitute the road base, binder and wearing courses.

In fact, fatigue cracking and rutting, which are the main forms of distress of this kind of pavements, clearly depend on the properties of the bituminous mixtures which, in turn, depend on the composition of the mixtures themselves.

Many experimental researches have been carried out to quantitatively evaluate the influence of the bituminous mixes composition on the behaviour of flexible pavements in service. The results of such researches enabled the formulation of prediction models which have proved very useful in the rational design of highway pavements.

Based on the above considerations, the present paper considers the possibility of using such prediction models in phase of bituminous mix design and during the analysis of the results of the quality controls carried out during the pavement construction. Thus, simple laboratory tests and the use of pavement performance prediction models could allow to add performance criteria to the standard bituminous mix design procedures and to evaluate the consequences of possible variations from the requirements.

2 Characterization of bituminous mixtures in terms of performance

2.1 Mix design
Standard procedures of bituminous mix design provide for performing laboratory tests to characterize (often in a very conventional way) the physical, mechanical and strength properties of the single components and of the mixture. The more frequently used mechanical tests are the Marshall test and the indirect tensile strength test. Their use allows for optimization of the composition characteristics of the mixture for the specific binder and aggregates available.

Such a procedure does not provide any information concerning the actual operational performance of the mixture. Its validity is based on the long-standing valuable records which couples the good performance of the mixture under traffic conditions to certain values of the laboratory measured quantities.

The fatigue behaviour and rutting prediction models presently available enable us, during the mix design phase, to have further information about the mix performance in service without having to resort to the more expensive and time consuming fatigue and static or dynamic creep tests.

In fact, prediction models are based solely on the knowledge of some parameters related to the volumetric composition of the mixtures.

Furthermore, these models can be very useful in judging the results of the quality controls performed during the construction phase. If a difference is observed during pavement construction between the composition characteristics of bituminous mixtures as defined in the mix design phase or prescribed in the contract specifications and those mixtures actually used, the consequences in terms of performance will be a variation (increase or decrease) of the pavement fatigue resistance and/or rutting.

According to present specifications, the above variations are normally settled by applying penalties, the value of which is often based solely on the quantity of materials layed down. They, therefore, bear no relation to the pavement actual performance.

The use of prediction models also gives in this case evident advantages as it allows for the introduction of performance criteria when determining any penalty which is to be levied.

2.2 Volumetric composition of mixtures
In order to evaluate the fatigue resistance and permanent deformation performance of a compacted bituminous mixture, the values of some parameters characterizing the mixture's volumetric composition have

to be determined.
These characteristics are:
- the aggregate mixture specific density (MVRA).
- the bitumen specific density (MVRB), whose value in most cases may be assumed to be equal to 1.02 gr/cm^3.
- the bituminous mixture specific density (MVRM) whose value, once MVRA and MVRB and the bitumen content are known, can be determined by the expression:

$$MVRM = \frac{100 + B}{(100/MVRA) + (B/MVRB)} \qquad (1)$$

- the bituminous mixture bulk density (MVAM), which is given as function of MVRM and mixture's air voids (Vv) by the expression:

$$MVAM = [1 - (Vv/100)] \, MVRM \qquad (2)$$

- the mixture's air voids percentage (Vv);
- the bitumen content given as weight percentage (B);
- the bitumen volumetric content (Vb), given by the expression:

$$Vb = 100 \left(1 - \frac{100}{[100/(100 + B)] \ (MVAM/MVRA)}\right) - Vv \qquad (3)$$

- the volume occupied by the aggregates (Va), given by the relation:

$$Va = 100 - (Vv - Vb) \qquad (4)$$

The values of the above listed quantities can be determined during construction by means of normalized laboratory tests carried out on samples of material taken from the stores and compacted according to a standard procedures or on cores or sample taken from the compacted layer.

The research work carried out during the last 20 years by European and American research agencies, using statistical analysis methods, have led to the formulation of regression expression relating the mixture composition parameters to the mixture dynamic modulus, to its fatigue life, and to the entity of permanent deformations which occur at the end of the pavement's service life, that is, after an established number of the standard axle repetitions.

Amongst the various available methodologies, reference is made in the present paper to that proposed by the Centre des Recherches Routieres (CRR) of Belgium, since it covers all aspects of bituminous mixtures' performance prediction and allows for a wide range of variability in the mixtures composition which encompasses the conditions most frequently encountered in practical use.

2.3 Dynamic modulus of the mixtures

The general relation proposed by CRR for the evaluation of the mixture dynamic modulus (E) as function of the load application frequence (fr) is [1]:

$$E (fr) = E_\infty \cdot R (fr) \tag{5}$$

where:

E_∞ : purely elastic component of the complex modulus; it characterizes the mixture behaviour at low temperatures and/or at high frequencies. It has been verified that value of this component (Mn/m^2) depends on the composition of the mixture according to the relation:

$$E = 3.56 + 10^4 \ [(Vb + Vv)]/Vb \ e^{(-0.1 \ Vv)} \tag{6}$$

R (fr) is an adimensional coefficient called "reduced modulus". It is a function of the frequency, of the ratio (Vb/Vv) and of the "consistency" of bitumen used; the latter can be expressed by means of the complex shear modulus of bitumen (G).

The general relation proposed for the calculation of R, when the ratio Vb/Vv varies between 3 and 12, is as follows:

$$\log R = A \log F (1 + 0.11 \log F) \tag{7}$$

where:

A: is a coefficient which takes into account the mixture composition, given by the following relation:

$$A = 1 - 1.35 \ [1 - e^{(-0.13 \ Va/Vb)}] \tag{8}$$

F: is the adimensional reduced shear modulus, given by the ratio $G/1000 \ (Mn/m^2)$.

2.4 Fatigue resistance

The composition of bituminous mixtures affects the pavement fatigue life both directly and indirectly. Directly, since mixtures of different compositions have different fatigue resistance under the same stress level. Indirectly, since the variation of the mixture mechanical properties (dynamic modulus, Poisson's ratio) due to the different volumetric compositions, involves a redistribution within the pavement of the stresses induced by applied loads and hence a variation of the stress level to which the mixture is subjected.

The general criterium for the evaluation of bituminous mixes fatigue life has been defined as follows [3]:

$$N \ (\leqslant 10\%) = No + N \tag{9}$$

where:

N ($\leqslant 10\%$): number of load applications which produce 2nd class fatigue cracks (as defined by AASHO test (2) on 10% of the surface of the wheel path;

No: number of load applications which produce the initial fatigue cracking [4,5];

ΔN: number of load applications which produce the reflection of the cracks at the pavement surface and their extension to 10% of the wheel path surface [3].

2.5 Permanent deformation

Referring to stress conditions far away from those causing the plastic rupture of the mixture, the relation between the accumulated plastic deformations (εp), the number of load applications (N), the frequency (fr), the stress conditions and the composition characteristics of the mixture can be given by the experimental formula [6,7]:

$$\varepsilon p = \frac{(\sigma 1 - \sigma 3)}{0.65 \ E \ F} \times \left[\frac{N}{fr \times 10^3}\right]^b \tag{10}$$

where:

$\sigma 1$: principal vertical stress;

$\sigma 3$: principal horizontal stress;

E: absolute value of the mixture's complex modulus;

F: parameter taking into account the volumetric composition of the mixture, according to the expression:

$$F = 5.5 \times 10^{-2} \ [1 - 1.02 \ (Vb/Vv + Vb)] \tag{11}$$

b: coefficient varying between 0.1 and 0.3.

2.6 Coefficients of relative performance

Supposing that the performance characteristics of bituminous mixtures corresponding to the contract specifications or defined during the mix design have the value of 100, it is possible to define the relative performance coefficient as the ratio between the effective performances of the mixtures used and those of the contract or design mixture.

Therefore, the following coefficients have been defined:

- coefficient of relative performance (CF) referring to the pavement fatigue life:

$$CF = \frac{Neff}{Nnom} \times 100 \tag{12}$$

where:

Neff: number of load repetitions supported up to fatigue rupture by the pavement, made with the actually laid mixtures;

Nnom: number of load repetitions which the pavement could support up to fatigue rupture if made with the design or contract mixtures.

154

- coefficient of relative performance (CR), referring to the pavement expected total rutting:

$$CR = \frac{\delta\,nom}{\delta\,eff} \times 100 \qquad (13)$$

where:

δ eff: total permanent deformation after 10^6 load repetitions on the pavement made with the actually laid mixtures.

δ nom: total permanent deformation after 10^6 load repetitions on the pavement made with the design or contract mixtures.

The values of total permanent deformation is given by the contribution of each single layer, according to the relation:

$$\delta = \sum_i p,i \; hi \qquad (14)$$

where:

$\varepsilon p,i$: unit deformation of layer "i";

hi: thickness of layer "i".

When evaluating Neff and .eff, not only the variations of the bituminous mixtures' composition characteristics with respect to the project ones, but also the variations of the other parameters which affect the behaviour of the pavement should be taken into account (foundation bearing capacity, effective thickness of the layers, etc.).

3 Cases Examined

On the basis of the considerations pointed out in the preceding paragraphs, an automatic calculation programme has been set up which allows calculation of the relative performance coefficients CF and CR, for given nominal characteristics of the design mixtures and pavement, as functions of the actual characteristics of the mixtures.

The conceptual flow chart of the finalized programme is shown in Figure 1.

The calculations carried out aimed at singling out the values of the coefficients CF and CR when the bituminous mixtures for road base, binder and wearing courses are affected by air voids (Vv) and bitumen content (Vb) values outside standards.

Those which more commonly appear in italian specifications for road works have been adopted as characteristics of the reference mixtures. Table 1 shows the range of acceptable values allowed by ANAS and AUTOSTRADE specifications.

Increase and decrease variations in the mixtures characteristics with respect to the Table 1 values have been taken into consideration in order to evaluate their consequences in terms of long-term behaviour of the pavement.

The values outside specifications of the composition characteristics examined are shown in Table 2.

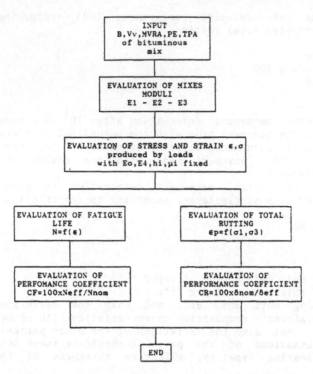

FIGURE 1: Procedure followed to determine coefficients of relative performance.

The calculations have been carried out with reference to two bitumen types, whose characteristics, meeting specifications in Table 1, are pointed out in Table 3.

Stresses due to traffic loads have been evaluated for a flexible pavement whose composition is shown in Figure 2.

The same Figure also shows the values, which remain constant, the resilient modulus of the sub-grade soil and of the sub-base unbound granular material, and of the Poisson's ratios assigned to the various layers.

In addition, calculations have been carried out for the following conditions:

- frequency of traffic load application equal to 10Hz, corresponding to a commercial vehicle speed of 25 km/h [8].
- reference temperature of the bitumen mixture: 20°C
- aggregate mixture specific density (MVRA) of the road base, binder and wearing courses equal to 2.70, 2.70 and 2.90 gr/cm³ respectively.

TABLE 1 : Range of variations usually allowed by standards in Italy

	ROADBASE COURSE	BINDER COURSE	WEARING COURSE
B	3.5 - 4.5	4.0 - 5.5	4.5 - 6.0
Vv	4.0 - 7.0	3.0 - 7.0	3.0 - 6.0
PE	60 - 80	60 - 80	60 - 80
RB	45 - 55	45 - 55	45 - 55

B = bitumen content percentage (ratio between bitumen and aggregates weights)
Vv = air voids percentage (ratio between air void and mix volume) .
PE = bitumen penetration (dmm)
RB = bitumen ring-ball point ('C)

TABLE 2 : Investigated values of bitumen content and air voids outside Italian standards.

	ROADBASE COURSE			BINDER COURSE			WEARING COURSE		
variat.%	+10	+20	+30	+10	+20	+30	+10	+20	+30
B	5.0	5.4	5.9	6.0	6.6	7.2	6.6	7.2	7.8
Vv	7.7	8.4	9.1	7.7	8.4	9.1	6.6	7.2	7.8
variat.%	-10	-20	-30	-10	-20	-30	-10	-20	-30
B	3.2	2.8	2.4	3.6	3.2	2.8	4.0	3.6	3.2
Vv	3.6	3.2	2.8	2.7	2.4	2.1	2.7	2.4	2.1

B = bitumen content percentage (ratio between bitumen and aggregates weights)
Vv = air voids percentage (ratio between air void and mix volume)

TABLE 3 : Investigated bitumen type

TYPE	PE	RB	IP	Eb(MPa)
1	60	45	-2.13	50
2	80	55	1.25	10

PE = bitumen penetration (dmm)
RB = bitumen ring-ball point ('C)
IP = penetration index
Eb = bitumen modulus in MPa evaluated according to Van der Poel graph for a pavement temperature of 20'C and a frequency of 10 Hz

4 Results

4.1 Dynamic modulus of the mixtures

The procedure described in paragraph 2.3 was used to calculate the E modulus of bituminous mixtures with reference to bitumen contents (B) ranging between 3% and 7%, to bitumen stifness modulus Eb of 50 and 10 MPa (see Table 3) and to mixture total voids (Vt) ranging between 11% and 19%. The results obtained are shown in Figures 3, 4 and 5 for the road base, binder and wearing courses respectively.
The B limits provided in the specifications are shown in the diagrams. In addition, in the upper part of each diagram, the

		stiffness moduli	Poisson ratio
WEARING COURSE	4 cm	E1 = variable	0.45
BINDER COURSE	8 cm	E2 = variable	0.45
ROADBASE COURSE	20 cm	E3 = variable	0.41
SUBBASE COURSE	40 cm	E4 = 125 MPa	0.30
SUBGRADE ///\\\///\\\///\ CBR = 5%		Eo = 50 MPa	0.40

FIGURE 2 : Pavement structure investigated

straight lines of equal value for residual air voids (Vv) of mixtures with varying B and Vt have been marked, thus indicating for each layer examined the field where the value of the variables Vv and B fall within the specifications limits.

Bituminous mixtures having higher moduli are obtained, as is clearly shown in Figures 3, 4 and 5, by using lower bitumen contents and lower residual or total voids.

The 20°C dinamic moduli of bituminous mixes having composition characteristics according to the specifications, vary within the following limits:

- road base course   4,100 MPa - 16,900 MPa
- binder course   2,500 MPa - 13,500 MPa
- wearing course   1,900 MPa - 10,000 MPa

Such quite wide variation ranges allowed by the current specifications has to be clearly taken into account during the pavement structural design phase.

Furthermore, in the mixture design phase, in addition to the usual criteria, reference should also be made to Figures 3, 4 and 5 which demonstrate the opportunity of choosing mixtures in which the B and Vv values are the lowest amongst those allowed according to the available materials.

4.2 Coefficients of relative performance to fatigue (CF)

To evaluate the influence of bitumen content and air voids of bituminous mixtures on overall pavement fatigue life, no.174 calculation conditions lying both inside and outside the contract

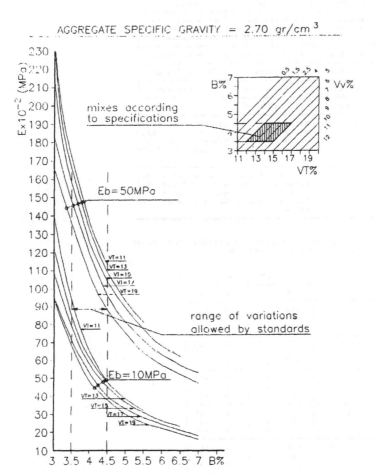

FIGURE 3: Roadbase course — mix stiffness as function of mix composition.

specification limits shown in Table 2 have been examined using the BISAR elastic multi-layer method.

In order to calculate the individual influence of each parameter, every calculation condition has been characterized by varying the mix composition characteristics of one layer at a time; for each one of these, the bitumen content and air voids has been varied separately. Tables 4 and 5 summarize the calculation conditions investigated.

As reference conditions for the calculation of Nnom, shown in the expression (12) specifications have been taken into account, at both their upper and lower levels, as detailed in Table 4. The results of the calculations performed for bitumen contents outside specifications for the road base, binder and wearing courses are

TABLE 4 : Type of investigated pavements

TYPE	ROADBASE	BINDER	WEARING
1	E3 = maximun	E2 = maximum	E1 = variable
2	E3 = minimum	E2 = minimum	E1 = variable
3	E3 = maximun	E2 = variable	E1 = maximum
4	E3 = minimum	E2 = variable	E1 = minimum
5	E3 = variable	E2 = maximum	E1 = maximum
6	E3 = variable	E2 = minimum	E1 = minimum

Ei = maximum,minimum : maximum or minimum stiffness value of i-th mix with variations of mix compositions inside the range allowed by Italian standards (see figures 3,4,5).

Ei = variable : mix stiffness variable with mix composition variations '(B and Vv) outside the range allowed by Italian standards according to table 5.

TABLE 5 : Mix composition variations outside the range allowed by Italian standards

REFERENCE MIX	VARIATIONS %			
	B	Vv	B	Vv
Bmax Vvmax Eb=10 Mpa	+10	0	0	+10
	+20	0	0	+20
	+30	0	0	+30
Bmax Vvmin Eb=10 Mpa	+10	0	0	-10
	+20	0	0	-20
	+30	0	0	-30
Bmin Vvmax Eb=10 Mpa	-10	0	0	+10
	-20	0	0	+20
	-30	0	0	+30
Bmin Vvmin Eb=10 Mpa	-10	0	0	-10
	-20	0	0	-20
	-30	0	0	-30
Bmax Vvmax Eb=50 Mpa	+10	0	0	+10
	+20	0	0	+20
	+30	0	0	+30
Bmax Vvmin Eb=50 Mpa	+10	0	0	-10
	+20	0	0	-20
	+30	0	0	-30
Bmin Vvmax Eb=50 Mpa	-10	0	0	+10
	-20	0	0	+20
	-30	0	0	+30
Bmin Vvmin Eb=50 Mpa	-10	0	0	-10
	-20	0	0	-20
	-30	0	0	-30

Bmin, Bmax = maximum or minimum value of bitumen content inside the range of Italian specifications

Vvmin, Vvmax = maximum or minimum value of air void inside the range of Italian specifications

Eb = stiffness modulus of bitumen (see table 3)

B% = percentage variation of bitumen content with reference to Bmax o Bmin

Vv = percentage variation of air void with reference to Vvmax o Vvmin

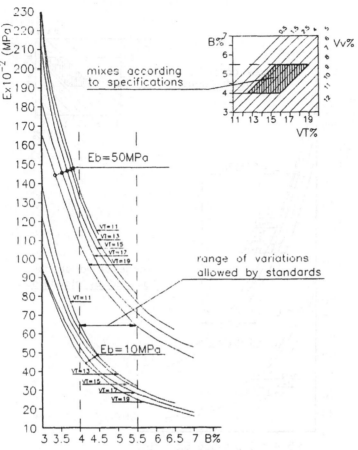

FIGURE 4: Binder course — mix stiffness as function of mix composition.

shown in Figures 6, 7 and 8 respectively.

Figures 9, 10 and 11 respectively show the results for similar variations of the residual air voids.

CF coefficient values of less than 100 indicate pavements with fatigue performances worse than those related to bituminous mixtures whose characteristics meet the requirements for each of the layers. CF values higher than 100, on the other hand, indicate pavements with better performances.

As shown in the above-mentioned Figures, the value of CF corresponding to a specific variation value of the bitumen content B or of the air voids Vv, is not unique, but may vary over a more or less wide range. The reason for this is that for each value outside standards of the parameter examined, the pavement fatigue life may vary according to

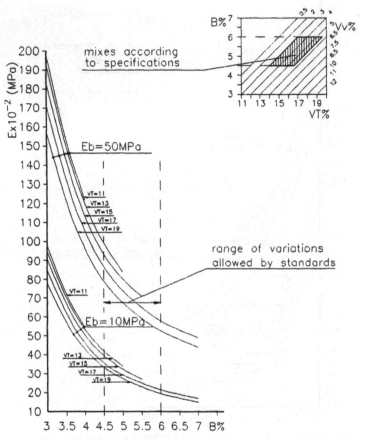

FIGURE 5:   Wearing course — mix stiffness as function of mix composition.

the values of the other parameters within the ranges allowed by the specifications. From Figure 6 it may be deduced, for example, that a 15% increase in bitumen content over the maximum value allowed for the road base course by the specifications, leads to performance coefficients CF of between 75% and 60% according to the residual air voids, the type of bitumen and the mechanical characteristics (elastic modulus) of the other layers forming the pavement.

In Figures 6 to 11 we tried to single out the influence due to the value assumed by the other parameters concerned.

Referring again to the above example, the CF can reach a value of between 60% and 67% if the bitumen employed for the road base has a higher stiffness modulus (harder bitumen) or a value of between 67% and 75% if it has a lower stiffness modulus.

Similarly, in Figure 7, which shows the variations of bitumen content in the binder course, it can be seen that a 15% increase of B enables CF to reach values of 83-95%; this variation range narrows if the other characteristics are considered, in particular:

- when the binder course is composed of harder bitumens (Eb = 50 MPa), the CF coefficient is in the order of 85-89% if the road base and wearing courses have higher performances (E3 = 16,900 MPa, E1 = 10,000 MPa), whilst it is in the order of 92-95% when the opposite holds, i.e., when the road base and wearing courses have lower performances (E3 = 4,100 MPa, E1 = 1,900 MPa);

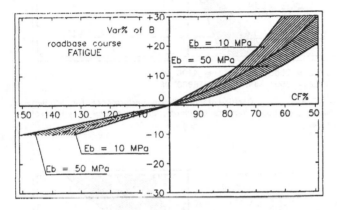

FIGURE 6: Coefficients of relative performance for bitumen content variations outside standards, with reference to pavement fatigue.

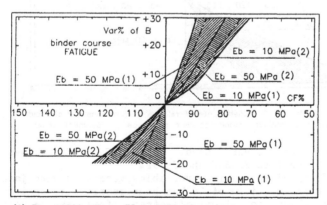

(1) E1 = 1900 MPa   E3 = 4100 MPa
(2) E1 =10000 MPa   E3 =16900 MPa

FIGURE 7: Coefficients of relative performance for bitumen content variations outside standards, with reference to pavement fatigue.

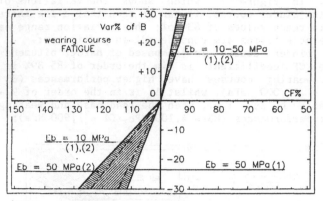

(1) E2 = 2500 MPa   E3 = 4100 MPa
(2) E2 =13500 MPa   E3 =16900 MPa

FIGURE 8:   Coefficients of relative performance for bitumen content variations outside standards, with reference to pavement fatigue.

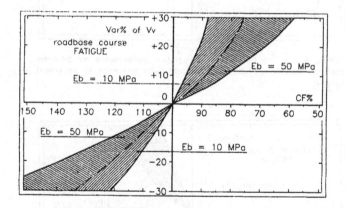

FIGURE 9:   Coefficients of relative performance for air void variations outside standards, with reference to pavement fatigue.

- Where, on the other hand, Eb=10 MPa, CF reaches values respectively equal to 83-85% and 89-92%.

Where the effects of the various variables concerned were jamed, only the total variation field of CF has been possible to be identified. The parameter which in almost all cases affects the results is the stiffness modulus of the bitumen.

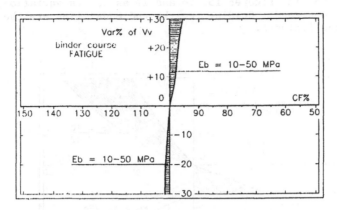

FIGURE 10   Coefficients of relative performance for air void
variations   outside   standards, with reference to pavement fatigue.

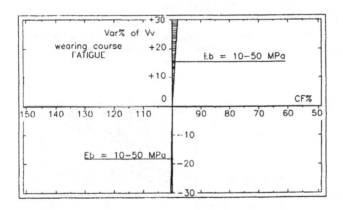

FIGURE 11:   Coefficients of relative performance for air void
variations outside standards, with reference to   pavement   fatigue.

4.3 Coefficients of relative performance to rutting

Similarly, CR values have been calculated according to the expression
(13), in order to analyze the influence of variations of the
parameters examined on the tendency to rutting of pavements.

The   results of the calculations performed are shown in Figures 12,
13 and 14, as far as variations outside specifications of bitumen
content in the road base, binder and wearing courses respectively are

concerned, and in Figures 15, 16 and 17 as far as variations of air voids are concerned. Reading remarks and modalities are the same as for the CF coefficients.

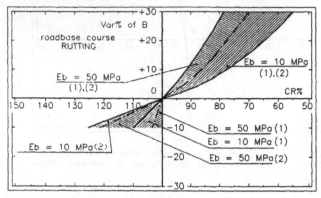

(1) E1 = 1900 MPa   E2 =2500 MPa
(2) E1 =10000 MPa   E2 =13500 MPa

FIGURE 12:   Coefficients of relative performance for bitumen content variations outside standards, with reference to pavement total rutting.

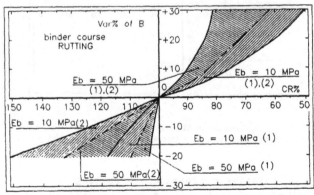

(1) E1 = 1900 MPa   E3 = 4100 MPa
(2) E1 =10000 MPa   E3 =16900 MPa

FIGURE 13:   Coefficients of relative performance for bitumen content variations outside standards, with reference to pavement total rutting.

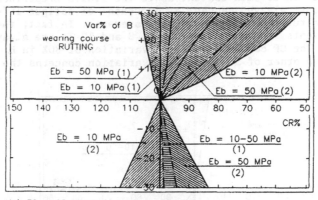

(1) E2 = 2500 MPa   E3 = 4100 MPa
(2) E2 =13500 MPa   E3 =16900 MPa

FIGURE 14:   Coefficients of relative performance for bitumen
content variations outside standards, with reference to pavement
total rutting.

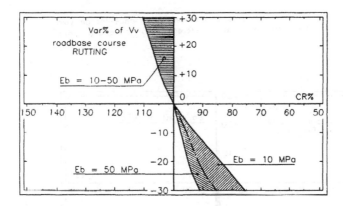

FIGURE 15:   Coefficients of relative performance for air void
variations outside standards, with reference to pavement total
rutting.

5 Analysis of the results

The results   shown in **Figures 6 - 17, with reference to the specific**
calculation conditions considered, allow the following considerations
to be drawn:

- An increase in bitumen content above the limits prescribed by the specifications always causes a decrease of fatigue life and an increase in rutting tendency in the pavements. In fact, the CF and CR coefficients shown in Figures 6-8 and 12-14 are always lower than 100. The CF coefficient for a variation of 30% in B, reaches values in the order of 50-60% if the variation concerns the road

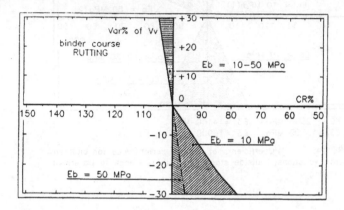

FIGURE 16·   Coefficients of relative performance for air void variations outside standards, with reference to pavement total rutting.

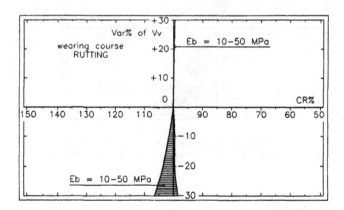

FIGURE 17:   Coefficients of relative performance for air void variations outside standards, with reference to pavement total rutting.

base, 70-90% if it concerns the binder course and in the order of 90-95% if it concerns the wearing course. CR values, on the contrary, always turn out to be very low for variations of B affecting anyone of the three layers considered (in the order of 50-80% for B variations of +30%), showing a high susceptibility to rutting of the pavement when the bitumen content increase, a susceptibility which is even more evident when B variations affect the surface layers. The reason for CF values less than 100 for B variations above the maximum limit prescribed can be found in the following consideration: the mixture modulus reduction, following a bitumen content increase (see Figures 3, 4 and 5) leads to an increase of the stresses induced in the pavement by the traffic loads. This produces a reduction in the allowable number of load repetitions to fatigue cracking, which completely cancels the beneficial effect of a greater bitumen content on the mixture fatigue resistance.

- An increase of air voids over the limits stated by the specifications produces a reduction of the pavement fatigue life only when such variations concern the road base (CF = 60-90% for B variations of +30%), whereas variations of Vv in the other layers do not produce appreciable effects. A Vv increase has, on the contrary, a beneficial effect on rutting. A greater tendency to rutting is found for air voids values lower than the minimum prescribed for road base and binder course mixtures (CR = 80-90% with a Vv variation of +30%). Such a tendency is not noticed for the wearing course.
- As far as fatigue life is concerned, the use of harder bitumens fundamentally involves a greater sensitivity of pavements under possible variations outside specifications of both bitumen content and air voids. As regards rutting, however, the contrary holds.
- Quality controls during construction are particularly important for the road base course. In fact, this layer has a determining influence on both fatigue life and rutting. An increase in the bitumen content over the maximum values allowed leads to CF and CR coefficient values up to 50-60%, which could result in a halving of the fatigue life and a doubling of accumulated permanent deformations. On the other hand, an increase in air voids over the maximum values allowed gives rise to a reduction in fatigue life as high as 60-70%, whilst a decrease below the minimum values gives rise to an increase in rutting in the order even of 20-25%.
- As regards the binder course, variations in the bitumen content can result in great troubles, especially as regards the pavement rutting (CR values can even reach 50-60%); slightly less, though still rather sensible, is its influence on fatigue life (CF maximum = 70-80%). Air voids control, on the other hand, is important for rutting alone, whilst it has no real influence on fatigue life.
- Finally, for the wearing course, only the bitumen content control should be carefully considered to control the pavement rutting.
- The results obtained prompt also the following consideration: the range of mixture composition characteristics normally admitted in the specification leads to very diversified mixture performances

and pavement behaviours over time.

This suggests the opportunity of carefully studying the composition of the bituminous mixes, according to the available aggregates, such as to always maintain the B and Vv values near the lower boundary of the allowed range. In any case, a reduction in the range width would appear to be appropriate.

The procedure developed for calculating the CF and CR coefficient values is suitable for defining performance criteria on which to base any penalty to be levied for works which do not correspond, either wholly or in part, to the contract specifications.

From this point of view, the numerical values shown in Figures 6 - 17 are only indicative being calculated in relation to a typical flexible pavement using well-defined calculation hypotheses. Therefore, it is not possible to generalize the results obtained without carrying out preliminary verifications with reference to each specific situation.

6 Conclusions

This paper examined the possibility of integrating the information concerning the properties of the bituminous mixtures for road base, binder and wearing courses obtainable by traditional laboratory tests, with that provided by the available experimental performance prediction models. This enables us to integrate the mix design with data concerning the mixture's performance under working conditions.

The application of the proposed criteria to a typical flexible pavement, taking into account the more frequent prescriptions of Italian contract specifications, gives rise to a series of general considerations.

The calculation of coefficients of relative performance (as defined by the relations (12) and (13)) for variations outside specifications in the road base, binder and wearing courses composition characteristics, enabled us to quantify the negative effects of increases in bitumen content and the air voids on the pavement fatigue life and on its rutting tendency (the latter limited to the effects of the bitumen content).

As far as rutting is concerned, a reduction of the air voids with respect to the prescribed value gives rise to negative effects.

The use of hard bitumens allows for mixtures with more limited permanent deformations, which are, however, more sensitive to variations in the bitumen content and air voids with regards to fatigue life.

The importance of a correct mix design and construction of the road base and binder course mixtures has been clearly shown; a variation in the composition characteristics of these two layers, in particular in that of the road base give rise to a reduction in the fatigue life or an increase in the pavement rutting of a considerable entity.

On the other hand, variations in the composition of the wearing course mixture may affect only the pavement permanent deformation; and only with regard to the bitumen content.

Finally, the range of variation intervals established in the specifications for the mixture' s composition characteristics leads to a very large diversification in the pavement performance, a fact which, whilst confirming the importance of an initial careful mix design of the bituminous mixtures and a careful quality control during their laying, should suggest the opportunity of a reduction in intervals allowed width.

The finalized procedure enables us to determine the entity of possible penalties to be levied when the road works do not conform to the contract specifications on the basis of performance criteria.

7 References

[1] Franken, L. (1977) Module des melanges bitumineux.
    Bitumes et Enrobes Bitumineux, N. Spec.
[2] The AASHO Road Test HRB Spec. Rep. 61 (1961-1962).
[3] Marchionna A. Fornaci M.G. and Malgarini, M. (1985)
    Modello di degradazione strutturale delle pavimentazioni,
    Autostrade 1/3.
[4] Verstraeten, J. (1972) Moduli and critical strains in
    repeated bending of bituminous mixes. Application to
    pavement design. **Proceedings, Intern. Conf. on the Struct.
    Design of Asphalt Pavement, London.**
[5] Verstraeten J. (1973) Loi de fatigue des melanges
    bitumineux. Influence des caracteristiques des bitumes
    et de la composition des melanges. **Rapport de Recherche**
    N. 161/JV C.R.R.
[6] Franken, L. (1977) Deformation permanentes observees
    en laboratoire et sur routes experimentales, C.R.R., **Coloque
    International sur la Deformation Plastique des Enrobes** , 29/30
    sett., Ecole Polytechnique Federale, Zurich.
[7] Franken, L. (1977) Permanent deformation law of bituminous
    road mixes in repeated triaxial compression. **Proceedings, Intern.
    Conf. on the Struct. Design of Asphalt Pavements** , University
    of Michigan, Ann Arbor.
[8] Klomp, A.J.G., Dormon, G.M. Stress distribution and dynamic
    testing in relation to road design, **Proc. 2nd Conf. Australian
    Road Research Board, Melbourne.**

12 MODIFIED MARSHALL TEST FOR DETERMINING THE SUSCEPTIBILITY OF BITUMINOUS MIXTURES AT ROAD SERVICE HIGH TEMPERATURES

F. DO PINO and J.M. MUÑOZ
Centro de Estudios de Carreteras, Madrid, Spain

Abstract
This work attempts to study the behaviour of the bitumens when forming bituminous mixtures at high temperatures of service in highway pavements. The used proceeding has been the Marshall method ASTM D 1559-82, modifiying the breakage temperature of the sample test. These temperatures oscillate from 45ºC up to 80ºC, from 5 to 5 degrees. In order to facilitate the study of the obtained results in relation to the thermic susceptibility of the tested mixtures, the ASTM Standard Test Method for Viscosity-Temperature Chart for Asphalts, ASTM D 2493, that relates the temperature and the absolute viscosity, has been adopted. The analysis of the obtained experimental results shows that bitumens with very similar penetration but different source, composition and nature present different rheological behaviour in the range of service temperatures. A marked change of rheological behaviour in all tested mixtures, about 55ºC, has been detected. This change is quantified by the slopes value calculated from the lineal regression equations.
Keywords: Bituminous Mixtures, Marshall Method, Thermic Susceptibility, Viscosity-Temperature Chart, Rheological Behaviour.

1 Introduction

This work attempts to study the behaviour of the bitumens when forming bituminous mixtures at high temperatures of service in highway pavements.

The results corresponding to the previous experiencies that have been the base of the present work formed part of the Spanish National Report sent to the 3rd International Symposium, RILEM, Belgrade 1.983.

A modification of the Marshall test has been used, to allow the determination in a simple way of the rheological properties of the bituminous mixtures, in the interval of the temperatures in which the most important characteristics of these materials are manifested fundamentally when employed in highway pavements.

The temperature about 60ºC represents this critical zone in which the phenomena of plastic deformation of these mixtures are produced, and constitute one of the actual and most important problems to set up in the bituminous materials.

The Marshall method, widely used in its actual normative, does not solve the set up problem, for which in this work some modifications to the mentioned method have been introduced in order to solve these difficulties.

In the Marshall test, as proposed as follows, some conditions that permit the study of the susceptibility to the temperatures of these materials during its period of service were established.

2 Experimental procedure

The used proceeding has been the Marshall method ASTM D - 1559 - 82, modifiying the breakage temperature of the sample test. The temperatures used oscillate from 45ºC up to 80ºC, both included, from 5 to 5 degrees.

In order to make manifest the possible differences of behaviour of the bitumens employed, mixtures of the same type of aggregates and identical aggregate composition, semidense type, has been used.

Sieve Size/passing,(%)
12,5mm	10mm	5mm	2,5mm	630μm	320μm	160μm	80μm
87,5	78,5	54,5	37,5	20,0	14,0	9,5	5,0

The bitumens employed are related together with the most important characteristics of the same in Table 1.

The binder content in all cases was 5,0% in aggregates, determinated as optimum for this type of mixture employed.

All the test were made in triplicate, and a total of 168 Marshall samples were tested.

Table 1. Hydrocarbon binder characteristics

Sample	Source	Nature	Refining Process	Additive	Pene- trat. (mm)	Soft Point (ºC)	Asphalt enes (%)
A	Middle Orient	Paraffi nic	Straight run	-	68	50,0	18,0
B	Venezue la	Asphalt naphte- nic	Straight run	-	66	51,0	14,5
C	Middle Orient	Paraffi nic	Partially oxidate	-	64	48,5	10,5
D	Middle Orient	Paraffi nic	Straight run	-	65	46,5	6,5
E	Middle Orient	Paraffi nic	Straight run	Sulphur	62	51,0	9,5
F	Middle Orient	Paraffi nic	Straight run	Polymer	70	53,0	10,0
G	Spain	Very Paraffi nic	Straight run	-	79	48,0	21,0

3 Results

The average bulk relative density of the 168 samples was 2,376 with a standard deviation, (σn-1), of 0,017.

The results obtained are shown in Table 2, in which are the mean stabilities of the specimens, identical intended to be identically the same, reached for each bitumen and each test temperature are pointed out.

Table 2. Test results

Temp (ºC)	Marshall Stability (kN)						
	A	B	C	D	E	F	G
45	22,8	22,0	22,4	22,2	22,2	22,8	19,8
50	18,6	19,2	19,7	17,5	19,7	17,4	17,7
55	16,7	15,8	15,9	16,2	17,5	18,2	14,2
60	15,7	15,2	15,6	17,0	15,5	15,9	14,9
65	14,2	15,8	16,0	15,5	14,9	14,5	14,5
70	14,2	13,6	12,1	13,6	12,5	14,4	11,7
75	12,3	12,6	13,9	15,2	13,5	13,0	13,3
80	13,9	10,5	12,4	12,9	11,6	13,3	11,3

The samples were tested in automatic press with XY graph recorder in which 1 mm in the graph correspond to 250 N (25kgf) of stability.

Though the study of the Marshall deformation determined in the tests of stability has not been the principal object of this work, it must be emphasized that statistical analysis of the results of the same, seems to mean a wanting of correlation among such deformation and the temperatures of the test. The statisticals calculated for this parameter have been

Samples number (n).............. 168
Mean deformation, ($\bar{X}$),mm....... 2,9
Standard deviation (σn-1),mm... 0,4
Linear regression equation,.... d(mm)=2,38+9,02·10^{-3}.t
Correlation coefficient........ 0,564

In order to facilitate the study of the obtained results in relation to the thermic susceptibilities of the tested mixtures, the ASTM Standard Test Method for Viscosity-Temperature Chart for Asphalts, ASTM D 2493, that relates the temperature and the absolute viscosity with the following equation, has been adopted as a most suitable way of graphycal representation and its later mathematical analysis.

$$\ln(\log \eta)=M\cdot\ln T + \ln B$$

175

where:

η = absolute viscosity in cP
T= absolute temperature (QC + 273)
M,B= constants.

In this work the absolute viscosity function has been substituted by the Marshall stability one.

The most significant characteristic noted upon representing in this graph, (Fig 1), the experimental obtained results in the Marshall test made at different temperatures, has made evident a clear change of slope, different for each mixture, to a temperature of about 55QC that indicates an important change in the rheological behaviour of the mixtures in relation with their susceptibility at those temperatures.

To study with more details this phenomenum it has been calculated, from experimental data, the lineal regression of the function considering two ranges: one from 45 to 55QC and another from 55 to 80QC. Once stablished the lineal equations,

$$ln(\log kN)=M \cdot \ln T + \ln B,$$

of each selected temperature range and of each mixture the points of intersection -the solutions - determining the abcisas (lnT) and the corresponding ordinates (ln(log kN)). Table 3 shows the obtained results in the regression analysis and their correlation coefficients.

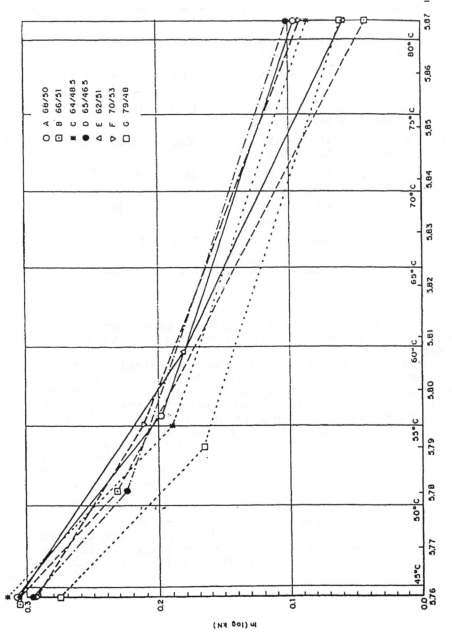

FIG. 1. STABILITY MARSHALL VS. TEMPERATURE

Table 3. Solution of the equations from experimental data

Sample	Temper. range	Slope −M	Correlat. coeffic.	Intersection (ºC)	(kN)
A	45–55	3,390	−0,990	55,7	16,16
	55–80	1,217	−0,835		
B	45–55	3,646	−0,992	50,9	18,24
	55–80	2,100	−0,910		
C	45–55	3,755	−0,984	54,9	16,22
	55–80	1,349	−0,763		
D	45–55	3,475	−0,967	51,0	17,84
	55–80	1,364	−0,810		
E	45–55	2,573	−1,000	59,7	15,82
	55–80	1,937	−0,929		
F	45–55	2,445	−0,786	54,9	17,34
	55–80	1,580	−0,944		
G	45–55	3,794	−0,976	53,6	15,14
	55–80	1,268	−0,760		

4 Conclusions

From the analysis of the obtained experimental results in this work the following considerations can be made:

- Then does not appear to be any correlation or dependency between the Marshall deformation and the temperature for a same bitumen content. This fact makes clear that the susceptibility to the plastic deformation due to the temperature in these conditions depends exclusivily on the internal friction of the mineral aggregate structure of the mixture; in practice, the observed deformation in the bituminous mixtures are promoted by an inadecuate content and type of bituminous material for the type of mixture used, which must be taken in account in the design and dosification of the same.

- Bitumens with very similar penetrations, but different source, composition and nature, present, when tested in mixtures of identical formulation, different rheological behaviour in the range of service temperatures.

- A marked change of rheological behaviour in all tested mixtures, about 55ºC, has been detected. This change is quantificied by the slopes value, M, calculated for the considered ranges of temperatures varying from -3,297, average value of the first stretch, up to -1,545, average value of the second one.
- The change of susceptibility (slope) calculated from the lineal regression equations is produced, in six of the seven bitumens studied, between 51 and 56ºC. Only in the bitumen modified with sulphur this temperature shifts up to about 60ºC.
- The temperature in which these phenomena are produce are of the same order of magnitude and very closed to that employed in most standard tests for determining the plastic deformation characteristics of the bituminous mixtures (Marshall, Wheel tracking and Gyratory) which implies a doubtful situation as to on which side of the sharp change zone of the slope a concrete mixture when tested is found.
- A good parallelism has been found between the conclusions of this work and that other the former work: "Float test a simple method to determinate the rheological properties of bitumes in the 45/75ºC range" made with the same bitumens, the results and conclusions of which were presented at RILEM Third International Symposium help in Yugoslavia (1983), though the obtained dispersions in the test results of the mixtures are significantly greater than in the corresponding binders, presumably due to the unavoidable heterogeneity of the mixture test samples yet identical intended.
- A subsequent investigation is considered necessary on the subject, in which new types of mixtures may be included and will prove to a more solid base to the preceding conclusions that allow the establishement of a normative test for the study of the temperature susceptibility of the bituminous mixtures.

5 References

Muñoz, J.M. and Do Pino, F. (1983). "Float test a simple method to determinate the rheological proper- ties of bitumen in the 45 - 75ºC range". Third International Symposium, RILEM, Belgrade, Yugoslavia.
Pap, I. (1984). "Discussion to the General Report: Testing of hidrocarbon binders". Third Interna- tional Symposium, RILEM, Belgrade, Yugoslavia.
Spanish National Report (1983). Third International Symposium. RILEM, Belgrade, Yugoslavia.

Suarez, L. (1989). "An approach to the thermal behaviour of the bitumens by use of the float test". 4th. Eurobitume Symposium. Vol. 1. Madrid. Spain.

Viscosity-temperature chart for asphalts, ASTM D 2493-68 (Reaprobed 1978).

13 A PROPOSED METHOD FOR MEASURING THE LATERAL DISPLACEMENTS DURING INDIRECT TENSILE TESTS ON ASPHALT BRIQUETTES USING LINEAR VARIABLE DIFFERENTIAL TRANSDUCERS

P.E. DUNAISKI and F. HUGO
Department of Civil Engineering, University of Stellenbosch, South Africa

Abstract
A new measurement system for measuring the lateral deformation during indirect tensile tests on asphalt-treated materials is described. The system consists of a light-weight aluminium frame with two linear variable differential transducers which is attached directly to the test sample. A very high resolution of less than $4,0 \times 10^{-5}$ mm is achieved resulting in a high measurement accuracy. The use of the system is demonstrated by describing the indirect tensile test with repeated load application for the determination of the stiffness of asphalt-treated materials as well as the indirect tensile test with a continuously increasing load for the determination of the strain-at-maximum stress of different test samples.
Keywords: Indirect Tensile Test, Asphalt, Displacement Measurements, Repeated Loads, Stiffness.

1 Introduction

The indirect tensile test (ITT) conducted on circular elements of asphalt treated mixes can be used to determine the tensile properties of these mixes. The tensile properties play a major role in determining the stiffness, the Poisson ratio, fatigue characteristics and the strain-at-maximum stress of this construction material.

During the ITT the increase in horizontal diameter of the circular element caused by a continuously increasing load or by repeated load impulses must be measured with great accuracy, since the deformations are very small. The load is applied in the vertical plane by means of two parallel loading strips which are machined to the same radius as the circular element. The lateral displacement is either measured by spring loaded displacement transducers attached to the loading frame on either side of the circular element and with the sensing probe touching the surface of the element or by a measuring device consisting of two cantilevered arms with strain gauges attached. Movements of the arms at the point of contact with the specimen have been calibrated with the output from the strain gauges (Hadley 1969). This measuring device is also attached to the loading frame.

During the load application, especially in the case of repeated load impulses, the circular test element undergoes a slight lateral rocking motion which can be attributed to various factors, which are the non-homogeneous nature of the material, a slight eccentric placement of the circular element in the testing machine or a slight misalignment of the load application strips. This lateral rocking motion causes a horizontal displacement of the circular element which can be greater than the horizontal deformation due to the applied load. This horizontal displacement is also measured by the two methods described above. The lateral deformation of the test sample is obtained by summing up the displacements measured by the two sensors. In the case of the repeated load impulses, where the amplitude of the load applied is between 10% and 50% of the expected failure load, the lateral increase in diameter of the circular element is in the order of 1×10^{-2} mm. For accuracy it is therefore advisable to try to eliminate any horizontal movement of the element during the testing. As this is very difficult to achieve, an alternative method would be to attach the lateral deformation measuring device directly to the test piece. In this way, the measuring device could follow any lateral movement of the test piece and only the increase in diameter of the circular element due to the applied vertical load is measured and the range of the measuring sensors can therefore be smaller and the resolution accordingly higher because no provision has to be made for the horizontal displacement due to the rocking motion of the sample. Likewise no inaccuracy results from irregularities on the sample surface.

In this paper such a device, which is directly attached to the test piece, is described.

2   Proposed measuring system

The basic idea of this measuring system is to attach a lightweight aluminium frame to the test piece a shown in figure 2.1.

Two contact beams are glued to the test piece by means of quick hardening two-component epoxy. The contact surface of the contact beam is machined to the same radius as the test piece. The cross beam, together with the longitudinal beam, is then attached to the contact beam by means of two wingnuts. The two HBM W2ATK LVDT's (Linear variable differential transformer) are attached on either side of the test piece with the sensing probe resting against the adjustable contact surface of the opposite longitudinal beam. The contact surface position is adjustable to set the LVDT to its zero position after setting up the measuring system.

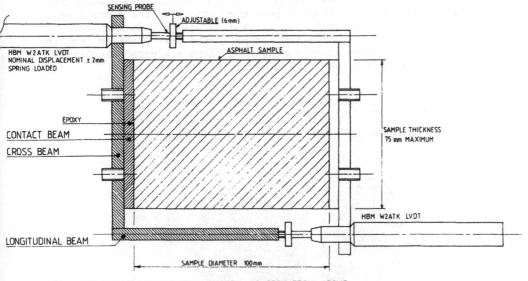

LATERAL DEFORMATION MEASUREMENT SYSTEM AS SEEN FROM ABOVE

Fig.2.1 Proposed measuring system

In figure 2.2 the measuring system is shown attached to a test piece between the parallel load application platens of the universal testing machine. The third LVDT to measure the vertical displacement can also be seen in figure 2.2.

Fig.2.2 Measuring system attached to test piece

183

The special jig, as shown in figure 2.3 is designed to ensure the correct position of the contact beam when attaching the contact beam to the test piece. With this jig it is also possible to mark the position of the load application strips on the test piece so that the load application plane and the measuring plane are perpendicular when the test piece is put into the universal testing machine.

Fig.2.3 Special jig to attach measuring system to test piece
Because only the contact beam is glued to the testing piece, as many test pieces can be prepared in advance as sets of contact beams are available.

The total lateral deformation due to the load is measured by each LVDT. The mean of the two readings is the actual lateral deformation.

The layout of the testing and data logging system is shown in schematic form in figure 2.4.

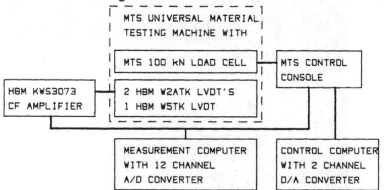

Fig.2.4 Schematic lay-out of control and data logging system

184

The control computer generates the load function for the repeated load impulses and controls the load application continuously via the control consol of the universal testing machine. The load applied is measured by means of the load cell, the lateral deformation is measured by means of the two HBM W2ATK LVDT's while the vertical displacement is measured by means of the HBM W5TK LVDT. The experimental data is then logged on the measurement computer via a 12 bit A/D converter using suitable software (Dunaiski 1986). The taking of the measurements after about 50 load cycles is triggered by hand. A window with a timespan of two seconds, that is two full loading cycles, will then be measured.

In figure 2.5 the time versus the digital value of the load function f(t) is shown. A digital value of 2048 (2^{11}) corresponds to an analog signal of 10 V. The load amplitude can then be set accordingly on the control console to obtain the specified load.

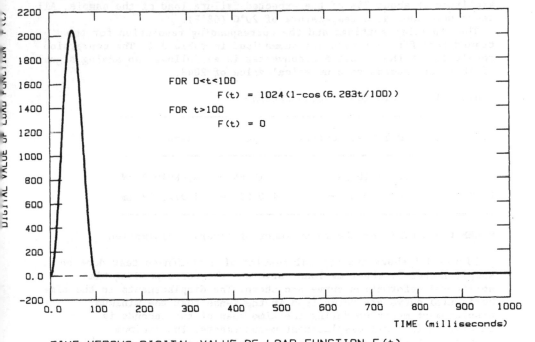

FOR $0<t<100$

$$F(t) = 1024(1-\cos(6.283t/100))$$

FOR $t>100$

$$F(t) = 0$$

TIME VERSUS DIGITAL VALUE OF LOAD FUNCTION F(t)

Fig.2.5 Graphical representation of load function

185

3   Applications of measurement system

Two applications of this measurement system will be demonstrated, i.e.
determination of the stiffness (MR) and strain-at-maximum stress.

3.1 Determination of the stiffness

Two series of seven tests each were done to determine the stiffness of
asphalt treated materials. The mix for both series was semi-gap graded
with a bitumen content of 5.5%. The bitumen for the first series had a
penetration of 60/70, while the bitumen for the second series had a
penetration of 80/100.

The stiffness was determined according to the method described in
ASTM D4123-82 (ASTM D4123-82 1987). The period of the load application
cycle was 1 second with the duration of loading 0,1 seconds and a
recovery time of 0,9 seconds. The sinusoidal load function had an
amplitude of about 25% of the expected failure load of the sample. All
tests were done at a temperature of 20°C (68°F).

The amplifier settings and the corresponding resolution for the
transducers for this test are summarised in table 3.1. The conversion
resolution of the 12 bit A/D converter is as follows: an analog signal
of 10 V corresponds to a numerical value of 2048.

Table 3.1   Resolution of transducers

Transducer	Amplifier setting	Range	Resolution
Load cell	10V $\triangleq$ 10 kN	$\pm$ 10 kN	$4,88 \times 10^{-3}$ kN
LVDT*	10V $\triangleq$ 0,08 mm	$\pm$ 0,08 mm	$3,91 \times 10^{-5}$ mm

* HBM W2ATK LVDT for the measurement of lateral deformation.

Figure 3.1 shows the typical results of a stiffness test done on
sample number 3. The time-load curve and the corresponding time-
horizontal deformation curve are shown. The displacements in the time-
horizontal curve are normalised in the sense that the minimum
displacement measured during the time span of two seconds is
subtracted from all displacement measurements. The minimum
displacement will therefore be zero. This approach is acceptable
because for the determination of the stiffness only the relative
displacements are used as indicated by H_I and H_T on the graph.

When the dynamic creep is to be measured by means of the indirect
tensile test, to determine input parameters for the VESYS analysis as
described by Kennedy (Kennedy 1977), the total deformation can also be
obtained.

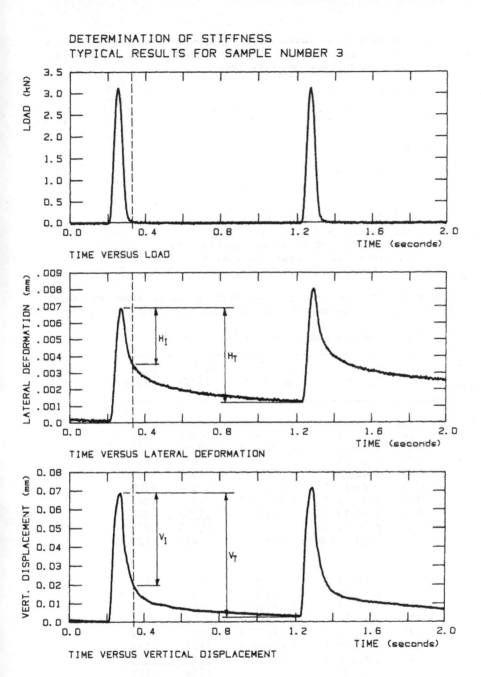

Fig.3.1 Results of repeated load test

Using the information as shown on the graph and assuming a Poisson ratio of 0,30 at a temperature of 20°C, the stiffness of the sample can be calculated as follows (Wallace 1980).

Instantaneous stiffness

$$M_{RI} = \frac{P}{H_I t} (0,273 + \nu) \qquad\qquad (3.1)$$

Total stiffness

$$M_{RT} = \frac{P}{H_T t} (0,273 + \nu) \qquad\qquad (3.2)$$

where
P =   amplitude of the repeated load (N)
t =   thickness of the sample (mm)
H_I =   instantaneous horizontal deformation (mm)
H_T =   total horizontal deformation (mm)
ν =   Poisson ratio

A summary of the results of the two test series is given in table 3.2. The parameters were calculated using the above equations.

Table 3.2   Instantaneous and total stiffness

Sample Number	P (N)	H_I (mm)	H_T (mm)	M_{RI} (MPa)	M_{RT} (MPa)
60/70 Penetration					
1	3550	$3,12 \times 10^{-3}$	$5,54 \times 10^{-3}$	10866	6115
2	3057	$2,73 \times 10^{-3}$	$4,59 \times 10^{-3}$	10694	6357
3	3120	$3,10 \times 10^{-3}$	$5,66 \times 10^{-3}$	9612	5264
4	3599	$3,61 \times 10^{-3}$	$6,73 \times 10^{-3}$	9521	5106
5	3311	$2,84 \times 10^{-3}$	$5,10 \times 10^{-3}$	11134	6200
6	3130	$3,01 \times 10^{-3}$	$5,68 \times 10^{-3}$	9931	5265
7	3472	$2,82 \times 10^{-3}$	$4,68 \times 10^{-3}$	11758	7083
		Mean Value		10503	5913
		Standard Deviation		839	728
80\100 Penetration					
8	3296	$3,83 \times 10^{-3}$	$7,01 \times 10^{-3}$	8218	4492
9	3252	$3,77 \times 10^{-3}$	$7,30 \times 10^{-3}$	8238	4252
10	3276	$3,98 \times 10^{-3}$	$7,19 \times 10^{-3}$	7861	4351
11	3291	$3,73 \times 10^{-3}$	$6,60 \times 10^{-3}$	8426	4672
12	3271	$3,84 \times 10^{-3}$	$7,82 \times 10^{-3}$	8135	3995
13	3286	$4,40 \times 10^{-3}$	$7,61 \times 10^{-3}$	7132	4123
14	3296	$4,86 \times 10^{-3}$	$8,74 \times 10^{-3}$	6477	3600
		Mean Value		7784	4212
		Standard Deviation		715	351

Using the information as shown on the graph in figure 3.1, the instantaneous and total Poisson ratio can be calculated as follows(ASTM D4123-82 1987):

Instantaneous Poisson ratio:

$$\nu_I \; - \; 3,59 \; \frac{H_I}{V_I} - 0,27 \tag{3.3}$$

Total Poisson ratio:

$$\nu_T \; - \; 3,59 \; \frac{H_T}{V_T} - 0,27 \tag{3.4}$$

where
H_I -   instantaneous horizontal deformation (mm)
V_I -   instantaneous vertical deformation (mm)
H_T -   total horizontal deformation (mm)
V_T -   total vertical deformation (mm).

Table 3.3 shows the results that were obtained when calculating the Poisson ratio. It is apparent that these are totally inconsistent with the normally expected values for asphalt at the test temperatures. When these values of the Poisson ratio are used to calculate the stiffnesses, the results are equally unacceptable.

Table 3.3   Instantaneous and total Poisson ratio

Sample Number	H_I mm	H_T mm	V_I mm	V_T mm	ν_I	ν_T
1	$3,12 \times 10^{-3}$	$5,56 \times 10^{-3}$	$3,35 \times 10^{-2}$	$7,08 \times 10^{-2}$	0,064	0,012
2	$2,73 \times 10^{-3}$	$4,60 \times 10^{-3}$	$4,21 \times 10^{-2}$	$7,08 \times 10^{-2}$	-0,037	-0,037
3	$3,10 \times 10^{-3}$	$5,68 \times 10^{-3}$	$3,98 \times 10^{-2}$	$6,58 \times 10^{-2}$	0,010	0,040
4	$3,61 \times 10^{-3}$	$6,76 \times 10^{-3}$	$6,10 \times 10^{-2}$	$8,48 \times 10^{-2}$	-0,058	0,016
5	$2,84 \times 10^{-3}$	$5,12 \times 10^{-3}$	$6,80 \times 10^{-2}$	$12,10 \times 10^{-2}$	-0,120	-0,118
6	$3,01 \times 10^{-3}$	$5,68 \times 10^{-3}$	$5,82 \times 10^{-2}$	$7,96 \times 10^{-2}$	0,101	0,218
7	$2,82 \times 10^{-3}$	$4,61 \times 10^{-3}$	$5,47 \times 10^{-2}$	$7,68 \times 10^{-2}$	-0,085	-0,055
8	$3,83 \times 10^{-3}$	$6,90 \times 10^{-3}$	$6,46 \times 10^{-2}$	$8,77 \times 10^{-2}$	-0,057	0,012
9	$3,77 \times 10^{-3}$	$7,31 \times 10^{-3}$	$6,83 \times 10^{-2}$	$10,43 \times 10^{-2}$	-0,072	-0,018
10	$3,98 \times 10^{-3}$	$7,29 \times 10^{-3}$	$6,85 \times 10^{-2}$	$9,05 \times 10^{-2}$	-0,061	0,019
11	$3,73 \times 10^{-3}$	$6,56 \times 10^{-3}$	$5,77 \times 10^{-2}$	$7,77 \times 10^{-2}$	-0,038	0,033
12	$3,84 \times 10^{-3}$	$7,78 \times 10^{-3}$	$7,20 \times 10^{-2}$	$10,25 \times 10^{-2}$	-0,079	0,002
13	$4,40 \times 10^{-3}$	$7,68 \times 10^{-3}$	$4,42 \times 10^{-2}$	$6,03 \times 10^{-2}$	0,087	0,187
14	$4,86 \times 10^{-3}$	$8,69 \times 10^{-3}$	$6,23 \times 10^{-2}$	$8,08 \times 10^{-2}$	0,010	0,116

It shoud be noted that the values of the Poisson ratio as calculated above using the vertical and the horizontal displacements are found to vary over a very large range. This is a well-known phenomenon. The authors believe that one of the reasons for this is the high stress in the compression zone close to the point of load application resulting in local shear failure. This failure causes excessive vertical deformation. To counter this effect it would be advantageous to use wider load application plattens. A 12 mm wide platten was used during the tests.

3.2 Determination of the strain-at-maximum stress

Strain-at-maximum stress and the corresponding energy requirements can be used to differentiate between mixes as to their ability to resist tensile stress (Hugo 1989). The proposed new method of measurement was therefore also used to determine this parameter.

The same samples which had been used to determine the stiffness of the material were used for the strain-at-maximum stress test.

For these tests the load was applied under displacement control with a rate of 50 mm/minute for the vertical displacement. These tests were also done at a temperature of 20°C.

The amplifier settings and the corresponding resolution for this test are summarised in table 3.4.

Table 3.4 Resolution of transducers

Transducer	Amplifier setting	Range	Resolution
Load cell	10V $\hat{=}$ 50 kN	± 50 kN	$2,44 \times 10^{-2}$ kN
LVDT*	10V $\hat{=}$ 0,8 mm	± 0,8 mm	$3,91 \times 10^{-4}$ mm

* HBM W2ATK LVDT for the measurement of lateral deformation

Figure 3.2 shows the time load curve of a strain-at-maximum stress test done on sample number 3.

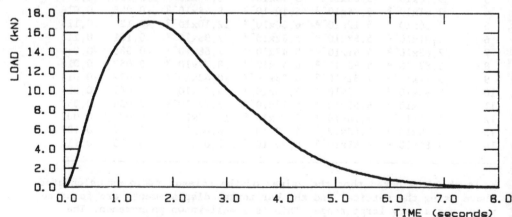

Fig.3.2 Time-load curve for strain-at-maximum stress test

Figure 3.3 shows the lateral deformation-load curve of a strain-at-maximum stress test done on sample number 3.

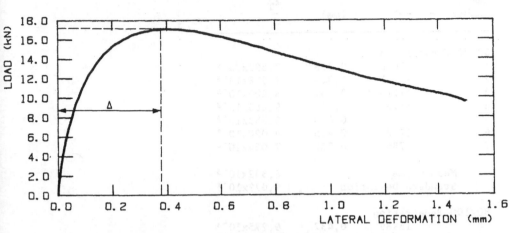

Fig.3.3 Lateral deformation-load curve

Using the information as shown on the graph the strain-at-maximum stress can be calculated as follows (Transportation 1975) assuming a Poisson ratio of 0,30 at 20°C:

$$\varepsilon_T = \Delta \left[\frac{0,00637 + 0,0191 \, \nu}{0,27 + \nu} \right] \tag{3.5}$$

where ε_T = strain-at-maximum stress
Δ = lateral deformation at maximum load (mm)
ν = Poisson ratio

A summary of the results of the two test series is given in table 3.5

Hugo et al. (Hugo 1989) determined the strain-at-maximum stress in a series of tests using the conventional measuring system. The standard deviation, expressed as a percentage of the mean value, for that series of tests was 17,8%. The standard deviation, also expressed as a percentage of the mean value, for the two series of tests using the proposed new measuring system for the strain-at-maximum stress is 10,9% and 8,8% respectively.

11

Table 3.5 Results of the strain-at-maximum stress test

Sample Number	P (N)	Δ (mm)	ε_T
60/70 Penetration			
1	18066	0,358	$7,593 \times 10^{-3}$
2	17261	0,369	$7,838 \times 10^{-3}$
3	17114	0,387	$8,205 \times 10^{-3}$
4	15991	0,427	$9,063 \times 10^{-3}$
5	12524	0,450	$9,553 \times 10^{-3}$
6	12622	0,425	$9,020 \times 10^{-3}$
7	17896	0,332	$7,054 \times 10^{-3}$
Mean Value			$8,332 \times 10^{-3}$
Standard Deviation			$9,072 \times 10^{-4}$
80/100 Penetration			
8	13989	0,437	$9,275 \times 10^{-3}$
9	13110	0,332	$7,054 \times 10^{-3}$
10	12329	0,419	$8,883 \times 10^{-3}$
11	13184	0,375	$7,969 \times 10^{-3}$
12	12231	0,388	$8,230 \times 10^{-3}$
13	12842	0,394	$8,361 \times 10^{-3}$
14	12354	0,369	$7,838 \times 10^{-3}$
Mean Value			$8,230 \times 10^{-3}$
Standard Deviation			$7,235 \times 10^{-4}$

4 Conclusion

This measurement system was developed to eliminate certain problems during indirect tensile tests on asphalt-treated materials. The following conclusions can be made:

Due to the fact that the measurement system is attached directly to the test specimen it can follow any movement of the specimen without affecting the measurement of the lateral deformation. The range of the LVDT can therefore be kept smaller with a correspondingly increasing resolution which enhances the accuracy of the measurement.

The total lateral deformation is calculated by taking the mean of the two transducer readings and not the sum as it is done with other measurement systems. This also increases the accuracy of the measurements.

The measurement system is also relatively inexpensive as all components are re-usable and any suitable data logging system can be used.

This measurement system is very easy to use. The process of glueing the contact beams to the test sample is simplified by the specially constructed jig. Together with dedicated software, which is instructing the user step by step, any indirect tensile test can become a routine testing procedure for even inexperienced laboratory assistants.

All components of the measurement system are robust. The working temperature range of the LVDT's as used in this system is from -20°C to +80°C. Any indirect tensile test under elevated temperatures can therefore be done using this system. The system is also small enough to fit into temperature controlled chambers in the frame of a universal testing machine.

The results as obtained by this measurement system are accurate and compare favourably with results obtained using other systems. Even electrical noise is almost not noticeable although a very high degree of signal amplification is used.

In general this measurement system for indirect tensile tests on asphalt treated materials performs very well and together with dedicated software this should help to make the indirect tensile test a routine procedure once the other sources of discrepancy alluded to in the paper have been addressed.

5   References

ASTM D4123-82 (1987) - **Test methods for indirect tension test for resilient modulus of bituminous mixtures**, American Society for Testing and Materials.

Dunaiski, P E (1986) **Program DLOAD, Measurements under dynamic loads, User's Manual**, Department of Civil Engineering, University of Stellenbosch.

Hadley, W O; Hudson, W R and Kennedy, T W (1969) **An evaluation of factors affecting the tensile properties of asphalt-treated materials**, Research Report 98-2, Center for Highway Research, University of Texas, Austin.

Hugo, F; Nachenuis, R (1989) **Some Properties of Bitumen-Rubber Asphalt and Binders**, Proceedings Association of Asphalt Paving Technologists, Technical Session, Volume 58.

Kennedy, T W (1977) **Characterization of Asphalt Pavement Materials using the Indirect Tensile Test**, Proceedings Association of Asphalt Paving Technologists, Technical Session, Volume 46.

Transportation Research Board (1975) **Test procedures for characterising dynamic stress-strain properties of pavement materials**, Special Report 162, Washington, D.C.

Wallace, K; Monismith, C L (1980) **Diametral modulus testing on non-linear pavement materials**, Proceedings Association of Asphalt Paving Technologists, Technical Session, Volume 49.

14 TESTING TENSION DISSIPATION OF ASPHALT PRISMS UNDER BENDING STRESS

P.K. GAUER
Ingenieurgesellschaft mbH, Regensburg, Federal Republic of
Germany

Abstract
Relaxation of asphalt is a very important property for the
asphalt's behavior at cold temperatures. In the author's
laboratory a device for testing the relaxation time of
asphalt prisms under bending stress has been developped.
The report shows the rheological background and the
practical meaning of relaxation, describes the testing
device and gives an example of an executed examination.
Keywords: Rheology, Relaxation, Bending Test of Asphalt
Prisms.

1 General Observations

1.1 Rheological theory of the Maxwell model
The term relaxation is derived from the field of rheology.
Rheology provides a comprehensive description of the
behaviour of materials with regard to Hookean elasticity
and Newtonian viscosity. The equations of the tension-
expansion behaviour are constructed from the basic equations depending on the combination of basic models through
superimposition of tensions or deformations.

Relaxation can best be described in pointing to the so-
called Maxwell body which stands for visco-elastic behaviour of materials. A Maxwell body consists of a serial
connection of a Newtonian damping element and a Hookean
feather which shows that the material law is composed of
the sum of deformations of the basic equations.

Hooke: $\tau = G \cdot \gamma$; $\gamma = \dfrac{\tau}{G}$ $\qquad$ (1), (1')

Newton: $\tau = \eta \cdot \dot{\gamma}$; $\dot{\gamma} = \dfrac{\tau}{\eta}$ $\qquad$ (2), (2')

Maxwell: $\dot{\gamma}_M = \dfrac{\dot{\tau}}{G_{fl}} + \dfrac{\tau}{\eta}$ $\qquad$ (3)

Index fl is added to the shearing modulus of elasticity to mark the characteristic elasticity of the liquid as an expression of the rigidity of the liquid. The equation of the Maxwell body is a differential equation of the type $y' = y$ and therefore has exponential character. In solving it, the parameters are considered constant so that they reappear as factors in the exponents of the e-function. The parameters indicate through their number the degree of the rheological equation

The equation (3) has the following solution:

$$\tau = e^{-\frac{G_{f1}}{\eta}} (\tau_0 + G_{f1} \int_0^t \dot{\gamma} \cdot e^{\frac{G_{f1}}{\eta} t} \, dt) \qquad (3)$$

If the deformation remains constant (i.e., $\dot{\gamma} = 0$) (3) becomes

$$\tau = \tau_0 \, e^{-\frac{G_{f1}}{\eta} t} \qquad (4)$$

The negative sign in the exponent means that $\tau = f(t)$ is gradually reduced according to the factor η/G_{f1}. This quotient is therefore called the relaxation time.

When $t = \eta/G_{f1}$, then the exponent becomes -1. The tension τ is then $0.368 \times \tau_0$. In a relaxation process which is measured in a Maxwell body this time mark is optimally suited to find the factor η/G_{f1} and therefore the specific relaxation time for this material.

It is self-evident that this time mark which can be easily found through experiments is also used to characterize relaxation processes which are observed in other than Maxwellian bodies.

Since the term relaxation time is defined through the quotient η/G_{f1} of the Maxwell model it is recommended that because of scientific exactitude one does not use the term relaxation time in non-Maxwellian behaviour but instead the general term tear-down time.

It should therefore be noted that this background is always presupposed in the following even when the term "relaxation time" is still used because it is more descriptive.

1.2 Relaxation of asphalts

Asphalts are in a rheological sense no two-parametric Maxwell bodies. The description of their material behaviour needs a complicated combination of all three basic models. According to our present knowledge at least five parameters are necessary. Therefore the relaxation behaviour cannot be described with the equation that is valid for the Maxwell bodies.

The comparison of the curve $\sigma/\sigma_0 = e^{-t/t_R}$ with the observed curve in figure 1.1 shows this very clearly. The curves intersect, with the exception of the trivial point at t = 0, only at t = t_R. Furthermore the observed relaxation function is considerably more curved than the e-function.

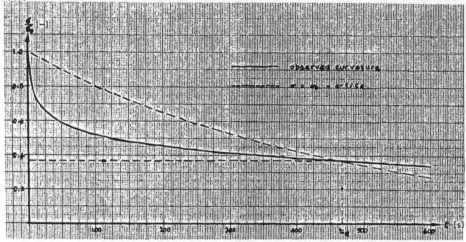

Figure 1.1. Comparison of the observed relaxation curve with the e-function $\sigma = \sigma_0 \cdot e^{-t/t_R}$

The reasons for this can be shown in model-like fashion through the following consideration:

1) The Hookean body does not relax because of the defined linearity between tension and expansion.

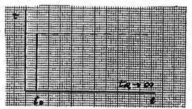

2) The Newtonian liquid relaxes immediately because of the relation $\tau = \eta \cdot \dot{\gamma}$ when $\dot{\gamma} = 0$

3) The Maxwell body relaxes as a combination of the Newtonian and Hookean body according to the e-function when $\dot{\gamma} = 0$. The relaxation curve moves from the Hookean toward Newtonian behavior.

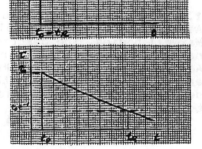

4) The St. Venant body "relaxes" immediately as well when the deformation is stopped.

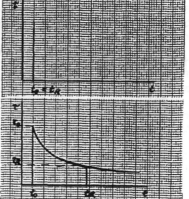

5) The relaxation curve of the asphalt must therefore move from that of the Maxwellian body toward that of the St. Venant body. Therefore it shows a stronger curvature than an e-function.

6) It has been shown that the relaxation behaviour can be described with an extraordinarily high degree of conformity through a logarithmic line, to the form σ = σo (A+B (ln t + 1)). Furthermore it can be shown that the value B with a certainty of approximately 80 % is dependent on the observed relaxation time t_R. The value A corresponds theoretically to the tension value at the time mark t = 0 and should therefore show the value 1. This condition can not always be observed especially in lower temperatures.

Nevertheless according to the present knowledge a general relaxation formula can be given through the equation

$$\sigma/\sigma_0 = 1 - [K] \frac{\ln (t + 1)}{t_R} \qquad (5)$$

The value K stands representatively for all parameters which are derived from regression analyses.

2 Practical Meaning of Relaxation

As with all other material characterising quantities of asphalt, relaxation is dependent on the current temperature. Relaxation occurs very rapidly at high temperatures. This fact explains both the high pliability of warm asphalt and the need to reduce the asphalt's tendency to become deformed during summertime use through stiffening additives.

As the temperature falls, so does relaxation ability. Sufficiently low temperatures can cause the asphalt to become practically quasi-elastic and incapable of relaxation.

With a negative temperature gradient, this can lead to inhibited contraction of an asphalt surface caused by friction and bindings to the material under the asphalt - all of which results in tension.

Tears or cracks result when the tension that grows from the edges to the middle of a road becomes greater than the asphalt's tensile strength.

The qualitative relationship between temperature and tensile strength and thermically induced tension is illustrated in the following sketch (fig. 2.1).

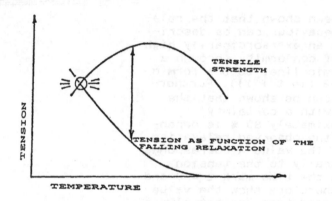

Figure 2.1. Relationship between tensile strength and thermically induced tension as function of the temperature

Tensile strength first grows as temperatures fall, then peaks and falls off again. The thermically induced tension rises as temperatures fall because of diminishing relaxation ability. This tension grows very gradually at first, then almost constantly.

When the tension curve meets the tensile strength curve, a break occurs. At temperatures higher than the breaking point, the double arrows represent the tension reserve, or the road's ability to absorb mechanically induced tension, i.e. from traffic.

The further the tension curve is moved to the left, the better an asphalt's low-temperature performance. The relaxation ability of a material is responsible for a large portion of any such movement to the left. Thus, relaxation ability is a very well-suited means of measuring the low-temperature behaviour of asphalt.

The difference in the temperatures at which various types of asphalt possess the same relaxation times can be determined through comparative experimentation.

3   Description of the Test

3.1 Purpose and Area of Application
This procedure is intended to establish the low-temperature behaviour of asphalt. To this end, tension dissipation of a predetermined base tension at various temperatures is tested.

The procedure may be performed on samples that are either produced in a laboratory or taken from asphalt layers that have already been installed.

The result is the time required until a remaining tension of 36.8% of the base tension is reached.

3.2 Short Description
Asphalt prisms are subjected to a defined bending stress at testing temperatures of less than 1 °C. This bending stress equals as a rule two-thirds of the asphalt's breaking point. Finally, the tension dissipation is charted and evaluated.

3.3 Terminology
Tension dissipation of asphalt is a product of the viscous characteristics of the bituminous binding agent. For purely viscoelastic materials, tension dissipation is expressed with the rheological term relaxation. The time required until a tension of 36.8% of the base tension is reached is a material constant and is called relaxation time.

With bituminous materials, relaxation is dependent not only on temperature, but also on the stress condition of the sample at the start of the experiment. The latter becomes apparent in that the asphalt's tensile viscosity falls as the sample is subjected to greater demands. Added to the list of relaxation-capable components of asphalt is its plasticity. Thus, the tension dissipation curve does not follow an e-function, but rather shows a stronger curvature; this curve can be depicted on a simple logarithmic scale as a straight line.

3.4 Equipment
Cooling device with container for refrigeration of
   cooling liquid to -30 °C.
Cryostat for refrigeration of asphalt samples.
Testing press with electromagnetic drive and controllable

feeding speed.

Four-point bending device for testing samples with dimensions 40 X 40 mm or 50 X 50 mm and length of 150 or 160 mm.

Cold chamber to hold the bending device for maintaining testing temperatures for the duration of the experiment.

Temperature monitor.

Electronic data storage or charting device.

Temperature charting system for several measuring points.

For production of laboratory samples:
Press for compaction of cylinders.
Pressure pot with diameter of 150 mm including footplate, stamp and depth gauge.
Diamond saw. (A)
Grinding machine. (A)

For production of samples from installed asphalt layers:
Devices marked with (A) above.
Devices according to DIN 1996, Parts 4,7,20.

3.5 Samples

3.5.1 Specifications
Asphalt prisms with cross-section dimensions of 40 X 40 mm or 50 X 50 mm and a length of 150 or 160 mm serve as samples. The samples are taken from cylinders with a diameter of 150 mm or produced in prism form according to DIN 18 555. The latter applies to mastic asphalt and similar materials. The cylinders can be taken from drilling samples from installed asphalt layers or can be produced in the laboratory. Drilling samples should have a layer thickness of at least 4 cm.

3.5.2 Production

3.5.2.1 Producing Cylinders in the Laboratory
Producing samples in the laboratory allows one to keep within the desired range of void content with an accuracy of ± 0.2 % volume.

The amount of mixture needed for each cylinder of 7 cm in height is determined mathematically taking into account the desired height and the corresponding volume, the mixture's density and the desired void content. Other factors that must be considered are:

- greater compaction in the center of the cylinder from which the prism will be extracted;
- wasted mixture during processing, accidents, etc.

The mixture, produced and climatised according to DIN 1996, Part 4, is put into the compaction mould and compacted with a press. A gauge is used to determine the final height of the sample. Once the final height is reached, the mould is left in the press for a while until the compaction-induced tension has dissipated.

Once the cylinders have cooled, the prisms are sawed out. The samples should be cut such that the width is exactly at the desired level and the height is 1 to 2 mm greater than desired. The exact dimensions are reached to 0.1 mm accuracy with a grinding machine.

With mastic asphalt mixtures, the prisms can be produced directly. The procedure is similar to the production of standard cubes, but instead of a cube mould, prism moulds are used according to DIN 18 555.

3.5.2.2 Preparing Samples from Installed Asphalt Layers
This process mirrors the preparation of laboratory samples, beginning with the sawing out of the prisms.

With layer thickness of less than 4 cm, the grinding machine may not be used. In such cases, it may be desirable to smooth out the storage area and where the load will be introduced.

3.6 Performing the Experiment

3.6.2 Determining the Breaking Point
After the desired temperature is reached in the cooling chamber, at least two samples are tested for bending strength. Due to the samples thermal inertia and the speedy completion of this process, it is not necessary to try to stabilise the testing temperature once the sample is inserted into the bending device, although this will be necessary for the following relaxation test.

The feeding speed of the press is to be set such that no relaxation takes place while the sample is under stress. Whether this condition is met can be observed visually in the stress-deformation curve. At a testing temperature of 0 °C a feeding speed of v=9mm/min has shown itself to be suitable; at -10 °C, v=6mm/min. Generally speaking, the desired elasticity is realised that much sooner the lower the temperature and/or the faster the stress demand is increased. On the other hand, the stress demand on the samples should be increased as slowly as possible so that later during the relaxation test the introduction of stress can be visually monitored.

3.6.3 Determining Relaxation Time
The two samples to be tested first are placed simultaneously in the refrigerated testing chamber. Sample 1 is inserted into the bending device, while sample 2 is set aside in an appropriate place. The temperature in the

chamber is altered when the door is opened. One must thus wait a while before starting the experiment until the temperature has stabilised. Then the experiment can be begun.

The temperature throughout the duration of the experiment should be recorded by a monitor. The temperature fluctuations in the chamber should be held to ± 0.1 K through the use of a suitable thermostat.

The sample is subjected to demands up to the pre-selected tension. The press drive then is stopped either manually or by a program. At this moment the relaxation process begins, which is charted on an x-y recorder. This chart, or the same data stored in a data storage program, is needed for later evaluation and analysis.

The experiment can be ended when the acute tension of the sample has been reduced to 30% of the beginning tension.

3.6.4 General

At least 8 prisms are recommended for each series of experiments. Of these, 2 are needed to determine bending strength and 5 for the relaxation test. The remaining sample serves as a backup for any repeated experiments that may be necessary if, for example, a sample were to break before the base tension were reached due to inhomogenity.

After the samples are produced, they are subjected to a volume determination according to DIN 1996, Part 7. The samples then can be dried to a constant weight at 30 C or at room temperature. Once they are dry, the samples' dry weight is used to determine density according to DIN 1996 Part 7.

The samples then are cooled to testing temperature by use of a cryostat. In general, a normal freezer can be used for this purpose, provided the freezer's factory-in-stalled temperature control has been replaced with a thermostat. In order to moderate the temperature change in the freezer, a sort of thermal ballast, such as a large amount of stone chips or a similar substance, can be used.

For a testing temperature of 0 °C, the samples must be left in the freezer for at least 3 hours. Further, it is advisable to establish the rate of cooling by recording the cooling of a dummy sample on a temperature recorder.

3.7 Evaluating the Results

The testing data can be evaluated either through a suitable PC program or manually by analysing the charted information.

It is convenient to divide the tension axis based on 100% of the base tension σ_o into steps of 5%, then to

chart the corresponding times for each step and to transfer
these onto a prepared table. It is sufficient to use the
75% mark as the first measured value and to stop at the
30% mark. The 36.8% value can be determined either directly
out of the curve or mathematically through the regression
equation:

$$\sigma/\sigma_0 = A + B \cdot \ln (t+1)$$

where $\sigma/\sigma_0 = 0.368$.

4 Example

The following is one example taken from many studies. It
demonstrates the possibilities of the testing method. It
shows the examination of different mixtures of mastic
asphalt 0/11.

4.1 Description of the analyzed mixture

4.1.1 Basic composition
Starting point for the examinations was a standard composition for mastic asphalt 0/11 mm according to TVbit 6/75.

In the following chart the basic values of the composition of the mixture as provided by the mixing plant are
compiled:

chippings	c. 53 %
thereof	18 % 8/11 mm
	15 % 5/8 mm
	20 % 2/5 mm
sand	c. 22 %
thereof	50 % natural sand and
	50 % crushed sand
filler	c. 25 %
thereof	the proportion of 2 %
	of Trinidad-Epuré
bitumen	6.8-7.0 % B 45
	including natural
	rock asphalt
softening point (ring and ball) of the binder	c. 64°C after extraction
identation depth	1.5-2.8 mm
mineral aggregates	basalt, natural sand

With the submitted samples of mineral aggregates a grading
curve was set up according to table 4.1, #1. The content of

binder including 2 % of natural rock asphalt was adjusted at 6.8%.

4.1.2 Variations

4.1.2.1 Constancy of quantity
The first variation of the basic composition with natural asphalt consisted of a replacement of the Trinidad-Asphalt through bituminous binder for road building and limestone filler. The composition of the aggregates is also recorded in table 4.1, #2.

The content of bitumen amounted correspondingly to 6.8%.

Table 4.1. Grading of the Examined Mastic Asphalts

Size of Sieves	Passing of Sieves		
	No.1	No.2	No.3
mm	%	%	%
45			
31,5			
22,4			
16	100,0	100,0	100,0
11,2	98,2	98,2	98,2
8	81,3	81,3	81,3
5	66,3	66,3	66,3
2	46,6	46,6	46,7
0,71	38,6	38,6	39,5
0,25	29,4	29,4	31,4
0,09	25,0	24,9	27,1

4.1.2.2 Constancy in viscosity
The second variation consisted in an equal-viscous replacement of the natural asphalt through the regulative of the relation between filler and binder.

As the characteristic of viscosity we took the softening point Wilhelmi of the ashapltic mortar as mixture of filler and binder. The softening point in question of the mixture of natural asphalt was reached when the relationship of the filler to the binder became F/B = 4.12. Accordingly for this variation with a constant binder content of 6.8 % the addition of foreign filler was increased while the sand component was reduced. The grading is also shown in table 4.1, #3.

4.2 Relaxation analyses

The relaxation tests were performed at temperatures of 0 °C and -10 °C. As the initial tension 2/3 of the tensile strength were adjusted. The results of the relaxation tests are grouped together in table 4.2.

Table 4.2. Tensile Strength and Relaxation Time

	Basic Composition		Equal Quantity		Equal Viscosity	
Temperature [°C]	0	-10	0	-10	0	-10
Tensile Strength $\beta_{b\,z}$ [N/mm²]	11,3	10,4	11,1	10,1	13,9	11,3
Relaxation Time t_R [s]	404	6640	147	2978	144	2840
Density of the Prisms [g/cm³]	2,553		2,552		2,561	

The chart in table 4.2 shows the mastic asphalts 1 and 2 which are identical in compositional quantity. They are practically identical in their tensile strength. Asphalt 3 which in recipe differs from the samples 1 and 2 through a higher filler content also shows a higher tensile strength.

Regarding the relaxation time both mastic asphalts 2 and 3 which were manufactured from bituminous binder for road building are practially identical. It should be noted here that the relaxation times are compatible with the results of asphalt concrete mixtures when B 45 is applied.

In comparison to the two other materials asphalt 1 which was manufactured by using Trinidad Natural Rock Asphalt shows in the mean a higher relaxation time by the factor of 2.7.

The graphic display in figure 4.1 showing the dependence of relaxation time on temperature in a semi-logarithmic scale indicates that asphalts 2 and 3 have the same relaxation time as the mastic asphalt 1 at a temperature that is 3 °C lower.

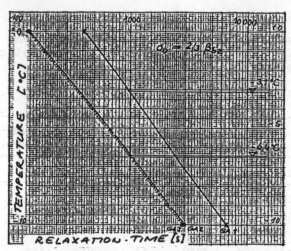

Figure 4.1: Relaxation time as function of the temperature

Therefore one can expect that with a mastic asphalt surface to which natural asphalt has been added to improve the thermal elasticity cold induced fissures will emerge more likely than with a mastic asphalt that does not contain natural asphalt.

Comparing the three mastic asphalt mixtures, mixture 3 in which a comparable viscosity to natural asphalt has been reached through a higher degree of filler should show a better behaviour in cold temperatures. This is founded in the fact that besides a smaller relaxation time this mixture also shows a higher tensile strength than comparable mixtures and therefore a larger tensile reserve in resisting to mechanically caused bending strengths.

5 References

Gauer, P.K. (1983) Prüfung des Kälteverhaltens von Asphalt durch Biegezugbeanspruchung. **Das stationäre Mischwerk,** 17, 205-208.

Gauer, P.K. (1984) Kälteverhalten von Asphalt mit synthetischer Kieselsäure. **Bitumen,** 46, 77-81.

Gauer, P.K. (1985) Relaxationsversuche an biegebeanspruchten Asphaltprismen bei niedrigen Temperaturen. **Bundesminister für Verkehr,** Forschungsbericht Nr. FA 07.115 G 83 E.

15 BENDING TEST OF ASPHALTIC MIXTURES UNDER STATICAL LOADING

J. JUDYCKI
Basrah University, Engineering College, Civil Engineering
Department, Iraq*

SUMMARY

   A method of bending testing of asphaltic mixtures and its
importance for evaluation of mixture resistance to cracking at
low temperature is presented in the paper. No clear correlation has been found between flexural strength and indirect
tensile stength. The flexural properties are more susceptible
to changes in mixture composition than indirect tensile
strength. Flexural strength is substantially dependent on
deformation speed during loading. The reduced variables method
has been proposed for evaluation of this relationship. Presented effect of type of mixture, content and type of bitumen,
filler content, mixture compaction, content of coarse aggregate, and elastomer additives on flexural strength, ultimate
strain and viscoelastic properties are of practical value. It
has been found that asphaltic mixtures are nonlinear viscoelastic materials and a theoretical model of nonlinearity is
presented. The bending test provides data suitable for
structural analysis of pavement structures.

1 INTRODUCTION

   Practical tests which have been currently used for design
of asphaltic mixtures do not sufficiently characterize their
behaviour at low temperatures. The performance of mixtures at
low temperatures has been judged from their certain physical
properties and composition but direct testing has rarely been
applied. As testing af asphaltic mixtures resistance to plastic deformation at high temperatures is relatively well developed there is still a need for a simple and reliable method
of testing their cracking resistance at low temperatures.

   A simple bending test of asphaltic mixtures has been worked
out at the Gdan̈sk Technical University (Ref. 1,2). The test
enables evaluation of flexural strength, ultimate strain and
viscoelastic properties of asphaltic mixtures at low temperatures. Testing does not require very sophisticated equipment
and may be performed at ordinary road laboratories. The paper
presents testing procedure, comparison of the bending test and
indirect tensile test, effect of deformation speed on flexural

---

* Permanently at the Gdansk Technical University, Poland

strength and relationships between various factors characterizing asphaltic mixtures and their flexural properties. Some results of the test have been applied for structural analysis of asphaltic pavements (Ref. 5).

2 IMPORTANCE OF BENDING TESTING OF ASPHALTIC MIXTURES AT LOW TEMPERATURES

The most frequent mode of asphaltic pavements distress which occurs at low temperatures are crackings which may be caused by one or some of the following factors:
(a) bending of asphaltic layers under wheel load;
(b) upward bending due to frost heaves;
(c) temperature tensile stresses in asphaltic layers induced by rapid intensive cooling,
(d) tensile stresses due to shrinkage of lower layers; especially on cement concrete or cement stabilized bases.

Bending of asphaltic layers is extremaly severe on pavements of steel bridges with orthotropic decks. Intensive bending which occurs under wheel load in asphaltic layers over cracks in bases contributes to propagation of reflective crackings. Much lesser but still significant bending may occur in asphaltic layers of pavements placed on subgrade soils if their stiffness is considerably greater than stiffness of lower layers. It usually happens in spring after subgrade thawing or in winter if temperature of asphaltic layers drops below 0 deg.C while moist subgrade is still unfrozen. Tensile stresses in asphaltic layers are responsible for low-temperature crackings and partially for reflective crackings. For these reasons, either bending or tension of asphaltic mixtures seems to be the most suitable method of testing their cracking resistance at low temperatures as such type of stresses occurs in asphaltic pavements.

The design of asphaltic mixtures according to the Marshall or similar methods may provide stability at high temperatures but their value in providing desirable resistance to cracking at low temperatures is rather limited. The interlocking of grains of aggregate is of the primary importance for the stability while properties of asphaltic mortar (mixture of bitumen, filler and fine sand) are the most important for the resistance to cracking. Therefore, no correlation has been found between the Marshall stability at 60 deg.C and flexural strength at - 20 deg.C (Fig. 1).

The principal purpose of bending testing of asphaltic mixtures is to improve their resistance to cracking which implies the following particular aims:
(a) to find the influence of mixture composition and compaction on flexural properties at low temperatures and to

improve design of asphaltic mixtures;

(b) to evaluate the real value of various "improving" additives to bitumen and asphaltic mixtures;

(c) to select the suitable mixture for severe bending or tension conditions (e.g. pavements on steel bridges, controlling reflective cracking, etc.);

(d) to evaluate the effect of substituting materials, such as waste materials from industry applied as fillers or aggregates;

(e) to provide data for structural analysis of asphaltic pavements.

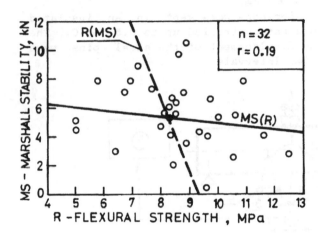

Fig. 1. Marshall stability at 60 deg. C and flexural strength at -20 deg.C

3 TESTING PROCEDURE

3.1 Preparation of specimens

Test specimens of asphaltic mixtures in form of beams 4*4*16 cm for fine and medium graded mixtures and 8*8*16 cm for coarse graded mixtures were compacted under statical pressure 15 MPa during 3 minutes. Temperature of the mixture during compaction was dependent upon the penetration of bitumen. Relative density of compacted specimens varied from 0.98 to 1.02 as compared with standard Marshall density. Greater density was achieved for greater bitumen content. In some cases Marshall compactor was used to compact asphaltic beams.

During compaction two holes were formed in the bottom of a beam in order to attach steel pins used for strain measurement. The pins were glued to the beam with epoxy resin. Before testing specimens were conditioned either in a water bath during 3 hours if testing temperature was above 10 deg.C or in a freezer during 18 hours if temperature was lower.

3.2 Testing

The testing assembly for asphaltic beams is presented in Fig. 2. The beam was tested either under constant load (creep test) or under constant-rate-of-deflection load. Strains at lower fibres of the beam were measured by means of especially designed displacement transducer. The transducer consists of a half-ring steel spring with two strain gauges, each one attached to one side of the spring. The spring is fixed in seats in the steel pins. The movable seat enables adjustment of required compression of the spring. When the beam is being bent the displacement of the pins is measured by the transducer with accuracy to 0.001 mm.

During testing the specimens were kept in an insulated chamber where temperature was controlled up to 0.5 deg.C. The bending force and the displacement of the steel pins were recorded at specified time intervals.

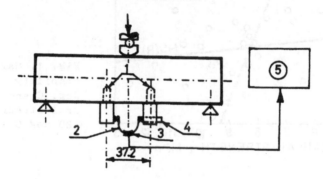

Fig. 2. Testing assembly for bending of asphaltic beam; 1 - steel pins, 2 - displacement transducer, 3 - strain gauges, 4 - movable seat, 5 - electronic recorder

3.3 Calculations of flexural properties of asphaltic mixtures

Stresses in the beam were calculated from the known formula of strength of materials. The average strain at the lower fibres of the beam, between two steel pins, were calculated from the following formula based purely on geometric considerations:

$$\varepsilon = (p/e) * c /(c + a) \tag{1}$$

where ε - average strain along maeasurement base e,
p - displacement of the steel pins,
e - measurement base; distance between axes of the steel pins,
c - half of the hight of the beam,
a - distance from the surface of the beam to the seats where the transducer is fixed.

3.3.1 Constant-rate-of-deflection test

Flexural strength and ultimate strains were determined as values corresponding to the highest load bearing capacity of asphaltic beam; either at the rupture at low temperatures or at the beginning of plastic flow at higher temperatures. Moreover, modulus of instantaneous elasticity E and viscosity coefficient η were determined from the approximation of experimental data with theoretical relationship derived for the Maxwell viscoelastic model:

$$\sigma(t) = d\varepsilon/dt * *(1 - \exp(1 - E*t/\eta)) \qquad (2)$$

where: σ, ε - stress and strain,
$d\varepsilon/dt$ - constant rate of strain,
t - time.

3.3.2 Creep test

Modulus of flexural stiffness and other viscoelastic properties of asphaltic mixtures were determined from the creep test. The stiffness modulus S(t,T) was calculated from the Van der Poel formula:

$$S(t,T) = \sigma/\varepsilon(t,T) \qquad (3)$$

where: σ - constant stress,
ε - strain as a function of time t and temperature T.

The viscoelastic properties were calculated from the approximation of an experimental creep curve with the following theoretical relationship derived for the Burgers model:

$$\varepsilon(t) = \sigma *(1/E1 + t/\eta1 + 1/E2*(1 - \exp(- E2*t/\eta2))) \qquad (4)$$

where: E1, E2, η1, η2 - viscoelastic parameters.

3.4 Accuracy of measurement and variability of test results

The accuracy of strain measurement was equal to 0.00002. Relative error of strain measurement did not exceed 5% with confidence level 95%. Flexural strength was measured with accuracy 0.15-0.25 MPa. The only disadvantage of the applied method was sligth creep of asphaltic mixture between the steel pins caused by springiness force of the transducer. This creep occured only at temperatures above 0 deg.C when mixtures exhibited greater creep compliance. Therefore, additional creep had been tested and measured strains were corrected to minimize its influence. It had been found that the steel pins attached to the beams did not affect the flexural strength of asphaltic mixture.

Variability of test results, presented in Table 1, resulted

from measuring errors, as well as, from nonhomogeneity of asphaltic mixtures and differences in temperature. The scatter of results was usually greater at higher temperatures. At lower temperatures greater stiffness of the mixture caused decrease of local stress concentration between grains what led to more uniform distribution of stresses and resulted in lesser scattering of results. The number of samples tested in one set was for practical reasons limited to six in the constant-rate-of-deflection test and to four in the creep test.

Table 1. Variability of bending test results

| Measured value | Coefficient of variability at temperature (deg.C), % | |
	10	-20
Flexural strength	17	10
Ultimate strain	16	12
Modulus of flexural stiffness	25	15

4 TESTED ASPHALTIC MIXTURES

Three types of asphaltic mixtures were tested, asphaltic concrete, asphaltic mastic and insulating mastic, the latter being used as a waterproofing layer on bridge pavements. Results of tests which are summerized in this paper have been obtained during last few years nad various paving mixtures were tested. The typical gradings of aggregates are presented in Table 2, however, grading of a particular mixture may sligthly differ.

Table 2. Gradings of tested mixtures

| Mixture | Percent passing sieve, mm | | | | | | | | |
	16	12	6.3	4	2	0.84	0.4	0.18	0.074
Asphaltic concrete	100	88	56	50	42	31	24	14	9.5
Asphaltic mastic		100	84	70	58	46	39	30	20
Insulating mastic				100	93	72	62	51	38

In most cases aggregate consisted of crushed basalt, crushed and natural sand and limestone filler. The following bitumens were applied: Polish paraffinic bitumens D200, D100, D70, D35 and also non-paraffinic Albanian A50 and German B65 and B45 (the numbers indicate standard penetration).

5 COMPARISON OF INDIRECT TENSILE TEST AND BENDING TEST

The indirect tensile test is widely accepted as a practical tool for characterization of asphaltic mixtures. The two tests, the indirect tensile test at deformation speed 50 mm/min and the bending test at deformation speed 1.25 mm/min were carried out on the same sets of samples of asphaltic concrete. Testing temperature varied from - 20 deg.C to + 40 deg.C. Three types of bitumens were used D200, D100 and A50. The content of bitumens varied over the range from 4% to 7.5 % out of the weight of asphaltic concrete.

The following conclusions may be drawn from the comparison of results of these two tests (Fig. 3):
(a) No clear correlation exists between the flexural strength and indirect tensile strenght of asphaltic concrete.
(b) The indirect tensile strength of asphaltic concrete is less susceptible to changes of temperature than the flexural strength.
(c) At low temperatures changes of bitumen content and type of bitumen have much smaller effect on the indirect tensile strength than on flexural strength.

The lack of correlation between the results of these two tests is likely to be caused by three following reasons;
(i)   different deformation speed appllied for these two tests;
(ii)   completely different stress distributions in tested specimens;
(iii) different role of mixture components in bearing the load.

Deformation speed appllied for bending (1.25 mm/min) was much lower than for indirect tensile test (50 mm/min). Relationship between strength, deformation speed and temperature is very complex (see next chapter) and must affect test results.

But it is more important that, in the indirect tensile test the specimen is subjected, both, to compression and tension acting in the same plane, and interlocking of grains of aggregate as well as cohesion of asphaltic mortar together withstand the combined stresses. Whereas, in the bending test the lower part of the beam is subjected only to tension and cohesion of asphaltic mortar along with adhesion of bitumen to mineral grains are of the primary importance to withstand the load, while interlocking of grains is less important. For the same reason, the flexural strength is more susceptible to changes of temperature and to changes of the type and content of bitumen. Due to greater susceptibility the bending test is more suitable to study the influence of various factors characterizing asphaltic mixture on its strength at low temperatures. Besides, the bending test more clearly indicates

defficiency   in mixture composition than the indirect tensile
test.

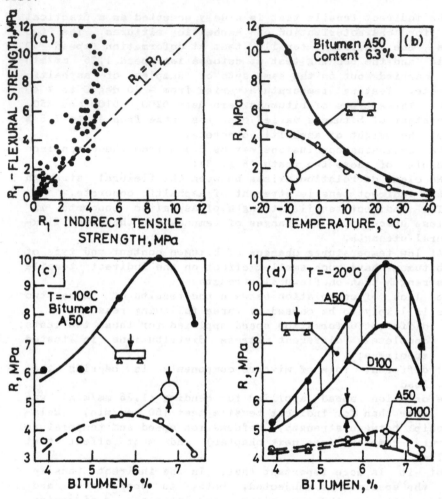

Fig.   3. Comparison of results of bending and indirect
tensile tests;   (a) strength at temperature from ~ 20 deg.C to
40 deg.C, (b) effect of temperature, (c) effect of bitumen
content on strength at -10 deg.C, (d) effect of type and
content of bitumen on strength at -20 deg.C

The stress distribution in a sample tested in the indirect
tensile test is very complex and does not resemble that one
which occurs in asphaltic layers of pavement structure. On the
contrary, the stresses induced in the bent beam are similar to
those in asphaltic layers bent under the wheel load. For that
reason, the flexural strength tested in laboratory is probably

more close to the actual strength of asphaltic layers of
pavements. Therefore, results from the bending test are likely
to be more suitable for evaluation of asphaltic mixtures and
for structural analysis of asphaltic pavements. The only undi-
sputable advantage of the indirect tensile test is the possibility of testing cores drilled from existing pavements
without a need of remoulding.

6 EFFECT OF SPEED OF DEFORMATION ON FLEXURAL STRENGTH OF ASPHALTIC MIXTURES

Asphaltic concrete which contained 6.3% of paraffinic bitumen D70 was tested with deformations speeds 1.25, 50 , 100
and 200 mm/min at various temperatures from - 35 deg.C to 40
deg.C. It was found that the deformation speed significantly
affected the flexural strength (Fig. 4).

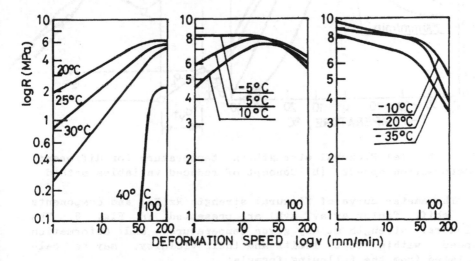

Fig. 4. Effect of deformation speed upon flexural strength of
asphaltic concrete at various temperatures

At temperatures above 20 deg.C when asphaltic concrete was at
a viscoelastic state the flexural strength increased with the
increase of deformation speed. On the contrary, at low temperatures, below -10 deg.C, when stiffness of the mixture was
high and viscous properties diminished the flexural
strength decreased with the increase of the deformation speed.
At intermediate range of temperature, above - 10 deg.C and
below 20 deg.C, the two patterns of mixture behaviour occured.
These complex phenomenon resembles behaviour of a viscoelastic
body at higher and a brittle body at lower temperatures.

The similarity of curves "flexural strength vs. temperature" for different deformation speeds (Fig. 5a) had suggested that they might be shifted to coincide one with another. The concept of, so called, "reduced variables method" was appllied, as shown in Fig 5b, and all experimental curves were shifted to one master curve.

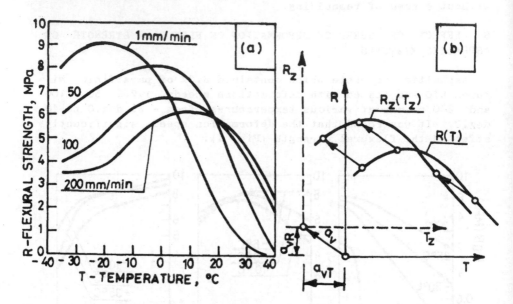

Fig. 5. (a) Flexural strength vs. temperature for different deformation speeds, (b) Concept of reduced variables method

The master curve of flexural strength Rz(Tz) and components of shift factor av(avT,avR) are presented in Fig. 6. The flexural strength at any given temperature and deformation speed, within test limits and their vicinity, may be calculated from the following formula:

$$R(T,v) = Rz(Tz) - avR \tag{5}$$

where: R(T,v) - actual flexural strength as a function of temperature T and deformation speed v,

Rz(Tz) - flexural strength in reduced coordinate system, as in Fig. 6,

Tz - reduced temperature:

$$Tz = T + avT \tag{6}$$

avT, avR - components of the shift factor av, dependent upon deformation speed v.

Good agreement has been found between measured flexural strength and calculated from reduced variables method; correlation coefficient r = 0.98.

It can be clearly seen that the flexural strength of asphaltic concrete for a given mode of loading, stress distribution and temperature is not a constant value but is dependent on the actual deformation speed and may vary up to three times for a high and low speed all other factors being equal. It has to be taken into consideration if any problem of pavement structure strength or fracture is analyzed. The speed of deformation which occur in an asphaltic layer of pavement under loading vary from very small for static or thermal loading to very high for dynamic loading from running vehicles.

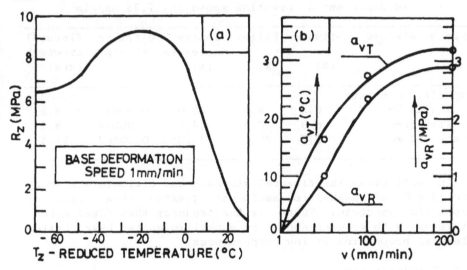

Fig. 6. (a) Master curve of flexural strength of asphaltic concrete for v = 1 mm/min, (b) Components of shift factor av

An approximate value of the deformation speed should be known to predict the possibility of pavement fracture. It seems that up now this phenomenon has not been sufficiently considered in structural analysis of pavement fracture.

7   INFLUENCE OF ASPHALTIC MIXTURE COMPOSITION AND COMPACTION ON FLEXURAL PROPERTIES

The influence of various features characterizing asphaltic mixtures was tested: type of mixture, type and content of bitumen, content of filler, mixture compaction, grading of aggregate and effect of elastomer additives. The majority of tests were carried out on medium graded asphaltic concrete (Table 2). Tests were conducted at variuos temperatures but the attention was focused on low temperatures at which bending of asphaltic layers occurs. Flexural strength and ultimate strain, whenever given throughout this chapter, has been tested at the deformation speed 1.25 mm/min, unless a different value is stated.

7.1 Type of asphaltic mixture

Typical results for three asphaltic mixtures: fine and medium graded asphaltic concrete and mastic asphalt are presented in Table 3.

Table 3. Typical flexural properties of asphaltic mixtures at
    T = -20 deg.C and deformation speed v = 1.25 mm/min

Type of mixture	Bitumen	Filler	Coarse aggregate	Ultimate strain	Flexural strength
	(%)	(%)	(%)		(MPa)
Asphaltic concrete					
– medium graded	6.2	9.5	58.0	0.00072	8.5
– fine graded	6.9	10.5	49.2	0.00074	9.5
Asphaltic mastic	8.0	20.0	42.0	0.00100	11.5

In most cases the flexural strength and ultimate strain at -20 deg.C of asphaltic mastic was greater than those of asphaltic concrete. There was the tendency that finer graded mixtures with greater amount of asphaltic mortar have better flexural properties at low temperatures.

7.2 Bitumen content

The typical relationships between properties of asphaltic concrete and bitumen content are presented in Fig. 7. There is a certain "optimal" content of bitumen for which the flexural strength reaches its maximum at low temperatures. It is asossiated, probably, with an optimal thickness of bitumen film coated grains of aggregate. If bitumen content is much lower than optimal the bond of grains is unsufficient and the flexural strength is low. If bitumen content is much greater than optimal the flexural strength decreases due to increased amount of, so called, "free bitumen" which is not sufficiently bound with grains of aggregate by adhesive forces. Cohesion and tensile strength of free bitumen is lower as compared with bound bitumen.

The optimal content of bitumen at maximum flexural strength at -20 deg.C is always close to that at maximum density and lower than that at maximum Marshall stability at 60 deg.C (Table 4).

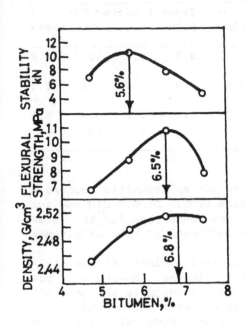

It has been found that the maximum flexural strength of asphaltic concrete occurs when almost all' voids in mineral aggregate are filled with bitumen, so there is no space for bitumen thermal expansion and voids ratio in the mixture does not exceed 1.1%. The usage of such mixtures would be impractical as they are susceptible to bleeding and plastic deformations in hot summers. If bitumen content is lower, so that voids content increases to a reasonable value 2.5% the flexural strength at -20 deg.C decreases by 7 to 16% as compared with the maximum value (Table 5).

Fig. 7. Effect of content of A50 bitumen on asphaltic concrete properties; Marshall stability at 60 deg.C, flexural strength at T = -20 deg.C and v = 1.25 mm/min and density

Table 4. Contents of bitumens corresponding to maximum values of flexural strength at T = - 20 deg.C and v = 1.25 mm/min, density and Marshall stability at 60 deg.C

Properties	Contents of different bitumens, %				
	D200	D100	D70	A50	B45
Flexural strength	6.3	6.4	6.5	6.5	6.8
Density	5.1	6.2	6.3	6.8	7.3
Marshall stability	4.8	5.3	5.3	5.6	6.3

Ultimate strain of asphaltic concrete at - 20 deg.C increased with the increase of bitumen content (Fig. 8). The influence of bitumen content on modulus of elasticity E was negligible if the content varied in practically accepted ranges. Scatter of E-value exceeded effect of bitumen content.

Table 5. Flexural strength of asphaltic concrete at T = - 20 deg.C and v = 1.25 mm/min

Properties	Type of bitumen				
	D200	D100	D70	A50	B45
Maximum flexural strength, MPa	10.4	9.2	8.5	10.9	9.8
Voids ratio at which maximum strength occurs, %	0.8	0.0	0.4	0.8	1.1
Flexural strength, MPa, if voids ratio = 2.5%	8.7	8.4	7.9	9.5	8.5
Decrease of strength, %	16	9	7	13	13

7.3 Type of bitumen

The bending tests were carried out on asphaltic concrete containing three paraffinic bitumens D200, D100, D70 and two non-paraffinic A50 and B45. It was found that for the same aggregate composition there is an influence of type of bitumen on flexural properties.

For the same bitumen content greater ultimate strain at - 20 deg.C was associated with softer bitumens. However, if the optimum bitumen content was selected according to the Marshall method; as an average value from contents corresponding to maximum stability, maximum density and voids ratio equal to 3%, the tendency was the same, but there were no distinct differences between ultimate strains for different bitumens and they varied from 0.00080 to 0.00086 at - 20 deg.C (see triangular signs in Fig. 8).

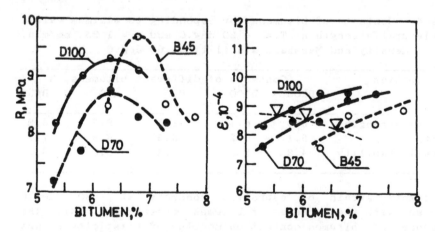

Fig. 8. Effect of type and content of bitumen on flexural strength and ultimate strain of asphaltic concrete at T = -20 deg.C and v = 1.25 mm/min

The type of bitumen does affect the flexural strength of asphaltic concrete but no clear relationship has been found for tested bitumens. In general, however mixtures with paraffinic bitumens had lower strength at low temperature than mixtures with non-paraffinic bitumens.

7.4 Mixture compaction

Tests were performed on medium graded asphaltic concrete designed according to Finnish specifications for wearing courses (Ref.4). The asphaltic beams were compacted by means of Marshall compactor with different number of blows. The flexural strength was measured at 10 deg.C and -25 deg.C at deformation speeds 1 mm/min and 50 mm/min. It was found that the compaction and voids ratio significantly affected flexural stregth (Fig. 9). The flexural strength decresed for poorly compacted mixtures. Effect of deformation speed, the same as described in chapter 6, was also very substantial.

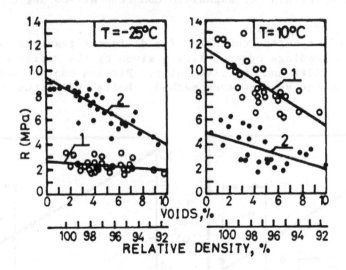

Fig.9. Effect of mixture compaction on flexural strength at 10 and -20 deg.C; 1 - v = 50 mm/min, 2 - v = 1 mm/min

7.7 Filler content

From rather limited number of tests on fillers it was found that for a given aggregate composition and bitumen content there was an optimum filler content at which the flexural strength and ultimate strain at -20 deg.C attained maximum values (Fig. 10). No clear effect of filler content on modulus of elasticity was found at -20 deg.C.

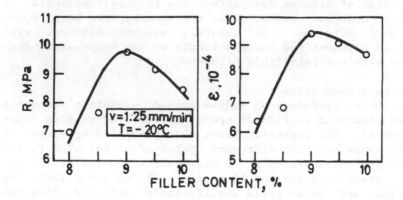

Fig. 10. Effect of filler content on flexural strength and ultimate strain of asphaltic concrete at -20 deg.C

7.8 Grading of aggregate

Samples of asphaltic concrete of 8 different gradings were tested. The gradings covered limits given by the Polish standard PN-74/S-96022 and their vicinity. Bitumen content was calculated from the surface area method. Polish D70 bitumen was applied.

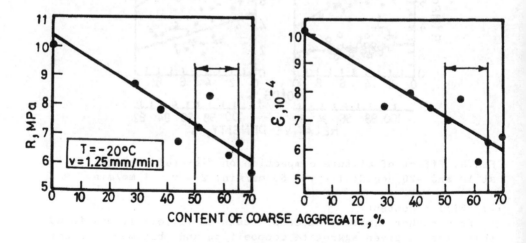

Fig. 11. Effect of coarse aggregate on flexural strength and ultimate strain of asphaltic concrete at T= - 20 deg.C and v = 1.25 mm/min

The asphaltic mortar composition was approximately the same for all tested mixtures and contained 8.5-8.8% of bitumen,

15-19% of limestone filler and 72-77% of quartzite sand. Crushed basalt was used as a coarse aggregate (retained on 2 mm sieve).

It had been found that there was tendency that the flexural strength and ultimate strain at -20 deg.C decreased for greater amount of coarse aggregate, as illustrated in Fig.11. The asphaltic mortar without crushed aggregate had the highest strength and the highest ultimate strain. Therefore, the asphaltic mortar may be considered as a governing factor in developing flexural strength of asphaltic concrete at low temperatures.

7.9 Flexural properties of asphaltic mixtures modified with elastomer

Tests were carried out on mixtures modified with carboxylated butadiene-styrene elastomer in form of "LBSK" latex. Asphaltic mixtures modified with the elastomer have the following properties as compared with conventional mixtures:
- greater flexural strength and ultimate strain at low temperature (Table 5);

Table 5. Flexural properties of modified and conventional asphaltic concrete at v = 1.25 mm/min

Temperature	Flexural strength (MPa)		Ultimate strain	
(deg.C)	convenional	modified	conventional	modified
25	1.0	1.7	0.02820	0.03300
10	6.9	7.5	0.01850	0.01790
0	11.4	13.0	0.00300	0.00282
-10	10.4	11.1	0.00126	0.00144
-20	9.8	10.1	0.00074	0.00083

- greater flexibility of mixtures at low temperatures; lower modulus of stiffness (Fig. 12);
- shorter relaxation time at low temperatures;
- greater resistance to plastic deformations at higher temperatures; greater modulus of stiffness and greater elastic recovery (Fig. 12);
- ageing process of modified mixtures was slower.

Laboratory tests have been proved by field trials. Very thin pavements (15 mm) made of conventional mixtures on very light steel bridge decks cracked badly after the first winter while elastomer-modified pavements did not indicated any sign of distress after three years of service. Results had been promising and modified pavements were applied with good result on

several bridges and fly-overs in Poland. The bending test was
very useful to select the suitable modifying agent and to find
out its influence upon asphaltic mixtures at low temperatures.

8 VISCOELASTIC PROPERTIES OF ASPHALTIC MIXTURES

8.1 Modulus of flexural stiffness and viscoelastic parameters

Moduli of flexural stiffness calculated from results of
creep test under bending are presented in Fig. 12. The time-
temperature superposition principle has been applied to obtain
master curve of modulus of flexural stiffness. Furthermore,
the master curve of modulus of stiffness was used for quasi-
viscoelastic analysis of pavement structure (Ref. 5). Viscoelastic parameters of asphaltic mixtures under bending are
presented in Ref. 1, 3 and 5.

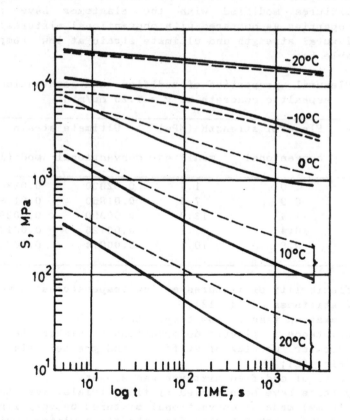

Fig. 12. Modulus of flexural stiffness of asphaltic concrete
(dotted lines - elastomer-modified mixture, continuous line -
conventional mixture)

8.2 Nonlinearity of viscoelastic behaviour of asphaltic-concrete

The bending tests were performed at various constant stress levels. It has been found that asphaltic concrete is a non-linear viscoelastic body and its modulus of stiffness is not only dependent upon time and temperature but also upon magnitude of acting stresses (Fig. 13).

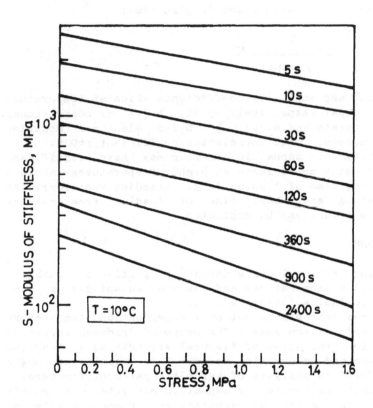

Fig. 13. Influence of stress level on modulus of stiffness of asphaltic concrete modified with the LBSK elastomer at 10 deg.C

The influence of stress level upon modulus of elasticity was very substantial at 10 deg.C. At lower temperatures nonlinear effects diminished. Nonlinearity of elastomer modified asphaltic conrete was more significant as compared with conventional material.

The following nonlinear viscoelastic model has been proposed to describe the behaviour of asphaltic concrete (Ref. 5):

$$\varepsilon(t) = \int_{-\infty}^{t} \Gamma \left[\sigma(\tau), t - \tau \right] * \frac{\partial \sigma(\tau)}{\partial \tau} * J(t - \tau) * d\tau \qquad (7)$$

where: σ, ε - stress and strain,
τ, t - time,
$\Gamma(\sigma,t)$ - nonlinearity coefficient.

It has been found that nonlinearity coefficient is the following function of stress and loading time:

$$\log \Gamma(\sigma,t) = \frac{t}{a + b*t} \sigma \qquad (8)$$

where a and b are empirical coefficients which at temperature 10 dec.C are equal respectively a=592, b=4.8 for conventional asphaltic concrete and a=157, b=2.8 for elastomer modified asphaltic concrete, with correlation coefficient r=0.9.

Nonlinearity of asphaltic concrete may have significant effect on pavement performance at higher temperatures (above 0 deg.C) and long time of loading (e.g. standing vehicles). At low temperature and short time of loading from running vehicles this effect may be neglected.

9 CONCLUSIONS

The bending test well characterizes properties of asphaltic mixtures at low temperatures and has some advantages as compared with the indirect tensile test.

The flexural stregth is not only dependent on temperature but also on deformation speed. The proposed "reduced variables method" enables evaluation of flexural strength as a function of temperature and deformation speed. This relationship ougth to be considered in analysis of asphaltic pavement fracture.

The flexural properties of asphaltic concrete are mostly dependent on composition of asphaltic mortar and contents of the mortar. Greater flexural strength at low temperature is associated with greater content of the mortar. There is an optimal bitumen content and optimal filler content at which flexural strength of asphaltic concrete reaches its maximum. The optimal content of bitumen is greater than that at maximum Marshall stability and close to that at maximum density of the mixture. The flexural strength significantly decreases for poorly compacted mixtures of high voids ratio.

Elastomer additives improve flexural properties of asphaltic mixtures and may be recommended for steel bridge pavements where bending is of primary importance as well as for preventing reflective cracks propagation.

Asphaltic concrete is a nonlinear viscoelastic body. Nonlinearity may affect pavement behaviour at long time of loading and at higher temperatures. For short-time loadings and low temperatures nonlinear effects may be neglected.

REFERENCES

1. JUDYCKI J.; Metoda badan wlasnosci reologicznych drogowego betonu asfaltowego, (A method of testing of rhelogical properties of road asphaltic concrete) Zeszyty Naukowe Politechniki Gdanskiej, Budownictwo Ladowe, 1976, No 254, p. 79-109,
2. JUDYCKI J.; Modul sztywnosci betonu asfaltowego, (Modulus of stiffness of asphaltic concrete), Archiwum Inzynierii Ladowej, Tom XXV, No 1/1984, p. 197-216
3. BORKOWSKI H., JUDYCKI J.; Reologia betonu asfaltowego i konstrukcji jezdni, (Rheology of asphaltic concrete and pavement structure), Zeszyty Naukowe Politechniki Gdanskiej, Budownictwo Ladowe, 1987, No 406, p. 111-141
4. EHROLA . E., JUDYCKI J.; Tivistamisen vaikutuksesta asfalttibetonin lujuuteen matalissa lampotiloissa (Effect of compaction on flexural properties of asphaltic concrete), Asfaltti, 1978, No 23, p. 4-9
5. JUDYCKI J; Drogowe asfalty i mieszanki mineralno-asfaltowe modyfikowane elastomerem, (Road bitumens and asphaltic mixtures modified with elastomer), Zeszyty Naukowe Politechnik Gdanskiej (To be published)

16 LABORATORY TEST FOR CHARACTERIZATION OF THE PERFORMANCE OF THIN ASPHALT LAYERS

Z. KUBANYI
Institute for Transport Sciences, Budapest, Hungary

Abstract
The long time performance of thin asphalt layers is influenced
by some important properties. The possibility of the
determination of some special characters of the used binder,
the correct composition of the aggregate part of mixes, the
compaction grade-durability connection were investigated.
Keywords: Thin asphalt layers, Modified binders, Durability,
Adhesion.

1 Introduction

The construction of thin asphalt layers for renovation of
existing surfaces became a wide ranging practice in many
countries because of economical respects. The technology was
promoted by the spreading utilization of the modified bitumens
too. According to the XVII. th Road World Congress 1983 the
thin layers are constructed of 2.5-4.0 cm laid asphalt
mixtures. They are not strengthening the pavement structures,
their part is only the improvement of surface characters so as
ageing, imperviousness to water and uneveness. The very thin
layers having a thickness of 0.5-2.5 cm are unable to improve
the uneveness they are suitable for the improvement of
characters related to the mixture properties. The asphalt
mixtures used for thin layers have a wide ranging variety in
their compositions in the used binder types and in the
aggregate parts.

Thin layers composed with conventional or modified binders
are constructed since 1985 in Hungary. Their properties were
investigated mainly by the test datas of experimental sections.
The type of mixtures used for thin layers where mainly dense
asphalt concretes but some mastix asphalt mixtures according to
ZTV bit where applied too.

Requirements for thin asphalt layers are the elasticity,
resistance against deformation, water imperviousness, adequate
roughness and all these for a long service time without no or
minimal maintenance. The above requirements can be fulfiled by

proper mix design, by use of suitable binder type, and by severe construction technology. The preliminary laboratory testing concerning these properties is important from the point of view of the performance of the layers. Test methods simulating the effect of traffic loads on the layers have great importance.

2. Binder

Important components of the asphalt mixtures for thin layers are the binders, nowadays mainly modified bitumens are used. The used binders are mostly elastomer modified types, for being able to guarantee the most important criteria of the durability of layers, the elasticity.

In the Institute for Transport Siences Hungary the following characters are tested:
- identifying tests according to Hungarian standard MSZ 3276
- penetration on different temperatures
- ductility on 25^{0}C and 4^{0}C
- elasticity tested in ductilometer on 25^{0}C and 4^{0}C
- flow test on 60^{0}C
- tube test on 180^{0}C
- viscosity

In case of modified bitumen the determination of the softening point -RB- and penetracion are to be used for identification. The informations about the type of modification and about the prospective perfomance in the asphalt mixtures can be obtained by ductility, elasticity, breaking point and flow property tests.

The term of utilization is tested by the tube test. The tube filled by the modified binder is cured on 180^{0}C for three days. After curing the tube is removed and the binder is tested to determine the rate of separation. The separation can be avoided by the correct modifing process, namely by correct mixing and by the gap size of the homogenizer. Further on the separation can be influenced by decreasing of the storage time.

The viscosity of the binder has a very important role in respect of the use in asphalt mixes, partly for the mixing process, partly for correct compaction. The equiviscosity 100 means the most favourable temperature for mixing and equiviscosity 1000 means the lowest temperature for effective compaction. These temperature values are determined by viscosity measurementes. The evaluation of viscosity test proves that the viscosity values determined on same temperatures are depending on the applied shearing rates, for the modified binders have a characteristic non Newtonian behaviour, therefore the testing should be carried out allways by applying the same shearing rate. Mixtures containing modified binders generally require 10-50^0 degree higher temperatures for manipulation than the conventionally mixes.

3 Aggregate

The other important constituent of asphaltic mixes is the aggregate part. The small layer thickness enables only the use of relatively small grain sizes, therefore the relation beetwee the crushed sand natural sand rate and the prospective performance of the layer was investigated. The other very important investigated relation was the influence of the compaction rate on the durability. These two problems are in close correlation. The evaluated data were obtained from laboratory tests and experimental sections.

In the laboratory asphalt mixtures were produced so that the rate of crushed sand natural sand was 100/0, 50/50 and 15/85, the specimen of the mixtures had equally 2,0 v% void contant. Mixtures having relativelly small void content have better impermeability and as higher binder dosing is possible the durability of the layer increases. The repetition number wuich is revealing of the durability was determined by dynamic splitting tests. The temperature of testing was +5ºC, the maximal applied load was the 40% of the maximal force determined by static tests. According to the test results the rate of crushed and natural sand has small influence on the repetition number. The test results proved that if the repetition number of the mix containing 100% crushed sand is 100%, the decrease in case of 50/50 was 6%, and that of the 15/85 was 15%.

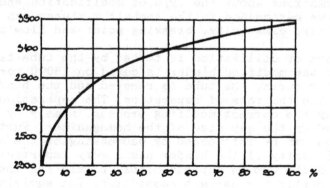

Fig.1. Crushed sand contant-load repetition relation

The stiffnes properties of conventional and in Hungary generally laid AB-12 type asphalt concret mixes and of modified bitumen containing thin layer mixes were tested. The investigation aimed to determine connection beetwen stiffness-durability and compaction rate-durability as well.

Instead of the conventional Marshall method carried out on 60ºC it was preferable for the testing of thin layer to determine the stiffness moduli on more temperatures. Cylinder

specimen were produced, the diameters were equall to the
Marshall specimen the heights were 3 cm. It was supposed that
this type of specimen is more available for modelling the thin
layers especially at lower temperatures. The tests for
stability and flow were carried out on 25, 40, 55 and 70ºC
temperatures. The test results presented a more limited
variation of stiffness moduli of thin layer mixes than that of
AB-12 type mix, as the temperature sensibility of modified
bitumen containing mixes is smaller than that of conventional
mixes.

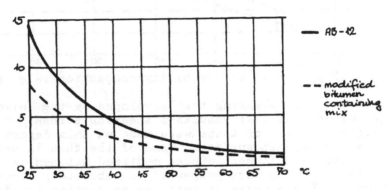

Fig.2. Stiffnes moduli-temperature relation

The other serial of tests aimed the modelling of compaction
grade - durability connection. Marshall specimen compacted
beetwen 92-100% rate were prepared. The specimen were tested
for the repetition number of dynamic splitting test. According
to the test data it can be stated that the thin layer mixes
containing modified bitumen have better compactibility
properties than the conventional mixes. The needed energy for
97% compaction grade is about 50% less than that for
conventional AB-12 mixes. The relation between compaction rate
- repetition number determined by dynamic splitting tests on
+5ºC temerature proved the considerable decrease of repetition
number, but the fall is smaller of thin layer mixtures.

The elasticity of the two type of mixtures were
characterized by the test data of static splitting tests
carried out on +5ºC. Test data proved that the elasticity
moduli of mixes containing elastomer modified bitumen having
the same splitt-tensile strength are 50% less than that of the
conventional mixes indicating better elastic properties.

4. Adhesion of thin layers

Important factor influencing the durability of thin layer is
the adhesion to the existing layer. The adhesion can be

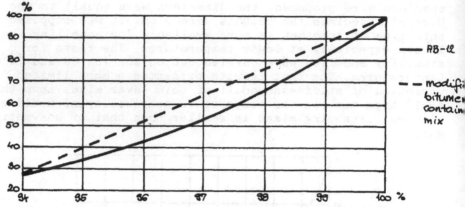

Fig.3. Fall of durability-compaction rale relation

determined by shearing test developed by the University of Karlsruhe. The test results a shearing stress - displacement diagram. Serial of tests were made on thin layers laid on existing asphalt surfaces. Some of the thin layers contained conventional bitumen, others modified bitumen as a binder. Some of the specimen were prepeared in laboratory other were core samples. Three types of test can be distinguished:
1. Modified bitumen containing thin layers on concrete surface
2. Thin layer containing conventional bitumen binder on conventional asphalt layer
3. Modified bitumen containing thin layers applied on conventional type of asphalt layer.
The test results are as follows
Pmax = shearing stress
Dmax = displacement
Pmax/Dmax = ratio
E = energy dispersion on the sheared surface
According to the evaluation of the serial test data the following can be stated:
- the adhesion is better on asphalt surface than on concrete
- the adhesion of asphalt layer containing modified bitumen is better than in case of normal bitumen containing asphalt layer
- the adhesion of asphalt wearing course laid on newly constructed base course is better, than that of laid on existing asphalt surface after spreading by bitumen emulsion. This statement is equally true in case of modified and conventional binder containing layers
- the best adhesion can be achieved between modified thin layer and existing concrete surface by spreading modified bitumen on the concrete surface.
Factors influencing the adhesion are the surface texture of

aggregates, the resistance of aggregate particles against embedding, the adhesion between binder and aggregate, the inner cohesion of binder, and the quality and quantity of spreaded adhesive material.

The test serie resulted minimal required values for the above mentioned cases, these are on 20°C as follows:
- minimal shearing stress for modified thin layers on concrete surface 10,5 kN,
- minimal shearing stress for modified thin layers on asphalt layers 18,0 kN,
- for conventional thin layer applied on conventional asphalt surface 15,0 kN.

5 References

Compactibility and some other properties of thin asphalt layer. Research Report. Institute for Transport Sciences Budapest Hungary 1987.

Comparison of new type binders in respect on utilization in construction technology. Research Report. Institute for Transport Sciences Budapest Hungary 1987.

Elaboration of test methods for modified binders. Research Report. Institute for Transport Sciences Budapest Hungary 1986.

Technological methods for layers produced by the used modified bitumens. Research Report. Institute for Transport Sciences Budapest Hungary 1986.

17 LE CONTROLE DE LA QUALITÉ DES ENROBÉS À L'AIDE DE LA MACHINE ASSERVIE D'ESSAIS RHÈOLOGIQUES

(Control of coating quality by mechanical rheological testing)

F. MOUTIER
Laboratoire Central des Ponts et Chaussées, Nantes, France
JL. DELORME
Laboratoire Régional de l'Est Parisien, Melun, France

Abstract
Bituminous materials have a highly pronounced viscoelastic
character. The only way to measure their mechanical
properties precisely is to subject them to well-defined
loading histories : references to tests at a "constant
rate" or with a "constant loading law" are common.
A special machine is needed to conduct the tests. It is
to meet this need that the Equipment Building Departments
of the LPC network have designed and built a
servocontrolled machine for rheological testing (MAER).
It incorporates the latest technologies ; in particular,
the servocontrol loop is digital, making it easy to adapt
the machine to the materials to be tested. It can also
carry out whole series of tests on a single specimen,
substantially improving the cost-effectiveness of the
machine.
It can be used on cylindrical specimens of widely
varying sizes, making it possible to anticipate its use to
characterize materials taken in situ. If the specimen
diameter must be less than 50 mm, a sample must be taken at
the plant outfeed and the material moulded and cored in the
laboratory.
The repeatability agreement or reproductibility is
excellent if the specimens tested come from the same
source: the fraction of error introduced by the operator is
negligible in this case, since all he does is place and
remove the specimens. The machine takes care of everything
else.
As for the use of the machine, all manipulations other
than the task mentioned above are reduced to entering data
in the control unit, and this can be done by an average
technician.
When this means of testing is used for quality control
of the binders of bituminous materials, it allows precise
identification of the binder, an important function given
the variety of binders available.

In addition, with a "constant" binder, a measurement of nonlinear behaviour serves to identify the behaviour of the mix.

This test, available at any laboratory that the machine, may now belong to the category of easy-to-use modern tests.

Résumé

Les matériaux bitumineux ont un caractère viscoélastique très prononcé. La seule façon de mesurer leurs caractéristiques mécaniques avec précision est de les soumettre à des histoires de sollicitation bien définies : on parlera ainsi d'essais à vitesse constante ou loi de sollicitation constante.

Pour exécuter ces essais, il faut disposer d'une machine spécifique. C'est pour répondre à ce besoin que les Services de Construction de Matériels du réseau des Laboratoires des Ponts et Chaussées ont conçu et réalisé une machine asservie destinée à l'exécution d'essais rhéologiques (MAER).

Elle fait appel aux technologies les plus avancées en particulier la boucle d'asservissement est numérique ce qui permet l'adaptation facile de la machine aux matériaux étudiés. De plus, elle est capable d'exécuter des séries d'essais sur une même éprouvette, ce qui améliore notablement la rentabilité de la machine.

Elle est capable de travailler avec des éprouvettes cylindriques de tailles très variables, ce qui permet d'envisager son usage pour caractériser les matériaux prélevés in-situ. Si le diamètre des éprouvettes doit être inférieur à 50 mm, il faut alors faire un prélèvement au "pied de la centrale" puis mouler et carotter le matériau en laboratoire.

Les résultats obtenus dans le cadre de la mesure du module pour tracer les courbes maîtresses présentent une bonne répétabilité.

L'accord inter-laboratoires ou reproductibilité est excellent si les échantillons testés proviennent de la même source : la part d'erreur apportée par le manipulateur est alors négliegable car celui-ci est réduit à monter ou démonter les éprouvettes. Tout le reste est géré par la machine.

Concernant son usage, les manipulations autres que la tâche citée précédemment se réduisent à entrer des données dans l'organe de commande et un simple technicien peut alors faire cette tâche.

Ce moyen d'essais appliqué au contrôle de la qualité des liants des matériaux bitumineux peut permettre une identification précise du liant, tâche importante compte tenu de la diversité de liant proposée par les entreprises.

De plus, à liant "constant" une mesure du comportement non linéaire permet d'identifier le comportement du mélange.

Cet essai, disponible dans tout laboratoire possédant la machine, pourrait répondre à la catégorie des essais modernes faciles d'emploi.

1 Introduction

La mesure précise des caractéristiques mécaniques des enrobés bitumineux novateurs destinés aux couches de bases est une nécessité - Cela permet d'une part, de les départager, d'autre part, d'évaluer le gain d'épaisseur que leur emploi peut entraîner. Ceci est particulièrement vrai en FRANCE pour les enrobés à base de bitume dur engendrant des modules élevés avec des performances à la fatigue équivalente à celle des enrobés à base de bitume traditionnel - Par ailleurs, le fait de promouvoir des produits novateurs à entrainer la nécessité de controler leur qualité in situ.

Ce besoin d'appréciation de la qualité des produits de chaussées noirs par la mesure de leurs caractéristiques mécaniques a obligé le LCPC à rechercher l'essai le mieux adapté.

Deux catégories d'essais répondaient au problème :
- l'essai de module complexe associé à l'essai de fatigue à déformation imposée,
- l'essai de traction directe développé par R. LINDER entre 1970 et 1974.

Après développement d'une machine spécifique pour répondre à l'essai de traction directe, il est apparu que l'emploi de cet essai, du fait de sa simplicité et de sa précision était le mieux placé pour satisfaire nos besoins de mesures rhéologiques à moindre coût.

Mais ceci n'a pas pu être atteint sans mal :
- Une exploration du comportement en traction des matériaux de chaussées noirs, puis une synthèse constituaient les contraintes préalables (travail de R. LINDER, 1970 - 1976).
- devant les difficultés relatives aux mesures des caractéristiques mécaniques, une machine d'essais spécifique devait être créée : c'est la machine asservie d'essais rhéologiques, dont la description a été faite dans une communication pour EUROBITUME 85.

La diffusion de cette machine et une formation systématique des utilisateurs permettent maintenant l'usage de la méthode d'essai pour l'évaluation des enrobés prélevés in situ.

Du fait que la méthodologie d'essai est implicitement exécutée par la machine d'essai, la reproductibilité est assurée, ce qui est une obligation quand on veut appliquer des seuils de choix de matériaux sur la totalité d'un territoire couvert par plusieurs laboratoires.

Après un rappel relatif aux propriétés rhéologiques des matériaux noirs et leurs conséquences sur les exigences en matière d'essais, les possibilités de la machine spécifique actuelle sont décrites, ainsi que les formes de dépouillement adoptées pour atteindre les informations concernant la courbe maîtresse et la non linéarité.

Ensuite le problème de la régression fatigue traction est abordé, enfin un exemple d'utilisation de la MAER comme moyen de contrôle des matériaux est développé.

2 Rappel des propriétés rhéologiques des matériaux de chaussées noirs

Pour les <u>faibles déformations</u>, le bitume confère aux enrobés bitumineux des propriétés viscoélastiques très accentuées. La rigidité entre autres dépend :

- de la température d'essai,
- de la loi de charge,
- du temps de charge.

Si un seul des paramètres cités ci-dessus est mal contrôlé, alors la répétabilité de l'essai est mauvaise et l'interprétation risque d'être "fantaisiste".

Les figures suivantes illustrent ces propriétés dans le cadre de l'essai uniaxial.

Si on soumet une éprouvette d'enrobé à des élongations du type $\epsilon = a_i t^n$ on obtient les réponses $F(t)$ suivantes :

a) pour $n = 1$

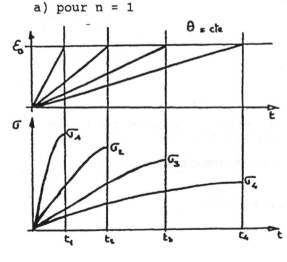

à même loi de charge, et même θ, σ_i est fonction du temps de charge t_i.

b) pour n = 1

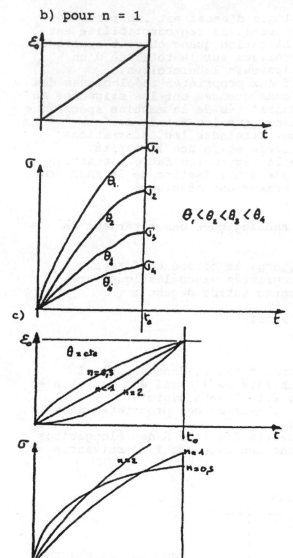

à même loi de charge,
et même temps de charge
σ_i est fonction de la
température de l'essai.

$\theta_1 < \theta_2 < \theta_3 < \theta_4$

c)

pour un même ϵ_0 atteint
au même temps t_0 la
contrainte est fonction
de la loi de charge.

Pour s'en convaincre, ci-dessous, quelques résultats
réels obtenus avec un enrobé bitumineux :

loi de déformation à vitesse constante a_i

Pour $\epsilon_0 = 10^{-4}$
temps de charge égal à 1 seconde

Effet de la température sur le module σ/ϵ_0

θ = 20 °C σ/ϵ_0 =   1 000 MPa
     10 °C   6 000 MPa
      0 °C   14 000 MPa
    - 10 °C   23 000 MPa

temps de charge : 100 secondes
Effet de la température
θ = 20 °C σ/ϵ_0 =   100 MPa
     10 °C   1 000 MPa
      0 °C   6 000 MPa
    - 10 °C   15 000 MPa
température égale à 10 °C
Effet du temps de charge pour atteindre ϵ_0
0,01 sec   15 000 MPa
 1 sec   6 000 MPa
 1 000 sec   400 MPa

loi de déformation du type $a_i t^n$
Effet de n sur la contrainte σ pour atteindre un même ϵ_0
en un temps t_0 :

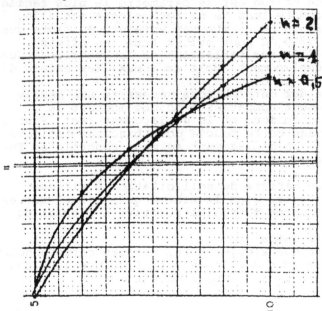

**On constate ainsi que le module est très affecté par la
température et le temps de charge à loi de déformation
identique.**

rapport de 23 en passant de -10 °C à 20 °C pour t = 1
sec.

239

rapport de 150 en passant de -10 °C à 20 °C pour t = 100 sec.

rapport de 10 à 20 °C pour t passant de 1 sec à 100 sec.
rapport de 1,5 à - 10 °C pour t passant de 1 sec à 100 sec.

On constate également que le module peut être affecté de 15 % lorsque n passe de 1 à 2.

Etablir un réseau d'isothermes ou une courbe maîtresse à partir d'un tel matériau dans de bonnes conditions de reproductibilité nécessite une très grande maîtrise des paramètres d'essais.

3 La maîtrise des paramètres d'essais - ses conséquences

Maîtriser les résultats avec ce type de matériau revient à maîtriser l'histoire des déformations, la température, le temps de charge et les conditions initiales de sollicitation :

- le temps de charge est le paramètre le plus facile à maîtriser, compte tenu de la technologie actuelle,
- la température nécessite une enceinte de haute qualité régulant au 1/10 °C pour éviter tous les effets secondaires liés aux variations de longueur des tiges de capteur de déplacement (voir en annexe les effets d'une régulation à ± 1 °C).
- l'histoire des déformations nécessite une boucle d'asservissement comparant à chaque instant le déplacement théorique et le déplacement réel. Pour des raisons d'usage cette boucle doit basculer facilement du mode contrôle en déplacement au mode contrôle en Force (et réciproquement).

L'exigence en matière de sollicitation à vitesse constante est résumée dans le schéma qui suit et qui est proposé dans la norme française relative à l'essai de traction directe.

Qualité de la rampe

Si t_i est tel que $\epsilon(t_i) = \epsilon_i$ (t_i = temps correspondant à l'élongation maximale de l'essai) il faut que pour tout temps t tel que b $t_i < t < t_i$ l'on ait :

$$\left| \frac{\epsilon (t_i) - a_i t}{a_i t} \right| < \frac{1}{100}$$

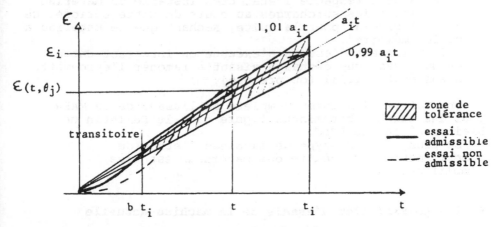

avec b fixé à 0,15 pour $a_i \geq 3.10^{-5}$ s^{-1}
0,05 pour $a_i < 3.10^{-5}$ s^{-1}

avec t_i temps d'essai élémentaire compris entre 0,5 et 1 000 secondes.

Les conditions initiales de stabilité avant chaque essai nécessitent d'être très scrupuleux quand au transport, montage, démontage des éprouvettes : c'est un matériau qui mémorise toutes les sollicitations, y compris celles non souhaitées. De plus, si on veut faire des séries d'élongation pour rentabiliser l'essai, les conditions de décharges devront elles-mêmes être très rigoureuses. On constate ainsi qu'une bonne exécution de rampe n'est pas une garantie de bons résultats.

Ce qui est vrai pour l'essai uniaxial est vrai pour tous les essais, y compris le pseudo essai de traction en compression diamétrale. Pour ce dernier essai les conditions initiales risquent d'être mal maîtrisées si l'axe de l'éprouvette est horizontal.

La synthèse de toutes ces exigences, plus celle de la rentabilisation (exécution de plusieurs rampes successives sans démontage de l'éprouvette de l'essai) a conduit à la réalisation de la machine asservie d'essais rhéologiques.

L'objectif a été atteint.

Celui qui a exécuté le même essai de Traction LPC à l'aide d'une presse quelconque et de la MAER a pu se rendre compte de l'extrême facilité avec lequel l'essai s'exécute pour la machine la plus récente :

- le nombre de montage et démontage est réduit, ceux-ci sont de plus facilités par la présence, en position manuelle, de pédales à actions proportionnelles,

- la machine respecte l'état très instable du matériau en limitant les surcharges au cours de cette période. Ce qui est crucial pour la suite, sachant que le matériau a de la "mémoire",
- l'empilement d'essais s'exécute en intégrant entre chaque essai une décharge visant à ramener l'éprouvette à son état initial (ou presque).

Une information assez complète de l'usage de la MAER pour les enrobés bitumineux figure dans le Bulletin de Liaison n° 142 du LCPC.

De plus, avec ce type de machines l'étude du comportement non linéaire des matériaux est très simplifiée.

4 Les possibilités d'essais de la machine actuelle

Celle-ci permet la constitution, par l'opérateur, de structures d'essais élémentaires en écriture BASIC.

Chaque essai élémentaire est une mini programme écrit en langage machine destiné à exécuter une tâche en contrôlant un paramètre d'arrêt à l'aide d'un seuil.

Les lois d'élongation de l'essai élémentaire de traction peuvent être quelconques, néanmoins, celles qui sont bien maîtrisées correspondent aux lois affines $\epsilon_i = a_i t^n$, avec a_i coefficient changeant d'un essai à l'autre (effet du temps de charge pour même déformation), n un coefficient définissant une classe d'essai donné (par exemple, pour n = 1 on a la classe d'essai à vitesse constante $(d\epsilon_i/dt = a_i)$).

D'autres lois d'élongation ou de charge existent pour répondre aux essais suivants :

- essai de fluage en compression,
- essai de fissuration sur éprouvettes préfissurées, mode I,
- essai de compression diamétrale,
- essai triaxial,
- essai cyclique, basse fréquence,
- essai impulsionnel périodique,
- essai de relaxation,
etc ...

Ceux-ci sont disponibles sur disquettes.

La capacité nominale de la presse est de 100 KN.

Les températures d'essai permettant une mesure suffisamment précise sont comprises entre - 10 °C et + 30 °C.

A vitesse constante a_i peut être compris entre 10^{-7}/s et 10^{-4}/s.

La gamme de matériaux caractérisables avec cette machine est très large : "du bitume pur à la grave ciment !".

5 Méthodes d'essais adoptées pour l'essai de traction directe LPC

L'essai de traction directe LPC est actuellement en cours de normalisation.

Pour prendre connaissance de la méthode d'essai adoptée on se reportera aux annexes où des extraits de ce projet de NORME figurent.

La MAER exécute l'essai de traction conformément au projet de NORME.

6 Estimation de l'erreur relative de mesure sur le module sécant – Caractère discriminant de l'essai

Le module sécant σ/ϵ calculé à partir de l'essai de traction directe dépend pour l'essentiel de quatre facteurs :
- le temps de charge,
- la température,
- la loi de charge,
- la déformation pour laquelle se fait le calcul.

D'autre part, il nécessite la mesure correcte de la force de réaction et de la déformation, et la connaissance précise de la base de mesure de la déformation ainsi que de la section de l'éprouvette.

En choisissant ce qui se fait de mieux en matière d'enceinte thermique et en respectant la loi d'élongation telle qu'elle a été décrite au paragraphe 3, l'essentiel des erreurs faites sur le module sécant sont liées aux mesures de F et Δl.

Sachant que :

$$S = \frac{\sigma}{\epsilon} = \frac{F.\Delta l}{s.l_0}$$

Le calcul d'erreur donne :

$$\frac{\delta S}{S} = \frac{\delta F}{F} + \frac{\delta \Delta l}{\Delta l} + \frac{\delta s}{s} + \frac{\delta l_0}{l_0}$$

Si on estime :
- $\delta F/F < 1/1000$ du fait du choix de la qualité du peson (classe 0,03).

- $\dfrac{\delta \Delta l}{\Delta l} \le \dfrac{1,5}{100}$ s = section de l'éprouvette

l_0 = base de mesure de déformation

et sachant que

$\dfrac{\delta s}{s} \# \dfrac{1}{500}$ et $\dfrac{\delta l_0}{l_0} \# \dfrac{1}{1\,000}$

alors :

$$\boxed{\dfrac{\delta S}{S} < \dfrac{2}{100}}$$

Cette erreur relative est compatible avec celle observée lorsque l'on teste une éprouvette "étalon" en alu pour mesurer la reproductibilité des machines de traction.

$\dfrac{\delta F}{F} < \dfrac{1}{100}$

Commentaires :
Cette résolution sur le module sécant, valable quelque soit le couple "temps de charge - température" tel que
- 10 °C < θ < + 30 °C
1 sec < t < 1 000 sec
permet de construire des isothermes ou isochrones de module sécant avec une bonne précision.
Ce qui permettra de mettre en évidence de faibles variations liées à la qualité du liant et de la formule globale.
On retiendra que cet essai présente un excellent pouvoir discriminant ainsi qu'une bonne reproductibilité si les calages métrologiques sont faits (vérifié par l'expérience).
Cet essai et la machine correspondante sont donc très adaptés pour faire du contrôle de qualité des caractéristiques mécaniques (voir paragraphe 9).

7 Le coût du matériel d'essai

Cette machine est disponible pour un prix voisin hors taxes de 850 KF.

Compte tenu de ses multiples possibilités et de son adaptation au cas des matériaux viscoélastiques, cette machine peut être considérée comme présentant un excellent rapport qualité - prix.

8 La régression fatigue traction - intérêts et problèmes

La notion de régression "fatigue traction" est officiellement apparue pour la première fois dans le rapport de recherche LPC écrit par R. LINDER en 1976 "Comportement en traction simple des enrobés hydrocarbonés" (n° 71 LPC Oct. 77).

A cette époque le fichier se composait de 18 groupes de valeurs (ϵ_{6F}, (1-Γ) ...). Pour expliquer ϵ_{6F} (déformation relative en fatigue entraînant une durée de vie moyenne de 10^6 cycles) deux paramètres étaient nécessaires : (1-Γ) ou non linéarité à 0 °C et 300 sec, S ou module sécant à 0 °C et 300 sec. La forme mathématique retenue était la suivante :

$$\hat{\epsilon}_{6F} = A_0 + A_1 (1-Γ) + A_2 S$$

et l'intervalle de confiance de ± 16.10^{-6}.

Ces essais étaient réalisés à l'aide de la presse de traction compression mécanique LCA.

Suite à la réalisation de la MAER, d'autres groupes de valeurs ont été disponibles.

Une étude statistique des groupes de valeur a mis en évidence les faits suivants :
- lorsque les paramètres de composition et compactage varient, toute chose égale par ailleurs (nature bitume et granulat), la non linéarité (1-Γ) explique <u>convenablement</u> la résistance en fatigue. L'intervalle de confiance est dans ce cas de ± 18.10^{-6} (2σ).

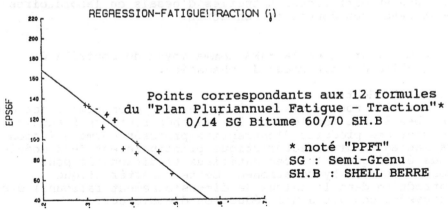

REGRESSION-FATIGUE!TRACTION (¡)

Points correspondants aux 12 formules du "Plan Pluriannuel Fatigue - Traction"*
0/14 SG Bitume 60/70 SH.B

* noté "PPFT"
SG : Semi-Grenu
SH.B : SHELL BERRE

- lorsque la nature du bitume seule change, $(1-\Gamma)$ reste constant !

Sachant que ϵ_{6F} est fortement altéré par la nature du bitume, la régression perdait en qualité de prédiction ; l'intervalle de confiance allait doubler, soit $\pm$ 32.10^{-6} (2σ).

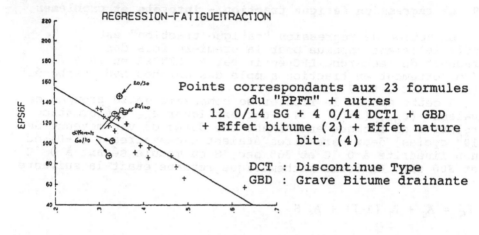

REGRESSION-FATIGUE!TRACTION

Points correspondants aux 23 formules du "PPFT" + autres
12 0/14 SG + 4 0/14 DCT1 + GBD
+ Effet bitume (2) + Effet nature bit. (4)

DCT : Discontinue Type
GBD : Grave Bitume drainante

Nul ne contestera l'intérêt d'avoir un résultat de fatigue à partir de quelques essais de traction spécifiques.

Néanmoins, la diversité des provenances de bitumes, pour une classe donnée n'est pas décodée par l'essai, ce qui jette le trouble sur son intérêt immédiat.

Est-ce que cette réalité est à rejeter au profit de celle mise en évidence à l'aide de la fatigue ?

Tant que la pertinence des essais (fatigue - traction) n'aura pas été testée, nul ne pourra répondre à cette question.

Une étude réunissant une expérience sur le manège de fatigue et différentes batteries d'essais en laboratoires pour lever ces doutes est en cours.

9 L'utilisation de la MAER comme moyen de contrôle de qualité des matériaux de chaussées.

9.1 Introduction

Depuis 1980 nous assistons à un développement important des enrobés à haut module dans les marchés routiers français. Ce sont des produits d'entreprise proposés comme solutions variantes. Leur caractéristique principale est leur module plus élevé que celui des matériaux traditionnels pour couche de base hydrocarbonée. Cette caractéristique introduite dans le calcul de dimensionnement rationnel des chaussées conduit à une réduction d'épaisseur.

Or cette valeur de référence est obtenue sur un matériau dans des conditions de laboratoire. Que deviennent ces performances avec les moyens de chantier et les variations éventuelles de qualité des constituants ? Face à ce problème, il a été proposé aux maîtres d'oeuvre de vérifier les performances in situ des matériaux proposés. Nous allons exposer le principe de la méthode appliquée et présenter un tableau de résultats.

9.2 Méthode - objectif

L'entreprise est responsable de son produit : elle propose une formulation, une nature de granulat, une nature de liant et un dosage en liant. Elle assure la fabrication et la mise en oeuvre par les moyens qu'elle choisit.

L'intervention du laboratoire du maître d'oeuvre devient un contrôle extérieur sur le produit fini.

Ce contrôle étant effectué sur prélèvement, il y a lieu de s'assurer de sa représentativité. A cet effet, on vérifie avant prélèvement la qualité du liant, la régularité de la fabrication et de la mise en oeuvre.

La teneur en vides des carottes est aussi un indicateur par rapport à la teneur en vides obtenue sur chantier.

L'objectif est d'obtenir sur le prélèvement des valeurs caractéristiques de module qui peuvent être comparées à quatre types de références :

1 - Référence "absolue" : les spécifications fixent une valeur de module uniaxial à 10 °C et 0,02 s.

Exemple : dans le marché d'enrobés spéciaux de la ville de Paris, les enrobés de la famille EHM (Enrobés à Haut Module) sont définis par une valeur de module S > 16 500 MPa - à 10 °C 0,02 s.

2 - Référence "relative" : la courbe maîtresse des modules uniaxiaux du matériau in situ est comparée à celle du matériau confectionné en laboratoire (en général avec des granulats de nature différente), qui a servi de base à la proposition de l'entreprise.

Exemple : les produits qui font l'objet d'un avis technique.

3 - Référence "analogique" : Dans le cas d'un produit nouveau, qui n'a pas subi l'ensemble des essais mécaniques, si les résultats de traction directe sont voisins de ceux d'un enrobé connu il est possible de considérer que le module en flexion à prendre en compte pour le dimensionnement est voisin de celui de l'enrobé connu.

4 - Référence "au module de dimensionnement" : Le module uniaxial à 15 °C - 0,01 s serait peu différent du module flexion 15 °C - 10 Hz. le module uniaxial 15 °C - 0,01 s est directement comparé à la valeur prise en compte dans le calcul de dimensionnement.

9.3 Processus d'essais

A) Prélèvements in situ - Prise d'échantillons

Le matériau est extrait par carottage grand diamètre de la couche de chaussée, dont l'épaisseur doit être au minimum de 9 à 10 cm. L'éprouvette d'essai est carottée selon l'axe "horizontal".

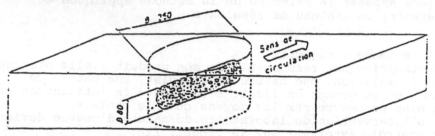

B) Mode opératoire

Chaque éprouvette est ensuite soumise aux divers chargements décrits dans le tableau 1, à l'aide d'une machine asservie d'essais rhéologiques (loi de déformation imposée, température d'exactitude 0,1 °C).

Cette série d'essais permet donc d'établir avec une grande précision les courbes maîtresses (Module en fonction du temps de chargement, à une température donnée). On appréhende ainsi le comportement du matériau dans des conditions très variées et on détermine les valeurs des modules caractéristiques définis au paragraphe 2.

Tableau 1 - conditions d'essais

Température d'essai	- 10°C	0°C	10°C	+ 15°C
Sollicitation	temps de charge	seuil de déformation	temps charge	déformation
	3s 10s 30s 100s 100s 300s	} 5×10^{-5} Pa	1s 3s 10s 30s 100s 300s	} 1.10^{-4}
		Non linéarité à 0°C		
	30s	10^{-4}		
	30s	2.10^{-4}		
	30s	3.10^{-4}		
	30s	4.10^{-4}		
	30s	5.10^{-4}		

9.4 Exemple de résultats

Famille de produit	A 1	A 2	A 3	A' 4	A 5	A' 6
Densité hydrostatique des carottes	2.472 (97 %)	2.511 (98,5 %)	2.48⁴ (97,8 %)	2.49 (97,9 %)	2.49 (99,3 %)	2.44 (97,4 %)
S (10°C, 0.02 s) MPa	17 800	19 500	19 500	17 300	19 050	18 300
S (0°C, 300 s) MPa	7 640	10 450	7 240	7 100	9 300	8 570
1 - r	rupt avant 5.10⁴	rupt avant 5.10⁴	0.39	0.16	0.25	0.28
S (15°C, 0.02 s) MPa	14 500	16 500				

Famille de produit	A 7	B 8	B' 9	C 10	D 11	E 12
Densité hydrostatique des carottes	(97,4 %)	2.43 (97,8 %)	2.43 (97,8 %)	2.412 (97,9 %)	2.449 (94,7 %)	2.52 (96,5 %)
S (10°C, 0.02 s) MPa	17 200	14 800	16 500	19 700	23 000	17 200
S (0°C, 300 s) MPa	8 600	4 300	6 300	11 300	9 000	7 400
1 - r		0.42	rupt avant 5.10⁴	rupt avant 5.10⁴	0.39	0.22
S (15°C, 0.02 s) MPa	13 000					

Famille de produit	F 13	G 14	G 15	G 16	G 17	G 18
Densité hydrostatique des carottes	2.364 (97,1 %)	2.55 (97,6 %)	2.56 (97,7 %)	2.58 (98,2 %)	2.5 (98,2 %)	2.5 (97,1 %)
S (10°C, 0.02 s) MPa	16 500	16 000	21 000	17 000	27 400	24 400
S (0°C, 300 s) MPa	8 000	4 400	11 250	4 870	11 500	12 900
1 - r	rupt avant 5.10⁴	0.1	0.17	0.14	0.18	0.35
S (15°C, 0.02 s) MPa						

Famille de produit	H 20	I 21	I 22	R 23	J 24
Densité hydrostatique des carottes	2.479 (97,8 %)	2.42 (98,8 %)	(94 %)		(98,4 %)
S (10°C, 0.02 s) MPa	18 300	18 800	15 000	10 300	18 000
S (0°C, 300 s) MPa	9 140	6 700	11 000	3 800	4 400
1 - r	0.39	0.25		0.29	0.42
S (15°C, 0.02 s) MPa	14 500				

Le tableau présente une série de valeurs de modules en traction directe, obtenus sur des matériaux appartenant à une même catégorie : ce sont des enrobés de couche de base pour lesquels un module de rigidité élevé est recherché.

A l'intérieur de cette catégorie, on distingue des familles désignées par une lettre, chaque famille est un type de produit proposé par une entreprise, défini par un fuseau granulométrique, par un liant dont les caractéristiques sont identifiées, et par une fourchette de teneur en liant.

Les autres paramètres tels que nature des granulats, conditions de fabrication et de mise en oeuvre interviennent dans les performances du produit ; l'essai de traction directe a permis, par exemple de détecter les effets suivants :

Famille A, essai 4.5 :

A formule et conditions de réalisation identiques, la composition du liant a été légèrement modifiée, ce qui a influé sur la compactibilité et la teneur en vides du matériau, donc sur les valeurs de module à 10 °C ; 0,02 s.

Essai 1.2 :

Sur un même chantier, une zone sur laquelle un défaut de compactage a été observé a conduit à un écart de 1800 MPa sur les valeurs de module.

L'essai n° 6 est une vérification des performances en laboratoire (même formule que l'essai n° 4). Les résultats sont équivalents à 1000 MPa près.

Entre **les essais 2 et 3,** la différence concerne les granulats (nature, forme, angularité, fines, etc ...) ; le module à 0 °C ; 300 s est plus faible avec le mélange n° 3. Dans la comparaison des essais 3 et 7, intervient une différence de teneur en liant (de 6 % à 5,8 %) qui induit un module plus faible à 10 °C ; 0,02 s et un module plus fort à 0 °C ; 300 s. (modification de la susceptibilité thermique).

Famille B : le matériau de **l'essai n° 8** est non conforme. Ses caractéristiques ont été modifiées pour conduire au matériau B' qui répond aux exigences prévues.

Famille C, D, E : ces 3 produits ont des squelettes granulométriques très différents, des dosages en liant divers, mais une nature de liant identique. On constate l'effet très important du facteur formulation-granulat sur les modules à 10 °C ; 0,02 s : environ 6 000 MPa.

Le produit F est conforme aux exigences du marché.

Le produit G est un procédé spécial pour lequel on a observé une grande dispersion de comportements en relation avec la qualité des additifs, des temps de murissement, des conditions de fabrication et de mise en oeuvre. G 15 est la référence de laboratoire, G 17 et G 18 ont des caractéristiques conformes à l'essai de laboratoire ; G 14et G 16, bien que conformes aux spécifications mettent en évidence un défaut de comportement.
H est conforme aux exigences du marché.
I contient une forte proportion de fraisat. I 22 est la confection de laboratoire. La réincorporation du fraisat en laboratoire simule difficilement le processus de chantier et induit des performances plus faibles.
R est une grave-bitume classique.
J satisfait les exigences à 10 °C ; 0,02 s. Mais le module à 0 °C ; 300 s est faible ; ce produit semble sensible à la température.

9.5 Conclusion

L'ensemble de ces résultats montre que l'essai de traction directe sur éprouvette prélevée in situ constitue une méthode simple et fiable pour caractériser ce type d'enrobés (sous réserve bien entendu de la représentativité du prélèvement). L'essai est répétable, et sensible aux facteurs de qualité du produit. (Propriétés intrinsèques, et paramètres de fabrication de mise en oeuvre).

C'est donc une méthode de contrôle extérieur qui permet de vérifier dans l'ouvrage réel les hypothèses de calcul du projet.

Cette méthode s'inscrit dans la démarche de calcul rationnel des chaussées. Elle fournit les données numériques intrinsèques sur matériau en place pour la vérification des hypothèses du modèle de calcul utilisé en France actuellement. On peut penser également que dans l'avenir, si les modèles de dimensionnement sont affinés par exemple par une meilleure prise en compte de la température, ou des lois de chargement, cette méthode est capable de donner des valeurs numériques nécessaires à la vérification des hypothèses. En effet les résultats d'essai couvrent une large gamme de températures de temps de charge et le mode de sollicitation est adaptable.

En outre, des exploitations complémentaires permettent sous certaines conditions de relier les résultats en traction directe à des estimations de résultats d'autres essais (fatigue par exemple).

Enfin, il est possible d'examiner le comportement du matériau sous des sollicitations particulières (comportement mécanique à basse température) ou de déceler des aptitudes au fluage par l'interprétation des pentes des courbes de module en fonction de la température.

10 Conclusions générales

La caractérisation du comportement des enrobés bitumineux dans le domaine des faibles déformations nécessite la réalisation de nombreux essais.

Ces essais doivent être très bien maîtrisés en température, temps de charge et histoire de déformation ou charge.

De même, il faut que les conditions initiales d'essais soient convenables, ce qui suppose que les conditions de montage et démontage d'éprouvettes à défaut d'être parfait (qu'est ce que la perfection dans ce cas) obéissent à une codification précise.

La **MAER** contient implicitement toutes ces qualités car elle a été conçue pour répondre spécifiquement au problème de la caractérisation mécanique de ce produit fortement viscoélastique qu'est l'enrobé bitumineux.

Son usage peut s'étendre aussi bien à la caractérisation des bitumes qu'au contrôle des matériaux bitumineux.

ANNEXE (Extrait du projet de la norme relative à l'essai de traction directe)

5.2 - But de l'essai
L'essai complet consiste :

- à établir la courbe maitresse de module sécant en fonction du temps de charge et de la température d'usage du mélange hydrocarboné,

- à calculer la perte de linéarité du même mélange à la température fixée dans les normes "matériaux".

But = recherche de $S(t, \theta_j)$ et $(1-\Gamma)$
        θ_j, t_i.

Pour établir la courbe maitresse, ou calculer la perte de linéarité, de nombreux essais de traction <u>élémentaires</u> à vitesse constante et température différentes sont nécessaires.

5.3 - Principe de l'essai élémentaire

5.3.1 - Nature du corps d'épreuve - obtention
C'est un cylindre orthogonal d'élancement supérieur à 2.
   Celui-ci est prélevé dans des plaques fabriquées en laboratoire (voir normes NF P 250-1, NF P 250-2, NF P 250-3), soit dans des plaques ou carottes prélevées sur chantier (voir normes NF P 250-1, NF P 250-2, NF P 250-3). Pour ce dernier, les dimensions des éprouvettes peuvent être adaptées à l'épaisseur des échantillons prélevés. Néanmoins, le diamètre minimal ne peut être inférieur à 50 mm.
   Le prélèvement du cylindre sera tel que l'axe du cylindre soit toujours horizontal et parallèle à l'axe de compactage.

   Pour améliorer la représentativité de l'essai, plusieurs corps d'épreuves provenant de plusieurs échantillons seront testés (minimum 4).

5.3.2 - Sollicitation du corps d'épreuve
La sollicitation n'est exécutée qu'après avoir stabilisé l'éprouvette.
   L'élongation maximale ϵ_i est fonction de la température θ_j de l'essai.
   Le temps t'_i d'exécution de l'essai pour $\epsilon = 100$ Mdef est fixé par l'opérateur.

La vitesse de la rampe de déformation est calculée à partir de l'expression suivante :

$$a_i = \frac{100.10^{-6}}{t'_i}$$

L'allure de la loi de déformation appliquée est la suivante :

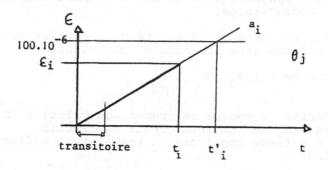

t_i, temps de charge réel de l'essai, est donné par

$$t_i = \frac{\varepsilon_i}{a_i}$$

5.3.3 – Réponse du corps d'épreuve – mesure du module sécant

Au cours de l'élongation une contrainte variable au cours du temps apparait. On l'enregistre ponctuellement (point noté o).

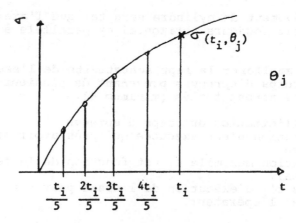

254

Cet enregistrement permet de calculer en particulier

la valeur $\sigma(t)$ pour le temps t_i et la température θ_j notée $\sigma(t_i, \theta_j)$ qui est la valeur recherchée pour estimer le module sécant à partir de l'expression:

$$S(t_i, \theta_j) = \frac{\sigma(t_i, \theta_j)}{\varepsilon_i}$$

5.3.4 - Nombre et nature des essais élémentaires à réaliser pour établir une courbe maitresse

Par éprouvette, au minimum trois températures θ_j et six élongations de temps de charge t'_i différents par température sont nécessaires.

Les ε_i choisis pour tous couples (t'_i, θ_j) sont seulement fonction de θ_j.

Le nombre d'exécution d'essais élémentaires est donc au minimum de 18.

Les caractéristiques des essais élémentaires sont résumés dans le tableau suivant :

	t'_1	t'_2	t'_3	t'_4	t'_5	t'_6
θ_1	ε_1	ε_1	ε_1	ε_1	ε_1	ε_1
θ_2	ε_2	ε_2	ε_2	ε_2	ε_2	ε_2
θ_3	ε_3	ε_3	ε_3	ε_3	ε_3	ε_3

L'interprétation est la suivante :
Pour une température θ_2 et un temps de charge t'_3 l'élongation pour laquelle la mesure de S sera faite est ε_2.
Pour cet essai, la vitesse de déformation relative sera

$$a_3 = \frac{10^{-4}}{t'_3}$$

le temps de charge réel

$$t_3 = \frac{\varepsilon_2}{a_3}$$

et le module donné par :

$$S(t_3, \theta_2) = \frac{\sigma(t_3, \theta_2)}{\varepsilon_2}$$

5.3.5 – Nombre et nature des essais élémentaires à réaliser pour calculer la perte de linéarité.

La perte de linéarité est une propriété mise en évidence lorsqu'on réalise des essais élémentaires aboutissant à des ε_k (k mis pour i dans ce cas) différents pour un même temps t_i de charge et une même température θ_j.

De ce fait, il faut choisir la température θ_j et le temps t_i puis exécuter au minimum 4 essais élémentaires par éprouvette et faire celà sur plusieurs éprouvettes.

Le tableau d'exécution peut être le suivant pour t_i, θ_j

ordre k	1	2	3	4
élongation	ε_1	ε_2	ε_3	ε_4

avec

$$\varepsilon_1 < \varepsilon_2 < \varepsilon_3 < \varepsilon_4$$

Les a_k correspondant sont toujours calculés avec

$$a_k = \frac{\varepsilon_k}{t_i}$$

et les modules recherchés $S_k(t_i, \theta_j)$ à l'aide de l'expression suivante :

$$\boxed{S_k(t_i, \theta_j) = \frac{\sigma(t_i, \theta_j)}{\varepsilon_k}}$$

(i et j peuvent être réduits à la valeur 1.)

5.7 – Expression du résultat
5.7.1 – courbe maitresse
5.7.1.1 – Calcul des modules ponctuels

Par éprouvette c et pour chaque couple de valeurs (t_i, θ_j) le module sécant est calculé :

$$\left[S(t_i, \theta_j)\right]_c = \frac{\left[\sigma(t_i, \theta_j)\right]_c}{\varepsilon_i}$$

5.7.1.2 – Calcul des caractéristiques de la distribution de S (S,CV)

Dans un tableau prévu à cet effet (annexe II) on reporte pour chaque éprouvette les modules correspondants, puis on calcule

$$\bar{S}(t_i, \theta_j) = \frac{\sum_{c=1}^{n} \left[S(t_i, \theta_j) \right]_c}{n}$$

$$CV(t_i, \theta_j) = \frac{1}{\bar{S}(t_i, \theta_j)} \sqrt{\frac{\sum_{c=1}^{n} \left(\left[S(t_i, \theta_j) \right]_c - \bar{S}(t_i, \theta_j) \right)^2}{n-1}}$$

5.7.1.3 – Traçage de la courbe maitresse

Sur un support (log, log) prévu à cet effet, on reporte les points correspondant aux différents résultats :
t_i en abscisse
$\bar{S}(t_i, \theta_j)$ en ordonnée.

On trace ensuite les arcs de courbe joignant ces points pour disposer des différentes portions de courbes maitresses.
En appliquant le principe de l'équivalence temps - température, qui correspond à une translation horizontale des portions de courbe maitresse θ_j, on trace la courbe maitresse à 10°C.

Voir exemple en annexe II.

5.7.2 – Perte de linéarité
5.7.2.1 – Calcul des modules ponctuels

Le module sécant est calculé par éprouvette c avec l'expression suivante :

$$\left[S_k(t_i, \theta_i) \right]_c = \frac{\left[\sigma(t_i, \theta_j) \right]_c}{\varepsilon_k}$$

5.7.2.2 – Calcul des caractéristiques de la distribution de $S(\bar{S}, CV)$

Dans un tableau prévu à cet effet (annexe 1) on reporte pour chaque éprouvette les modules correspondants, puis on calcule

$$\bar{S}_k(t_i, \theta_j) = \frac{c = \sum_{1}^{n} \grave{a} n \left[S_k(t_i, \theta_j) \right]_c}{n}$$

$$CV_k(t_i, \theta_j) = \frac{1}{\bar{S}_k(t_i, \theta_j)} \sqrt{\frac{c = \sum_{1}^{n} \grave{a} n \left(\left[S_k(t_i, \theta_j) \right]_c - \bar{S}_i(t_i, \theta_j) \right)^2}{n - 1}}$$

5.7.2.3 – Calcul de la perte de linéarité (provisoire)

A l'aide de tous les résultats, on exécute une régression parabolique entre les S_k et ϵ_k, les variables expliqués et explicatives étant respectivement S_k et ϵ_k.

Si on pose

$$(S_k)_c = A_0 + A_1 \epsilon_k + A_2 \epsilon_k^2 + \text{résidu } (k, c)$$

Alors la perte de linéarité est donnée par

$$(1 - \Gamma) = \frac{\hat{S}_4(t_i, \theta_j) - \hat{A}_0}{\hat{A}_0}$$

<u>à condition</u> que le minimum de la parabole estimé ne soit pas entre

$\bar{S}_3$ et $\bar{S}_4$.

Dans le cas contraire, la régression sera linéaire de modèle
$$(S_k)_c = A_0 + A_1 \epsilon_k + \text{résidu } (k, c)$$
et $(1 - \Gamma)$ sera donné par la même expression.
Voir exemple en annexe II.

MODULE SECANT EN TRACTION DIRECTE

Rappel des conditions d'essais choisis

Nature, comportement matériau : ~~très-rigide~~ / ~~rigide~~ / normal / ~~souple~~ / ~~très souple~~

temps de charge t'_i pour 10^{-4} : 1 / 3 / 10 / 30 / 100 / 300 / ~~1-000~~ / ~~3-000~~

température d'essais θ_j : - 10 / 0 / 10 / ~~20~~ / ~~30~~ / ~~40~~

Déformation relative ϵ_i conforme au tableau   oui - ~~non~~

si non valeurs de ϵ_j en fonction de θ_j

	t_i			
θ_j	ϵ_i			

MODULE SECANT EN TRACTION DIRECTE

Résultats (en GPA)

Ep = Eprouvette

$\theta_1 = -10$

	1	3	10	30	100	300	1000	3000
Ep 1		22.8	20.3	18.1	15.9	13.8		
Ep 2		23.6	21.0	18.8	16.5	14.4		
Ep 3		22.7	20.2	18.2	16.0	13.8		
Ep 4		21.5	18.9	16.7	14.2	12.1		
Ep 5								
Ep 6								
$\bar{S}$		22.7	20.1	18.0	15.6	13.5		
100 CV		3.8	4.3	4.9	6.3	7.4		

$\theta_2 = 0$

	1	3	10	30	100	300	1000	3000
Ep 1	12.8	10.6	8.3	6.4	4.7	-		
Ep 2	13.5	11.2	8.9	6.8	4.9	3.5		
Ep 3	13.7	11.3	8.9	6.9	5.0	3.6		
Ep 4	13.2	11.1	8.8	6.8	5.1	3.7		
Ep 5								
Ep 6								
$\bar{S}$	13.3	11.1	8.7	6.7	4.9	3.6		
100 CV	2.9	2.8	3.2	3.3	2.9	3.2		

MODULE SECANT EN TRACTION DIRECTE

Résultats (en GPa)

Ep = Eprouvette

$\theta_3 = 10$

	1	3	10	30	100	300	1000	3000
Ep 1	5.8	4.0	2.6	1.7	1.0	0.6		
Ep 2	5.9	4.2	2.7	1.7	1.0	0.6		
Ep 3	6.4	4.5	2.9	1.9	1.1	0.6		
Ep 4	5.8	4.1	2.7	1.6	1.1	0.7		
Ep 5								
Ep 6								
$\bar{S}$	6.0	4.2	2.7	1.7	1.0	0.6		
100 CV	4.8	5.3	4.6	7.8	5.1	5.9		

$\theta_4 =$

	1	3	10	30	100	300	1000	3000
Ep 1								
Ep 2								
Ep 3								
Ep 4								
Ep 5								
Ep 6								
$\bar{S}$								
100 CV								

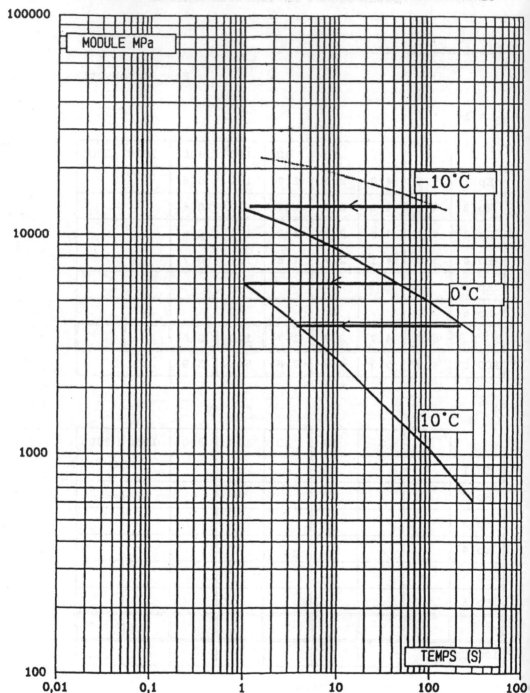

ELEMENTS DE COURBE MAITRESSE A DIFFERENTES TEMPERATURES

MODULE MPa

−10˚C

0˚C

10˚C

TEMPS (S)

COURBE MAITRESSE DU MODULE. à 10°C

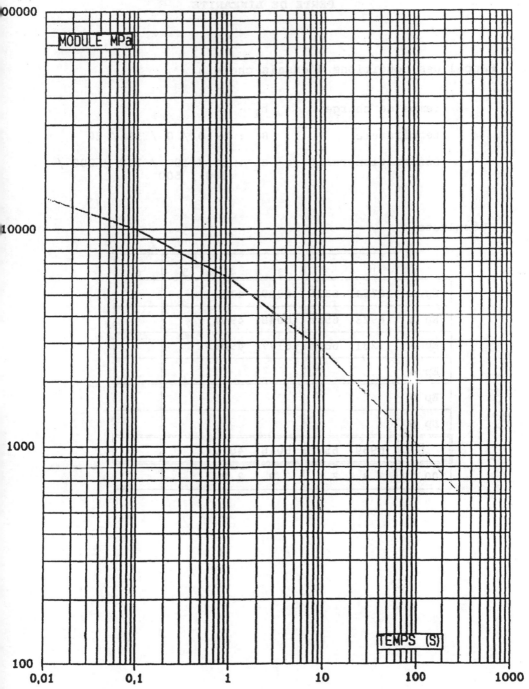

MODULE MPa

TEMPS (S)

PERTE DE LINEARITE

Rappel des conditions d'essais choisis

temps de charge t_i : 30

température θ_j : --$\cancel{10}$ / 0 / $\cancel{10}$ / $\cancel{20}$

Déformation relative ϵ_k : 50 / 100 / 200 / $\cancel{300}$ / $\cancel{400}$ / 500

Résultats

	1	2	3	4	5	6
Ep 1	6764	6326	5829	4723		
Ep 2	6923	6585	6048	4902		
Ep 3	7281	6804	6177	4986		
Ep 4						
Ep 5						
Ep 6						
$\bar{S}$	6989	6572	6018	4870		
100 CV	3.8	3.6	2.9	2.8		

PERTE DE LINEARITE

Régression :

Modèle choisi : $\boxed{\text{X}}$ $S_i = A_0 + A_1 \, \epsilon_k + A_2 \, \epsilon_k{}^2$

$\boxed{}$ $S_i = A_0 + A_1 \, \epsilon_k$

Valeurs de A_i : $\hat{A}_0 = \quad 7,340.10^3 \qquad s_{y/x}$

$\hat{A}_1 = - \ 7,920.10^6$

$\hat{A}_2 = \quad 5,940.10^9$

$\hat{S}_5 = \quad 4,865.10^3$

Perte de linéarité

$$(1 - \Gamma) = \frac{\hat{A}_0 - \hat{S}_5}{\hat{A}_0} = \qquad \boxed{0,337}$$

COURBE DE PERTE DE LINEARITE

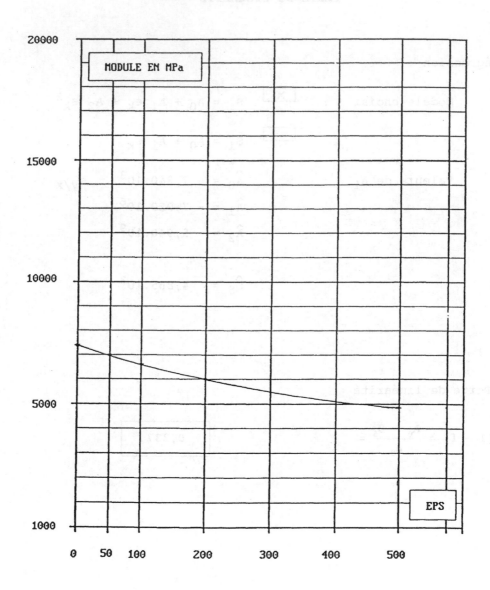

18 THE EVALUATION OF THE CONSTRUCTION MATERIAL ASPHALT BY THE 'CARINTHIAN MODEL'

H. PIBER
Amt der Kärntner Landesregierung, Klagenfurt, Austria

Abstract
This paper described a new testing method by unique loading.
Keywords: Testing method, Equipments, Specimens, Description,
Evaluation, Requirements, Practical application, References.

1 Introduction

Since about a decade the road administration in Carinthia (Austria) has
been evaluating road construction materials and pavements with a model
developed by the department 17 - BT (construction techniques) of the
state goverment.
 Testing methods used in asphalt-mechanics up to now have the
disadvantage that it is impossible to seperate "elastic" from "plastic"
deformation behaviour. The basis of this (so called) "Carinthian Model"
is a testing method which determines the characteristic features of the
material asphalt completely.

2 Range of application

This method is used for the determination of the deformation modulus of
asphalt. It allows to analyse the resistance to deformation, in
consequence to determine the quasielastic part of deformation behaviour
and the dynamic "elastic" deformation modulus for structural design of
pavements.

3 Laboratory testing equipment and specimens

The testing instrument is a 100 KN compression tester. It's precision requires class one according to DIN 51 223. The controllable feedtravel was chosen with 1.27 mm/min. From this speed results a speed of strain in the test specimen, which is equivalent to the speed of lateral strain under the rolling wheel at 25 km/hr. Thus relation to practice is given.

Force and time are recorded in a diagram.

The asphalt specimens tested are of cylindrical shape with a diameter of 100 mm and with a height of 60 mm. They can be marshall-specimens or asphalt cores. The maximal particle size is 22 mm. The faces of the cylindrical specimens must be flat, parallel and polished. The testing temperature could be optional but practice has been showing that tests at -20°C, +10°C and +30°C are very effective.

The specimens are stored in a waterbath or in refrigerating plants until they arrive at a constant temperature for testing in their center. A thermostat regulates the temperature with a precision of $\pm$1°C.

4 Way of experiment

Two specimens must be examined at every testing temperature. For at least one hour these specimens lie in a waterbath or in a refrigerating plant. After that one of those is put between the both pressure-plates centrically. For 80 seconds it is deformated with a feedtravel of 1.27 mm/min.

The electronic recording unit prints a force-time-diagram. The scale of the diagram must be adjusted to a reading accuracy of $\sim$0.1 KN for force and 1 second for time. (Look at figure 1.)

5 Description of the method

The total deformation of asphalt consists of an "elastic" and a "plastic" part. Between both the limit is flowing. It is necessary to find that point in which predominantely elastic deformation changes into predominantely plastic deformation. The method is based on the thesis that deformation of every material requires energy. Against this deformation energy every material opposes its deformation resistance energy. The more energy is necessary to deform the material the bigger is the deformation resistance.

We proceed on the assumption that in the elastic phase the increase of deformation energy is growing unproportionally for every unit of time. Therefore it is possible to determine the point between the elastic and plastic phase. The so calculated energetic deformation modulus is determined for the "elastic" range. Therefrom the dynamic elastic modulus can be calculated. During the test the course of the deformation-energy can be found out. At the same time the change in the deformation-energy's increase is calculated. The energy activated until the maximum possible increase of deformation resistance capacity

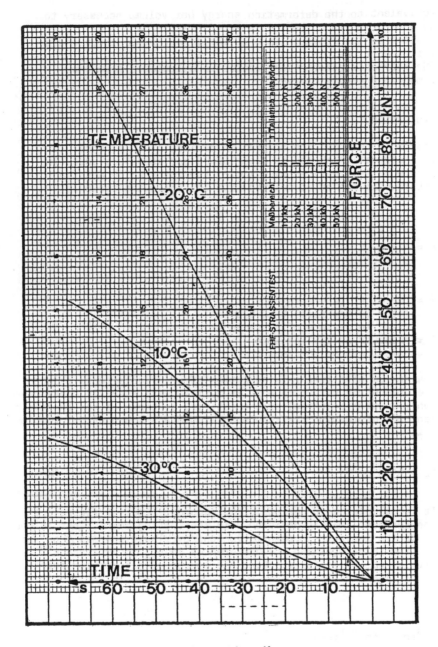

Fig. 1. Force-time-diagram

is equivalent to the deformation energy per volume necessary to cause a deformation up to the limit of elastic behaviour. This is equivalent to the fatigue limit of the granular structure under elastic conditions.

The energetic deformation modulus is calculated from the deformation-energy's course.

6 Evaluation

The force is read off the diagram on periods of 10 seconds.

The deformation energy is ascertained per structure unit. The structure unit is defined with 1 mm³. The force which acts on this structure unit can be calculated with the following forumula:

$$F = \frac{P \times 4}{D^2 \times \pi} \times \frac{H}{D}$$

F = force per structure unit
P = force of the force-time-diagram
   (read in periods of 10 seconds)   (N)
D = diameter of the specimen   (mm)
H = height of the specimen   (mm)

The energy of a time related action of a variable force related to the volumetric structure unit is defined with $F^2/2$. In consequence the unit related potential of the material in terms of the energetic concept of deformation resistance at a constant speed of deformation also is

$$a_t = F^2/2 \qquad\qquad (N/s)$$

It is calculated in periods of 10 seconds. (Look at figure 2.)

The fatigue limit is positioned in the turning point. That is the limit to the increase in deformation-energy. These changes $\triangle a_{tn}$ are calculated in determining the differences in a_{tn} in 10 seconds intervals, eg. by differentiating the course of deformation energy. (Look at figure 3.)

$$\triangle a_{tn} = a_{tn} - a_{tn-1}$$

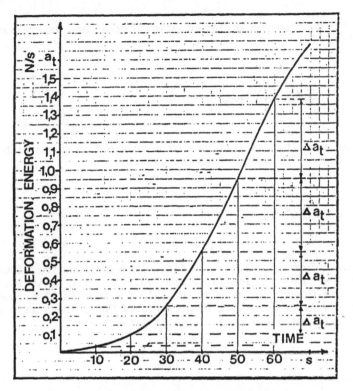

Fig. 2. Deformation energy-time diagram

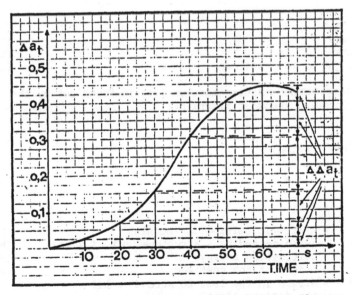

Fig. 3. Increase of deformation energy-time-diagram

271

After that by a second differentiating process one determines the turning point "S".

$$\triangle \triangle a_{tn} = \triangle a_{tn} - \triangle a_{tn-1}$$

The "change of fhe deformation-energy-changes" is given with regard to double periods (20 seconds) to 0.1 exactly.

$$\triangle \triangle = 20 \times \triangle \triangle a_{tn}$$

This values form a curve, which first increases and then decreases. After that the turning point is determined. The point is indicated in the force-time-diagram in seconds. Usually the time is 35 seconds for asphalt specimens at a testing temperature of 30°C. The strain is calculated from speed and time. (Look at firgure 4).

$$\xi = \left| \frac{v \times t}{H} \right| \times 100 \qquad\qquad (\%)$$

ξ = strain	(%)
v = speed	(mm/s)
t = time	(s)
H = height of the specimen	(mm)

The energetic deformation modulus (Ve) is found out by the following formula: (Look at figure 5).

$$Ve = \frac{\left(a_{tS} - a_{tS-20} \right) \times H}{\triangle t \times v}$$

Ve	= energetic deformation modulus	(Ns)
	(with regard to a structure unit of 1 mm³)	(Ns/mm³)
a_t	= deformation energy	(N/s)
H	= height of the specimen	(mm)
$\triangle t \times v$	= strain in the last 20 seconds until the fatique limit	

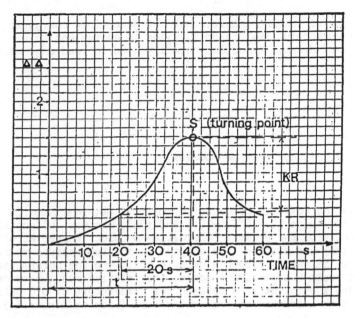

Fig. 4. "Change of the deformation energy changes"-curve

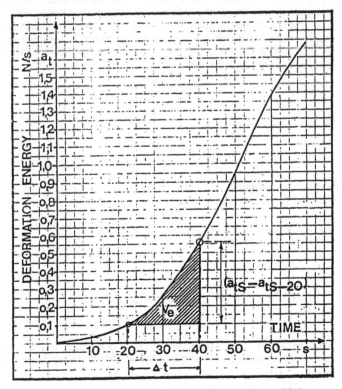

Fig. 5 Energetic deformation modulus (Ve)

Comparative testing showed, that the determined energetic deformation modulus (Ve) taken 10 times is equivalent to a conventional dynamic E-modulus normally used in road engineering.

$$V_d = 10 \times V_e \qquad\qquad\qquad\qquad (N/mm^2)$$

V_d = dynamic deformation modulus.

The testing temperature could be optional but practice has been showing that tests at -20°C, +10°C and +30°C are very effective. If these results of the dynamic deformation-modulus (tested at different temperatures) are plotted in a diagram versus temperature where temperature is shown through a decadic scale and the deformation-modulus through a logarithmic-scale, the relation-modulus-temperature will follow a straight line between the temperature of 0°C until to 35°C. At both ends the curve flattens out. (Look at figure 6.).

At the temperature of more than +35°C aggregate friction, eg. the friction strength, determines the deformation behaviour. Granulometry, properties of aggregates as broken faces angularity shape, strength, type and content of binder define the level of the friction strength. It can be backcalculated from the dynamic deformation-modulus (VdB 30°) and the friction activated between the particles of the granular skeleton with following approximation.

$$VdR = VdB\ 30° \times \frac{1}{1.5 + KR}$$

VdR = friction strength
KR = particle friction

The friction is evaluated from the change in deformation energy increase. (Look at figure 4)

$$KR = 20 \times \left(\triangle\triangle a_{tS} - \triangle\triangle a_{tS-20} \right)$$

Between 0°C an +35°C the cohesive strength of the binder defines the dynamic deformation modulus. The quality of binder bridges depends on

274

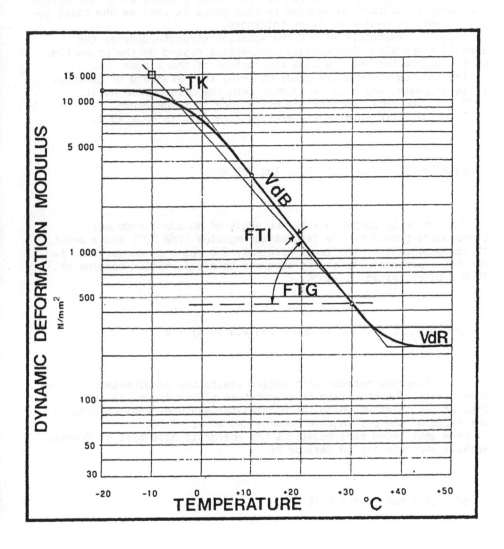

Fig. 6. Modulus-temperature-gradient

the type and content of the binder and on the properties of aggregates. The dynamic deformation modulus in this range is used as the basis for the structural design of layer thickness.

Below a temperature of 0°C the adhesion strength defines the asphalt properties. The adhesion properties depend on the properties of the aggregates and the type and content of the binder.

The modulus-temperature gradient (FTG) is a standard of quality. It is an angle, which is calculated form the deformation moduli (VdB 30°, VdB 10°) and from the difference in temperature.

Then the difference of temperature form 20°C is equated 1.

$$\text{tg FTG}_{(IST)} = (\log \text{VdB } 10° - \log \text{VdB } 30°)$$

This angle is compared with the angle of an ideal modulus-temperature-gradient. The deformation modulus (VdB 30°) and a point of convergence form the ideal gradient. The point of convergence is defined experimentally and is adequate to a deformation-modulus of 15 000 N/mm² tested at -10°C.

$$\text{tg FTG}_{(SOLL)} = (\log 15\ 000 - \log \text{VdB } 30°C) : 2$$

The difference between both angles yields the modulus-temperature-index (FTI). The criterion of temperature (TK in figure 6) is this theoretical point at which the modulus-temperature-gradient crosses the value of the deformation modulus tested at -20°C.

From well known considerations the energetic potential of dynamic asphalt strength may be defined by

$$BF \ldots° = VdB \ldots° \times \varepsilon\ /2$$

$$HF \quad = VdB\ -20° \times \varepsilon\ /2$$

BF = cohesive potential of the asphalt structure   (N/mm²)
HF = adhesion potential of the asphalt structure   (N/mm²)
ε = strain amount of deformation   (%)

7 Requirements

The actual quality of layed asphalt is evaluated from asphalt cores.
The asphalt layers must meet a sery of requirements on quality that
the pavement performs well for a long time.

The loading case - "heat period" - is relevant to deformation. The
traffic load and its speed are factors which define the friction
strength. The traffic intensity is calculated analogous to the
Austrian specification RVS 3.63.

$$NLW = DTLV \cdot R \cdot V \cdot S \cdot Ä \cdot T$$

NLW = design load
DTLV = daily number of commercial vehicles
R   = direction factor
       traffic uniformly distributed in both directions   R = 0.5
V   = traffic lane factor
       one or two lanes in a direction   V = 1.0
       three lanes in a direction   V = 0.9
S   = wheel track factor
       width of the lane: < 3.0 m   S = 1.0
                           3.0 - 3.5 m   S = 0.9
                           > 3.5 m   S = 0.6
Ä   = traffic equivalence factor
       normal case: DTLV > 2000   Ä = 1.0
                    1000 - 2000   Ä = 0.8
                    < 1000   Ä = 0.5
T   = number of days which have a temperature
       more than 25°C

The traffic speed is subdivided into three characteristic ranges:

Table 1. Speedfactor in dependence on traffic speed

Speedfactor RLF	traffic speed
1.4 1.0 0.8	< 10 km/hr 10 - 40 km/hr > 40 km/hr

The required friction strength can be estimated:

Wearing course

$$VdR_{SOLL} = 100 \times RLF \times NLW^{1/4}$$

Bituminous roadbase

$$VdR_{SOLL} = Vdp \times ((ABV \times NLW^{1/4})^{1/4} \times VdR_{IST} : Vdp)^{1/3}$$

$$Vdp = 500 \times RLF \times ABV \times (ABV \times NLW^{1/4})^{1/4}$$

$$ABV = 1 - \frac{(1 - h^3)}{(a^2 + h^2)^{1.5}}$$

Vdp = dynamic deformation modulus of an ideally elastic isotropic half space in a distance from the upper surface
ABV = axial load relationship (Look at figure 7)
h = distance from the upper surface to the centre line of the course
a = radius of loading area (150 mm)

The modulus-temperature-index (FTI) is comparable to the penetration-index of a binder.
Positive FTI-values mark a steeper increase of the modulus depending on the temperature. What means, that crack formation brittleness thermal cracking is favoured.
Negative FTI-values mark a flatter increase of the modulus depending on the temperature. What means that crack formation is excluded.
The loading case - "cold period" - is relevant to crack formation.
The criterion of temperature (TK) is a feature of durability of asphalt courses too. It marks the critical temperature range in which the maximum asphalt strengthes - the adhesion strengthes - are activated. In construction practice this critical temperature depense on the climatic conditions (0°C for Carinthia).
Higher temperatures mark an inflexible asphalt course. Critical thermal stresses occur more often and the service life of the course is shorter.
Lower temperatures mark an asphalt which is able to relax. Critical structure stresses occur less often and because of this the service life is getting longer.
The loading case - "freeze-thaw period" - is relevant to the

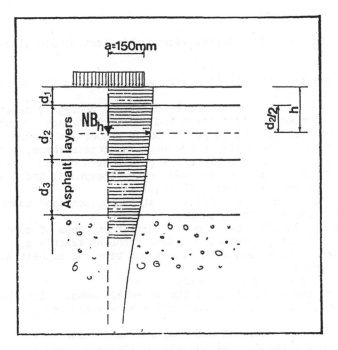

Fig. 7. Axial load relationship (ABV)

the structural design of the pavement structure.

The critical load of the pavement structure occurs at the bottom of the asphalt courses in the freeze and thaw period.

The bearing capacity of the subsoil or subgrade is very little due to water absorption and the asphalt is - as everybody knows - rigid due to the low temperatures.

At this time deflection measurements are carried out. In cooperation with an elastic multi-layered system ration between the subgrade, the subbase and the asphalt courses are calculated. In this structural analysis the asphalt strengthes are replaced by the earlier described dynamic deformation modulus (VdB). The occuring tensile stresses (energies) at the bottom of the asphalt course can be determined with a design system. This result is compared to the cohesive strength (BF) (energetic potential) of the asphalt structure. The design load (NLW) is considered too.

The road administration in Carinthia (Austria) developed her own structural design of pavement structures. This topic can not be described here because it would be too extensive.

8 Practical application

In comparison to nominal or target values this test method allows
for checking on

- the optimisation of asphalt mixes, with regard to the desired
  deformation modulus
- the influence on the behaviour by different aggregate materials
  and their proportions (friction and adhesion strength, kind of
  aggregates with respect to petrographic properties, proportion of
  crushed faces, angularity roundedness aggregate shape, grading)
- the behaviour of certain binders e.g. standard bitumen (type)
  special binders (modified bitumen, natural asphalt) and admixtures
  (adhesion promoting agents, fillings)
- the suitability of sands (friction strength, adhesion capacity
  of crushed sands or round sands)
- the material specific behaviour of different types of asphalt
  (asphalt concrete, pervious macadam, granular materials,
  stabilised by treatment with bituminous binders, asphalt after
  repave, remix)
- the behaviour of asphalt courses
- as well as the suitability of the material asphalt also the
  behaviour of pavement constructions can be evaluated.

The assessement of material and construction failures is
facilitated and the standard of evaluation improved. Consequently
technically optimised proposals for reconstruction are possible. At
the structural design of new pavements there is more possibility to
consider the quality of construction materials in detail.

The road administration of Carinthia can make reference of a long
period of practical use and application of this testing method and of
many positive experiences.

9 References

Füreder, H. (1983) Verformungsmechanik und Verformungsfestigkeit
    der Oberbaukonstruktion von Asphaltstraßen. Oberbaudimensionierung-
    Oberbaubewertung. Dissertation an der TH Graz
Füreder, H. (1987) Qualitätsparameter der Asphaltsichten. Amt der
    Kärntner Landesregierung, unveröffentlicht.
Meier, H. Eisenmann, J. und Koroneos, E. Beanspruchung der Straße
    unter Verkehrslast, Forschungsarbeiten aus dem Straßenwesen,
    Heft 76.
RVS 3.63 - Bautechnische Details - Oberbau.

19 INVESTIGATION ON HIGH AND LOW TEMPERATURE BEHAVIOUR OF ASPHALT BY STATIC AND DYNAMIC CREEP TESTS

M. SCHMALZ, R. LETSCH and M. PLANNERER
Prüfamt für bituminöse Baustoffe und Kunststoffe, Technical
University, Munich, Germany

Abstract
Investigations on the deformation behaviour of a bitumen
B 65 and two polymer modified bitumens were carried out at
extreme temperatures. The low temperature behaviour was
derived from static creep tests with bituminous mortars at
temperatures between -10 and -40 °C. The resistance
against deformation at high temperatures was determined by
dynamic loading creep tests with Marshall specimens at 50
and 60 °C. The results allowed a clear distinction of the
three bitumens and showed the effects of the modification
with polymers. This difference was not revealed by the
standard test methods.
Keywords: Bitumen, Polymer modified Bitumen, Asphalt,
Creep, Cracking, Temperature Behaviour, Dynamic Loading,
Rutting.

1 Introduction

In Germany asphalt pavements on roads and bridges and also
asphalt waterproofing membranes are in service for many
decades. Although practical experience with asphalt is
very large and a great deal of research has been conducted
on the behaviour of asphalt at extreme temperatures, low
temperature cracking and high temperature rutting still
seem to be unsolved problems.

It is well known, that the properties of bitumen have a
major influence on the temperature behaviour of asphalt
pavements or waterproofing membranes. Practical experience
has been gained with various types of polymer modified
bitumen (PmB). Most of the polymers added increase the
softening point ring and ball. Therefore it is possible to
use bitumen with higher penetration values which improve
the low temperature behaviour of the asphalt mixture.

Unfortunately the standard test methods for bitumen and asphalt do not show this improvement in a satisfying manner / 1 /. New testing methods have to be developed. One of them could be long term testing (creep) of bituminous mortars or asphalt at different temperatures. The results of such a test allow conclusious on the elastic, viscous and viscoelastic components of the deformation. These components may be used as criterion for the prediction of the resistance against deformation at high temperatures and fracture susceptibility at low temperatures.

In the past investigations have been made using static and dynamic creep tests to predict the permanent deformation of asphalt mixtures. They were all carried out at normal and elevated temperatures / 2-8 /.

This paper presents low temperature static and high temperature dynamic creep tests and shows the results of a comparative study on three different kinds of bitumens for asphalt pavements.

To predict the low temperature behaviour of asphalt it is very important to know the low temperature properties of the bitumen used in the mixture.

Bitumens are viscoelastic materials. With falling temperatures the behaviour of bitumen changes from viscous and viscoelastic to almost elastic. Viscous and viscoelastic behaviour are the causes for stress relaxation which reduces the danger of cracking. By the standard test methods (e.g. breaking point by Fraaß) no precise defined physical properties of bitumen can be measured.

2 Static creep tests at low temperatures
2.1 Scope
Unfilled bitumens showed at normal temperature too big viscous deformations so that the creep tests had to be carried out with bituminous mortars. The mix design followed the specifications of the German TV bit 6/75 / 9 / for asphalt-mastix:

Bitumen content	15 % by weight
aggregate: washed lime sand	0/2 mm
ratio: natural sand / crushed sand	50/50

The mortars were prepared using three bitumens, an unmodified one (B65) and two polymer modified ones (PmB1 and PmB2). The standard specifications are given in table 1.

Table 1. Standard specifications of the bitumens

Method of Testing	DIN		B 65	PmB1	PmB2
Penetration	52 010	1/10 mm	50	79	65
Softening Point	52 011	°C	53,0	54,5	54,0
Breaking Point	52 012	°C	-9,0	-11,5	-20,0
Ductility at 25 °C	52 013	cm	24	>100	20
Difference between Softening Point and Breaking Point		K	62	66	74

The mortars were mixed for 3 minutes at 170 °C and poured into a steel form treated with releasing agent. The dimensions of the specimens were 20x20x80 mm³. The influence of the aggregate on the stress distribution within the specimens is very small / 6 /, as long as the maximum grain size does not exceed 1/10 of the smallest dimension of the specimen. The tests were carried out in loading frames shown in figure 1.

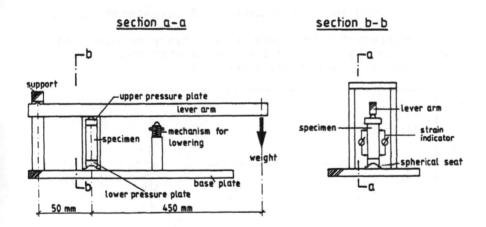

Fig. 1. Loading frame for creep tests / 10 /

The deformations measured by two strain indicators were recorded continuously during the loading and unloading periods. The tests were conducted at temperatures between -10 and -40 °C. The temperature deviations within a dummy specimen were measured to ± 0,1 K.

2.2 Test results

Examples of the time-depending deformations under load and after unloading are given in figure 2.

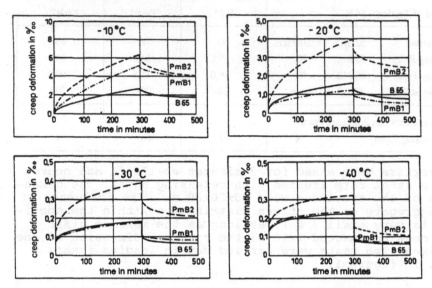

Fig. 2. Creep and recovery curves of 3 bituminous mortars

The three different components of the creep deformation were calculated by applying the 4-parameter model by Burgers (figure 3) using equation (1):

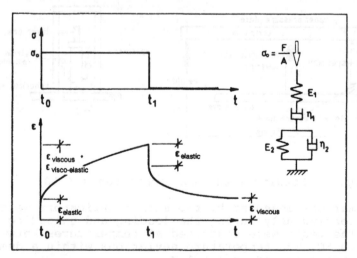

Fig. 3. Deformation according to the Burgers model

$$\varepsilon_{(t)} = \frac{\sigma_0}{E_1} + \frac{\sigma_0}{\eta_1} \cdot t + \frac{\sigma_0}{E_2} \cdot [1-e^{-\frac{E_2}{\eta_2} \cdot t}] \qquad (1)$$

The calculation of the parameters E_1, E_2, η_1 and η_2 was carried out using a computer curve-fitting program (Gauß-Newton algorythm) which gives the best least-square fit to a xy-dataset. Figure 4 shows these parameters versus temperature.

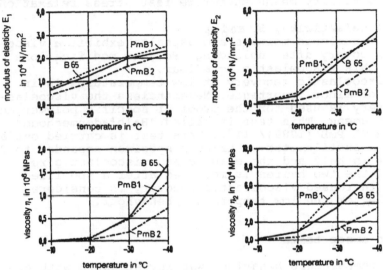

Fig. 4 Parameters E_1, η_1, E_2 and η_2 of the 3 bituminous mortars between -10 and -40 °C

2.3 Discussion
The properties of bituminous mortars and asphalts depend on temperature, stress, speed of loading and type of bitumen.

The standarized test methods (penetration, softening point and breaking point) are only valid for a spezified condition at one single temperature. To estimate the time- and temperature depending deformations it is necessary to know the materials's properties as modulus of elasticity and viscosity over a wide temperature range.

2.3.1 Modulus of elasticity, E_1
The modulus of elasticity increases almost linearly between -10 and -40 °C up to more than 20 000 N/mm². There is a distinct difference in the moduli of elasticity with PmB2 having the smallest and PmB1 the highest one. With decreasing modulus the stresses induced by hindered or im-

posed deformations decrease. That means that the risk of cracking at low temperatures is also lowered by a smaller modulus of elasticity.

2.3.2 Viscosity, η_1
The viscosities calculated for the mortars with B65 and PmB1 are almost the same while the one for mortar with PmB2 is only half as big over the whole temperature range. Stresses in bituminous mortars and asphalts decrease due to stress relaxation caused by viscous flow of the bitumen. Low viscosity means therefore fast stress relaxation.

2.3.3 Viscoelasticity, E_2 and η_2
Bitumen, bituminous mortars and asphalts exhibit a viscoelastic recovery after unloading. Polymer modifications increase this viscoelastic behaviour.

The viscoelastic recovery at low temperatures is in practice of minor importance. Nevertheless the viscoelastic recovery is used for the proof of specific polymer modified bitumens. This test is called "Halbfadenmethode" according TL PmB (1989)/ 11 /. This test is carried out at 25 °C in a ductilometer. A briquet specimen expanded to 20 cm is cut in half and the elastic and viscoelastic recovery of the two halfes is measured after a specified time. An increased viscoelastic recovery is considered favourable for the low temperature behaviour.

3 Dynamic creep tests
3.1 Scope
The high temperature behaviour was investigated with asphalt concrete. The maximum grain size was 8 mm, the binder content 6,2 % by weight. Marshall specimens prepared with the same bitumens used for the low temperature creep tests were subjected to dynamic loading according figure 5 at 50 and 60 °C.

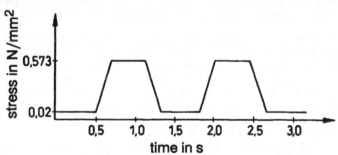

Fig. 5. Dynamic load function

3.2 Test results
Figure 6 shows the dynamic creep curves at 50 and 60 °C of the 60 mm high Marshall speciments in relation to the number of load cycles.

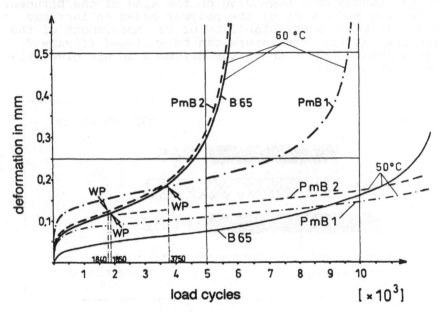

Fig. 6. Deformations of Marshall specimens at 50 and 60 °C vs. load cycles

3.3 Discussion
Due to consolidation of the aggregates within the specimens, the initial deformations are very high. After this period the speed of deformation decreases to an almost constant value. The dynamic loading leads to a point of inflection (WP) after a number of load cycles specific for every material. A progressive increase of the deformations follows.

The evaluation of the creep curves at 60 °C gives a distinct difference in deformation speed at the point of inflection. The number of load cycles at the point of inflection was nearly the same was for the Marshall specimens with B65 and PmB2 and only half as big as for the PmB1-specimen. For the deformation speed an inverse relation was observed:

B 65 and PmB2: $\varepsilon = 2,8.10^{-5}$ mm per load cycle
PmB1 $\varepsilon = 1,5.10^{-5}$ mm per load cycle.

The results show, that the specimens prepared with PmB1 have the greatest resistance against irreversible deformation (e.g. rutting) at high temperatures.

287

4 Conclusions

The results of the low and high temperature creep tests show that polymer modified bitumens cover a wider temperature range in their application for pavements than an unmodified bitumen B65. Depending on the kind of the bitumen and the type and amount of the polymer added an increase in the high temperature stability or an improvement of the flexibility at low temperature can be achieved (figure 7).

The standard test methods for bitumens do not give evidence of this fact.

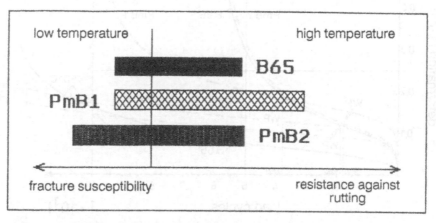

Fig.7. Comparison of the high and low temperature behaviuor of a B65 and two polymer modified bitumen (PmB1 and PmB2)

5 Literature

/ 1 / Kolb, K.H.: Die Beurteilung von polymermodifiziertem Bitumen und polymermodifiziertem Asphalt im Laboratorium. Diss. TH Darmstadt 1984

/ 2 / Arand, W.: Zum Einfluß tiefer Temperaturen auf das Ermüdungsverhalten von Asphalten. In: Straße und Autobahn 34 (1983) H. 10, S. 424-431

/ 3 / Huschek, S.: Der Kriechversuch. In: Straße und Verkehr 62 (1976) H.4, S. 134-142

/ 4 / Huschek, S.: Die Beurteilung des Verformungswiderstandes bituminöser Mischungen durch den Kriechversuch. In: Bitumen 42 (1980) H. 2, S. 44-48

/ 5 / Jongeneel, D.J.,Haugh, L.: Creep Testing of Asphalt
Mixes. Results of an Interlaboratory Study of Laboratory Apparatus and Test Procedures. In: 3. International Eurobitume Symposium, Den Haag, Sept.
1985, S. 294-300

/ 6 / Kraß, K.: Kriechuntersuchungen an zylindrischen
Asphaltprobekörpern. Veröffentlichungen des Insti-
tutes für Straßenbau und Eisenbahnwesen der Universität Karlsruhe, Heft 5/1971

/ 7 / Van de Loo, P.J.: Creep Testing, a Simple Tool to
Judge Asphalt Mix Stability. Ann. Meeting Ass. of
Asphalt Paving Technologists, Williamsburg (1974),
S. 253-285

/ 8 / Peffekoven, W.: Der Kriechversuch. In: Bitumen 45
(1983) H. 1, S. 15-23

/ 9 / Bundesminister für Verkehr, Abt. Straßenbau:
Technische Vorschriften und Richtlinien für den Bau
bituminöser Fahrbahndecken (1975), TV bit 6/75

/ 10 / Letsch, R.: Über das Verformungsverhalten von
Epoxidharzen und Epoxidharzmörteln bei stationären
und instationären Temperaturen. Diss. TU München
1983

/ 11 / Forschungsgesellschaft für Straßen und Verkehrs-
wesen: Technische Lieferbedingungen für Polymer-
modifizierte Bindemittel (Entwurf 1989), TL PmB
(1989)

20 UTILISATION DE LA TRACTION DIRECTE POUR L'ÉTUDE DE LA VALEUR STRUCTURELLE DES ENROBÉS BITUMINEUX

(Use of direct tension for studying the structural qualities of coated bitumens)

J.P. SERFASS
Screg Routes, Saint-Quentin-en-Yvelines, France
S. VAN-BELLEGHEM
Recherche-Technique-Entreprise, Bonneuil-sur-Marne, France

La prévision du comportement mécanique des enrobés bitumineux dans les structures de chaussée impose de connaître leur réponse sous chargement et leur tenue en fatigue sous sollicitations répétées. La méthode rationnelle de dimensionnement utilisée en France fait appel aux notions de module complexe et de tenue en fatigue en flexion alternée. Les essais correspondants sont représentatifs et bien calés par rapport aux comportements en place, mais ils sont sophistiqués et nécessitent un matériel spécifique et coûteux.

Depuis le début des années 80, il existe un nouvel essai de traction directe, conçu par le Laboratoire Central des Ponts et Chaussées, et un nouveau matériel, relativement simple, pour l'évaluation structurelle des enrobés.

Le Centre Technique du groupe SCREG Routes, R.T.E., pratique couramment cet essai de traction directe depuis 6 ans, parallèlement aux essais de module complexe et de fatigue classiques.

La présente communication analyse les résultats obtenus en traction directe sur divers types d'enrobés et les compare à ceux des essais "lourds" de référence, évoqués plus haut.

1 DESCRIPTION DES ESSAIS

1.1 L'essai de traction directe

1.1.1 Principe de l'essai

L'essai consiste à appliquer des efforts de traction sur éprouvettes cylindriques, sans aller jusqu'à la rupture. Il permet la détermination d'un ensemble de caractéristiques mécaniques, notamment le module sécant et la perte de linéarité.

Les éprouvettes sont obtenues par carottage dans des plaques.

On applique à l'éprouvette des déformations dont la loi de variation en fonction du temps est de la forme :

$Eps = a.t^n$, avec :

a = coefficient correspondant aux vitesses

n = coefficient propre à la machine

On effectue d'abord des essais dans le domaine des petites déformations (pas d'endommagement de l'éprouvette) pour définir des modules qui sont fonction du temps de charge et de la température, puis un essai à 0°C dans un domaine de grandes déformations, pour en déduire une "perte de linéarité" (perte de rigidité définie par rapport au comportement viscoélastique linéaire).

L'ensemble du processus opératoire exécutable sur une seule éprouvette
(Ø = 80 mm et H = 200 mm), tel qu'il a été défini à l'origine par le L.C.P.C.,
comporte les séquences suivantes :
- Modules sécants

Températures en °C   Temps de charge en secondes
20, 10, 0 et -10°C   1, 3, 10, 30, 100, 300

Les déformations maximales, Eps max, programmées par la machine pour les
différentes températures, tiennent compte de la teneur et de la pénétration du
bitume de l'enrobé.

Ces paramètres, ainsi que le temps de charge et les températures sont rentrés
dans le programme de pilotage de l'essai.
- Non-Linéarité

Température	Temps de charge	Déformations successives
0°C	30 ou 300 sec.	50, 100, 200, 500 microdef

Le temps de charge, la température et les déformations successives sont les
paramètres à programmer pour exécuter l'essai.

1.1.2 La Machine Asservie d'Essais Rhéologiques (MAER)

La presse MAER est représentée par le Graphique 1. La machine est totalement
pilotée par un micro-ordinateur Apple.

La machine d'essai (MAER) comportant un asservissement extensométrique,
permet d'effectuer des essais à vitesse de déformation constante, et d'assurer une
période de recouvrance de manière à ce que les contraintes se relaxent entre
chaque essai.

Graphique 1 : PRESSE M.A.E.R.

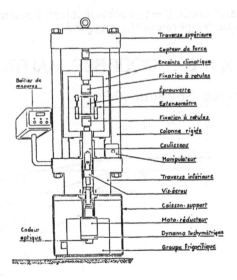

1.1.3 Interprétation des résultats

1.1.3.a Module sécant uniaxial

Pour chaque température, on détermine une série de modules sécants à amplitude
de déformation constante. Ces modules sont fonctions du temps uniquement :

$$S_{(t)} = \frac{\sigma(t)}{\varepsilon_0} \quad \text{avec :}$$

$\sigma(t)$ = contrainte longitudinale
$\varepsilon_{(o)}$ = déformation longitudinale imposée

Le niveau de déformation imposé, ε_0, est très petit, de manière à n'infliger qu'un endommagement négligeable à l'éprouvette.

Le schéma de principe de la détermination du module est illustré par le Graphique 2.

Graph 2 : ESSAIS DE MODULE

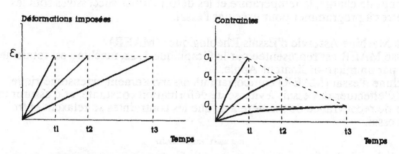

A partir des différents essais, il est possible d'obtenir l'ensemble des courbes maîtresses pour chaque température (Graphique 3).

Graph 3 : COURBES MAITRESSES

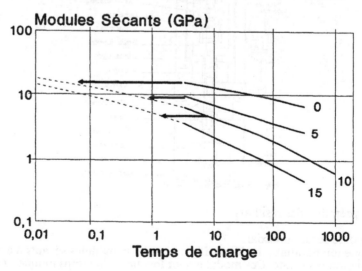

1.1.3.b Théorie de la non-linéarité - Prévision de la tenue en fatigue

Dans le domaine des très petites déformations (Eps $< 10^{-4}$), les enrobés bitumineux peuvent être considérés comme viscoélastiques linéaires. Ceci n'est plus le cas dans le domaine des grandes déformations, où il y a une "perte de linéarité" qui est définie comme une perte relative de rigidité en fonction de la déformation pour un temps fixé :

Γ : est appelé facteur de non-linéarité

1-Γ: est le perte de linéarité

Pour caractériser 1-Γ, on réalise la suite des amplitudes de déformations précisées plus haut. On définit alors la suite des modules sécants :

$$S_{(\varepsilon i)} = \frac{\sigma_i}{\varepsilon_i}$$

d'où on déduit $S_{(\varepsilon = 0)}$ par extrapolation.

La perte de linéarité est : $1 - \Gamma) = \dfrac{S(\varepsilon = 0) - S^{(\varepsilon = 5.10^{-4})}}{S_{(\varepsilon = 0)}}$

Le schéma de principe de la détermination de la perte de non-linéarité (1-Γ) est donné dans le Graphique 4.

Graph 4 : ESSAI DE NON-LINEARITE

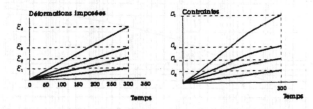

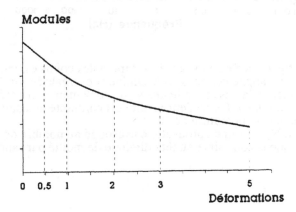

1.2 L'essai de module complexe en flexion alternée - Rappel des points essentiels

Rappelons brièvement que l'éprouvette, de forme trapézoïdale, encastrée à sa base, est sollicitée sinusoïdalement à son extrémité libre.

Si l'on applique à cette éprouvette une déformation sinusoïdale, la contrainte résultante est de même fréquence, mais déphasée par rapport à cette déformation.

Le module complexe est :

$$E^* = \frac{\sigma}{\varepsilon} = \frac{\sigma}{\varepsilon_0}e^{i\varphi} \qquad (\varphi = \text{angle de déphasage})$$

Les conditions d'essai concernent la température (de -10°C à +40°C) et la fréquence (1, 3, 10, 30 Hz), la déformation étant constante.

L'interprétation des résultats se base sur la représentation graphique de la courbe maîtresse du module (Graphique 5).

Graph 5 :
COURBE MAITRESSE DU MODULE COMPLEXE

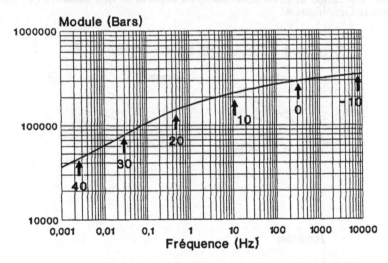

1.3 L'essai de fatigue par flexion alternée - Rappel des points essentiels

La simulation de la fatigue des enrobés bitumineux est réalisée sur des éprouvettes trapézoïdales collées à la base et sollicitée sinusoïdalement à une fréquence de 25 Hz. La déformation en tête de l'éprouvette est constante au cours de l'essai. L'essai s'effectue à la température de 10°C.

La durée de vie Nj pour une éprouvette correspond au nombre de chargements nécessaire pour que la contrainte en tête diminue de moitié par rapport à sa valeur initiale.

La connaissance de la durée de vie de plusieurs lots d'éprouvettes soumis à des déformations différentes permet de déterminer la droite de fatigue de l'enrobé (Graphique 6).

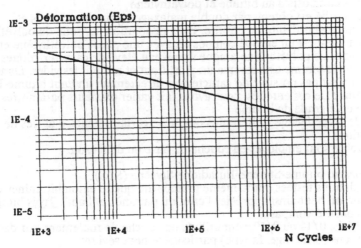

Graph 6 : FATIGUE EN FLEXION ALTERNEE A DEFORMATION IMPOSEE
25 Hz - 10 C

L'équation de la droite de régression est de la forme :

$$\mathcal{E} = \mathcal{E}_6 \, (N/10^6)^p$$

avec $\mathcal{E}_6$: déformation admissible pour 10^6 cycles
    p : pente de la droite de fatigue
    N : nombre de cycles

2 LES DIFFERENTS TYPES D'ENROBES ETUDIES

Dans le cadre de ses études et recherches, le laboratoire de R.T.E. a testé en traction directe plusieurs types d'enrobés classiques et spéciaux.
   Les différents enrobés étudiés se répartissent ainsi :

2.1 Bétons bitumineux classiques
Nous avons pris, à titre de référence, un béton bitumineux 0/14 semi-grenu, type Recommandation SETRA-LCPC, au bitume 50/70.

2.2 Enrobé à haut module au bitume très dur
Il s'agit d'un enrobé 0/14, de granularité semi-grenue et à très haute teneur (6,3%) de bitume très dur (pénétrabilité à 25°C égale à 10). Ce type de formulation est utilisé en couche de base ou de liaison.

2.3 Enrobé à haut module au bitume additivé

L'additif employé est d'origine bitumineuse et se présente sous forme pulvérulente. Il peut être incorporé soit dans le bitume, en donnant un liant prêt à l'emploi, soit directement dans l'enrobé par double malaxage.

Cet additif augmente considérablement le module du liant, donc de l'enrobé. L'enrobé étudié est un 0/14 semi-grenu, à très forte teneur en liant (6%).

Les enrobés de cette famille s'utilisent en couche de base ou de liaison.

2.4 Enrobés composites au bitume et polyéthylène

Le type d'enrobé est obtenu par double malaxage : granulats chauds d'abord, bitume ensuite. Il y a dissolution partielle du polyéthylène et le matériau final est un enrobé composite, avec une phase polyéthylène et une phase bitume. Le polyéthylène amène un niveau élevé de caractéristiques mécaniques; le bitume agit comme un plastifiant et remplit les vides. Dans ce procédé, il n'est pas possible de caractériser directement un liant bitume-PE.

Pour l'étude de ces enrobés, nous avons fait varier plusieurs paramètres :
- l'origine et le grade du bitume
- le type de polyéthylène (différentes origines : certains neufs, d'autres de récupération)

Ces enrobés sont employés en couche de base ou de liaison.

2.5 Enrobés au bitume-Styrène-Butadiène-Styrène (SBS)

Les formulations testées ont la même granularité que le béton bitumineux au bitume pur; on peut ainsi apprécier l'effet du paramètre "liant". <u>Trois</u> bitumes-SBS différents ont été comparés.

Ces enrobés, 0/14 ou 0/10, sont utilisés en couche de roulement, sur des sections à sollicitations (orniérage, fatigue) particulièrement sévères.

2.6 Enrobés au bitume et fibres

Les fibres sont d'origine minérale (roche ou verre). Elles sont très fines et courtes. Leur action est double : d'une part, grâce à leur surface spécifique très importante, elles fixent une quantité appréciable de bitume, d'autre part, elles ont un effet de micro-armatures dans le mortier.

Les enrobés correspondants ont donc une teneur en liant particulièrement élevée : 7,2% pour la formulation 0/10 étudiée ici.

Les enrobés au bitume-fibres s'utilisent en couches de roulement minces.

Le détail des formulations des enrobés étudiés figure dans le Tableau 1. Les caractéristiques des liants sont regroupées dans le Tableau 2.

TABLEAU 1 : FORMULATIONS ETUDIEES

TYPES ENROBES	CLASSIQUE	BITUME TRES DUR	BITUME + ADD SOLIDE	BITUME + P.E		BITUME + S.B.S	+ FIBRES
GRANULARITE	0/14	0/14	0/14	0/14	0/10	0/14	0/10
NATURE	DIORITE	CALCAIRE	CALCAIRE	CALCAIRE		DIORITE	DIORITE
GRANULOMETRIE							
14mm	95	92	92	96	100	95	100
10	72	72	72	73	96	72	90
6,3	50	55	55	50	54	50	49
2,0	32	32	32	30	34	32	32
0,08	8	8	8	6	7	8	13
LIANT + ADDITIF	5,6	6,3	6,0	6,1	5,6	5,6	7,2
LIANT N°	1	6	7	1-2-3-4-5-6-8		9-10-11	1
ADDITIFS	---	---	E	A-B-C-D		---	F

TABLEAU 2 : CARACTERISTIQUES DES LIANTS

LIANT N°	1	2	3	4	5	6	7	8	9	10	11
NATURE	BITUME	BITUME	BITUME	BITUME	BITUME	BITUME DUR	BITUME ADDITI SOLIDE	BITUME-PE PRET A L' EMPLOI	BITUME S.B.S	BITUME S.B.S	BITUME S.B.S
PENE à 25°C	60	45	25	58	35	10	11	31	40	45	48
PENE à 40°C	299	203	100	278	170	29	39	112	129	162	223
T.B.A (°C)	48	53	62	51	55	87	75	63	110	90	63
I.P P40°/25°C	-1	-0,6	0	-0,8	-0,9	+1,8	+0,6	+0,5	+1,1	+0,5	-0,7
Pt de FRAASS	-11	-10	-8,5	-12	-10	-5	-1,5	-10	-9	-9	-12

ADDITIFS	A	B	C	D	E	F
NATURE	P.E de RECUPERATION	IDEM	IDEM	P.E NEUF	ADDITIF BITUMINEUX PULVERULENT	FIBRES MINERALES

3 MESURES DE MODULE EN TRACTION DIRECTE - COMPARAISONS AVEC LE MODULE DYNAMIQUE EN FLEXION

L'ensemble des résultats obtenus est regroupé dans les Tableaux 3 et 4.

TABLEAU 3 :   MODULES EN T.D et FLEXION ALTERNEE
ENROBES COMPOSITES AU BITUME-POLYETHYLENE

FORMULE N°	1	2	3	4	5	6	7	8	9	10	11
LIANT N°	1	2	3	6	1	1	1	1	8	4	5
PENETRABILITE	60	45	25	10	60	60	60	60		58	35
BITUME-PE (%)	6.1	6.1	6.1	6.1	6.1	6.1	6.1	5.6	6.3	6.1	6.1
TYPE de P.E	A	A	A	A	B	C	D	A		A	A
C%	97,9	97,2	97,5	96,8	97,6	96,9	96,8	96,0	98,7	97,0	96,9

M.A.E.R - T.D (RESULTATS DES MODULES SECANTS EN GPa)

Sm 10°C-0,02s	21,0	23,0	23,7	25,0	20,5	17,6	17,3	19,0	15,8	18,0	21,5
Sm 15°C-0,02s	17,4	20,0	21,0	22,0	18,4	16,3	15,3	16,0	13,1	16,5	18,5

FLEXION ALTERNEE (MODULES COMPLEXES E* en GPa)

E* 10°C-10Hz	21,2				19,3				20,9	17,3	19,8	21,5
E* 15°C-10Hz	17,6				16,2				16,6	13,9	16,4	18,1

TABLEAU 4 :   MODULES EN T.D et FLEXION ALTERNEE
AUTRES FORMULATIONS

FORMULE N°	12	13	14	15	16	17	18
TYPES D'ENROBES	CLASSIQUE	BITUME TRES DUR	BITUME + ADD SOLID	BITUME + S.B.S	BITUME + S.B.S	BITUME + S.B.S	BITUME + FIBRES
LIANT N°	1	6	7	9	10	11	1
PENETRABILITE	60	10	11	40	45	48	60
BITUME-ADDITIF (%)	5,6	6,3	6,0	5,6	5,6	5,6	7,2
TYPE ADDITIF	--	--	E	--	--	--	F
C%	96,8	98,2	98,0	95,4	95,9	95,9	98,0

M.A.E.R - T.D (RESULTATS DES MODULES SECANTS EN GPa)

Sm 10°C-0,02s	7,5	21,1	23,8	13,9	13,5	13,0	12,6
Sm 15°C-0,02s	5,8	18,2	22,0	9,5	9,1	8,5	9,5

FLEXION ALTERNEE (MODULES COMPLEXES E* en GPa)

E* 10°C-10Hz	7,2	21,4	27,2		10,9		11,5
E* 15°C-10Hz	5,4	17,9	22,8		8.5		8,5

3.1 Béton bitumineux classique

Nous ne développerons pas ici les résultats obtenus sur enrobés classiques, qui ont fait l'objet de diverses publications (voir la bibliographie).

Notons simplement que les valeurs trouvées sur la formulation 0/14 étudiée comme référence à R.T.E. confirment que l'essai en traction directe a un bon pouvoir prédictif du module dynamique en flexion tel qu'il est rentré dans les calculs de dimensionnement :
- Module dynamique en flexion à 15°C - 10 Hz : 5,4 GPa
- Module sécant en T.D. à 15°C - 0,02 s : 5,8 GPa
- Module dynamique en flexion à 10°C - 10 Hz : 7,2 GPa
- Module sécant en T.D. à 10°C - 0,02 s : 7,5 GPa

3.2 Enrobé à haut module au bitume très dur

3.2.1 Adaptation du mode opératoire

Notre expérience des essais de traction directe sur enrobés à haut module nous a très vite conduits à modifier le mode opératoire.

Nous avons en effet constaté que l'application de déformations relatives de 10^{-4}, avec les temps de charge de la méthode standard, entraîne, pour ce type d'enrobé, un niveau d'endommagement excessif. Celui-ci se traduit, de manière générale, par des valeurs apparentes de module faibles, voire par des ruptures d'éprouvettes lors des essais aux températures les plus basses (0 et 5°C). Par ailleurs, le mode opératoire "standard" ne permet pas, avec ces enrobés, de retomber sur les valeurs du module complexe mesuré en flexion alternée.

A la suite de ces observations, nous avons, en collaboration avec le L.C.P.C. et le Laboratoire d'Angers, modifié le mode opératoire "standard", à savoir :
- choix d'un comportement supposé de l'enrobé testé avant paramètrage de l'essai :
  . "Très souple"
  . "Souple"
  . "Normal" (essai standard)
  . "Rigide"
  . "Très rigide"

Cette procédure permet à la machine MAER de programmer les déformations relatives, les temps de charge, le type de mors (avec ou sans jeu) et de peson à utiliser en fonction de la rigidité de l'enrobé.

En comportement "rigide" ou "très rigide", le programme impose une déformation relative réduite de moitié (50 microdef au lieu de 100 microdef) et un temps de charge divisé par 2.

Inversement, en comportement "souple" ou "très souple", le programme sélectionne la déformation relative double (200 microdef au lieu de 100) et un temps de charge double également.

Il est à noter que d'autres laboratoires ont procédé à des modifications analogues du mode opératoire.

3.2.2 Résultats

Moyennant les modifications exposées ci-dessus, l'essai de traction directe permet de prédire correctement le module complexe "de dimensionnement" des enrobés au bitume très dur. Voici un exemple de résultats typiques :
- Module dynamique en flexion à 15°C - 10 Hz : 17,9 GPa
- Module sécant en T.D. à 15°C - 0,02 s : 18,2 GPa
- Module dynamique en flexion à 10°C - 10 Hz : 21,4 GPa
- Module sécant en T.D. à 10°C - 0,02 s : 21,1 GPa

Les courbes maîtresses de module en T.D. et en flexion alternée sont données dans les Graphiques 7a et 7b respectivement.

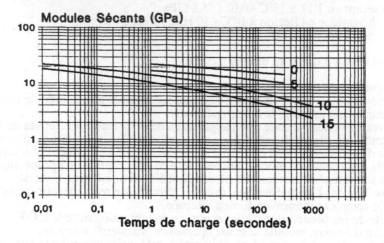

Graph 7a : TRACTION DIRECTE - M.A.E.R COURBES MAITRESSES ENROBES AU BITUME TRES DUR

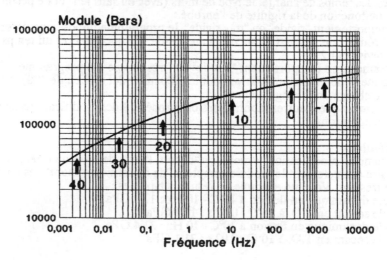

Graph 7b: COURBE MAITRESSE DU MODULE COMPLEXE ENROBES AU BITUME TRES DUR

3.3 Enrobé à haut module au bitume additivé

Les adaptations au mode opératoire décrites en 3.2.1 s'imposent pour ce type d'enrobé.

La valeur de module uniaxial en T.D. apparaît comme une bonne prédiction du module complexe :
- Module dynamique en flexion à 15°C - 10 Hz : 22,8 GPa
- Module sécant en T.D. à 15°C - 0,02 s : 22,0 GPa

A 10°C, l'écart trouvé entre les deux modalités est relativement faible (moins de 15%). Ils convient de noter que ce type d'enrobé atteint des niveaux de rigidité très élevés et que la dispersion des mesures a alors tendance à augmenter.

3.4 Enrobés composites au bitume et polyéthylène

Ces enrobés ont des modules élevés. Les adaptations du mode opératoire de traction directe au paragraphe 3.2.1 sont donc nécessaires pour leur évaluation.

Les résultats présentés ici sont extraits d'une importante étude, qui a établi l'influence de divers paramètres, tels que l'origine et la pénétrabilité du bitume, l'origine et la forme du polyéthylène, leurs dosages, etc. Cette étude a confirmé que l'essai de traction directe rend bien compte des variations de composition (granularité, teneur en liant, consistance du bitume, etc.).

A titre d'exemple, le graphique N° 8 illustre la loi de variation du module uniaxial en fonction de la pénétrabilité du bitume, ceci avec le polyéthylène couramment utilisé.

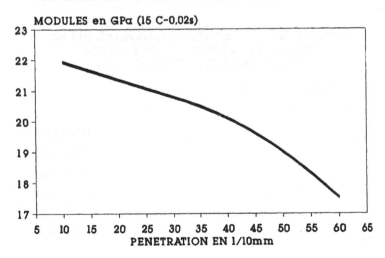

Graph 8: ENROBES COMPOSITES BITUME-PE INFLUENCE DE LA PENETRABILITE DU BITUME

D'une façon générale, on a trouvé pour ces enrobés au bitume et polyéthylène une très bonne corrélation à 15°C entre module sécant en T.D. à 0,02 s et module dynamique en flexion à 10 Hz (Graphique 9). Par contre, à 10°C, la corrélation n'est que moyenne (Graphique 10).

Graph 9: ENROBES COMPOSITES
CORRELATION ENTRE T.D (15 C - 0,02s)
ET FLEXION ALTERNEE (15 C - 10Hz)

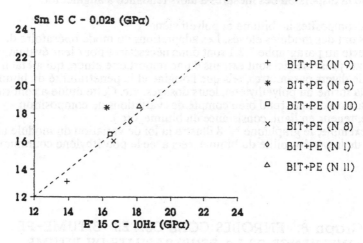

Sm 15 C - 0,02s (GPα)

+ BIT+PE (N 9)
* BIT+PE (N 5)
□ BIT+PE (N 10)
× BIT+PE (N 8)
◇ BIT+PE (N 1)
△ BIT+PE (N 11)

E° 15 C - 10Hz (GPα)

Graph 10: ENROBES COMPOSITES
CORRELATION ENTRE T.D (10 C - 0,02s)
ET FLEXION ALTERNEE (10 C - 10Hz)

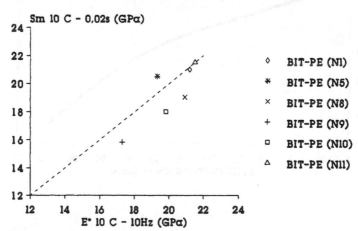

Sm 10 C - 0,02s (GPα)

◇ BIT-PE (N1)
* BIT-PE (N5)
× BIT-PE (N8)
+ BIT-PE (N9)
□ BIT-PE (N10)
△ BIT-PE (N11)

E° 10 C - 10Hz (GPα)

Voici un exemple de résultats obtenus sur un enrobé composite au bitume et polyéthylène le plus couramment utilisé :
- Module dynamique en flexion à 15°C - 10 Hz : 17,6 GPa
- Module sécant en T.D. à 15°C - 0,02 s : 17,4 GPa

On notera au passage que la technologie du liant bitume-polyéthylène préparé à l'avance donne des résultats de module nettement inférieurs à ceux du double malaxage.

3.5 Enrobés au bitume-SBS
Les adaptations au mode opératoire décrites en 3.2.1 ont conduit à réaliser les essais de T.D. en choisissant le comportement "souple".

La valeur du module uniaxial en T.D. donne à 15°C une bonne prévision du module complexe :
- Module dynamique en flexion à 15°C - 10 Hz : 8,5 GPa
- Module sécant en T.D. à 15°C - 0,02 s : 9,1 GPa

L'écart entre les deux modalités est relativement faible (<10%). Par contre, à 10°C, l'écart est notable. Il convient de noter que l'enrobé étudié présente des niveaux de rigidité plus élevés que ceux d'un béton bitumineux au bitume pur.

3.6 Enrobés au bitume et fibres
Comme pour les enrobés au bitume-SBS, la modification du mode opératoire a permis de réaliser l'essai de T.D. avec un comportement supposé "souple".

Là aussi, cette procédure permet d'obtenir une assez bonne prédiction du module complexe :
- Module dynamique en flexion à 15°C - 10 Hz : 8,5 GPa
- Module sécant à 15°C - 0,02 s : 9,5 GPa

Le Graphique 11 regroupe tous les résultats obtenus sur les différents enrobés étudiés. Il confirme la validité de la prévision du module "de dimensionnement" à partir de l'essai de traction directe pour des types très divers d'enrobés.

On notera que la modalité qui donne la meilleure corrélation avec le module complexe à 15°C - 10 Hz est celle à 15°C - 0,02 s (et non pas 10°C - 0,02 s, comme il avait été proposé précédemment par le L.C.P.C.

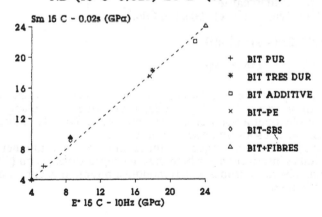

Graph 11: CORRELATION ENTRE
T.D (15 C-0,02s) ET E° (15 C-10Hz)

4 ESTIMATION DE LA TENUE EN FATIGUE PAR LA NON-LINEARITE - COMPARAISON AVEC LA FATIGUE EN FLEXION

L'ensemble des résultats est récapitulé dans le Tableau 5.

TABLEAU 5 : ESTIMATION FATIGUE EN T.D et FLEXION ALTERNEE
AUTRES FORMULATIONS

FORMULE N°	12	13	14	1	5	8	9	10	11	15	18
TYPES D'ENROBES	CLASSIQUE	BITUME TRES DUR	BITUME + ADD SOLID	BITUME + POLYETHYLENE						BITUME + S.B.S	BITUME + FIBRES
LIANT N°	1	6	7	1	1	1	8	4	5	9	1
PENETRABILITE	60	10	11	60	60	60	31	58	35	40	60
BITUME-ADDITIF (%)	5,6	6,3	6,0	6,1	6,1	5,6	6,3	6,1	6,1	5,6	7,2
TYPE ADDITIF	--	--	E	A	B	A	--	A	A	--	F
C%	96,8	98,2	98,0	97,9	97,6	96,0	98,7	97,0	96,9	95,4	98,0
M.A.E.R - T.D (RESULTATS DES Eps6 en 10-6)											
1 - Gamma	0,298	0,304	0,300	0,435	0,339	0,316	0,344	0,441	0,392	0,403	0,388
Eps6 corrélation (1)	155	248	255	162	219	210	155	165	222	113	122
Eps6 corrélation (2)	135	133	134	88	121	129	119	86	103	99	104
Eps6 corrélation (3)	140	139	140	96	127	135	125	94	110	106	111
FLEXION ALTERNEE (RESULTATS DES Eps6 en 10-6)											
Eps6 (Flexion)	135	148	139	118	139	102	105	131	124	165	184

Trois estimations différentes de la déformation relative en flexion ε_6 à partir de la non-linéarité ont été successivement proposées par le réseau technique des Laboratoires de l'Equipement pour les enrobés semi-grenus classiques (Références bibliographiques), à savoir :

(1) $\varepsilon_{6T} = 10^{-6} [240 - 445 (1 - \Gamma) + 12,4 S]$
où 1-Γ est la perte de linéarité
S est le module sécant à 0°C et 300 s (en GPa)

(2) $\varepsilon_{6T} = 10^{-6} [238 - 345 (1 - \Gamma)]$

(3) $\varepsilon_{6T} = 10^{-6} [239 - 330 (1 - \Gamma)]$

Il est clair que la première régression proposée n'est pas satisfaisante, dès qu'on sort de l'échantillonnage initial. D'après elle, il suffirait en effet d'augmenter le module d'un enrobé pour améliorer sa tenue en fatigue. Dans la réalité, ces deux caractéristiques varient souvent en sens contraire ...

Par contre, il apparaît que, lorsque le liant est uniquement bitumineux, nos quelques résultats s'inscrivent assez bien dans les régressions (2) ou (3). Cette constatation englobe nos enrobés à haut module à bitume très dur ou à additif d'origine bitumineuse.

Il convient toutefois de noter que, à la suite d'un important plan d'expérience, le L.C.P.C. a établi que la perte de linéarité ne dépend pratiquement pas de la structure du bitume (liée au brut utilisé et à la méthode de raffinage), alors que ce facteur a une influence importante sur la fatigue en flexion.

Par ailleurs, aucune des corrélations proposées pour les enrobés au bitume pur ne peut être étendue à un quelconque des enrobés contenant un additif non bitumineux, qu'il s'agisse de polyéthylène, de SBS ou de fibres.

De plus, les résultats obtenus montrent qu'aucune corrélation valable ne peut être établie entre déformation admissible en flexion et perte de linéarité en traction directe.

Pour les autres types d'enrobés (SBS, fibres), les résultats sont trop peu nombreux pour qu'on puisse savoir s'il existe des corrélations significatives.

CONCLUSIONS

D'une manière générale, l'essai de traction directe rend bien compte des variations de composition d'un enrobé. Cet essai, avec détermination du module sécant, s'avère un moyen fiable de prévision du module complexe "de dimensionnement".

La méthode apparaît valable non seulement pour les enrobés au bitume pur, mais aussi pour les enrobés spéciaux (additifs étudiés : pulvérulent bitumineux, polyéthylène, SBS, fibres).

Un aménagement de la procédure d'essai est nécessaire pour l'évaluation correcte des enrobés à haut module.

Par contre, la prévision de la tenue en fatigue par détermination de la perte de linéarité n'est pas possible dans l'état actuel des connaissances.

On notera enfin que le matériel va évoluer et voir ses possibilités étendues à d'autres essais, comme le fluage, la relaxation, les sollicitations cycliques, les essais triaxiaux et d'autres. Cette évolution apportera un intérêt supplémentaire au matériel actuel.

Bibliographie

- L. FRANCKEN - "Module complexe des bétons bitumineux" - Bulletin de Liaison des Laboratoires des Ponts et Chaussées - Spécial V - 1977
- F.H. DOAN - "Les études de fatigue des enrobés bitumineux" - Bulletin de Liaison des Laboratoires des Ponts et Chaussées - Spécial V - 1977
- R. LINDER - "Application de l'essai de traction directe aux enrobés bitumineux" - Bulletin de Liaison des Laboratoires des Ponts et Chaussées - Spécial V - 1977
- Rapport national français - "Essais mécaniques pratiques de formulation et de contrôles des enrobés bitumineux" - RILEM Belgrade - 1983
- R. LINDER, F. MOUTIER, M. PENET, F. PEYRET - "La machine d'essais rhéologiques asservie (MAER - LPC) et son utilisation pour l'essai de traction directe" - EUROBITUME La Haye - 1985
- F. MOUTIER - "Etude statistique de l'effet de la composition d'enrobés bitumineux sur le comportement en fatigue et le module complexe" - EUROBITUME Madrid - 1989
- J.P. SERFASS, P. BENSE - "Enrobés au bitume-polyéthylène : formulation et performances mécaniques" - EUROBITUME Madrid - 1989
- "Enrobés bitumineux - Perspectives d'avenir" - Journées d'information du L.C.P.C. - 12 - 13 décembre 1989 d'information

21 PRACTICAL MEASUREMENTS OF BITUMINOUS LAYER MODULI IN PAVEMENT STRUCTURES

A.A. SHAAT
Civil Engineering Department, The Queen's University, Belfast,
Northern Ireland

Abstract
Bituminous layer moduli play a principal role in the evaluation of
pavement serviceability and in the design of pavement overlay layer.
These data can be measured with various laboratory and field
techniques. This paper presents a theoretically rigorous approach for
backcalculating bituminous layer elastic moduli from the results of
non-destructive testing for measuring pavement deflection bowl under
load. A FORTRAN computer program has been coded, called EPLOPT, to
implement the method on a VAX computer. The author carried out an
extensive programme of field deflection measurements on several parts
of the highway network and performed laboratory testing and
theoretical analysis to evaluate the stiffness moduli of the
in-service bituminous layers. It is concluded that the combination of
the measured deflection data using both the Deflectograph and the
Falling Weight Deflectometer with the theory of Equivalent Layer
Thicknesses is much more efficient and field representative than
current laboratory methods.
Keywords: Flexible, Pavements, Mechanistic Design, Pavement Testing,
Nondestructive Testing, Laboratory Testing, Backcalculation,
Equivalent Layer Thickness, Pavement Deflections.

1   Introduction

In the analytical design of pavement structure there are two major
aspects of material properties which have to be considered. Firstly,
the stress-strain characteristics of the material used in the various
layers of the pavement structure need to be known so that analysis of
stress and displacement of the deformable medium due to external
environmental and loading conditions, can be performed. Secondly, it
is essential to define the mechanisms which lead to loss of
serviceability and eventual pavement failure. Each distress mechanism
must be described in terms of stresses, strains or deflections, so
that it can be incorporated into an analytical design procedure. The
materials used in the flexible and rigid pavement constructions fall
essentially into two categories, bound materials (e.g., bituminous or
Portland cement concrete), and unbound materials (e.g., crushed rock
or gravel).

For the last 30 years the Transport and Road Research Laboratory

(TRRL) has been studying experimental pavements in the laboratory and monitoring full scale road tracks, which has shown the modes of failure of flexible pavements and established how the various bituminous materials behave in the pavement.

The test methods used for collecting data on pavement material properties which used in evaluating the structural condition of a pavement are either destructive or nondestructive. The difference between them is normally dependent on physical disturbance of the pavement materials. The pavement destructive evaluation methods depend on field sampling and laboratory testing is not only expensive and time consuming but also lacks the consideration of materials, loading and environmental conditions variability. Also, it is difficult to select a laboratory test procedure which models exactly the field stress conditions, primarily because these are not known with certainty. In addition, a time-dependent strength loss occurs after sampling in pavement materials due to the relief of confining stresses. These factors must be considered when comparing the laboratory measured material parameters with those estimated using field tests. Most important, current destructive test methodologies severely affect the traffic movement and cause a lot of inconvenience to the road users.

Fortunately, in recent years, great strides have been made in the development of equipment that can rapidly and nondestructively collect data on which an estimate of the pavement material properties can be made with a reasonable degree of accuracy. Of particular importance in evaluation of flexible pavements was the development of the Falling Weight Deflectometer (FWD) and the British version of the Deflectograph. These devices measure the surface deflections produced from stationary impulsive load or from rolling wheel load, respectively. Simultaneously, researchers worldwide have been developing empirical and analytical models that could backcalculate the elastic moduli of pavement layers utilising the shape of deflection bowl. This is thus based on an overall 'System Approach'. Than analogies and differences between the Component Approach and the System Approach are illustrated in Figure 1.

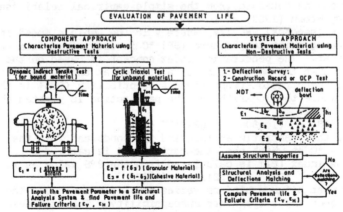

Fig. 1. APPROACHES FOR EVALUATION OF PAVEMENT LIFE

2   BITUMINOUS MATERIAL

Bituminous mixes are visco-elastic in their behaviour as are their bitumen binders. Bituminous materials are influenced by bitumen binder properties and aggregate characteristics. Some of the important variables influencing the properties of the bituminous mix are: loading time, temperature, bitumen binder properties, binder content in the mix, aggregate type and graduation, density of mix and air voids in the mix. Because of this, the response to stress application can be more completely described by a complex dynamic modulus, i.e.., the stress-strain relationship is dependent upon time of loading (t) and temperature of mix (T), as follows:

$$S_{(t,T)} = (stress/strain)_{t,T}$$

in which, S is the mix stiffness. In an elastic analysis, the elastic component of the complex dynamic behaviour is separated from the viscous component and referred to as the resilient or elastic modulus (E).

Bituminous mixes are essentially linear viscoelastic with a resilient modulus which is independent of stress level, and which decreases considerably with increasing temperature or decreasing loading time.

Two procedures are presented to determine bituminous mix stiffness.

2.1 Estimation of Bitumen and Mix Stiffness from Component Analysis

The determination of the bitumen stiffness according to the Van der Poel stiffness theory requires, the time of loading, the softening point temperature minus test temperature and penetration index of the bitumen binder. For the design of pavement materials the time of loading is related to the typical traffic speed, e.g. for speeds of 50-60 kph the time of loading is approximately 0.02 seconds. Where speeds are higher the time of loading is reduced. A reasonable estimation of loading time for bituminous layers between 100 and 350mm thickness can be obtained from the simple empirical relationship proposed by Brown (1980) as follows:

t (seconds) = 1/V, where V is the vehicle speed in kph.

Softening point temperature (SP) °C is determined from the ring and ball test. The penetration index concept is based on the assumption that the penetration of a bitumen at the SP temperature is about 800. Penetration-temperature susceptibility (PTS) is the slope of a line where the logarithm of the penetration is plotted against the temperature:

PTS = (log800 - logP)/(SP - T)

where P is the penetratin at 100g and 5 seconds and T is the temperature °C at which the penetration test is carried out. The penetration index (PI) is calculated from the PTS value as follows:

PI = (20 - 500PTS)/(1 + 50PTS)

The PI of most bitumen binders varies from -2.6 to +8.0. The lower PI, the higher the temperature susceptibility. The stiffness of the bitumen, using the above mentioned parameters, can be estimated using the well known nomograph developed by Van der Poel (1954).

As an alternative to using the Van der Poel nomograph, a simplified equation (Ullidtz, 1979) can be used to calculate binder stiffness extracted from recovered cores:

$$S_b = 1.157 \times 10^{-7} \, t^{-0.368} \, 2.718^{-PI} \, (SP - T)^5$$

where S_b is the bitumen stiffness (MPa), t is the time of loading in seconds (0.01 < t < 0.1), PI is the penetration index of recovered bitumen (-1 < t < +1), SP is the softening point ring and ball in °C [20 < (SP - T) < 60] and T is the bitumen temperature °C.

The PI and SP are calculated from the following equations:

PI = (20SP + 500LogP - 1951.55)/(SP - 50LogP + 120.15)
SP = 99.13 - 26.35LogP

where P is the binder penetration measured at 25°C. To measure the penetration of the extracted bitumen, 100g of it has to be recovered and then subjected to the standard penetration test at 25°C. This quantity has proved to be difficult and time-consuming to extract from a core and it is also expensive to do so.

The sliding plate viscometer (Fig. 2) was used to determine the viscosity of bitumens extracted from different cores at temperature of 40°C (Shaat, 1989a). This increased test temperature was used as it was anticipated that the majority of bitumens to be examined would be quite hard (aged bitumen), and it was not possible to measure their viscosity at lower temperature, such as 21°C used by the Asphalt Institute, since the recommended load that could be applied in the

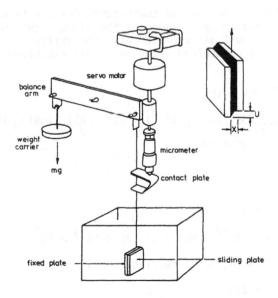

Fig. 2. SCHEMATIC DIAGRAM OF THE OPERATING MECHANISM OF A SLIDING PLATE MICROVISCOMETER (FROM HUET AND ISTA, 1985)

309

microviscometer was too small to develop a constant shear rate. Thus a relationship was established between penetration as measured at 25°C and the viscosity of the same bitumens at 40°C using the microviscometer (McKibbin, 1987):

$$P = 16264.557/(\log V)^{3.96}$$

where, P is the penetration at 25°C, and V is the viscosity at 40°C (Poises) measured by the microviscometer. The above equation has a correlation coefficient (R^2) of 0.987 and valid for penetrations at 25°C between 10 and 530 for non-blown bitumens.

The procedure for determining the bitumen and mix stiffness from laboratory component analysis was to separate the bitumen from the aggregate for each core recovered from each site and perform a series of tests according to the following steps:

(1) The bulk density of the core was measured.
(2) Representative samples of mix were taken from different parts of the surfacing layer.
(3) A solvent was added to each sample for 2 hours.
(4) A filtration process was carried out on the dissolved sample to remove aggregate and fines.
(5) The solvent mixture (solvent + bitumen) were heated to evaporate the solvent.
(6) A sieve analysis and relative density test were carried out on the recovered aggregate, fines and filler.
(7) Sliding plate microviscometer tests at 40°C were carried out on the recovered bitumens.
(8) The binder content and air void content were determined in each mix.

With the stiffness modulus of the bitumen derived from the above method and the volumetric proportions of the mix determined by measurement of the actual mix constituents, the stiffness of the bituminous mix can be determined using the shell equations (Bonnaure et al, 1977):

For $5 < S_b < 1000$ (MPa)

$$\text{Log } S_m = \frac{(S1 + S4)}{2} \log(S_b.10^{-8}) + \frac{(S4 - S1)}{2} \log(S_b.10^{-8}) + S2$$

For $1000 < S_b < 3000$ (MPa)

$$\log S_m = S2 + S4 + 2.096(S3 - S2 - S4) \log(s_b.10^{-9})$$

where:

$$S1 = 0.6 \log[(1.37V_b^2 - 1)/(1.33V_b - 1)]$$

$$S2 = 8 + (5.68 \times 10^{-3} \times Vg) + (2.135 \times 10^{-4} \times Vg^2)$$

$$S3 = 10.82 - [(1.342(100 - Vg)/(Vg + Vb)]$$

$$S4 = 0.758(S3 - S2)$$

To calculate the percentage volume of bitumen (V_b) and that of aggregate (V_g), for substituting in these equations, from the commonly specified values of percentage weight of bitumen (W_b), percentage of aire voids (Va), and specific gravities of bitumen (SG_b) and aggregate (SGg) the following equations are used:

$$Vb = \frac{(1 - V_a)}{1 + \frac{SG_b}{SG_g}\left(\frac{100}{W_b} - 1\right)} \times 100$$

$$V_g = [1 - (V_a + V_b)] \times 100$$

where V_a, V_g and V_b are practical volumes of air, bitumen and aggregate respectively.

Based on this theoretical developments, a computer program (DYSTIF) was developed to compute the stiffness modulus of bitumen binders and bituminous mixes. Typical bitumen and mix stiffnesses, computed by DYSTIF for several recovered cores from the highway network in N. Ireland are given in Table 1.

Table 1. Results of Mix Component Analysis Programme.

Site No.	Core Density Mg/M^3	Void in Mix (%)	Binder Content (%)	Viscosity Poises x 10^6	Pentr-ation	Bitumen Stiff (MPa)	Mix Stiff (MPa)
6	2.284	1.93	8.2	0.692	15	43.2	3050
7	2.404	0.00	8.0	0.035	41	13.0	1415
9	2.320	1.11	7.7	0.200	22	28.5	2606
10	2.324	0.94	8.0	0.179	23	27.3	2381
11	2.386	0.91	8.0	0.439	17	38.5	3125
13	2.288	3.26	7.6	0.416	18	37.0	2794
15	2.344	1.84	6.9	1.197	13	53.5	4783
17	2.300	3.77	8.3	0.485	17	40.3	2477

2.2 Estimation of Mix Stiffness from System Analysis

The System Approach involves the backcalculation of material properties based on nondestructive testing (NDT) on the full scale pavement by measuring the deflection response to a known applied load. The determination of mix stiffness using NDT techniques involves the following steps:

(a) Application of an NDT load of known magnitude, geometry and duration.

(b) Measurement of the response deflection bowl and layer thicknesses in the pavement structure.

(c) Backcalculation of the material properties (stiffnesses) of the layers using a selected analysis model (e.g. elastic theory).

In the evaluation process used in the system approach, the response of the pavement is measured and the material properties are then backcalculated. On the basis of these properties, a "theoretical" response is determined and compared with the measured response. The structural evaluation of a pavement by the System Approach is considered as an inverse design process. Among the different load responses, only the surface deflections are easily measurable by means of nondestructive field testing methods. These methods have many advantages over laboratory testing programmes, e.g. the nondestructive nature of the test, the full-scale in-situ loading of the materials, and the inexpensive and rapid testing relative to that based on Component Approach.

These advantages make the system approach an economic and useful tool for measuring the material properties in-situ with minimum inconvenience to the traffic and road users.

Deflection Measuring Equipment
The current deflection measuring equipment using NDT techniques are classified under three groups: (a) Rolling Wheel Techniques, Fixed Point Techniques and Wave Propagation Techniques. A review of these methods has been presented elsewhere by Shaat (1989a). The procedure for bituminous mix evaluation presented herein is based on the measurement of deflection bowl produced from a Fixed Point Technique well known as the Falling Weight Deflectometer (FWD).

The principle of the FWD, schematically shown in Figure 3, is that a mass with a given weight drops from a certain height on a number of parallel rubber springs that are mounted on a 300 mm diameter circular footplate. The load pulse generated in this way aries depending on the falling weight and drop height. For road pavements a typical load level between 40 and 60 kN is normally used. The duration of the load pulse is of the order of 30 msec which is dependent on the geometry of the pavement structure, rubber springs and the stiffness of the loading plate. The deflections are measured by means of six velocity transducers (Geophones) mounted on a beam which is lowered automatically with the loading plate. The beam places the six goephones at locations up to 1270mm from the centre of the loading plate.

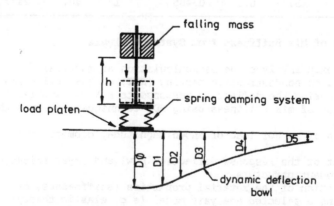

Fig. 3. SCHEMATIC OF FALLING WEIGHT DEFLECTOMETER

The information from the geophones are converted to the actual vertical displacements and is fed into a microcomputer based control and registration unit.

The FWD load technique has a number of advantages over other NDT techniques, the most significant being that the load pulse and stress levels are broadly comparable with normal lorry traffic, also that no reference level is needed to measure the deflections. The main disadvantages of the FWD are, the loading duration produced by the impulse load is constant with depth of pavement, and the FWD cannot take measurements at sufficiently close spacings.

Deflection Measurements for Structural Diagnosis

In order to detect localised defects in the highway network it is important to use a high-output deflection measuring device which can provide the maintenance engineer with global view at the structural condition of the tested pavement in terms of the variability of the maximum deflection.

The Deflectograph (Shaat, 1988) is NDT device which realistically represents the moving design wheel load. The Deflectograph has the ability to take continuous measurements, one deflection measurement on the edge and in the centre line every 3 metres, therefore providing nearly continuous monitoring of the highway network structural condition. Whereas, with the FWD, to cover the same length of pavement, one would have to settle for one measurement every 50 to 80 metres, coving only one line. Therefore, the first step in the structural diagnosis is based on the deflection survey using the Deflectograph. The defected areas are located from the high deflection values and a more detailed examination of such areas is carried out using the FWD. The sole objective of this method is to establish an order of priority for the maintenance of the sections of the network and to confine the use of the FWD. Having measured the deflection bowls and the layer thicknesses, the determination of the stiffness moduli from these information becomes solving an inverse design problem.

3 Backcalculation of In-Situ Pavement Elastic Moduli

The determination of in-situ pavement moduli based on measured performance parameters such as the deflection bowl is a field of growing interest for maintenance engineers since it involves NDT techniques of multi-layered systems. No direct theoretical solution exists giving the elastic moduli for the various layers in a pavement structure from measured surface deflection bowls. Such a procedure has to be an inverse technique that is opposite to the conventional approach which calculates the deflection bowl from given layer moduli. The inverse procedure is necessarily an iterative or trial and error process.

The extensive findings of the parametric study and sensitivity analysis (Shaat, 1989a) of 300 typical pavement structures resulted in the formulation of an automatic iterative methodology for determining a set of layer elastic moduli based on the best fit technique using least squares theory. This methodology relies on generating theoretical deflection bowls using simplified elastic theory (MELT)

(Peattie and Ullidts, 1981). The initial values of assumed moduli are
then changed using a procedure of successive correction in order to
obtain a best fit to the deflection bowl measured by the Deflectograph
or the FWD. This has been achieved using the finite elements model
DEFPAV (Shaat et al, 1988), and the layer moduli were determined from
the measured deflection bowls by a successive trial and error
technique. This involved choosing a set of initial modulus values,
computing the resulting absolute deflection bowl from these moduli
adjusting this bowl to the recorded Deflectograph deflections, and
then repeating the entire procedure until the computed deflection bowl
matched the measured one within the required accuracy. A disadvantage
of this method was found to be the length of time required to
establish a good element diagram and prepare the vast amount of data
for this trial and error procedure. This is considered impractical
for a maintenance engineer to handle especially when the procedure is
to be used with a high output deflection measuring equipment such as
the Deflectograph. Therefore a new structural model (DUAL) was
developed based on the Method of Equivalent Layer Thickness (MELT)
(Shaat, 1989). Confidence in using DUAL was gained by checking its
results against those from the elaborate and expensive finite elements
model DEFPAV. However it must be emphasised that for very special
sites (e.g. Airports) where a very detailed analysis is required, the
DEFPAV model may need to be used.

As an alternative to the finite element progam DEFPAV, which uses
the classical elastic theory, the multilayer elastic structure is
transformed into an equivalent semi-infinite space which allows the
use of Boussinesq's equation for stress, strain and deflection
computation. This transformation is achieved by using the method of
equivalent thicknesses.

The basic assumption is that the stresses, strains, and
deflections below a layer will be unchanged as long as the flexural
stiffness of the layer remains constant. The flexural stiffness is a
function of the cube of the thickness of the layer and its modulus of
elasticity, (E) and Poisson's ratio (m). If two layers of moduli E_1
and E_2 and thicknesses h_1, h_2, respectively, are to have the same
flexural stiffness then:-

$$h_1^3 \times E_1/(1 - m_1^2) = h_2^3 \times E_2/(1 - m_2^2)$$

Using this principle, a two-layer system may be transformed to a
semi-infinite space, provided that layer 1 is replaced by a thickness
(h_e) of material having the properties of the semi-infinite space, so
that:

$$h_e = h_1 \times E_1 (1 - m_2^2)/E_2(1 - m_1^2)$$

Boussinesq's equations are used to compute the stresses, strains
and deflections at the underside of interface. The method is
approximate and the results obtained will therefore deviate from those
using the more rigorous and fundamental layered elastic theory (e.g.
the finite element program DEFPAV). For better agreement with the
DEFPAV values, a correction factor, c, is introduced (Shaat, 1989a).
The equivalent thickness of the (n - 1) layers above layer n may then
be calculated from:

$$h_{e'n} = c \times \sum h_i \times [E_i(1 - m_n^2)/E_n(1 - m_i^2)]^{0.33}$$

It should be kept in mind that the correction factor (c) improves agreement with the finite element program DEFPAV only, but not necessarily with other methods or with actual responses in real pavement systems.

The simplicity of the MELT method makes it possible to use interactively, and to analyse the several thousand measurements of deflection bowls taken by the Deflectograph daily.

The analytical model (DUAL) was further developed and incorporated as a subroutine in a self-iterative computerised system called EPLOPT to determine the layer elastic moduli that provide the minimum error between measured deflection bowl and the computed theoretical bowl.

4   Matching Algorithms

The objective is to determine the set of elastic moduli (E_1, E_2, E_3) that will provide the best fit between the series of computed deflections (DC_1, DC_2, ..., DC_9) and the measured ones using the Deflectogrph or FWD (DM_1, DM_2, ..., DM_9). This was accomplished by incorporating the inter-relationship found in the parametric study between the surface deflection bowl parameters and the pavement layer elastic moduli, i.e.:

$$DC_j = f_j (E_1, E_2, E_3)$$

The discrepancies between the theoretical deflections (DC_j) and measured deflections (DM_j) at a location j are given by the error function (E_j) as follows:

$$E_j = DM_j - DC_j$$
$$= DM_j - f_j(E_1, E_2, E_3)$$

The total error function (TEF) over the interval 1 to n (where n = 9 for Deflectograph and 6 for FWD) can be expressed as a function of the elastic moduli E_1, E_2, E_3 of the various component layers of a pavement structure, i.e.:

$$TEF = \sum [DM_j - f_j(E_1, E_2, E_3)]$$

The values of elastic moduli can be chosen to minimise the total error function (equation 3) but this would be unsatisfactory since unacceptably large positive and negative values of error (E_j) could occur within the deflection bowl parameters which, nevertheless, partially cancel each other in the total error function (TEF). A more satisfactory approach would be to calculate the squared value of error for each deflection parameter in the bowl (over the interval j = 1 to n). Also, since the error in the various measured deflection bowl parameters (DM_j) can vary significantly from one parameter to another, a weighting function (W_j) should be introduced for each of the n deflections in the total square error function (TSEF), i.e.:

$$TSET = \sum_j [W_j \, DM_j - f_{jj}(E_1, E_2, E_3)]^2 \bigg/ \sum_j w_j^2$$

and by applying the partial differentiation with respect to each unknown modulus a set of linear simultaneous equation is obtained and is solved by the standard Gaussian Elimination Method to obtain values of unknown elastic moduli. The computed modulus values are input to the DUAL model to compute the theoretical deflection bowl parameters and the mismatching deflection error between the measured and the theoretical bowl is found. This is calculated and compared with the user-assigned permissible tolerance (e.g. 5%). At this point it is said that one cycle of iteration has been completed. The procedure for successive correction to the derived modulus values from the first cycle of iterations is continued following similar steps to those described above, until the difference between the calculated and measured deflections in the bowl are within a predetermined tolerance value. Then the final set of moduli is obtained representing the effective in-situ moduli of the tested pavement system corresponding to the test load, temperature and season. The final set of moduli are adjusted to a standard temperature (20 $^\circ$C), loading frequency (10 Hz) and design season (March). These corrections were developed in this investigation and described elsewhere (Shaat et al, 1989b). The non-linearity of the subgrade is not considered yet in the EPLOPT backanalysis Program. This can however be done manually beforehand. The steps used in the evaluation procedure using EPLOPT are shown schematically in the simplified flow chart given in Figure 4.

5   Comparison of Modulus Values Measured by Component Analysis and Backcalculated by System Analysis

The stiffness modulus of the bituminous layer for each test section was backcalculated using the evaluation program EPLOPT from in-situ FWD deflection bowl measurements. The results of the backcalculations are summarised in Table 2 comparing them with those measured in the laboratory using the component analysis discussed in section 2.

Table 2.   Comparison of Backcalculated and Measured Moduli.

Site No.	Backcalculated Modulus (E_b) (MPa)	Laboratory Measured Modulus (E_m) (MPa)	Ratio (E_b)/E_m
6	3070	3050	1.01
7	3676	1415	2.59
9	1529	2606	0.59
10	3659	2381	1.54
11	2635	3125	0.84
13	3529	2794	1.26
15	3609	4783	0.76
17	3882	2477	1.56
	Average Ratio (E_b/E_m) =		1.25

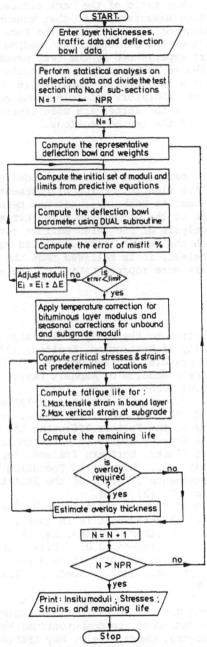

START.

Enter layer thicknesses.
traffic data and deflection
bowl data

Perform statistical analysis on
deflection data and divide the test
section into No. of sub-sections
N = 1 ⟶ NPR

N = 1

Compute the representative
deflection bowl and weights

Compute the initial set of moduli and
limits from predictive equations

Compute the deflection bowl
parameter using DUAL subroutine

Compute the error of misfit %

Adjust moduli
$E_i = E_i \pm \Delta E$ — no — is error < limit

yes

Apply temperature correction for
bituminous layer modulus and
seasonal corrections for unbound
and subgrade moduli

Compute critical stresses & strains
at predetermined locations

Compute fatigue life for:
1. Max. tensile strain in bound layer
2. Max. vertical strain at subgrade

Compute the remaining life

is overlay required? — no

yes

Estimate overlay thickness

N = N + 1

N > NPR — no

yes

Print: Insitu moduli; Stresses;
Strains and remaining life

Stop

Fig. 4. SIMPLIFIED FLOWCHART FOR EVALUATION OF PAVEMENT
LIFE BASED ON DEFLECTION BOWL ANALYSIS

317

It is noted from Table 2 that a poor correlation exists between the two sets of moduli. The ratio of the backcalculated modulus from the FWD in-situ deflection measurements to that measured in the laboratory is in the range of 1.01 to 2.59. The fact that this ratio is not unity is a result of the different FWD loading frequencies (in the range 8.5 - 22.8 Hz) experienced by the test pavement sections when subjected to the FWD as compared with the constant frequency (loading time) used in the laboratory. Thus, while there seems to be some disagreement between laboratory determined and backcalculated moduli, it is believed that the latter are more representative of the actual field conditions of the tested sections.

6. Conclusion

In this paper a material parameters assessment model has been developed to determine the elastic modulus of pavement layers from in-situ nondestructive testing technique. It has been demonstrated that the elastic modulus of the bituminous mixes estimated from the laboratory component analysis of recovered surface cores do not confirm, generally, with the in-situ backcalculated values at the same core location. Nevertheless, it is believed that the in-situ backcalculated moduli are more representative of the actual field conditions.

7   References

Brown, S.F., "An Introduction to the Analytical Design of Bituminous Pavements". University of Nottingham, England, 1980.

Bonnaure, F., Gest, T., Gravois, A., and Ugs, A., "A New Method of Predicting the Stiffness Modulus of Asphalt Paving Mixtures". Assn. of Asphalt Paving Techs., Vo. 46, 1977.

Huet, J., and Ista, E., "Viscosity/Penetration Relationship of Bitumen at 25°C". Proc. 3rd Eurobitume Symp., The Hague, Sept. 1985.

McKibbin, D.M., "A Study of the Factors Affecting the Performance of Dense Bitumen Macadam Wearing Course in Northern Ireland". Ph.D. Thesis, University of Ulster, Northern Ireland, Sept. 1987.

Peattie, K.R., and Ullidts, P., "Simplified Computing Techniques for Analysing Flexible Pavements". Proc. of the Institute of Civil Engineers, London, Vol. 71 (2), 1981.

Shaat, A.A., "Evaluation of Highway and Airport Pavement Life Based on Nondestructive Structural Assessment Techniques". Ph.D. Thesis, The Queen's University, Northern Ireland, May 1989(a).

Shaat, A.A., Farouki, O.T., Ferguson, J.D., "Evaluation of Pavements in Northern Ireland Based on Nondestructive Structural Assessment Techniques". Proc. 1st Intr. Conf.: Roads and Traffic 2000, Vol. 2, Berlin, Sept. 1988.

Shaat, A.A., Farouki, O.T., Ferguson, J.D. and McCullough, L.M., "Use of Nondestructive Test Results in Evaluation of Simplified Pavement Systems". Proc. 4th Int. Conf. on Computational Methods and Experimental Measurements, Capri, Italy, May 1986(b).

Ullidtz, P., "A Fundamental Method for Prediction of Roughness, Rutting and Cracking of Pavements". Proc. Ass. Asph. Pav. Technol., (48), 1979.

Van der Poel, C., "A General System Describing the Visco-elastic
   properties of Bitumens and its Relation to Routine Test Data".   J.
   Appl, Chem., (4), 1954.

319

22 THERMIC EFFECTS FOR RUPTURE TESTS AT INDIRECT TENSILE TEST ON ASPHALT CONCRETE

G. TESORIERE and S. MARINO
Road Construction Institute, Palermo University, Italy
S. CANALE
Hydraulics, Transport and Roads, Rome Universty, Italy

Abstract
This research is made to put in evidence some special aspects of rupture test at indirect tensile test on asphalt concrete, with special reference to the temperatures and to the curing and conditioning times. Keeping in mind the results obtained we must point out that the best temperatures to represent the rupture effects are between 20–25°C. so to exalt the elastic component of the sample and to reduce the inevitable mistakes which may arise with the application, for rupture tensions, of a formula valid only for quite elastic materials.

1 Results of previous research

Recently we have been interested to analize the various factors which influence the rupture test at indirect tensile test on asphalt concrete for binder in view of the introduction of the relative rules in Italian Specifications.

This research is made to know the following aspects:
– definition of the highest value of rupture tension in function of the ratio between the percentage of bitumen ($b = 100 \, Pb/(Pi+Pb)$) and the specific surface of aggregate (Σ);
– influence of bitumen penetration in the range 80/100;
– influence of rupture speed;
– height of the sample;
– presence or absence of listel;
– compaction degree through the measurement of the volumetric mass;
– repeatability of results.

The aggregates used to manufacture the concrete were constituted by calcareous rock for crushing (with filler obtained by the same aggregates) which granulometry was included in A.N.A.S. (Azienda Nazionale Autonoma delle Strade) grading envelope for binder.

For all the tests we have put attention to adopte a precise

methodology for what concerns aggregates and bitumen pre-heating temperatures and also for samples conditioning before submitting them to rupture.

Prelimnary operations, followed in previous tests and which we refer to, are resumed in table n° 1.

TABLE 1. Conditions followed in manufacturing preliminary operations

Manufacturing of asphalt concrete

preparation	aggregates	T = 110°C for 12 h
	bitumen	T = 110°C for 30'
pre-heating	aggregates	T = from 110 to 177° for 1 h
	bitumen	T = 135°C for 1 h
aggregates mixing		T = 177°C for 5'
heating of bitumen		T = from 135°C to 177°C for 15'
aggregates-bitumen mixing		T = 152°C + 5°C for 5'

Samples manufactoring

blows per face	n° 50
curing	T = 20°C for 12 h
conditioning	T = 25°C for 6 h

The unitary load of rupture was determined appling the formula obtained in the hypothesis of perfectly elastic, isotropous and omogeneous material: $\sigma_R = \dfrac{2P}{\pi DH}\left[N/mm^2\right]$

beying : P the load in N

   D the diameter in mm.

   H the height in mm.

The average dimensions of D and H of each sample have been obtained through statistical considerations of normal distribution of measurement taken (8 heights and 8 diameters of which 4 on one base and 4 on the other one).

Furthermore, each value reported is the average (statistic) value of 4 samples.

The rupture speed has been maintained equal to 0.85 mm/s (Marshall press) but we have made also checks at a speed of 0.42 mm/s (Hubbard Field).

While making the tests, by means of transducers, we have measured vertical deformations (rupture) and lateral deformations.

The first results obtained put in evidence some interesting aspects which may be synthetized as follows:

a) the function $(\sigma R, b/\Sigma)$ comes out with a bell shape, flattened towards the highest values di σR which are obtained for percentages of bitumen between 3.828 and 4.237%;

b) σR value is influenced, in addition to bitumen percentage, by the variation of penetration value, even if for limited ranges: when penetration increases rupture tensions goes from 1.15 N/mm^2 (pen = 82-85 dmm) to 0.85 N/mm^2 (pen = 100 dmm);

c) the volumetric mass remains always within restricted limits $(2.33 \div 2.42 \cdot 10^3 \cdot \text{Kg/m}^3)$ with a porosity between 6.6 and 8.3%;

d) lowering rupture speed (from 0.85 mm/s to 0.42 mm/s) σR value decreases (viscous effet);

e) repeatability results to be between 0.09-0.11, under the same penetration of bitumen and other characteristcs, so it's possible to have variations of σR values of about 10% (with Marshall press it's possible to obtain variations up to 15-20%);

f) the influence of sample height is relative and remains restricted into repeatability);

g) an higher compaction (75 blows per face) increases the σR values of about 13%;

h) the absence of listels determines a dispersion in the values;

i) the vertical deformation is limited (0.5 mm); the lateral one is absolutely insignificant.

All these results persuade us to continue these experiences to study the influence of temperature and conditioning times, which in the first researches were kept scrupolously constant.

2 Considerations on the thermic effects in the tests

In our previous memory we put in evidence a certain perplexity about the definition of σR tension, reached in elastic field, also for viscous-elastic materials like asphalt concrete and we explained the choise of curing temperature (20°C. for 12h) and the choise of conditioning temperature (25°C. for 12h) before the test, because we thought that in such a way we could avoid viscosity effects.

At this purpose, from bibliography, we learnt that, for indirect tensile test, 4 temperatures were suggested: -10, 25, 45 and 65°C., in a range of temperatures which seems of excessive wideness.

As a matter of fact, very low temperatures exalt elastic properties of the mixture, bringing it close to a vitreous structure; very high temperatures which reach or exceed the value of bitumen softening point (p.a.), certainly exalt the viscous component.

We thought, therefore, to further examine, in this sector, under the same asphalt concrete and test characteristics of previous research: granulometry of aggregates, type and percentage of bitumen, modality of test mixing, presence of listels, press speed (0.85 mm/s) and changing only the temperature and conditioning time.

The aggregates, always constituted by calcareous rock for crushing, have been mixed in such a way to maintain the same granulometric curve of previous tests (A.N.A.S. grading envelope for binder); the bitumen used has a penetration 85 dmm with softening point p.a. = 49°C. and IP = 0, equal to a type with which we have worked in previous tests.

As we have already observed, the highest σ_R value swings in a very restricted range of bitumen percentage, having kept constant the specific surface of aggregates, the mixture have been all prepared with b = 3.94% (b/Σ = 0.40 with a specific surface Σ = 9.853) (mm^2/Kgp).

We point out that, while in previous researches the conditioning was always obtained in a stove, in these new experiences we have used a thermostatic tank reducing conditioning time to 2 h only for operative reasons, after we could be sure that there wasn't valuable variation in the results (control tests).

We say again that we wanted to define also the repeatability we may reach with test at high temperature (40°C.), on 30 samples, following very well tested modalities.

3 Test results

It is interesting to report exactly the material characteristics with which the sample have been made:
—Aggregates:
calcareous nature (quarries of Palermo area)
L.A. (type B) = 25.7%
E.S.   = 89.08%
granulometry:

Jig UNI	Undersize %	Checkoff%
Sieve ASTM		
UNI 25	100	--
UNI 15	83	17
UNI 10	65	18
UNI 5	45	20
ASTM 10	33	12
ASTM 40	15	18

ASTM 80   10   5

ASTM 200   6   4

  filler   6

                                           100%

−Bitumen:

for refining, it is indicated 80/100

penetration at 25°C. = 85 dmm

softening p.a. = 49°C.

penetration index IP = 0

b percentage = 3.94%

−Samples:

compaction: 50 blows per face with Marshall hammer

average values for all tests:

average diameter between mm 101.60 - 10180

average height between mm 64.77 - 66.35

ΔD = Sv/D (%) included between 0.73 - 1.94

(Sv = vertical slipping)

−Temperatures:

curing in thermostatic stove for 12 h at 20°C.

conditioning in thermostatic tank for 2 h at 10-18-25-30-40 and 50°C.

    The average values of results obtained, taken at various conditioning temperatures are reported in table n° 2.

TABLE 2. Average   values   obtained   in   the   tests   in   function   of conditioning temperature

T (°C)	D (mm)	H (mm)	P (N)	σR (N/mm^2)	Volumetric mass (kg/cm^3)	porosity %	Sv·R mm^3/N	ΔD %
10	101,60	66,36	26.340,00	2,487	2,271	7,93	0,44	1,08
18	101,64	64,84	14.641,43	1,414	2,278	7,65	0,52	0,73
25	101,65	65,20	11.640,54	1,118	2,290	7,16	0,98	1,08
30	101,70	63,85	5.691,03	0,558	2,290	7,16	2,41	1,33
40	101,65	64,77	2.593,89	0,247	2,280	7,57	4,45	1,08
50	101,80	65,35	1.525,50	0,146	2,270	7,97	13,56	1,94

    In   the   previous   test   (bitumen,   pen 85   dmm) with conditioning temperature at 25°C (in thermostatic stove) for 6 hours, we obtained the following results: highest value σR = 1.1036 N/mm^2, γ = 2.34 · 10^3 kg/m^3, p = 8.0960%, with values assimilables to the ones above reported for conditioning temperature in tank for 2 hours.

    As a matter of fact, this result is due to the fact that the temperature used for curing (20°C for 12 h) is very close to the test temperature (25°C) so a time reduction has no influence. However we thought to maintain for all tests the conditioning time limited to 2 h.

The diagram of Fig.1 (ordinate σR, abscissa conditioning temperature) allows us to make some interesting consideration:

a) in the range between the temperature 10 - 25°C the highest σR tension varies in a linear way, so we must believe that the elastic component is predominant.

b) over the 30°C the function bends with a parabolic course; the σR values pull considerably down reaching at 50°C a minimum of 0.146 N/mm^2; this shows that over 30°C the viscous component prevails;

c) the vertical deformations referred to the oR value increase accordingly to the increase of temperature while the lateral deformations, almost null for temperatures up to 30°C, start to be valueable over the 30°C still remaining within values not bigger than one tenth of mm;

d) the repeatability at 40°C, obtained on 30 samples (all the relative indications are reported in table n° 3) gives the value r = 0.07, this result may seems in contrast with the value r obtained for test temperature of 25°C (0.09 - 0.11), but if we make reference to the percentage variation with regard to the value obtained for $\sigma Rmax$ we obtain that, in case of test at 25°C with σR = 1.1 N/mm^2 the variation $\pm$ 0.11 is only of about 10% of $\sigma Rmax$ while at 40°C with $\sigma Rmax$ = 0.24 N/mm^2 the variation $\pm$ 0.07 is of about 30%, therefore with a large dispersion.

TABLE 3. Repeatability tests for conditioning temperature
T = 40°C for 2h

n°	(N/mm^2)	n°	(N/mm^2)	n°	(N/mm^2)
1	0,247	11	0,260	21	0,236
2	0,280	12	0,199	22	0,261
3	0,242	13	0,220	23	0,283
4	0,229	14	0,273	24	0,220
5	0,220	15	0,254	25	0,283
6	0,185	16	0,265	26	0,206
7	0,259	17	0,247	27	0,215
8	0,262	18	0,220	28	0,251
9	0,273	19	0,252	29	0,265
10	0,282	20	0,243	30	0,273

A similar consideration has to be done about the values of vertical slipping, which have to be referred to the highest value σR; fig. 2 reports the function (Sv/σR,T) relative to the different conditioning temperatures from which we notice that the value Sv/σR become more

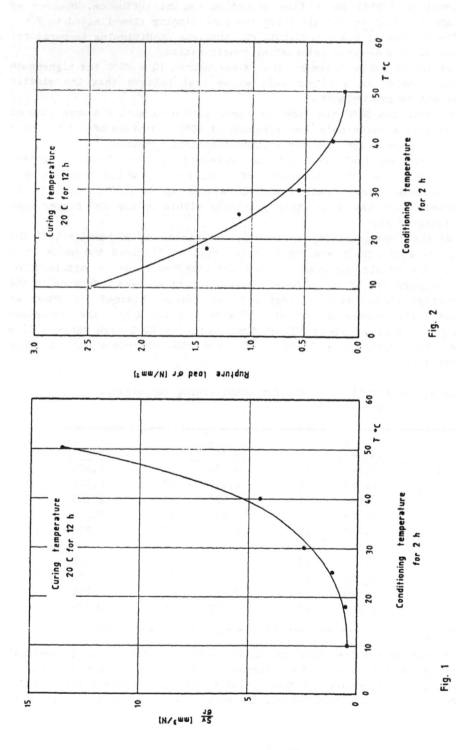

Fig. 2

Curing temperature
20 C for 12 h

Conditioning temperature
for 2 h

Rupture load σ_r [N/mm²]

Fig. 1

Curing temperature
20 C for 12 h

Conditioning temperature
for 2 h

$\frac{\sigma_r}{SV}$ [mm³/N]

intensive for high temperatures (over 30°C) while stays on values which may be considered very close for temperatures between 10 and 30°C.

We already saw that reducing the conditioning time to 2 h for temperature of 25°C the oR values remain unchanged, in this case the sample submitted for 12 h to conditioning of 20°C reached after 2 h the thermic conditions relative to 25°C.

It was foreseeable a different behaviour when different curing and conditioning temperatures were taken: for example: curing at 20°C and conditioning at 40°C.

Remaking the experience with a conditioning temperature of 40°C but bringing the time at 4 hours instead of 2 h, it has been obtained a oR value = 0.233 N/mm^2 lower than the value reached with duration conditioning equal to 2 hours (0.247 N/mm^2).

This states that adopting a very high conditioning temperature in comparison with curing temperature more time is needed because the sample be representative of the chosen temperature.

Conclusions

This research put in evidence the function of conditioning temperature (limited to 2 hours in thermostatic tank) on results which may be obtained in rupture tests at indirect tensile test for asphalt concrete samples, binder type.

It was specially noticed that at temperatures lower than 20-25°C an effect of mass stiffening comes out; the stress-deformation curve has the characteristics shape of a stiff material and with the instrument used it is obtained a sort of blow-up which gave hard time to the dynamometric ring.

For higher temperatures (30 - 40 - 50°C) the viscous component prevails and, in percentage, the results dispersion increases.

Therefore it seems to us confirmed that the most indicative values are obtained adopting the following temperatures:
- curing in thermostatic stove at 20°C for 12 h.
- conditioning in thermostatic tank at 25°C for 2 h.

Acceptable variants may be proposed about temperatures which we have worked at, for example choising for curing and conditioning the same temperature (25°C).

We confirm, however, that the values of rupture at indirect tensile test obtained in different laboratories may be comparable only if the same manufacturing, test, and curing and conditioning temperatures modalities have been strictly followed.

Note
The authors thank Engineers L. Simone and A. Abbagnato for the help given during the execution of laboratory tests.

References

Tesoriere G. and Simone L., (1988) Considerations on rupture tests at indirect tensile test on asphalt concrete samples for binder. Magazine "Rassegna del Bitume"

Huet J., Recherche d'un essai global pour le control de la qualité des bètons hydrocarbonès: Compressions diamétrale. Centre de Récherches Routières, Bruxelles: Das Stationare Hischwerk, n. 4/73.

Shapiro S.S. and Wilk M.B. (1965) An anlysis of variance test for normality. Biometrika, vol. 52, 3-4, 1965.

D'Orsi R. and Giannattasio P. (1978). Repeatability and reproducibility of a test methodology. XVIII Road National Meeting, Taormina, Italy

PART THREE
TESTS WITH REPEATED LOADING

23 MASTIC ASPHALTS MODIFIED BY SCRAP RUBBER

K. BEDNAR
CVUT Praha, Fakulta Stavebni, Thakurova, Praha,
Czechoslovakia

Abstract

In the first part of paper ecologic, economic and technical reasons for the utilisation of waste rubber in road - building industry are given. The solution starts from the hypothesis that bounds resulting from the vulcanization of the rubber polymer network are damage by the mechanical and thermal treatment, i.e. by the homogenization with asphalt. In the second part of the paper the possibility of transferring results of experimental work into practice is documented on the basis of building the whole series of field experiments.

Keywords: Asphalt-rubber, Mastic Asphalt, Modification, Temperature Susceptibility, Low Temperature Behavior, Fatigue Charakteristics, Resistence to Deformation.

1 Rubber-waste utilization problem

In Czechoslovakia, similarly to the case of great many of industrially developed countries, a vast amount of waste rubber occurs, becoming a significant problem. There are at least two reasons why to solve it, namely:
- environmental protection,
- economic reasons.

According to the data provided by the Barum and The Salvage Materials national enterprises, the waste rubber amounted to $8.4*10^5$ tons in the year 1980, $10.0*10^5$ tons in 1985, and the expected value for the year 1990 amounts to $11.4*10^5$ tons. It is debeaded tyres that take the lion´s share of this waste.

Various methods of reclaimation of debeaded vehicle tyres are known both abroad and in Czechoslovakia. The matter is to choose and put into practice the economically optimal variants, also justifiable from the point of view of environmental protection. In the prevailing oppinion held in the sixties, it was assumed that the direct combustion in special furnaces and/or cement kilns was the only method capable of solving the problem of economical disposal of the waste. Currently it has been demonstated that higher economic effects can be achieved by using waste rubber as a raw material in applications both within and outside the rubber industry.

The utilization of debeaded tyre waste as a reclaim or rubber powder in the rubber industry is only of a limited range. Among applications outside the rubber industry, practical usage has been made in the building industry and the road-building industry (cf. Fig. 1). Apart from seal courses, sealing compounds, and surface dressigs for various purposes, particular attention is given to the application of waste rubber in the road building. Road trials with rubberized binder lait in the U.S.A. in the sixties have brought a considerable enhancement in the service life of road surface, particularly in heavy-duty places.

The advanteges of the modified bitumenous mixtures, demonstrated in a wide range of road trials abroad, include:
- extension in service life of the treatment,
- wider elastic range of the modified binder,
- easier maintenance (lower costs),
- savings in quality asphalt,
- higher stripping resistance,
- higher coefficient of friction,

Using rubber waste as a substitution for bitumen provides neither heat from the combustion process for the energy incorporated. However, a certain improvement of bitumen and substitution for the same are achieved, in view of the service life of the treatment and the

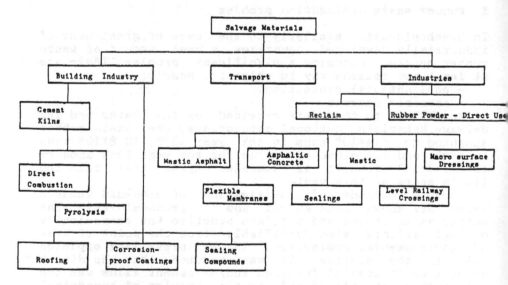

Fig. 1 Possibilities of utilization of reclaiming rubber in national economy

technology employed on up to 1 : 1 scale, which yields up to 210 kJ/g. This application is exceptionally promising for the utilization on both the great and the small scale. It provides the possibility of processing all tyres, and seems to be the best solution as far as the viewpoint of energy conservation or recovery is concerned.

A summarized survey of the possibilities of energy recovery in terms of kJ/g for individual modes of application of discarded tyres is presented in Table 1. Analysing the Table we can draw the conclusion that using rubber shreds for the asphalt modification, the energy recovered is 6 times higher than in the case of combustion; 2.3 to 6.9 times higher than in manufacture of chemicals and fuels, and 1.8 times higher than in manufacture of chemicals and fuels, and 1.8 timas higher than when used as reclaim.

Although upgrading the accuracy of above mentioned values calls for further technological research and for the construction and long-run observation of road trials, nowadays it is possible to state positively that the yeild of recovered energy for the rubberized asphalt is higher than for the other avalaible technologies of waste rubber processing.

The most difficult problem, not yet satisfactorily tackled worldwide, is the disintegration of tyres to shreds of particular grading. Considering exceptional toughness of the material, primarily with the steel cord tyres, all the equipment used as yet work with low afficiency. The price of the granulation product is considerably high.

Regardless the fact that the path to the large-scale utilization of discarded tyres is pretty difficult to materialize, it is neccessary even under the conditions of our country to verify the possibility of the economical processing of the waste rubber in the road-building industry, particulary in the manufacture of bitumenous mixtures.

The possibilities of modification of bituminous mixtures using rubber shreds are considerably confined by the following economic requirements:
- neccessity of rubber shreds being produced at specified particle-size range, below 0.5 mm, on a machinery of Czechoslovak make,
- asphalt modification shall not considerably increase the energy consumption being commonplace in the manufacture of bitumenous mixtures,
- the modification shall facilitate the application of lower viscosity asphalts, subject to all quality requirements specified by Czechslovak Standard,
- saving in asphalt and/or mineral filler shall amount to 10% at least.

Table 1 — Scrap Tyre Uses — Energy recovered (kJ/g) (Ref. 2)

Use	Process	Scale and potential	Energy recovered (kJ/g)			Comments
			Process	Later combustion	Total	
Combustion	Whole tyres	1	35	-	35	special furnace req.
	Shredded tyres	1, 2	35	-	35	existing furnace
Chemicals & Fuels	Pyrolysis (fuel products only)	1	25.6-34.9	-	30.25*	pilot plant shut
	Pyrolysis (recover carbon black)	1	44.2-53.5	8.7	57.55*	experimental
	Depolymerisation (subst. oil & carb.b.)	2	51.2	40.1	91.3	
Re-use in new rubber products	Reclaim	2	79	35	114	
	Microwave devulcanization	2	93	35	128	
	Shreds (cryogenic)	2	93	15	108	
	Shreds (gould)	2	86	15	101	
Asphalt modification		1, 2	210	-	210	most promising option
Whole tyres use	Reefs, crash barriers	2	-	35	35	negligible potential

Scale and potential : 1 all scrap tyres consumable
2 applicable on small scale

※ on the average

2 Modification of asphalt by rubber shreds

A research project (Ref. 3) was conducted at the Faculty of Civil Engineering, Technical University of Praque, resulting in a draft for interim specifications for the manufacture of mastic asphalt with rubber shreds admixture.

First it was neccesary to follow the physico-chemical properties of the mixture of asphalt and rubber shreds. It was demostrated that partly the rubber dissolves in the asphalt solution, and parlty is in colloidal state.

During homogenization process the rubber particles swell to two to three times their original volume.

An important indicator of the properties of the asphalt and rubber shreds mixture is the degree of dispersion of rubber in bitumen. It is determined by:
- type of binder used,
- type of rubber and its particle-size range,
- method of homogenization of the mixture.

The requirements ganerally imposed upon the properties of used materials include:

ruber - soluble or well-swelling in low-molecular
      hydrocarbon of asphalt,
    - suitable particle-size range(as fine as possible
    - higher tensile strenght,
    - wider elastic range.

bitumen - high contens of maltenes,
    - high degree of dispersion incorporation of
      rubber

resulting mixture - temperature invariable,
              - compatible,
              - hot-storage resistant (degradation
                of the rubber must be avoided)

The following tasks resulted the above mentioned requirements for practical verifications:
- to identify the suitability of a partioular type of rubber shreds for asphalt modification,
- to devise a testing method simple enough to be applicable for job tests of asphalt - rubber shreds mixtures employing instruments avalaible in every road laboratory,
- to suggest the optimum mixture of asphalt and rubber (selected type) when defining the duration and temperature of mixing for a before-stated type of bitumenous treatment,
- to verify whether the applications to service conditions of the results of laboratory investigation is possible,
- to verify the possibility of asphalt being saved when using modified binder.

On investigation into the problem of suitability of a particular type of grading, optimum percentage of comminuted scrap rubber as an asphalt modifier, and of homogezation procedure, the following conclusions can be drawn:

- establishing the penetration index is a sufficient method for selecting the suitable type of comminuted scrap rubber for asphalt modification,
- conventional tests without modifications, commonly used with asphalt binders, cannot be employed as conclusive methods for modification by comminuted rubber,
- test conditions and test methods (modified conventional or non-conventional) have to be precisely defined and strictly adhered to,
- homogenization procedure is a decisive factor for the efficiency of the modification by comminuted rubber of asphalt,
- duration and temperature of homogenization are important technological factors which determine resulting properties of the blend. The appropriate determination of them forms the basis for applicability of bitumen and comminuted rubber blend in practice,
- viscosity measurement depends on properties of bitumen type and elasticity of rubber compatibility between bitumen type and elasticity of rubber, and degree of completion of the reaction between bitumen and rubber,
- due to rubber additive, the viscosity changes irreversibly, at higher temperatures (250°C) the effect of rubber additive decreases, and significant changes may cause softening of bitumen and act in a way similar to that of fluxation oils,
- ductility of the blend of bitumen and comminuted rubber is always lower than that of pure bitumen; the method is sensitive to homogenization temperature.

Suitability of a particular type of comminuted rubber and its grading for bitumen modification was observed using variation of penetration index, cf. Fig. 2, Fig. 3. Observations were made for 7 types of comminuted rubber, currently produced to cover the needs of rubber industry, and 1 type of shreds, manufactured according to our requirements from scrapped conveyer belts from open pit coal mines. The effect of comminuted rubber upon the performance of modified bitumen depends on a large number of factors. Regardless the duration of heating and method of homogenization the effect would be dependent on the magnitude of contact area between the rubber shreds and the bitumen, on the properties of the bitumen itself, but especially on the properties of the rubber shreds. While the magnitude of contact area can be estimated quite well from the rubber shreds grading curve, the properties of the comminuted rubber itself cannot be established beforhand. We do not know chemical constitution of shreds since various materials mixtures are used, neither we know

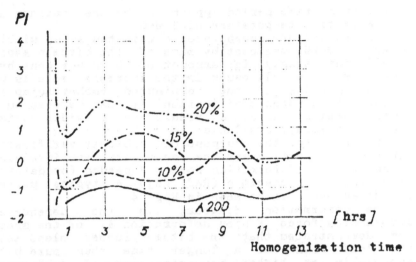

Fig. 2 PI changes A 200 asphalt modified by "Karkasa" crumb

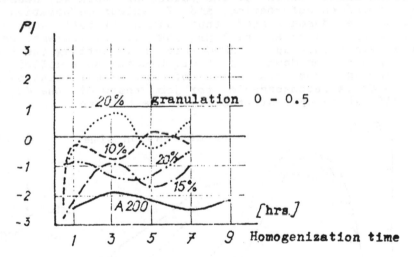

Fig. 3 PI changes of A 200 asphalt modified by
"tread-runner" crumb

the degree of vulcanization or devulcanization.
Experiments conducted on preparation of rubberized bitumen
suggest that according to the values of penetrtion index
(PI) achieved, and related properties of bituminous
mixtures, the "Karkasa" comminuted rubber display the
highest increase in PI and favourably affects the
performance of the modified bitumen (Ref. [1],[5]).

Durations of homogenization within the interval of 1 to
13 hours have not provided any significant shift against
the values obtained during a heating extending over 3

hours. This time period appers to be the optimum at the homogenization temperature of 180°C.

The ammount of incorporated comminuted scrap within the range of 10-20 per cent by mass of the bitumen manifests itself increasingly. The ammount of 15-20 per cent by mass shows itself in all cases as the increase role is played by the grading of the comminuted rubber scrap in the process of bitumen modification. It can be assumed that finer comminution could result in a wider variety of comminuted scraps available.

Adjusted for the purpose of job-test verification of bitumen modification was the ductility test involving the measurement of deformation force F needed for deformation Δl of the test specimen (Ref. [6]). A typical pattern of the dependence can be seen in Fig. 4.

The deformation diagram, i.e. the dependence of deformation force F upon deformation Δl of the specimen, has demonstrated that the bitumen/rubber blend sustains the load F_{max} for a longer time than pure bitumen, especially at higher concentrations of rubber crumb. Taking measurements of the values of work W_p, needed for the specimen deformation, and of maximum deformation force F_{max}, has demonstrated that values of both work W_p and force F_{max} increase with growing rubber content. Raising the temperature up to 220°C results in cutting the time of homogenization down to 1 hour. Despite some deficiencies, following from assessment of individual relations, the method is recommendable for job tests of the degree of modification by crumb rubber of the bitumen.

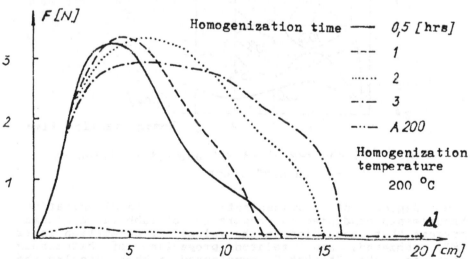

Fig. 4 Relation between deformation force F and
extension l of the specimen. "Karkasa"
modif. A200 asphalt - 25% by mass

3   Rheological properties of modified bitumen

In order to verify the assumptions of the favourable effect of rubber crump upon bitumen modification, the evaluation of rheological properties of the optimal blend has been carried out (Ref. [7]). It consist in the determination of the phase shift between force and deformation in the dynamic establishing of complex modulus G. Viscoelastic materials like bitumen exhibit the phase shift 0° to 90°, according to the sort of bitumen, temperature of testing, and frequency. Under extreme conditions, the bitumen properties approach those of elastic or viscous materials. The value of the phase shift is a component of the viscosity. The delay of deformation after force in the viscoelastic region can be explained by the transformation into thermal energy of a part of the mechanical energy responsible for the deformation.

$$\text{tg } \varphi = \frac{G''}{G'}$$

G" - imaginary component of complex modulus
G´ - real component of complex modulus

It follows from Fig. 5 that the behaviour of the A 200 bitumen without modification resembles that of a newtonian liquid, the temperature of 60°C and lower values of circular frequency (1 rad.s⁻¹) provided. At about 10°C and higher (20 rad.s⁻¹) frequency the behaviour of the bitumen is entirely elastic. At the temperature below -10°C the behaviour of the bitumin is entirely elastic at a bulk of vibration frequencies employed. Thus the temperature of -10°C is to be considered to be a limit of transition into elastic state.

Fig. 6 shows that the behaviour of the A-200 bitumen modified with 20 per cent by mass of rubber crumb is quite different. At the temperature of 60°C the blend does not pass into viscous state, and its behaviour at -30°C is not quite elastic, either. For the pure A 200 bitumen the viscoelastic range was found to cover the interval of 70°C, whereas for the modified bitumen it extended over more than 90°C. This offers an evidence that the modification favourably affects the viscoelastic properties of the bitumen.

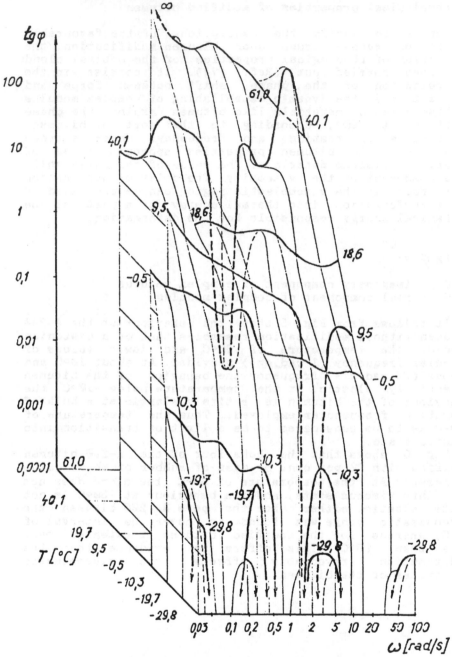

Fig. 5 Influence of Temperature T and Frequency ω on
the tg φ for A 200

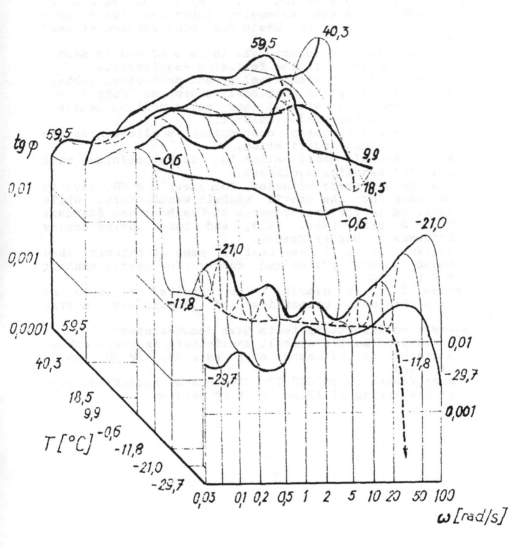

Fig. 6   Influence of Temperature T   and Frequency ω on
         the   tg φ for A 200   modified by   "Karkasa"
         (20% by mass of mix)

4 Rubber-crumb modificatin of bituminous mixtures

Following the evaluation of basis parameters of rubber/bitumen systems, extensive laboratory tests were carried out to form the basis for construction of road trials.

To begin with, the solutin was to be confined to mastic asphalt and subjected to the following requirements:
- to employ the conventional ground pulverised rubber, manufactured by the national enterprise "Rudy Rijen" at Otrokovice, without any subsequent mechanical, thermal, or chemical treatment,
- no increase in energy consumption is allowed for the mastic asphalt modification,
- mastic asphalt modification has to be performed with existing handling equipment,
- also softer-grade bitumens, for exemple A 80, have to be used for the mastic asphalt manufacture, while adhering to the requirements of Czechoslovak Standart CSN 73 6150 (Ref. [4]), and thus saving scanty bitumens of higher grades,
- to save 10 to 15 per cent by mass of bitumen when using ground pulverised rubber for mastic asphalt modification.

Tables 2 to 5 display the results of tests of physico-mechanical properties of rubberised mastic asphalt.

Summarized laboratory tests have demonstrated:
- rubberized mastic asphalt manufacture is practicable,
- softer-grade bitumens, such as A-80, AP-80, can be employed,
- specifications of CSN 73 6150 can be satisfied, even at the saving of 10 per cent of bitumen.

Table 2 - Physical and mechanical properties of fine mastic asphalts — FMA (mixtures and check tests)

Composition of mixture / Trial mix series:	1	2	3	4	5	6	7	8	9	10	11	12	13
Binder — Specification designation	AP - 25	AP - 25	AP - 25	AP - 25	AP - 25	AP - 25	AP - 25	AP - 25			A - 80	A - 80	A - 80
Softening point R.B. (°C)		62,2				61,5					47,6		
Penetration (mm^{-1})		25				26					83		
Aggregates — Content (% by mass of total mix)	8,6	8,5	8,5	8,5	8,6	8,6	8,6	8,6	8,5	8,5	8,4	8,4	8,4
Coarse aggregate (% by mass t. mix)	42,0	41,7	41,7	41,7	42,0	41,8	41,8	41,8	36,6	36,6	36,3	36,3	36,3
Fine agg. 0–4 (dtto)	22,9	22,7	22,7	22,7	22,9	22,8	22,8	22,8	27,4	27,4	27,2	27,2	27,2
Mineral filler (dtto)	26,5	26,3	26,3	26,3	26,5	26,4	26,4	26,4	27,5	27,5	27,3	27,3	27,3
Rubber powder (dtto)	–	0,8	0,8	0,8	–	0,4	0,4	0,4	–	–	0,8	0,8	0,8
Temperature of homogenization (°C)	218	225	225	225	211	215	224	224	225	228	210	215	230
Time of homogenization (hours)	1,5	3,0	3,0	4,5	–	1,5	3,0	4,5	–	0,5	1,5	3,0	4,5
Static indentation test (mm) (1)	3,0	2,26	0,63	0,12	3,85	1,02	0,41	0,17	5,26	5,72	5,27	2,58	0,44
Increment – static indentation test (2)	0,35	0,61	0,12	0,03	0,60	0,22	0,10	0,05	0,97	1,30	0,93	0,53	0,11
Compressive strength R_k – 22°C (MPa) (3)	4,63	5,93	9,24	11,56	6,70	9,12	11,51	13,40	3,93	5,06	5,87	7,78	13,07
Flexural strength R_{122} – 22°C (MPa) (4)	3,66	5,76	9,56	7,34	6,70	9,86	11,32	11,67	3,11	4,85	5,75	8,54	12,58
Flexural strength R_{10} – 0°C (MPa)	16,42	15,83	7,86	7,58	14,80	13,39	6,86	9,27	15,50	16,01	17,43	14,42	11,91
Quality coefficient $Q_1 = R_{122}/R_k$	0,79	0,97	1,034	0,635	1,0	1,081	0,983	0,871	0,79	0,958	0,98	1,098	0,963
Quality coefficient $Q_2 = R_{122}/R_{10}$	0,22	0,364	1,21	0,968	0,452	0,736	1,650	1,258	0,201	0,303	0,330	0,592	1,056
Forming property space	–	3	2–3	5(6)	4–5	5	5	5	1–2	5	1–2	2–3	3–4
Rubber powder (% by mass of binder)	–	10	10	10	–	5	5	5	–	–	10	10	10
meets Czech. Standard ČSN 73 61 50		X				X						X	

LEGEND : 1) 5 cm^2 / 515 N, 40 °C, 30 min. /
2) 5 cm^2 / 515 N, 40 °C, 30 - 60 min /
3) Test conditions according to Appendix 1, ČSN 73 61 50
4) Test conditions according to Appendix 2. ČSN 73 61 50
5) evaluated according to Ref. 17, (cf. Legend below Table 3)

Table 3. - Physical and mechanical properties of fine mastic asphalts - FMA (mixtures and check tests)

Trial mix series:	14	15	16	17	18	19	20	21	22	23
Composition of mixture — Binder										
Specification designation	A - 80				A - 200				A - 200	
Softening point R.u.B. (°C)	47,6				40,6				40,6	
Penetration (mm⁻¹)	83				144				144	
Aggregates										
Content (% by mass of total mix)	8,5		8,4		8,6		8,5		8,4	
Coarse aggregate (% by mass t. mix 4-8)	34,8		34,2		34,7		34,2		33,9	
Fine agg. 0-4 (dtto)	27,5		27,0		27,5		26,9		26,7	
Mineral filler (dtto)	29,2		28,8		29,2		28,8		28,5	
Rubber powder (dtto)	-		1,6		-		1,6		2,5	
Temperature of homogenization (°C)		220	225	170	201	182	192	180		207
Time of homogenization (hours)		1,5	3,0	4,5	-	1,0	2,0	3,0	1,0	2,0
Static indentation test (mm) (1)		11,54	6,67	4,17	21,82	14,10	13,74	14,64	18,19	25,0
Increment - static indentation test (2)		4,50	2,64	0,81	6,04	4,48	6,05	4,84	4,24	6,4
Compressive strength R_k - 22°C (MPa) (3)		3,18	3,97	4,17	1,97	2,09	2,10	2,10	1,85	2,03
Flexural strength R_{122} - 22°C (MPa) (4)		2,65	3,64	4,37	1,25	2,16	2,48	3,00		1,19
Flexural strength R_{10} - 0°C (MPa)		15,46	19,0	11,56	13,29	11,15	12,53	14,51		12,35
Quality coefficient $Q_1 = R_{122}/R_k$		0,83	0,92	1,05	0,63	1,03	1,18	1,43		0,58
Quality coefficient $Q_2 = R_{122}/R_{10}$		0,17	0,19	0,24	0,09	0,19	0,20	0,21		0,10
Forming property space		1	1	5	0(7)	5	3-4	2-3	5	4
Rubber powder (% by mass of binder)			20				20		30	
meets Czech Standard ČSN 73 61 50										

Note: *Check mixture No.18*

LEGEND - cont. ; 5) 0 - fluid mixture, 1 - mixture suitable for hand spreading in thickness of 2 - 2,5 cm, for machine spreading unsuitable; 2 - suitable for hand and/or machine spreading in thickness of 3 cm, spreading in thickness of 2 - 2,5 cm difficult, 3 - suitable for machine spreading in thickness of 3 cm, and hand spreading in thickness of 4 cm, hand spreading in thickness of 3 cm difficult, 2 - 2,5 cm practically impossible, 4 - machine spreading in thickness of 3 cm and hand spreading in thickness of 4 cm difficult, hand spreading in thickness of 2 - 3 cm practically impossible.

6) marking (5) indicates unworkable mixture

7) marking (0) indicates absolutely fluidic mixture

8) rubber powder - pulverized passenger car tyres

Table 4 - Physical and mechanical properties of fine mastic asphalts - FMA (mixtures and check tests)

		Trial mix series:	24	25	26	27	28	29	30	31	32	33	Requirements of Czechoslovak Standards ČSN 73 61 50 or corresponding foreign Standards
Composition of mixture	Binder	Specification designation					λ - 80						
		Softening point R.B. (°C)					49,0						
		Penetration (mm⁻¹)					91,0						
		Content (% by mass of total mix)	7,74	7,5	7,3	7,1	7,55	7,35	7,15	6,95	6,76	7,15	
	Aggregates	Coarse aggregate (% by mass t. mix) 4-8	44,83	44,90	45,0	45,10	44,93	45,04	45,15	45,25	45,36	45,15	1 - 4
		Fine agg. 0-4 (dito)	22,87	22,9	23,0	23,1	28,43	28,49	28,56	28,63	28,70	28,56	
		Mineral filler (dito)	23,79	23,9	24,0	24,00	18,34	18,38	18,43	18,47	18,51	18,43	
		Rubber powder (dito)	0,77	0,75	0,7	0,7	0,75	0,74	0,71	0,70	0,67	0,71(8)	rec.max.0,5(0,6)
Temperature of homogenization (°C)			212	234	236	227	228	233	233	230	242	228	
Time of homogenization (hours)			3	3	3	3	3	3	3	3	3	3	
Static indentation test (mm) (1)			4,26	3,89	4,16	3,8	6,03	3,75	3,32	2,51	2,98	2,11	1 - 4
Increment - static indentation test (2)			1,01	0,98	0,91	0,90	1,29	0,65	0,65	0,57	0,47	0,41	rec.max.0,5(0,6)
Compressive strength R_k - 22°C (kPa) (3)			5,94	6,85	5,69	6,51	4,68	5,28	7,01	5,81	7,32	7,67	3,5 - 8,0
Flexural strength R_{122} - 22°C (kPa) (4)			4,65	5,81	4,88	5,56	3,93	5,19	6,01	5,23	7,12	6,91	3,0 - 7,0
Flexural strength R_{10} - 0 °C (kPa)			14,24	18,36	19,43	17,79	17,48	17,40	17,35	15,18	15,94	14,13	14 - 18
Quality coefficient $Q_1 = R_{122}/B_k$			0,78	0,85	0,86	0,85	0,84	0,98	0,86	0,9	0,97	0,9	rec. 0,8 - 1,0
Quality coefficient $Q_2 = R_{122}/R_{10}$			0,33	0,32	0,25	0,31	0,22	0,30	0,35	0,34	0,45	0,49	max. 0,5
Forming property space			2-3	2-3	1-2	5	3	1	1-2	4	4-5	5	
Rubber powder (% by mass of binder)			10	10	10	10	10	10	10	10	10	10	
meets Czech. Standard ČSN 73 61 50			10 x	x	x			x	x	x	x		

Table 5 - Physical and mechanical properties of FMA (mixtures and check tests)

		Trial mix series:	34	35	36	37	38	39	40	41
Composition of mixture	Binder	Specification designation	AP - 80							
		Softening point R.B. (°C)	50,1							
		Penetration (mm⁻¹)	72							
		Content (% by mass of total mix)	7,94	7,54	7,44	7,24	6,95	7,54	7,35	7,15
	Aggregate	Coarse aggre. 8-16 (% by mass t.m.)						12,84	12,82	12,90
		Coarse aggre. 4-8 (dtto)	45,63	45,84	45,90	46,01	46,18	33,00	33,09	33,17
		Fine aggregate 0-4 (dtto)	27,38	27,49	27,55	27,62	27,70	22,92	23,00	23,03
		Mineral filler (dtto)	18,25	18,35	18,36	18,40	18,47	22,94	23,00	23,03
		Rubber powder (dtto)	0,80	0,78	0,75	0,73	0,70	0,76 [9]	0,75 [9]	0,72 [9]
Temperature of homogenization (°C)			230	235	240	240	242	240	245	240
Time of homogenization (hours)			1,5							
Static indentation test (mm) (1)			3,77	3,63	4,56	2,70	1,86	3,34	2,55	1,63
Increment - static indentation test (2)			0,48	0,41	0,63	0,53	0,32	0,63	0,53	0,22
Compressive strength R_k - 22°C (kPa) (3)			3,83	3,66	3,80	5,30	7,02	7,79	8,76	7,65
Flexural strength R_{122} - 22°C (MPa) (4)			3,36	3,07	3,40	4,42	5,42	6,91	7,90	6,85
Flexural strength R_{10} - 0°C (MPa)			16,00	15,52	14,85	16,98	15,10	14,97	18,05	15,38
Quality coefficient $Q_1 = R_{122}/R_k$			0,877	0,899	0,895	0,834	0,772	0,887	0,902	0,895
Quality coefficient $Q_2 = R_{122}/R_{10}$			0,210	0,198	0,229	0,260	0,359	0,461	0,427	0,445
Forming property space			2	2	2	2-3	3-4	3-4	4	5
Rubber powder (% by mass of binder)			10							
meets Czech Standard ČSN 73 61 50			x	x	x	x	x	x	x	x

LEGEND - cont. : (9) rubber powder from belt conveyers

5 Road Trial Construction

On the basis of laboratory results the verification of properties of rubberized mastic asphalts was initiated in road trials. Five road trials have been laid in total. Three of them have been carried out in co-operation with the Stuctural Engineering Research and Development Institute, Ostrava, and two have been laid in co-operation with the Praque Road and Water-Service Constructions, national enterprise.

Physico-mechanical properties of fine mastic asphalt in road trials are presented in Tables 6 and 7 (Ref. 8, 9, 10, 11).

The main conclusions that can be drawn from evaluation of road trials with fine mastic asphalt with ground rubber additive include:

- most suitable for modification is the A-80 bitumen,
- with harder grades of bitumens the effect of modification is not so much demonstrative,
- as to the workability of fine mastic asphalt it has been demonsrated that the maximum of rubber additive amounts to 20 per cent by mass of bitumen content,
- the result of the modification is chiefly influenced by the temperature. At high temperatures of homogenization (above 260°C) the usability time interval for the modified mastic asphalt reduces, the optimum duration of homogenization is 1 to 2 hours at temperature of 230 to 260°C,
- in service conditions it is possible to manufacture a modified mixture of fine mastic asphalt making use of soft grade bitumens corresponding to Standards' requirements,
- application to service conditions of the results of laboratory research is possible,
- in service conditions a reduction of bitumen content in mixtures of fine mastic asphalt with comminuted rubber additive is possible providing that the aggregate grading curve should be adjusted accordingly.

Table 6 - Physical and mechanical properties of FMAs in field trials - "Ostrava" (mixtures and check tests)

Composition of mixture	Field trial series:	1a	0	1b		2	3
Binder	Bitumen kk	A 80				A 80	A 80
	Softening point R.B. (°C)	46,1				48,1	48,1
	Penetration (mm^{-1})	102				67,0	64
Aggregates	Bitumen content (% by mass of total mix)	8,4				7,6	7
	Coarse aggre. (% by mass t.m.)	37,7				39,55	28,3
	Fine aggregate (ditto)	22,5				28,45	45,0
	Mineral filler (ditto)	29,7				22,72	18,0
	Rubber (ditto)	0	1,7	1,7	1,68	1,68	1,7
	Homogenization temperature (°C)	212	260	245	290	230 – 280	240 – 290
	Homogenization time (hours)	0	4	0	1	0,5	1,5
	Static indentation test (mm)	unmeasurable	unmeasurable	0	3,08	3,68	Unmeasurable, too a plastic mixture
	Increment – static indent. test (mm)	unmeasurable	unmeasurable	0	0,64	0,73	
	Compressive strength R_k – 22°C (MPa)	2,59	1,8	3,27		4,40	
	Flexural strength R_{122} (MPa)	1,38	0,71	2,46	0,61	3,57	
	Flexural strength R_{10} (MPa)	15,00	12,09	16,64	9,83	14,67	
	Flexural strength R_{1-15} (MPa)	11,50	–	–	12,25	–	
	Quality coefficient Q_1	0,53	0,394	0,75		0,811	
	Quality coefficient Q_2	0,092	0,06	0,15	0,062	0,244	
	Quality coefficient Q_3	0,12	–	0,05	0,05	–	
	Forming property	1	1	1	1	3 – 4	
	Rubber content (% by mass of bitumen)	20				20	25

Table 7 - Physical and mechanical properties of FMAs - field trials "PSVS" [+] (mixtures and check tests)

Field trial series:	4		5			Requirements of ČSN 736150 (Ref 4)
Bitumen Mk	AP - 65		A - 80 + AP-25	A - 80 + AP-25	A-80+AP-25	
Softening point R.B. (°C)	49			51	51	
Penetration (mm^{-1})				50	50	
Bitumen content (% by mass of total mix)	8,07			7,7	7,9	
Coarse aggre. (% by mass t.m.)	43,38			48,6	48,4	
Fine aggregate (ditto)	20,75			21,1	21,00	
Mineral filler (ditto)	26,16			21,9	21,9	
Rubber (ditto)	1,64			0,7	0,8	
Homogenization temperature (°C)	220 - 240		260 - 270		250 - 260	
Homogenization time (hours)	3	5	0,5	2	1,5	
Static indentation test (mm)	5,03	5,08	2,21	2,16	2,69	1 - 4
Increment - static indent. test (mm)	0,72	0,98	0,17	0,15	0,38	rec. max. 0,6
Compressive strength R_k - 22°C (MPa)				4,39	4,38	3,5 - 8,0
Flexural strength R_{122} (MPa)	3,39	3,25		3,51	3,40	3,5 - 7,0
Flexural strength R_{10} (MPa)	15,74			18,66	18,51	14 - 18
Flexural strength R_{1-15} (MPa)	13,94			12,60	13,69	
Quality coefficient Q_1	0,96			0,8	0,78	rec. 0,8 - 1,0
Quality coefficient Q_2	0,21			0,19	0,18	max. 0,5
Quality coefficient Q_3	0,23			0,28	0,25	
Forming property	2 - 3			4	5	
Rubber content (% by mass of bitumen)	20		10			

+) Czech abbreviation for the Prague Road and Water Service Constructions

6   Observations of Deformational Characteristics at Low Temperature

The measurements consist in establishing deflections and/or creep curves of console-shaped specimens loaded statically at the temperatures of -5°C and -20°C. It is assumed that the behaviour of the bitumenous mixture is that of the rheological Burghers model where the stress-strain relation is defined as

$$\varepsilon_{(t)} = \sigma \left[\frac{1}{E_1} + \frac{t}{\lambda_1} + \frac{1}{E_2} \left(1 - e^{-\beta t} \right) \right], \quad \beta = \frac{E_2}{\lambda_2}$$

Established from the creep curves found by measurements are the following rheological constants
   - moduli of elasticity E_1 and E_2,
   - factors of normal viscosity λ_1 and λ_2.

The evaluation has been carried out for the mixture manufactured on laying the road trial No. 5. Presented in Fig. 7 are the results of creep curves measurements from which the inclination to develop cracks at low temperatures can be dependably inferred. To complement the graph it should be noted that the mixtures of fine mastic asphalt, modified by rubber crumb and sulphur, exhibit the same behaviour of creep curve at -20°C as that of a common fine mastic asphalt mixture (within the graph´s scale the

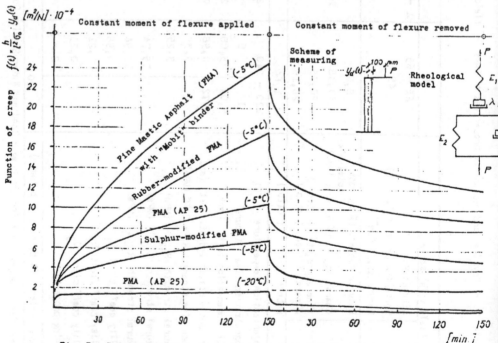

Fig. 7   Deformation time curve of bituminous materials under low temperature
- constant moment of flexure

differences would be unobservable).

From the results obtained by measurements the following partial conclusions can be drawn:

- with rubberized mixtures of fine mastic asphalt, the anticipated better deformational compliance at temperatures of about -5°C was demonstrated,
- mixtures of fine mastic asphalt, modified by 10 per cent by mass of crumb rubber in AP-65 bitumen content, do not exhibit at the temperature of -20°C any properties different from those of commonly produced fine mastic asphalts.

7 Evaluation of Mixture's Performance at Higher Temperatures

To evaluate the performance of a mixture at higher temperatures, and/or its creep compliance under loading, the creep test under simple compression has been employed. In foreign countries, results of that test have been used to anticipate rutting.

The principle of the test consists in measuring vertical deformations of cylindrical speciments loaded by simple compression.

The results of the test of the mixture produced for the road trial No. 5 are plotted in Fig. 8, for the test temperature of 40°C and loading of in 0.1 N/mm². No mandatory criteria are avalaible for the assessment of mastic mixtures from the point of view of the results of creep tests under simple compression. The assessment then should be based on comparison with previously conducted tests of mastic asphalts.

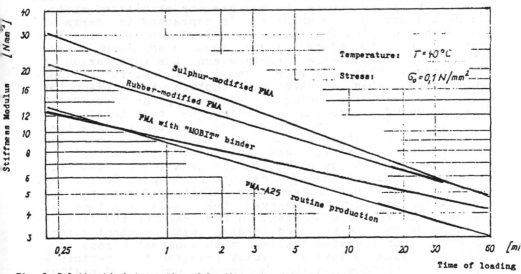

Fig. 8 Relationship between time of loading and modulus of stiffness

351

From the results obtained by measurement the following partial conclusions can be drawn:
- time pattern of stiffness modulus of rubberized fine mastic asphalt turns out to be favourable, as compared with common sorts of fine mastic asphalts. This especially applies to individual values of stiffness moduli, whereas the values of the decreases in stiffness moduli, remain in effect unchanged,
- also in comparison with "MOBIT" - modified fine mastic asphalt the rubberized fine mastic asphalt seems to be better as far as individual values of stiffness moduli are concerned.

Considering the performance of the fine mastic asphalt with crumb rubber additive in general, i view of the creep test under simple compression and at the temperature of 40ºC we can refer to the fact that the stiffness of that mixture is level with harder-grade mastic asphalt surfacings, tested by the date.

8 Resistance against Developing Plastic Deformation

In order determine the resistance of bituminous mixture against developing plastic deformations, the following tests have been carried out:
- fatigue tests,
- repeated wheel-tracking tests.

8.1 Fatigue Tests

The material under testing is subjected to repeated stressing, at least one component of which (stress, strain) is time-variable. During the test the material is physically ageing up to the rupture or stabilized state. The durability of the material is expressed in terms of the number of loading cycles required for the material rupture to take place under test conditions agreed beforehand. The material rupture and the termination of the fatigue test is very easy to identify at low temperatures when a clear-cut rupture takes place. At higher temperatures, owing to viscous properties of the material being tested, establishing the test termination is more difficult. In such a case, as the fatigue test termination we refer to the point of time at which the value of specimen´s stiffness modulus reduces to half the initial value. The specimen is stressed in the test rig exerting a cycle force of constant amplitude. During the test the phase shift between the exciting force and the deformation produces is observed. The results obtained are evaluate in Wöhler´s diagram. In the case of the fatigue tests performed by repeated loading with constant stress the number of repeated loading with constant stress the number of repeated loading cycles required for specimen´s rupture is plotted on the horizontal axis, and maximum stress amplitude on the vertical. For various temperatures a system of parallel isotherms of the dependence $\log \tilde{\sigma}$ versus $\log N$ is obtained, cf. Fig. 9 (Ref. [13], [14]).

The material fatigue of rubber crumb/mastic asphalt mixture is a complex process, in the course of which the mixture´s physical and mechanical properties gradually change due to repeated stress, and which results in failure of various character.

With growing number of repeated loading the absolute value of the complex modulus (modulus of stiffness) reduces, namely the higher the action stress, the faster the effect of the viscous component of the mixture upon the whole of its properties increases.

At the temperatures of +40°C and +30°C within the range of doubling maximum deformation, the rubber/mastic asphalt specimen did not fail a rupture. In the bulk of cases the increase in deformation was regular and uniform in the course of entire test.

The higher the stress, the lower the number of repeated loadings needed for failure to take place; for both rupture and the failure defined by agreement, whichever the case is. With temperature decreasing and stress being equal, the number of repeated loadings needed for specimen´s failure increases.

The fatigue tests with rubber crumb/mastic asphalt mixtures have provided evidence that a straight-line relationship between $\log \tilde{\sigma}$ and $\log N$ in Wöhler´s diagram is valid, with constant temperature and stresses within the range used. At the same time it is obvious that individual isotharms can be regarded with sufficient accuracy as paralell.

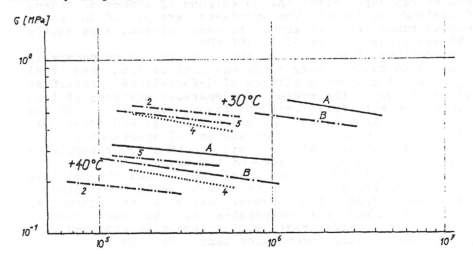

Fig. 9 Comparison of fatigue tests results ;   — induced stress
           N   — traffic (No of axles)

In view of the fatigue test, the mixture containing 20 per cent of rubber from road trial No. 2 appears to be the worst. To achieve the half value of the stiffness modulus calls for unproportionally lower number of repeated loadings than with rest of mixtures. For exemple, applying stress of 0.214 MPa, the specimen from road trial No.2 requires $6.3*10^4$ repeated loadings, whereas the specimen from road trial No. 4 as $2.88*10^5$ repeated loadings.

In so far as we assess the results of fatigue tests in comparison with the earlier conducted tests on commonly used materials we may conclude that they are comparable. It is to be noted that the binder content in the A-mixture has amounted to 8.5% (1.75% "Selenice" and 7% B-65), and in the B-mixture to 8.3% (AP-25). By contrast, for rubberized mastic asphalts specimens we have (i) binder content 7.6% (A-80) in road trial No. 2, (ii) 8.07% (AP-65) in road trial No.4, and (iii) 7.9% (A-80 + A-25 - penetration 50, R.B. 51) in road trial No. 5. In view of this fact the results of the fatigue tests are very favourable.

8.2 Whell-tracking test

The deformational properties of rubberized mastic asphalts have been investigated again on specimens made from the mixtures which has been laid on road trials, and on laboratory-tested mixtures with the rubber crumb produced from scrap conveyer belts. For the sake of comparison, several samples from production of the national enterprise "Prague Road and Water-Service Consructions", Division 6 at Stredokluky, have also been tested.

Employed for evaluation is the longitudinal profile area of rut depth after the corresponding number of passes of testing equipment. The specimens are placed in water bath at temperature of 40°C. The wheel contact area equals 7.68 cm² at the stress of 0.7108 MPa

As follows from Fig. 10, the results can be generally assessed as satisfactory. Only one mixture from road trial No. 5 exhibits lower qualities of deformational properties and falls beyond the range of comparable mixtures of fine mastic asphalt from routine production. With FMA the deformations grow faster in the beginning, and after some 1000 passes. Upon visual inspection of specimens it has been found out that with rubberized FMAs minor expulsion of the mixture ocurs on sides of the track.

The mixtures from road trial No. 2 and the mixture No. 40 (cf. Table 5) - fine mastic asphalt modified by crumb rubber from scrap conveyer belts can be regardet as very good They are comparable to the best results achieved in coarse-grade mastic asphalts produced with A-25 binder (static indentation test), Ref. 16.

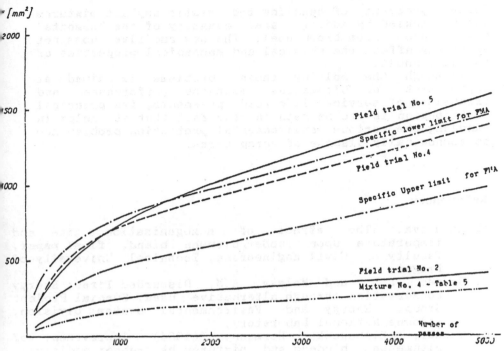

Fig. 10 Wheel-tracking test results - relation between number of passes and longitudinal profile area

9 Overall evaluation of mastic asphalt modified by scrap rubber

It follows from the submitted study that using rubber crumb for modification of asphalts is practicable in the conditions of our country, and represents a valuable asset in the energy budget.

On the other hand, it should be noted that the verification has carried out only for the technology of production and laying of mastic asphalts with the "Karkasa" rubber crumb. Had we introduced the mentioned technology on full scale with, for exemple, "Prague Road and Water-Service Constructions", hardly 2000 tons of rubber crumb could be handled annually, considering total production of 10^5 tons of mastic asphalts per an.. This implicates that our further effort should be directed at the technology of asphaltic concretes, macro surface dressings, flexible membranes, and sealings.

For practical application in the production of modified mastic asphalts it is neccessary to complement the mixing plant with the homogenization equipment. Its task is speed up the homogenization process, and to reduce to minimum the operation stresses of the precoating equipment.

355

Certain problems of handling the mastic asphalt mixtures can be tackled by adding a small quantity of the "Romonta" wax produced from brown coal. The wax additive does not adversely affect the physical and machanical properties of mastic asphalt.

Although the solving these problems is aimed at improvement of bitumenous mixtures performance and extension of service life road pavements, its principal contribution is to be seen in the fact that it helps in handling the serious environmental protection problem how to economically dispose of scrap tyres.

References

1. Micanova, The effect of homogenization time and temperature upon rubber/bitumen blend, **final paper,** Faculty of Civil Engineering, Technical University of Prague,
2. Gaines, L.L. and Wolsky, A.M. Discarded Tires: Energy conservation through alternative uses, Special Project Group, Energy and Environmental System Division, Argonne National Laboratory,
3. Bednar, Kloboucek, Hermann, (1984) Improvement of bitumenoes binders and mixtures by rubber additives, **final report of research project P-12-526-01,** Faculty of Civil Engineering, Technical University of Prague,
4. Czechoslovak Standart CSN 73 6150 "Mastic asphalt for pavements"
5. Musil, (1984) The effect of homogenization time and temperature upon rubber/bitumen blend, **final paper,** Faculty of Civil Engineering, Technical University of Prague,
6. Hrib, (1984) Evaluation of Rubber-modified bitumen, **final paper,** Faculty of Civil Engineering, Technical University of Prague,
7. Kalabinska, Pilat, Wyszynski, (1985) Rheological properties of bitumens with rubber reclaim, Warsaw Polytechnics,
8. Vorlicek, (1984) The effect of rubber additive on physico-mechanical properties of mastic asphalts, **final paper,** Faculty of Civil Engineering, Technical University of Prague,
9. Lausman, (1984) The effect of rubber additive on physico-mechanical properties of mastic asphalts, **final paper,** Faculty of Civil Engineering, Technical University of Prague,
10. Hanzik, Vanek, (1983) Verification of potential processing of rubber crump in bitumenous mixtures production, **summary of laboratory results,** Prague Road and Water-Service Constructions,

11. Hanzik, Vanek, (1984) Verification of potential processing of rubber crump in bitumenous mixtures production, **summary of laboratory results**, Prague Road and Water-Service Constructions,
12. Novak, (1985) The effect of rubber additive on physico-mechanical properties of mastic asphalt, **final paper**, Faculty of Civil Engineering, Technical University of Prague,
13. Konecna, (1985) Fatigue of rubberized mastic asphalts, **final paper**, Faculty of Civil Engineering, Technical University of Prague,
14. Luxemburk, F. and Lehovec, F. (1975) Investigation into measurements of dynamic modules, and establishment of relationship between number of loading repetitions and service life of bound pavement materials, **final report of research project P-12-256-074-00-037**, Technical University of Prague,
15. Vanek, (1985) Technical experience gained in Prague Road and Water Service Constructions, upon introducing production of modified mastic asphaplts for municipal communications, **Advanced Pavement maintenance Review**, Technical Centre of CSVTS, Bratislava,
16. Masek, (1985) Macro surface dressings, **final paper**, Faculty of Civil Engineering, Technical University of Prague,
17. Hanzik, (1975) Contribution to determination of properties of pavement mastic asphalts, with evaluation of suitability of AP-25 bitumen, **doctoral thesis**, Technical University of Prague,

24 A LABORATORY STUDY OF THE EFFECTS OF BITUMEN CONTENT, BITUMEN GRADE, NOMINAL AGGREGATE GRADING AND TEMPERATURE ON THE FATIGUE PERFORMANCE OF DENSE BITUMEN MACADAM

M.J. BRENNAN
Department of Civil Engineering, University College,
Galway, Ireland
G. LOHAN
Cold Chon (Galway) Ltd., Ireland
J.M. GOLDEN
Environmental Research Unit, Department of the Environment,
Dublin, Ireland

Abstract
Since 1985, several studies have been carried out at
University College Galway to investigate the fatigue
performance of bituminous materials in repeated uni-axial
tension-compression. The bituminous specimens are
compacted in standard cylindrical moulds using a double
acting static load. The most recent study on dense
bitumen macadam under constant stress loading conditions
would on balance suggest a tentative conclusion that
increasing temperature has a beneficial effect on the
initial strain fatigue life relationship. The same trend
was observed to occur when the stiffness of the mix was
decreased (by using either a higher test temperature or a
softer bitumen). For the limited range of variables
considered, nominal aggregate grading, test temperature
and bitumen grade were found not to significantly affect
the positions of the stress fatigue life relationships.
Increasing bitumen content in the range of from 3.4 to
4.6% was found to significantly improve the positions of
both the initial strain and stress fatigue lines.

1 Introduction
The acid test for proving the performance of a new
roadmaking material is to lay it in a full-scale trial
under real traffic conditions and monitor its progress
over a long period of time. Alternatively, the lengthy
time scales that are associated with such tests can be
reduced by using special test tracks in which roadmaking
materials can be subjected to repeated heavy wheel loads
at much shorter intervals than occur under traffic. Such
facilities and, indeed, full scale trials are so expensive
that they are mainly used for studies of strategic
importance. Researchers in small institutions engaged in
exploratory research on new materials are often limited to
studying behaviour using proprietary laboratory testing

358

machines. This is the case at University College Galway
where research is ongoing on the fatigue performances of
modified bituminous materials. In order to assess the
performances of new products it was necessary at the
outset to establish the performances of standard products
using the available testing facilities. This article
presents an account of the most recent study on dense
bitumen macadam.

2 Scope

The objectives of the study were to compare the
stiffnesses and fatigue lives of three different bitumen
macadams, namely : a 28mm nominal size roadbase material
manufactured using a 100 pen bitumen ; a 14mm nominal size
wearing course material manufactured using a 100 pen
bitumen ; and a 14mm nominal size wearing course
manufactured using a 200 pen bitumen.

Both the 28 and 14mm materials manufactured using the
100 pen bitumen were made up at three bitumen contents :
4.6, 4.0, and 3.4% bitumen. These are the upper, middle
and lower limit of the specified range for a 28mm material
; and the middle, lower limit and below the specified
range for a 14mm material, British Standards Institution
(1973). The 14mm material manufactured using 200 pen
bitumen was made up using a bitumen content of 4.6%
bitumen only.

It was necessary to test the materials at the ambient
laboratory temperatures in Winter and Summer. This made
it possible to compare results at two test temperatures :
$16^{o}C$ and $22^{o}C$. Accordingly, an opportunity was provided
to investigate the "strain criterion" i.e. that the
applied tensile strain controls the fatigue life and that
the effect of temperature can be accounted for by its
effect on the stiffness of the specimen, Pell (1962) and
Pell and Cooper (1975).

3 Test method

Laboratory moulded cylindrical specimens of the bituminous
materials were tested in repeated uniaxial loading using a
computer controlled Dartec 9500 20 kN servo hydraulic
actuator. The tests were carried out under controlled
stress conditions as this mode of loading has better
reproducibility and shorter test durations than the
controlled strain mode, Celard (1977).

A sinusoidal loading form with zero mean stress was
applied to the specimens. The frequency of 10 Hz that was
used corresponds to an average vehicular speed of 65 km/h.
The applied stresses were chosen to cause failure between
5,000 and 300,000 load applications. In this way, most

tests were completed within a normal working day and under reasonably uniform temperature conditions.

The initial strains and stiffnesses were recorded after 200 loading cycles. Failure was defined by the complete fracture of the specimen.

4 Preparations of the test specimens

The specimens were produced using standard 100mm diameter cylindrical moulds, British Standards Institution (1970). Each specimen was manufactured using a mass of 3200 g of aggregate. The aggregate, the bitumen and the moulds were heated in a thermostatically controlled oven for approximately six hours, two hours and one hour, respectively. The temperatures were regulated at 120°C using a 200 pen bitumen and at 140°C using a 100 pen bitumen. These temperatures are within the ranges recommended for manufacture by the British Standards Institution, British Standards Institution (1973).

The aggregate and the bitumen were mixed in the pre-heated bowl of a 12 litre capacity Hobart mixer for one-and-a-half minutes. Preliminary trials established that the rotation of the blade of the mixer would eject aggregate retained on the 20mm B.S. sieve from the bowl during mixing. For this reason, it was necessary to omit the 20 to 28mm fraction from the 28mm nominal grading. Accordingly, the grading that was adopted was slightly out of specification above the 20mm sieve : 100% passed the 20mm sieve instead of 95%. The particle size distributions of the 14 and 28mm gradings used in the study are presented in Figure 1.

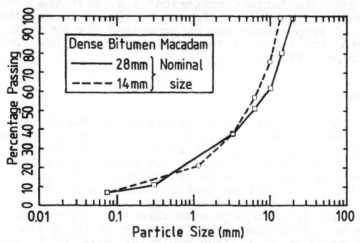

Fig. 1. Grading curves for the 14 and 28mm nominal aggregates

The hot mixture was placed in three layers in the mould.
Each layer was hand tamped 60 times with a steel tamping
rod. A simple collar was attached to the top of a mould
to accommodate loose material before compaction.

The material was compressed using a double acting
static load of 100 kN that was maintained for five
minutes. This load has been found to produce uniform
specimens of dense bitumen macadam, hot rolled asphalt and
grave-emulsion with voids contents that compare
satisfactorily with typical in situ values : a load of 125
kN has been required to adequately compact asphaltic
concrete, Brennan and Sweeney (1986) and Brennan and
O'Kelly-Lynch (1988). After a specimen had cooled
overnight in the mould, it was extruded using an hydraulic
jack. A compacted specimen had a height of approximately
182mm. The specimens were left at room temperature for
one month before testing.

The final preparations were made on the eve of a test.
Firstly, steel plates with 20mm high collars were glued to
the ends of a specimen with epoxy resin in a jig that
ensured that they were perpendicular to its longitudinal
axis. These end plates served the purpose of transmitting
the loads from the fatigue testing machine to the
specimen. Secondly, linear variable displacements
transducers were mounted diametrically opposite on the
specimen. The locating nuts for mounting the transducers,
spaced 100mm apart, were simply bonded to the specimen
with epoxy resin. The transducers were used to record the
strains in the course of a test.

As a specimen neared the end of its fatigue life, it
was necessary to disconnect the displacement transducers
in order that they might not be crushed by coming loose
due to the large vibrations that were transmitted.

Table 1. Mean densities and voids contents of the
specimens together with the coefficients of variation
(C.V.)

Nominal size	Bitumen content	Density (g/cm^3)		Voids content (%)	
		Mean	C.V.	Mean	C.V.
14mm	4.6%	2.337	0.31%	7.2	4.1%
	4.0%	2.354	0.42%	7.4	5.2%
	3.4%	2.331	0.63%	9.1	6.1%
28mm	4.6%	2.351	0.32%	6.6	4.6%
	4.0%	2.362	0.37%	7.0	6.1%
	3.4%	2.353	0.29%	8.3	3.2%

The mean densities and mean voids contents of the compacted specimens together with the coefficients of variation are given in Table 1. The lower voids content for the 28mm nominal size material is possibly attributable to the fact that its grading curve is closer to a Fuller's curve than that of the 14mm material.

5 Experimental results

The mean stiffnesses that were obtained for the six combinations of bitumen content and nominal aggregate grading using the 100 pen bitumen at 22°C are given in Table 2.

Excepting the 28mm material with 3.4% bitumen, there is a trend for stiffness to increase with the bitumen content for the range under consideration.

The effects of bitumen grade and test temperature on the mean stiffnesses of the specimens with 14mm nominal aggregate and 4.6% bitumen are summarised in Table 3. An average increase in stiffness of the order of 50% was obtained by decreasing the test temperature or by using the harder bitumen.

Estimates of the stiffnesses were derived from the characteristics of the materials using the equations developed by Heukelom and Klomp (1964), and Ullidtz(1979). For predicting the stiffness of the bitumen, it was assumed that the softening point increased by 10% and that the penetration decreased by 35% during mixing. The ratio of estimated to measured stiffness ranged from 0.8 to 2.0.

Both initial strain fatigue life and stress fatigue life relationships were established for each material by carrying out linear regressions of the logarithm of fatigue life, N_f, on the logarithms of the initial strain, ε, and the applied stress, σ, respectively. The parameters of the fatigue lines that were obtained are given in Table 4. The coefficients of correlation, R, ranged from 0.79 to 0.97 for the initial strain fatigue lines and from 0.56 to 0.92 for the stress fatigue lines.

Table 2. Mean stiffnesses of the specimens using 100 pen bitumen at 22°C.

| | | Nominal aggregate grading | |
		28mm	14mm
Bitumen	4.6	4100 MN/m²	3000 MN/m²
content	4.0	2700 MN/m²	2800 MN/m²
(%)	3.4	2800 MN/m²	2500 MN/m²

Table 3. Mean stiffnesses of the specimens with 14mm nominal aggregate and 4.6% bitumen.

| | | Test temperature | |
		22°C	16°C
Bitumen	200 pen	1800 MN/m^2	2700 MN/m^2
grade	100 pen	3000 MN/m^2	4200 MN/m^2

Accordingly, a better degree of correlation was obtained with the initial strain fatigue lines.

Two criteria were used to test the significance of the difference between two fatigue lines. The principal test was the standard statistical technique for checking whether the slopes and the intercepts of the regression lines are the same at the 5% confidence level, Neter and Wasserman (1974). In the event that two lines, found to be statistically different, crossed or appeared to be close when plotted, a second test was carried out to establish the extent of overlap of the 90% confidence

Table 4. The parameters of the fatigue lines.

| Nominal aggregate grading, bitumen content, bitumen grade and test temperature | | | | Fatigue model | | | |
| | | | | $N_f = C (1/\varepsilon)^m$ | | $N_f = K(1/\sigma)^n$ (kN/m^2) | |
Nom. size	% bit.	Bit. grade	Test temp.	C	m	K	n
14mm	4.6	200	22°C	1.29×10^{-7}	2.88	3.95×10^{11}	3.08
14mm	4.6	100	22°C	2.15×10^{-8}	2.97	1.52×10^{18}	5.76
14mm	4.0	100	22°C	1.20×10^{-7}	2.73	1.13×10^{33}	12.63
14mm	3.4	100	22°C	7.43×10^{-11}	3.43	3.08×10^{12}	3.75
28mm	4.6	100	22°C	3.95×10^{-8}	2.87	1.18×10^{18}	5.61
28mm	4.0	100	22°C	2.92×10^{-12}	3.78	3.06×10^{18}	6.36
28mm	3.4	100	22°C	1.21×10^{-16}	4.68	4.07×10^{14}	4.87
14mm	4.6	100	16°C	5.21×10^{-9}	3.06	1.52×10^{15}	4.38
14mm	4.6	200	16°C	5.62×10^{-8}	2.88	1.13×10^{14}	4.09

bands of the regression lines.

At bitumen contents of 4.6 and 4.0%, the different aggregate gradings were found not to affect the positions of either the initial strain or stress fatigue lines. At a bitumen content of 3.4%, the initial strain fatigue lines were different but the stress fatigue lines were not.

Increasing bitumen content improved the positions of both the initial strain and stress fatigue lines for the range of bitumen contents considered. The beneficial effect was greater for the stress fatigue lines. The increase in bitumen content from 3.4 to 4.6% extended the strain fatigue lives at least twofold whereas the stress fatigue lives were extended at least tenfold. The stress fatigue lines reflect the additional benefit of increased stiffness that accrues by increasing the bitumen content.

The comparison of the fatigue lines with 100 and 200 pen bitumen at the two test temperatures showed that grade of bitumen has a significant effect on the initial strain fatigue lines. The mix with the softer bitumen had a longer fatigue life. However, the effect was not significant for the stress fatigue lines. Thus, the reduction in stiffness that resulted from using the softer bitumen cancelled the improvement in fatigue life at a given initial strain.

The initial strain fatigue lines for the materials with 200 pen bitumen tested at the two test temperatures are shown in Figure 2 ; those for the materials with 100 pen bitumen are shown in Figure 3. Although the fatigue lines at $16°C$ and $22°C$ are close, the differences between them are statistically different. These results do not support the strain criterion.

Figures 2 and 3 also show the 90% confidence bands of the initial strain fatigue lines. The areas that are crosshatched indicate the extent to which the confidence bands overlap. Consequently, fatigue lines could be drawn for each mix that would be independent of test temperature. Such lines would support the strain criterion. It is pertinent to note that differences in the ratios of fatigue lives of the order of six have been accepted in promoting the strain criterion. On this basis for making comparisons, the results of the study could reasonably be interpreted as supporting it.

Test temperature was found not to significantly affect the positions of the stress fatigue life relationships. If the strain criterion applied, one would expect that the stress fatigue lines would have a greater dependence on temperature than the initial strain fatigue lines. Thus, although the evidence is not definitive, the results obtained would seem on balance to show that increasing temperature has a beneficial effect on the initial strain fatigue life relationship.

The trend for the initial strain fatigue life

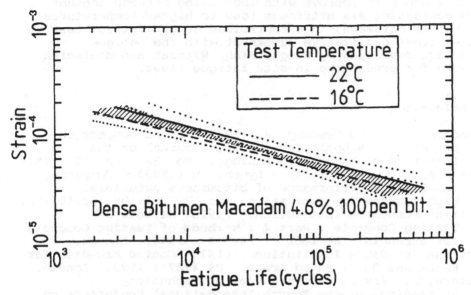

Fig. 2. Effect of temperature on the initial strain
fatigue life relationship for the 100 pen bitumen.

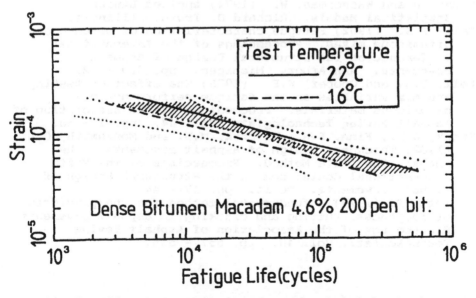

Fig. 3. Effect of temperature on the initial strain
fatigue life relationship for the 200 pen bitumen.

relationship to improve with increasing bitumen content and decreasing mix stiffness (due to higher temperatures or softer bitumens) in a controlled stress fatigue test is in at least qualitative agreement with the fatigue equation developed by Shook, Finn, Witczak and Monismith (1982) for predicting in situ fatigue lives.

6 References

Brennan, M.J. and Sweeney, A. (1986) Duriez compaction procedure. **Ashpalt Technology.** Journal of the Institute of Asphalt Technology, No. 33, pp. 12 - 15.

Brennan, M.J. and O'Kelly - Lynch. W (1989). Comparing the fatigue performance of bituminous materials. **Highways and Transportation.** No. 2, Vol. 36, pp.10 -13.

British Standards Institution. (1970) **Methods of Testing Concrete : Part 4 : Methods of Testing Concrete for Strength. BS 1881 : Part 4 : 1970.** London.

British Standards Institution. (1973) **Coated Macadams for Roads and Other Paved Areas. BS 4987 : 1973.** London.

Celard, B. (1977) Esso road design technology. **Proceedings of the Fourth International Conference on the Structural Design of Asphalt Pavements.** Ann Arbor. Michigan. pp. 249 - 268.

Heukelom, W. and Klomp. A.J.G. (1964) Road design and dynamic loading. **Proceedings of the Association of Asphalt Paving Technologists,** Vol. 33, pp. 92 - 123.

Neter, J and Wasserman, W. (1974) **Applied Linear Statistical Models.** Richard D. Irwin. Illinois.

Pell, P.S. (1962) Fatigue characteristics of bitumen and bituminous mixes. **Proceedings of the International Conference on the Structural Design of Asphalt Pavements.** Ann Arbor. Michigan. pp. 310 - 323.

Pell, P.S. and Cooper, K.E. (1975) The effect of testing and mix variables on the fatigue performance of bituminous materials. **Proceedings of the Association of Asphalt Paving Technologists,** Vol. 44, pp. 1 - 33.

Shook, J.F., Finn, F.N., Witczak, M.W. and Monismith, C.L. (1982) Thickness design of asphalt pavements - the Asphalt Institute Method. **Proceedings of the Fifth International Conference on the Structural Design of Asphalt Pavements.** Delft. pp. 17 - 44.

Ullidtz, P. (1979) A fundamental method for the prediction of roughness, rutting and cracking in asphalt pavements. **Proceedings of the Association of Asphalt Paving Technologists,** Vol. 48, pp. 557 - 586.

25 L'ESSAI DE MODULE COMPLEXE UTILISE POUR LA FORMULATION DES ENROBÉS

(Complex modulus testing for coating formulation)

J.J. CHAUVIN
Centre D'Études Techniques De L'Équipement de Bordeaux,
France

Résumé
Cet article décrit une méthode de mesure dynamique du module de rigidité par flexion deux points ; les éprouvettes, de forme trapézoïdales sont obtenues par sciage des plaques compactées en laboratoire ou extraites d'une chaussée. L'essai est réalisé à 6 températures et pour chaque température à 4 fréquences ; il est entièrement géré par micro-ordinateur : pilotage, mesures, exploitation, édition des résultats. A travers un plan d'expérience, cet essai a permis de mettre en évidence l'influence sur la valeur du module de caractéristiques comme la teneur en bitume, la granulométrie, l'énergie de compactage. Associé à la tenue en fatigue (définie comme étant la déformation relative conduisant à une rupture du matériau pour un million de chargement), le module permet de caractériser le matériau en terme d'épaisseur de dimensionnement et ainsi de mieux apprécier l'importance respective des différents paramètres de formulation.
Mots-clés : Enrobés, Essais, Formulation, Laboratoire, Module complexe, Micro-informatique, Fatigue, Dimensionnement.

1 Introduction

Au cours des dernières années se sont développés pour les besoins de la construction routière des enrobés très divers, tant dans leur formulation, que dans leur emploi : enrobés pour couches de base, pour couches de roulement épaisses ou couches d'entretien minces, enrobés drainants ... dans chaque cas il est nécessaire d'optimiser la composition et de définir l'épaisseur d'usage : la qualité des matériaux, la nature et le dosage en liant, le pourcentage de vides, autant d'éléments dont il faut évaluer l'importance car ils conditionneront le coût, mais aussi la durée de vie de la chaussée.

Parmi l'ensemble des essais permettant d'évaluer les caractéristiques mécaniques des enrobés bitumineux, la mesure du module dynamique de rigidité constitue une donnée intéressante ; il s'agit en effet d'une caractéristique utilisable dans un calcul rationnel de dimensionnement; la variation du module liée à des variations de formulation et associée à la tenue en fatigue pourra directement s'exploiter en terme d'épaisseur, chose difficilement réalisable avec les essais technologiques traditionnels.

Les deux méthodes les plus répandues pour l'évaluation du module dynamique sont celle en traction - compression et celle en flexion ; - c'est cette dernière méthode qui est employée en France.

2 Principe de l'essai

Il est basé sur la méthode de l'impédance mécanique qui consiste à établir le rapport force sur déplacement.

Une éprouvette à profil trapézoïdal d'épaisseur constante, encastrée à sa base est sollicitée sinusoïdalement à son extrêmité libre (Figure 1). Les vibrations sinusoïdales en flexion sont destinées à réaliser une simulation du trafic.

Si on applique à cette éprouvette visco-élastique une contrainte de la forme $\sigma = \sigma_0 \exp(i\omega t)$, la déformation relative qui en résultera sera de même fréquence mais déphasée par rapport à cette contrainte : $\varepsilon = \varepsilon_0 \exp i(\omega t - \varphi)$

Le module complexe est donc :

$$E^* = \frac{\sigma}{\varepsilon} = \frac{\sigma_0}{\varepsilon_0} \exp(i\varphi) \tag{1}$$

σ : contrainte, σ_0 contrainte max

$\varepsilon = \frac{\Delta \ell}{\ell}$ déformation relative, ε_0 valeur max

$\omega = 2\pi f$: pulsation, f : fréquence

φ : angle de déphasage ; sa valeur rend compte du caractère visqueux ou élastique du matériau ($\varphi = 0$ pour un corps parfaitement élastique).

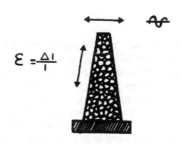

FIGURE 1 . PRINCIPE DE L'ESSAI

La forme de l'éprouvette est destinée à obtenir une zone de sollicitation homogène dans la partie centrale.

3 Préparation des éprouvettes

Les enrobés expérimentés proviennent de deux sources principales :
a) sur chaussée : on prélève une plaque d'enrobés dans le sens de la circulation à l'aide d'une scie circulaire équipée d'une lame à diamants
b) fabrication en laboratoire : après la reconstitution pondérale de la formule d'enrobés bitumineux à étudier, on procède au malaxage dans un malaxeur grande capacité, puis au compactage ; cette opération est réalisée avec un compacteur à pneus de laboratoire permettant de réaliser des plaques de 600 x 400 mm ayant 120 mm d'épaisseur.
Ces plaques sont ensuite découpées en 4 tronçons de 250 x 150 mm ; chaque tronçon comportera un numéro de plaque et permettra de réaliser 5 éprouvettes également numérotées.
Une plaque permettra la réalisation de 20 éprouvettes **(Figure 2 et 3).**

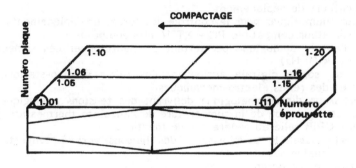

FIGURE 2   DECOUPAGE DE LA PLAQUE EN 4 TRONCONS

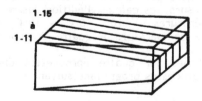

FIGURE 3 : DECOUPAGE DE CHAQUE TRONCON EN 5 EPROUVETTES

Toutes les éprouvettes devront être soigneusement mesurées et pesées, afin de définir leurs caractéristiques géométriques et physiques :
    . il convient en effet d'expérimenter des séries d'éprouvettes homogènes en compacité
    . les caractéristiques géométriques sont nécessaires pour transformer la mesure de flèche en valeur de déformation relative.

4   Machine d'essai

4.1   Description (Photos 1 et 2)
La chaîne de mesure est constituée des élèments suivants :
* Une enceinte thermostatée (- 10°C à + 40°C) contenant :
    . 1 socle rigide en acier mécano - soudé
    . 4 vibreurs de 18 daN
    . 4 capteurs de force (piezo)
    . 4 capteurs de déplacement (inductifs)
* Une armoire électronique comprenant :
    . 1 synthétiseur générateur de fonction
    . 4 amplificateurs de puissance (500 W)
    . 4 amplificateurs de force
    . 4 amplificateurs de déplacement
    . 8 afficheurs numériques pour le contrôle des forces et déplacements
* Un micro-ordinateur compatible PC - XT 16 bits équipé de :
    . 1 carte d'entrée analogique permettant d'échantillonner des signaux sinusoïdaux (< 50 Hz)
    . 1 carte de sortie digitale actionnant des relais électro-statiques opto-isolés et des relais électro-mécaniques
    . 1 convertisseur digital-analogique délivrant des tensions analogiques (commande et régulation de la température de l'enceinte thermostatée)
    . 1 interface GPIB relié au générateur de fonctions
    . 1 carte sortie parallèle pour assurer des liaisons vers 4 convertisseurs digitaux/analogiques
    . 1 imprimante et un écran
    . 1 lecteur de disquettes
    . 1 disque dur

4.2   Fonctionnement (Figure 4)
L'essai est entièrement géré par le micro -ordinateur qui assure la pilotage, la régulation, la mesure, les calculs, l'édition des résultats.
Il est réalisé à 6 températures : - 10°C, 0°C, + 10°C, + 20°C, + 30°C, + 40°C et pour chaque température à quatre fréquences : 30, 10, 3 et 1 Hz.
Les mesures sont réalisées sur quatre éprouvettes disposées dans la même enceinte selon le processus programmé suivant :

PHOTO 1 . VUE GENERALE DE LA MACHINE D'ESSAI

PHOTO 2 . EPROUVETTES D'ESSAI

1 affichage de la première température (- 10°C) et attente de 4 heures

2 au bout de 4 heures : connection du vibreur 1

3 mise à zéro de l'amplificateur de déformation 1

4 envoi de la première fréquence (30 Hz)

5 mesure de la force

6 mesure du déphasage

7 répétition des mesures (4 fois) et mise en mémoire des moyennes

8 passage à la fréquence suivante (10 Hz) et réalisation des séquences 5 à 7

9 mêmes séquences (5 à 7) aux deux dernières fréquences (3 et 1 Hz)

10 deconnection du vibreur 1 et connection du vibreur 2

11 séquences 3 à 9 sur vibreur 2, déconnection et connection du vibreur 3

12 séquence 3 à 9 sur vibreur 3, déconnection et connection du vibreur 4

13 séquence 3 à 9 sur vibreur 4 déconnection

14 affichage de la deuxième température (0°) et attente de 4 heures

15 répétition des séquences 2 à 14 à 0°C, + 10°C, + 20°C, + 30°C et + 40°C - arrêt des mesures

16 édition des résultats sur imprimante et classement dans un fichier "résultats" du disque dur du micro-ordinateur.

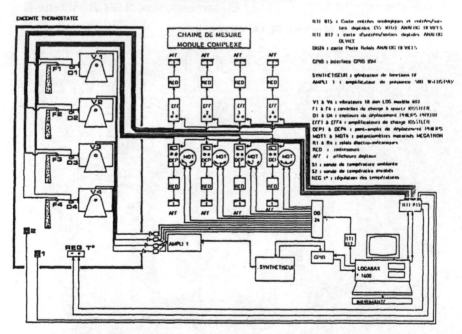

FIGURE 4 . SCHEMA DE LA CHAINE DE MESURE

Dans sa configuration actuelle, cette machine d'essai permet également de réaliser des essais de fatigue avec temps de repos (par salves) ; c'est ce qui justifie la présence dans l'armoire électronique des quatre amplificateurs de puissance ; dans ce cas, les quatre voies fonctionnent simultanément : on programme l'amplitude de sollicitation, la longueur des trains d'onde et celle du temps de repos entre chaque train d'onde.

5 Présentation des résultats

5.1 Tableau des valeurs
Après avoir mis en mémoire les valeurs de la force, du déplacement et de l'angle de déphasage à chaque fréquence et chaque température, le micro-ordinateur effectue le calcul du module complexe E^*

$$E^* = \sqrt{E_1^2 + E_2^2} \quad \text{et} \quad \varphi = \text{arc tg} \frac{E_2}{E_1} \quad (2)\,(3)$$

$$E_1(\omega) = \gamma \frac{Fo}{Zo} \cos \varphi + \mu M \omega^2 \quad (4)$$

$$E_2(\omega) = \gamma \frac{Fo}{Zo} \sin \varphi \quad (5)$$

les différents symboles employés ayant la signification suivante :
E_1 partie réelle, E_2 partie imaginaire du module
Fó force appliquée, Zo flèche correspondante
μ facteur dépendant de la géométrie de l'éprouvette ($\mu = 0,135$)
M masse de l'éprouvette
ω pulsation ($\omega = 2\pi f$), f : fréquence
γ facteur dépendant des dimensions de l'éprouvette

$$\gamma = \frac{12\,L^3}{b(ho-hi)^3} \left[(2 - \frac{hl}{2ho}) \frac{hl}{ho} - \text{Log} \frac{hl}{ho} - \frac{3}{2} \right]$$

L (hauteur de l'éprouvette)
ho (grande base)
hl (petite base)
b (épaisseur)

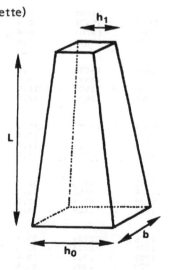

FIGURE 5 . DIMENSIONS DE L'EPROUVETTE

373

Le **Tableau** 1 est un exemple de présentation des résultats ; il est accompagné de différents graphiques illustrant le comportement du matériau :
* courbes isochrones **(Figure 6)** donnant pour chaque fréquence la variation du module en fonction de la température
* courbes isothermes **(Figure 7)** donnant pour chaque température la variation du module en fonction de la fréquence
* courbe maitresse **(Figure 8)** : cette courbe permet d'évaluer le module à des fréquences extérieures à la gamme des fréquences expérimentées, elle est basée sur l'équivalence entre la température et le temps ; elle s'obtient par décalage des différentes courbes isothermes par rapport à la courbe obtenue à 10°C qui reste fixe ; l'opération s'effectue directement sur l'écran du micro-ordinateur
* relation déphasage - module dite courbe de Black **(Figure 9)** caractérisant la structure bitumineuse du matériau.

TABLEAU 1 . EXEMPLE DE PRESENTATION DE RESULTATS

RESULTATS MODULE COMPLEXE
GB 0/14 NOUBLEAU 6.2% BIT.DUR 10/20

GB 0/14 NOUBLEAU
6.20% BITUME DUR 10/20
COMPACTAGE MOYEN-FAB.LABO
EXPERIMENTATION MANEGE DE FATIGUE

compacite moyenne = 97.00

T	F-Hz	E1-MPA	E2-MPA	E*-MPA	TETA
-10	30	25023	2143	25116	4.9
-10	10	24385	1563	24440	3.7
-10	3	23345	1391	23391	3.5
-10	1	22358	1917	22442	4.9
0	30	21116	2505	21265	6.8
0	10	20026	2080	20136	5.9
0	3	18580	2105	18701	6.4
0	1	17054	2272	17207	7.6
10	30	15850	2785	16094	10.0
10	10	14282	2441	14491	9.7
10	3	12364	2436	12608	11.2
10	1	10511	2591	10829	13.9
20	30	11432	3104	11849	15.2
20	10	9736	2673	10098	15.4
20	3	7718	2493	8114	17.9
20	1	5959	2376	6417	21.8
30	30	7162	2838	7706	21.6
30	10	5535	2439	6053	23.9
30	3	3899	2005	4384	27.3
30	1	2658	1604	3104	31.1
40	30	3290	2138	3924	33.1
40	10	2230	1575	2731	35.3
40	3	1334	1057	1702	38.4
40	1	808	720	1082	41.8

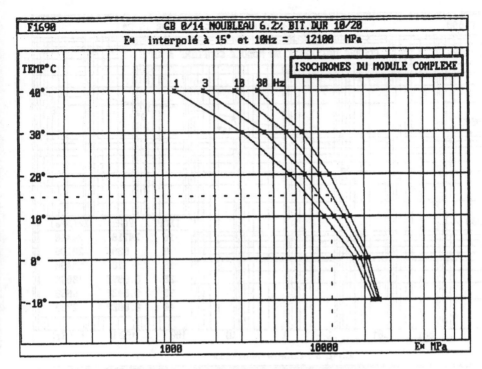

FIGURE 6 . VARIATIONS ISOCHRONES DU MODULE

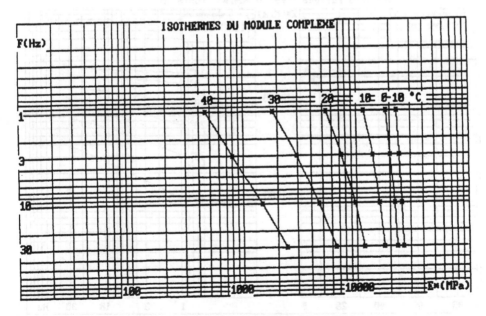

FIGURE 7 . VARIATIONS ISOTHERMES DU MODULE

375

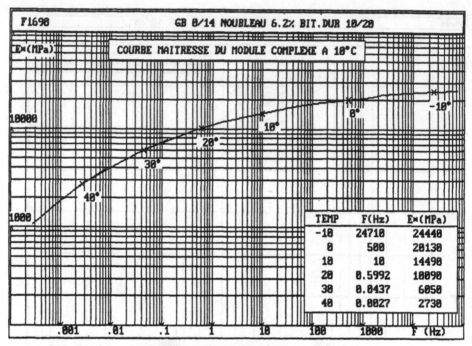

FIGURE 8 . COURBE MAITRESSE

TEMP	F(Hz)	E*(MPa)
-10	24710	24440
0	500	20130
10	10	14490
20	0.5992	10090
30	0.0437	6050
40	0.0027	2730

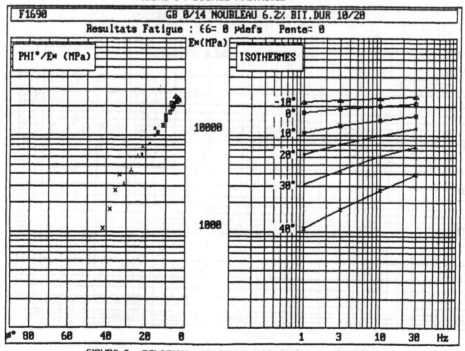

FIGURE 9 . RELATION MODULE-DEPHASAGE (Courbe de BLACK)

L'ensemble de ces graphiques permet de caractériser de façon complète une formule ; de façon usuelle (notamment pour les calculs de dimensionnement), tous les renseignements fournis ne sont pas utilisés : la valeur couramment employée est le module à 15°C et 10 Hz : 15°C représentant la température moyenne en France et 10 Hz une fréquence de sollicitation proche de celle des véhicules ; c'est cette valeur qui sera prise en compte dans les exemples de résultats qui suivent.

6 Exemples de résultats et conséquences pratiques

6.1 Variation du module en fonction de la teneur en liant et de l'énergie de compactage

Un plan d'expérience visant à déterminer l'influence de la teneur en bitume et de l'énergie de compactage sur le module et la tenue en fatigue d'une formule d'enrobé standard a été réalisé.

- . matériaux : diorite, granulométrie 0/14 (identique pour toutes les formules)
- . bitume : pénétration 60/70 ; dosages (3,6), (4,2), (4,8), (5,4) et (5,7 %)
- . énergie de compactage : trois niveaux : faible (f), moyen (M) et fort (F).

La gamme de compacité ainsi recouverte allait de 87 % à 98 % ; les résultats ont conduit aux conclusions suivantes :

Pour une énergie de compactage donnée (Figure 10) :

l'augmentation de module lorsque la teneur en liant varie est de l'ordre de 1000 à 1500 MPa ; le gain passe par un optimum correspondant à une teneur en liant d'autant plus faible que l'énergie de compactage est élevée ; l'optimum est par ailleurs d'autant plus marqué que l'énergie de compactage est forte.

Pour une teneur en liant donnée (Figure 11) :

l'augmentation de module lorsque l'énergie de compactage augmente est de l'ordre de 2500 à 5000 MPa ; le gain en proportion est plus élevé lorsque la teneur en liant est faible.

En définitive, la valeur du module d'un enrobé sera très sensible à l'énergie de compactage, plus qu'à la teneur en liant.

Les résultats obtenus mettent également en évidence la difficulté d'apprécier les bonnes performances en module d'un enrobé à la seule vue de son niveau de compacité. A titre d'exemple, pour une compacité de 93% (couramment mesurée sur chantier), le module peut varier de 7500 MPa à 11500 MPa ; une faible valeur correspondra à un dosage en bitume élevé et à une faible énergie de compactage et une valeur élevée à un dosage en bitume faible et à une forte énergie de compactage.

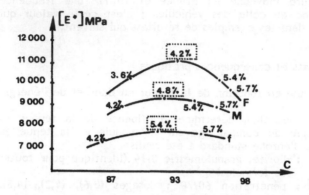

FIGURE 10 . EFFET DE LA TENEUR EN LIANT SUR LA VALEUR DU
MODULE A TROIS NIVEAUX D'ENERGIE DE COMPACTAGE:
f (faible) M (moyen) F (fort)

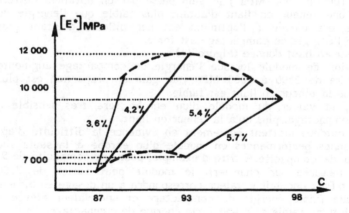

FIGURE 11 . EFFET DE L'ENERGIE DE COMPACTAGE A DIFFERENTES
TENEURS EN LIANT

Toutes les formules expérimentées en module ont fait parallèlement l'objet d'essais de fatigue ; cet essai est réalisé à déformation imposée et donne la relation entre déformation appliquée (ε) et durée de vie (N) exprimée en nombre de cycles **(Figure 12).** On s'intéresse plus particulièrement à la déformation correspondant à un nombre de 10^6 chargements à la rupture que l'on nomme ε_6 et qui caractérisera le matériau en ce qui concerne son comportement à la fatigue.

Les facteurs correspondant à une amélioration de la tenue en fatigue (augmentation de la valeur de ε_6) sont différents de ceux qui améliorent le module ; la tenue en fatigue en particulier est très sensible à la teneur en bitume. Ces deux grandeurs (module et tenue en fatigue) étant prises en compte au niveau du dimensionnement, il est dans ces conditions difficile de comparer la qualité de deux enrobés lorsqu'elles varient simultanément, d'où l'idée de les associer dans un même indicateur qui permettra de réaliser des comparaisons.

6.2 Indicateur global de performance (IQE)
. Principe

L'indicateur utilisé est issu d'un modèle simple de structure de chaussée: on dispose les différents enrobés à comparer sur un même support de module Eo et on les soumet à la même charge Po ; chaque enrobé est caractérisé par la valeur de son ε_6 et de son module E ; on calcule pour chaque enrobé l'épaisseur **(h)** telle que la déformation à la base de la couche soit égale à la valeur de ε_6 correspondant à la formule **(Figure 13).**

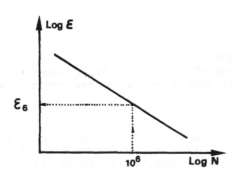

FIGURE 12 . RESULTAT D'EN ESSAI DE FATIGUE

Cette épaisseur **(h)** caractérisera chaque formule ; elle va différer d'une formule à l'autre selon la valeur de ε_6 et E et pourra servir de comparaison ; on la nomme **IQE** (Indice de Qualité Elastique) ; plus elle est faible, plus la formule est performante.

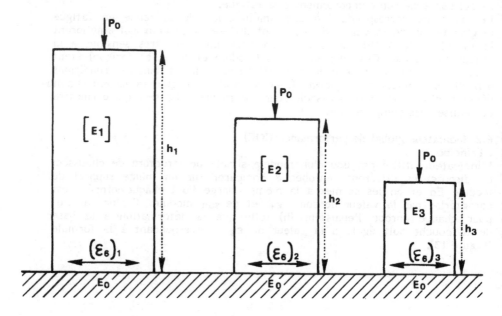

FIGURE 13 . IQE (h1, h2, h3) principe

. **Exemple d'utilisation de l'IQE**
Lorsque les seuls paramètres de variation sont la teneur en liant et l'énergie de compactage, on obtient le résultat présenté sur la **Figure 14.**
Trois zones apparaissent :
- faibles compacités : **(< 92)**
   Cette zone correspond à de faibles teneurs en liant et/ou à une faible énergie de compactage ; l'enrobé est de mauvaise qualité et cette qualité est très sensible aux variations de compacité (dûes aux variations de teneur en liant et d'énergie de compatage).
- compacités usuelles **(92 < C < 96)**
   Dans cette zone, la qualité est peu sensible à la compacité, il y a en fait des phénomènes de compensation entre les variations de ε_6 et de E :

$$\varepsilon_6\uparrow + \quad E \downarrow \# \quad \varepsilon_6 \downarrow + \quad E\uparrow$$

- fortes compacités (> 96)
 la qualité s'amèliore mais cette zone correspond à des formules sensibles à l'ornièrage.

Ces résultats sont particulièrement importants et montrent tout l'intérêt qu'il y a à considérer **simultanément** le module et la tenue en fatigue. On peut ainsi identifier la zone de compacité où la formule est le moins sensible aux variations de composition.

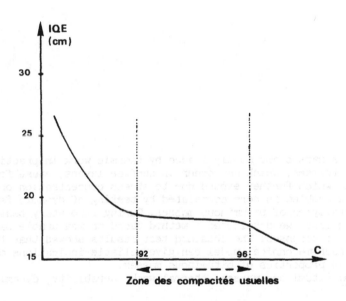

FIGURE 14 . VARIATION DE LA QUALITE MECANIQUE D'UN ENROBE, LORSQUE LA TENEUR EN LIANT ET L'ENERGIE DE COMPACTAGE VARIENT

26 USE OF THE CYCLIC FATIGUE METHOD TO DETERMINE THE DURABILITY OF BITUMINOUS MASSES

N. DENIĆ and R. KRSTIĆ
Institute for Testing of Materials of SRS, Belgrade, Yugoslavia

Abstract
Bituminous masses are mostly loaded by thermic work in practice. The highest stresses, which can occur on surface layers, cause formation of cracks which further expand due to stress concentration on their ends. The problem is here correlated by testing of dynamic fatigue of various types of bituminous masses. During laboratory testing of masses fatigue, we developped a method based on measurable parameters of frequent stresses. The obtained test results showed that the testing of this macr-mechanical model can give reliable indications of changes of certain properties of bituminous masses.
<u>Keywords</u>: Bituminous masses, thermal work, durability, dynamic fatigue.

1 Introduction

The use of bituminous masses as construction materials implies more thorough studies of dependance of their physical, mechanical and rheological properties on weather conditions than with classical in-organic construction materials. Durability of these materials must be determined with the knowledge of influence of weather conditions to macro-mechanical properties of these composite materials.

The testing method for fatigue of bituminous masses has been introduced as macro-mechanical, rheological method for investigation of changes of their micro-mechanical properties due to ageing. To develop the method suitable for this characterisation of the material we introduced some simplifying assumptions. We analysed the causes of fatigue which were empirically determined.

2 Fatigue of bituminous masses under influence of ambient temperatures

Changes of ambient temperatures have most important influence on the composite material during its production and further constructive application. Practically, masses fatigue in hydroinsulation system appears as crack created by alternate stretching and shrinking. Such dilatations caused by natural temperature frequences determined choice of macro-mechanical model for simulation of this physical process in nature.

Breakage due to fatigue is announced by appearance of initial crack in matrix which, as a rule, appears on the point with failure in structure of material or with some surface damage made during sample treatment or as a consequence of physical ageing of matrix without mass convection due to sudden variations of ambient temperature. Thermal shock can cause mechanical breakage caused by various dilatations. Sudden variation of ambient temperature creates temperature differences in the depth of layers resulting with different dilatation of external and internal side of layers. With high amplitude thermal shocks the effect of one temperature extreme can cause cracking which corresponds to low-cyclic fatigue in macro-mechanical model. In case of considerably lower amplitude cycles the damages occur through time which corresponds to high-cyclic fatigue model.

3 Testing methods

During testing of bituminous masses durability to temperature changes by means of fatigue model, the starting point was the life of material expressed by time needed for development of crack i.e. number of cycles to the breakage of test sample. The development of crack is influenced by numerous variables such as mechanical, geometrical and production influences as well as ambient properties. Main variables taken into consideration in development of this method are as follows:

Mechnaical varibales:
a) maximum stress,   max.
b) cyclic stress amplitude,
c) relation of minimum and maximum stress in cycle, R
d) frequency of cyclic load, f
e) cyclic load wave shape (for constant load amplitude)
f) load interaction (with variable amplitude load)
g) stress status
h) delayed stress.
Geometrical variables:
a) size of crack and its relation to sample dimensions
b) crack shape (geometry)
c) presence of neighbouring cracks.
Production variables:
a) mass contents
b) microstructure and coloidal structure
c) technological processing
d) mechanical processing
e) distribution of additives

Ambient variables:
a) temperature, T
b) ambient type - gas, liquid, etc.
c) corrosive effect degree.

Cyclic fatigue tests shown here were made on mass samples of the dimensions 170x15xd mm, at the temperature of 20°C, velocity of v=100 mm/min and force F limits of 25%, 50% and 75% of breaking force. All tests were made on universal testing machine - Zwick.

4 Test results and conclusions

The tests were made on samples of blown bitumen, polymerbitumen with atactic polypropilene(APP) with or without addition of filler (F). Only four characteristic cases are shown here for samples marked M6, BIT200/APP, M16, IB100/40, M17, IB85/25, M5, IB85/25+F. Table 1 shows the results of testing of main properties of these samples, while the results of testing of cyclic fatigue are graphically shown on figure 1

Table 1.   Results of testing of main properties of tested masses

Mass type	PK (^{o}C)	FRAS (^{o}C)	Penetration 1/10 mm (^{o}C)				
			0	10	15	25	40
M6, BIT200/APP	130	-15	6	14	22	31	46
M16, IB100/40	103	-24	6	14	19	33	65
M17, IB85/25	88	-19	6	10	14	24	72
M5, IB85/25+F	93	-12	6	11	17	24	43

The same method was used for testing of mass with addition of thermoplastic elastomer SBS with limited deformation varying between 50% - 300% from breaking elongation.

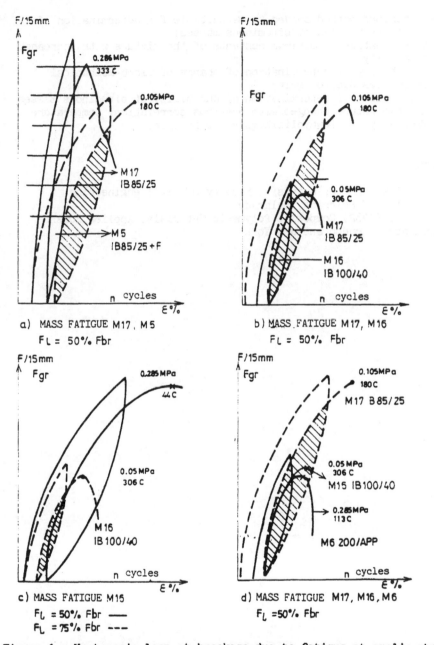

Figure 1 – Hysteresis loop at breakage due to fatigue at cyclic stretching of masses at velocity v=100 mm/min, T=20°C, a) influence of filler to fatigue, b) influence of bitumen structure, c) influence of value of limiting force at stretching, d) influence of primary matrix type

The described method proved to be suitable for determination of the following properties of bituminous masses:

1) Determination of optimum contents of the mixture with increased durability
2) Possibility of determination of change of micro-mechanical properties due to ageing
3) Possibility of determination of durability of bituminous masses at temperature changes with previous correlation: temperature frequency - lowcyclic/highcyclic fatigue.

5 Literature

Denić, N. (1990) Doctor`s thesis, Faculty of civil engineering, University of Belgrade, Belgrade

Sheldon, R.P. (1982) Composite Polymeric Materials, Applied Science Publishers, London and New York

27 NOUVELLE APPROCHE DU COMPORTEMENT DES ENROBÉS BITUMINEUX: RÉSULTATS EXPERIMENTAUX ET FORMULATION RHÈOLOGIQUE

(New approach to behaviour of coated bitumens: experimental results and rheological characteristics)

H. DI BENEDETTO
Laboratoire Géomatériaux, École Nationale des Travaux Publics de l'État, Vaulx-en-Velin, France

Résumé

Nous présentons, tout d'abord, les essais traditionnellement effectués sur les mélanges bitumineux en les situant dans un cadre général. Cette présentation montre la nécessité de réaliser des essais "multi dimensionnels".

Une étude sur l'anisotropie visant à préciser la représentativité des échantillons obtenus en laboratoire est ensuite proposée.

Les résultats d'une campagne d'essais triaxiaux (dimension 2) qui ont permis de caractériser l'évolution du comportement mécanique d'un enrobé bitumineux avec la compacité, la contrainte de confinement, la vitesse de déformation axiale, la température sont, ensuite, proposés.

Enfin, le modèle rhéologique de type incrémental développé au Laboratoire Géomatériaux est introduit. Des comparaisons entre des simulations à l'aide du modèle et des résultats de la campagne d'essais triaxiaux sont également présentés.

Mots clés : Anisotropie, Essais Mécaniques, Enrobés Bitumineux, Lois de Comportement, Rhéologie, Essai Triaxial.

1 Introduction

L'évaluation des contraintes et/ou des déformations en divers points au sein de l'ouvrage sert de base à l'ensemble des méthodes actuelles de dimensionnement des chaussées. Or, les méthodes numériques ou analytiques utilisés pour cette évaluation supposent, en général, que les matériaux constitutifs ont un comportement élastique linéaire isotrope. Cette dernière hypothèse est une approximation grossière puisque les couches de roulement constituées d'enrobés bitumineux, les couche de base traités aux liants hydrauliques et les sols supports ont des comportements fortement non linéaires et irréversibles au sein desquels existent également, parfois des phénomènes visqueux.

Il apparaît donc qu'une meilleure évaluation des champs de contrainte et de déformation au sein d'une chaussée qui est nécessaire à l'amélioration du dimensionnement, impose une meilleure description du comportement des matériaux constitutifs. Nous présentons dans cet article quelques résultats d'une recherche en cours au Laboratoire Géomatériaux (LGM) de l'Ecole Nationale des Travaux Publics de l'Etat, dont l'objectif est de développer une loi de comportement pour les matériaux traités aux liants hydrocarbonés. Ces matériaux étaient, en effet, restés en marge des développements observés pour les autres géomatériaux, tels que les sols, les bétons hydrauliques, les roches, en ce qui concernent la description de leur propriété mécanique. Cette recherche s'appuie sur l'utilisation d'un appareillage permettant d'effectuer des essais triaxiaux de révolution. L'utilisation plus systématique de cet essai, correspond, à notre sens, à une nouvelle voie dans le champs des essais mécaniques en vue du dimensionnement et du contrôle des mélanges bitumineux.

Afin d'expliquer notre choix, nous présentons ci-dessous, une analyse et une classification des différents essais mécaniques ainsi que leurs utilisation pour les mélanges bitumineux. Les essais mécanique de laboratoire utilisés pour les matériaux bitumineux (ou les autres géomatériaux), peuvent être classés en deux grandes catégories :

- les essais "non homogènes" qui correspondent à des essais de structures. Ces essais sont interprétés de manière totalement empirique ou déterminent une caractéristique de l'ouvrage grâce à l'utilisation de lois de similitude (modèle réduit). Cette détermination s'effectue, en général, en considérant une propriété d'usage donnée, Bonnot (1983).
- les essais "homogènes" qui sont les seuls à fournir des éléments directement exploitables pour la détermination de la loi de comportement du corps.

Nous proposons de classer en trois catégories les essais homogènes réalisés sur les mélanges. Ces trois classes sont identifiables lorsque l'on considère les valeurs de la norme de la déformation et celles du nombre de chargement appliqués (Figure 1).

- Pour un faible nombre de chargement et des déformations de quelques pourcents, les essais sont des essais de déformabilité, le comportement observé est alors fortement non linéaire.
- Pour des chargements comprenant quelques centaines de cycles et des déformations "faibles" ($< 10^{-4}$) le comportement est considéré, en première approximation, viscoélastique linéaire.
- Lors de chargement de plusieurs milliers de cycles et des déformations faibles les phénomènes d'endommagement apparaissent, le matériau se "fatigue".

Ces trois classes d'essais peuvent donc d'être qualifiées respectivement d'essai de déformabilité, de détermination des caractéristiques linéaires et de fatigue.

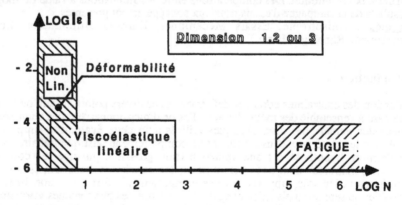

Figure 1 : Essais homogènes sur les enrobés (température 20°C)
ε déformation   -   N nombre de chargement

Actuellement, les études sur les enrobés privilégient les essais de détermination des caractéristiques linéaires et de fatigue.

D'autre part, chacun de ces essais peut être considéré comme mono ou multi dimensionnels selon que l'on applique une ou plusieurs sollicitations indépendantes. Par exemple, les essais de traction-compression simple pour lesquels seule la contrainte (ou la déformation axiale) est imposée correspondent à des essais de dimension 1. Ces essais qui ne traduisent pas le confinement latéral (existant par exemple dans une chaussée) sont, bien sûr, insuffisants pour fournir une description de ce phénomène. Nous donnons quelques exemples d'essais ainsi que leurs fréquences d'utilisation pour les mélanges bitumineux dans le tableau 1.

Tableau 1 : Catégorie d'essais homogènes selon la dimension
de l'espace des sollicitations

DIMENSION DE L'ESSAI	EXEMPLE D'ESSAIS	UTILISATION POUR LES MELANGES BITUMINEUX
1	* Compression - traction simple	Trés fréquente
	* Torsion de poutres	Faible
	* Cisaillement (cission)	Faible (Assi (1981))
	* rhéomètre "tournant"...	Fréquente pour les bitumes
2	* Triaxial de révolution	Faible
	* Presse à cisaillement giratoire	Fréquente
	*	
3	* Triaxial vrai	Inéxistante
	* Torsion compression - traction	Inéxistante
	de cyclindre creux confinés	(1 exemple sur bitume, Addala (1989)
	*	Di Benedetto (1986) (1987)
4	* Direct shear cell, Arthur et al (1981)	Inéxistante
5 et 6	Pas d'essai connu	

Une analyse de ce tableau montre qu'une évolution des pratiques actuelles consiste, dans un premier temps, à utiliser :

- des essais de dimension 1, assurant des rotation des axes principaux (freinage, efforts tangentiels sur chaussés, ...) ;
- des essais triaxiaux de révolution (dimension 2), qui introduisent le confinement lattéral.

Cette rapide synthèse des orientations possibles et des pratiques actuelles en vu de la détermination des caractéristiques mécaniques des mélanges bitumineux, explique la raison pour laquelle notre choix s'est porté vers la réalisation d'essais de déformabilité à l'appareil triaxial de révolution. Signalons que ces essais fournissent également aux très petites déformations des caractéristiques linéaires.

Nous avons, d'autre part, également cherché à quantifier l'anisotropie liée au mode de fabrication de ce type de matériau.

Dans cet article, nous présentons successivement : le matériau testé ainsi que le dispositif expérimental utilisé, les résultats expérimentaux obtenus sur la classe de chemin de sollicitation choisie, la loi de comportement développé ainsi que des comparaisons entre des similations numériques et des résultats expérimentaux.

Les conventions de signes adoptées sont celles de la mécanique des sols où les compressions et les contractions sont comptées positivement.

2 Le matériau et le dispositif expérimental

2.1 Le matériau
Le matériau utilisé est un mélange d'un liant hydrocarboné et d'une formule 0/6

reconstituée. Le bitume (ELF-FEYZIN) est de classe 60/70, de température bille anneau T_{BA} = 48°C et d'indice de pénétrabilité Ip = -0,122. La teneur en bitume choisie est de 7,3 ppc.

2.2 Essais de caractérisation de l'anisotropie

Afin de valider la représentativité des essais réalisés sur éprouvettes cylindriques, une campagne expérimentale, visant à préciser l'anisotropie induite par le mode de fabrication des enrobés bitumineux, a été conduite. On constaste, en effet, que lors de la mise en place des enrobés trois directions peuvent être distinguées : la direction de compactage, la direction de roulement du compacteur et la direction perpendiculaire aux deux précédentes. La représentativité des résultats résultant d'essais sur des éprouvettes obtenues en laboratoire par rapport à celles carottées perpendiculairement ou parallèlement à la surface de la chaussée nécessite donc d'être confirmée.

Nous avons réalisé au Laboratoire, Laradi (1990), Di Benedetto et Laradi (1989), des essais de traction-compression simples, à vitesse de déformation constante, sur des éprouvettes cubiques découpées dans des plaques d'enrobés (Figure 2) obtenues à l'aide du compacteur de plaques type L.P.C. (Le matériel LPC (1988)). Les éprouvettes cubiques, de compacité (Cp = 93,6 %), ont été sollicitées selon les 3 directions I, II et III, correspondant respectivement au sens de compactage, de roulement des roues du compacteur, perpendiculaire aux deux précédents. Les caractéristiques des 50 essais réalisés par N. Laradi sont précisées Tableau II.

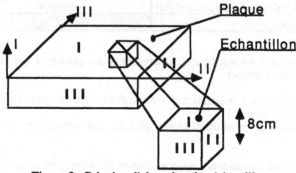

Figure 2 : Principe d'obtention des échantillons ;
I : Direction de compactage ; II : Direction de roulement ; III : Direction perpendiculaire

Tableau II : Caractéristiques des essais T = cycles en traction, C = cycles en compression, compacité Cp = 93,6 ± 0,5 %, température T = 23°C ± 1°C

Vitesse de déformation (% / m n)	Direction de sollicitation		
	I	II	III
0,05	3C	-	3C
0,25	3C + 2T	-	4C + 2T
1	5C + 2T	3C + 2T	2T
4	2C + 3T	-	3C + 2T
16	2C + 3T	-	2C + 2T

Les déformations dans les directions I, II et III qui sont, à priori, toutes différentes, ont été mesurées à l'aide successivement de deux dispositifs :
- des jauges de déformations,
- deux capteurs mis au point au laboratoire et présentés au paragraphe II-3.
Les positions sur l'échantillon des systèmes de mesure sont indiquées figure 3.

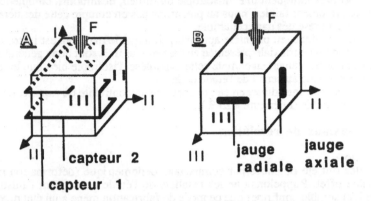

Figure 3 : Disposition adoptée : A capteur, B jauges de déformation (sens d'écrasement I)

Ces deux dispositifs ont permis de couvrir une large gamme de déformation (quelques 10^{-6} à quelques 10^{-2}).

Une analyse plus détaillée des résultats obtenus est présentée dans Di Benedetto et al (1989) et Laradi (1990). Nous n'exploitons dans les développements qui suivent, que les résultats relatifs aux contraintes de rupture.

Soulignons, néanmoins, qu'une des conclusions intéressantes qui ressort de l'étude est que la direction I correspond à un axe de symétrie de révolution pour le matériau.

Cette observation tend à confirmer que la confection d'éprouvette dans des moules cylindriques est représentative de l'état du matériau après compactage.

Les valeurs des contraintes de rupture obtenues en traction (signe opposé) et en compression pour les différentes vitesses de déformation sont reportées figure 4.

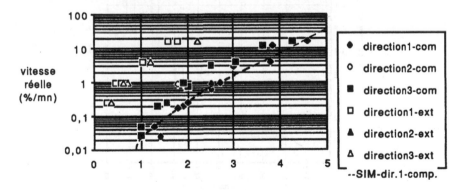

Figure 4 : Contrainte de rupture (MPa) en compression et en traction (*-1)

Nous constatons sur cette figure que l'écart relatif entre les valeurs absolues des contraintes de rupture lors de sollicitation dans la direction I et la direction II (ou III) est compris entre 10 et 25 %. Le signe de l'écart est inversé entre la compression et la traction : plus grande résistance dans le sens I en compression et moindre résistance dans ce même sens en traction.

Les écarts observés indiquent une anisotropie du milieu, néanmoins, compte tenu de leurs valeurs relativement faibles, nous ne prendrons pas en compte cette dernière dans les développements présentés dans cet article.

Nous supposerons donc, en première approximation que le matériau est isotrope. Les résultats de la campagne triaxiale présentée au paragraphe II-3. sont donc généralisés au cas tridimensionnel en prenant en compte cette hypothèse d'isotropie bien que le sens de sollicitation corresponde au sens I de l'étude ci-dessus.

Il semble que l'erreur introduite, en ce qui concerne la contrainte de rupture, peut être considérée comme faible (inférieure à 25 %).

2.3 Essais triaxiaux de révolution

a) Présentation

Les éprouvettes ont été obtenues par compactage oedomètrique (déformation radiale nulle) à double effet. Rappelons que les résultats de l'étude concernant l'anisotropie (paragraphe 2.2) semble confirmer que ce mode de fabrication mène à un état proche de celui existant in situ.

Nous avons observé qu'il était possible d'obtenir, par ce mode de compactage, une compacité proche du maximum de densité que l'on peut obtenir en considèrant uniquement le squelette granulaire. Ce résultat montre bien le rôle de lubrifiant joué par la phase bitumineuse.

Trois groupes de compacité : faible (Cp = 89 %), moyenne (Cp = 93,6 %) et forte (Cp = 96,6 %) ont été obtenues. Les éprouvettes sélectionnées pour la campagne triaxiale se trouvent dans des fourchettes de compacité de plus ou moins 1/2 point (Figure 5).

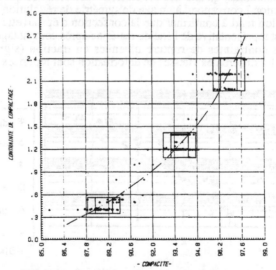

Figure 5 : Représentation des échantillons dans un
graphique compacité (%) contrainte de compactage (KN/cm²)

Le dispositif expérimental se compose des différents organes suivants : la presse, la cellule triaxiale, le système de mise en pression, les systèmes de mesure (capteurs de déplacement, capteur de force, alimentation, amplificateurs, ...), l'enceinte thermique, le système informatique d'acquisition et de traitement des données.

Nous ne présentons pas ces divers organes, sauf le capteur de mesure de déplacement axial et radial dont l'originalité mérite d'être soulignée.

Nous avons utilisé, pour la mesure des déformations axiale et radiale, un seul capteur qui a été conçu au laboratoire. Ce capteur est présenté sur la figure 6. La mesure, à l'aide de jauges de déformation montées en pont complet, de la flexion des lamelles métalliques permet d'apprécier des variations des déplacements axiaux et radiaux de l'ordre de 2 microns. Nous évaluons la précision globale de la mesure à $\pm 10^{-2}$ mm. Le capteur est positionné dans le tiers central environ de l'échantillon. La mesure des déformations par ce capteur permet de s'affranchir des effets de bord induits par le frettage.

Nous avons observé expérimentalement des différences atteignant 25 % entre la déformation axiale déterminée à partir des mesures du déplacement du piston de la presse et celle obtenue a l'aide de notre capteur, l'introduction d'un tel capteur nous semble donc très importante.

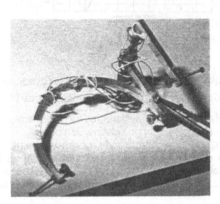 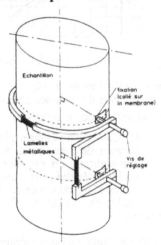

Figure 6 : Vue et principe du capteur de mesure des déformations axiale et radiale
(nouveau système/1989)

b) Résultats

L'objectif des essais est de permettre une calibration aisée de la loi constitutive que nous proposons tout en assurant une meilleure connaissance du comportement des enrobés. Ceci suppose la prise en compte de l'influence des déformations instantanées, différées et de la température. Nous avons pour cela choisi une classe de sollicitation en conciliant les contraintes de l'appareillage et les objectifs de l'étude.

La sollicitation choisie pour la calibration, correspond à des cycles de charge-décharge à vitesse de déformation axiale, pression de confinement et température constantes. Trois pressions de confinement (0 MPa, 0,2 MPa, 0,4 MPa), trois températures (5°C, 23°C, 41°C) et quatre vitesses de déformation axiale (0,25 %/mn, 1 %/mn, 4 %/mn, 16 %/mn) ont été imposées pour les trois classes de compacité.

Nous présentons, figure 7, un exemple de sortie graphique pour 1 essai réalisé à une contrainte de confinement de 0,4 MPa, une compacité de 93,8 %, une température de 23°C et une vitesses de déformation axiale de 16 %/mn. 130 essais, environ, ont été effectués.

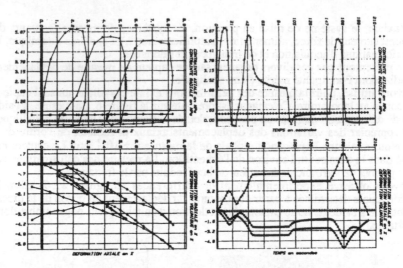

Figure 7 : Exemple de résultat expérimentaux

c) Interprétation et analyse

L'allure générale des courbes obtenues ainsi que l'évolution qualitative de la réponse avec les paramètres vitesse de déformation axiale ($\dot{\varepsilon}_1$), contrainte de confinement (σ_3), compacité (Cp) et température (θ) est représentée figure 8, dans les axes déformation axiale - déviateur de contrainte et déformation axiale - variation relative de volume.

Nous remarquons, en premier lieu, l'aspect non linéaire du comportement qui comprend des irréversibilités instantanées que l'on peut observer, par exemple, au changement de signe de la vitesse de déformation axiale et des irréversibilités visqueuses qui se manifestent par des courbes distinctes pour des vitesses de déformation axiale différentes.

Les autres points suivants méritent d'être soulignés :

- Dans les axes déformation axiale - déviateur de contrainte :

. lors de la charge, pour une déformation donnée, la contrainte axiale est d'autant plus importante que la vitesse de déformation est grande, la contrainte de confinement grande, la compacité élevée et la température faible ;
. lors de la décharge, les phénomènes observés en charge sont transposables si l'on considère la valeur absolue du déviateur de contrainte et des temps "suffisamment éloignés" du point d'inversion.

- Dans les axes déformation axiale - variation relative de volume :

. lors de la charge, le matériau est contractant puis dilatant ;
. en décharge, le matériau est soit immédiatement dilatant, soit immédiatement contractant. Lorsqu'elle apparaît, la phase de dilatance initiale est très faible. Si la décharge est maintenue suffisamment longtemps, le matériau devient dilatant.

Chacune des courbes présentées est qualitativement équivalente à celle obtenue sur d'autres matériaux granulaires, par exemple les sables, Mohkam (1983), les argiles,

Doanh (1983), les bétons hydrauliques, Torrenti (1987), le matériau de Schneebeli, Cambou (1979), les billes de verre, Bouvard (1982)...

Nous pensons que cette <u>similitude de comportement se retrouve pour l'ensemble des matériaux ayant un squelette granulaire</u>. Le liant éventuel modifie quantitativement la réponse et confère le cas échéant des propriétés visqueuses. Dans le cas des bétons bitumineux, le liant a un comportement visqueux très marqué et est d'autre part très sensible à la température.

Nous analysons, dans les paragraphes suivants, les caractéristiques à la rupture en compression, les caractéristiques à la rupture en extension ainsi que le phénomène de relaxation en extension et en compression. Une étude des variations de volume et des modules d'Young sécants et tangents a été présentée par Di Benedetto (1987). Bard et al (1989) propose une analyse à l'appareil triaxial des modules d'Young et coefficients de Poisson complexes pour les petites transformations. ($|\varepsilon| < 10^{-4}$).

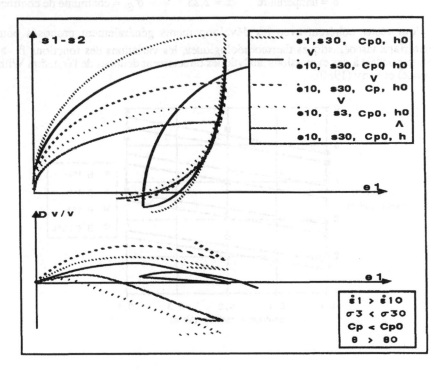

Figure 8 : Allure générale des courbes obtenues, évolution par rapport à la variation des paramètres : vitesse de déformation axiale ($\dot{\varepsilon}_1 = \dot{e}_1$),

contrainte de confinement ($\sigma_3 = s_3$), compacité (Cp) et température (θ)

Soulignons que des résultats fournissant la variation de volume sur les mélanges bitumineux sont rares. Cette variation correspond à un phénomène important qui est à l'origine des évolutions des contraintes perpendiculaires et des déformations parallèles à la surface de la chaussée. Une étude plus approfondie et systématique des coefficients de Poisson correspond, à notre sens, a une évolution nécessaire des essais mécaniques sur les enrobés bitumineux.

Rupture en compression

Nous nous sommes principalement attachés à trouver une expression analytique de la contrainte axiale de rupture en fonction des variables de sollicitation. Nous avons considéré que la valeur de la contrainte au 1er pic en charge ou au palier lorsque ce pic n'apparaît pas est celle qui précise le mieux la contrainte de rupture du matériau.

L'analyse des résultats nous amène à exprimer la contrainte de rupture (σ_{1p}) du matériau par la formule suivante :

$$\sigma_{1p}/\sigma_0 = \alpha\ \sigma_3/\sigma_0 + \beta(Cp, \theta))\ \ln[(\dot{\varepsilon}_1 + \delta(\theta))/\dot{\varepsilon}_0)] + \gamma(Cp, \theta) \quad (1)$$

où $\sigma_0 = 1$ MPa $\dot{\varepsilon}_0 = 1$ %/mm   Cp = compacité

θ = température $\alpha = 2,25$ σ_3 = contrainte de confinement

Pour rester cohérent avec les développements généralement proposés pour les matériaux viscoélastiques thermorhéologiques, les variations des fonctions β, γ et δ correspondent à des expressions analytiques directement déduites de l'équation Williams, Lendel et Ferry (1980).

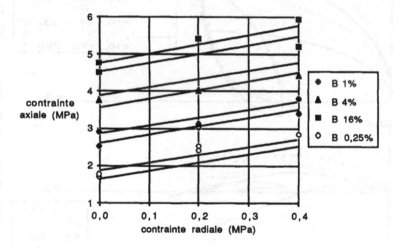

Figure 9 : Contrainte axiale de rupture en compression ; points expérimentaux-courbes déduites de l'équation (1) (Cp = 93,6 ± 0,5 %; θ = 23°C)

L'ensemble des comparaisons entre les résultats fournis par cette équation et les points expérimentaux montrent qu'elle décrit relativement bien le phénomène observé (figure 4 et 9).

En fait, une légère divergence est observée pour la température la plus élevée et la vitesse la plus basse, ainsi que pour la température la plus faible et la vitesse la plus grande. Cette divergence pourrait s'expliquer par la difficulté que pose pratiquement lors de ces essais le maintien des valeurs de sollicitation.

Remarquons que si nous cherchons un critère de rupture de type Mohr-Coulomb, l'angle de frottement obtenu (φ) est indépendant de la compacité et de la température. Il se déduit du paramètre α par l'équation : $tg(\varphi/2 + \pi/4) = \sqrt{\alpha}$

Nous trouvons $\varphi = 23°$.

Les résultats en compression impliquent que tout critère de rupture isotrope "conique" convient.

En supposant un critère de rupture de type Mohr-Coulomb, l'angle de frottement obtenu est identique pour l'enrobé bitumineux quelles que soient la compacité, la vitesse de déformation axiale et la température considérées lorsque le déviateur de contrainte est positif. Seul le sommet du cône - c'est-à-dire la cohésion - évolue avec ces paramètres.

Rupture en extension

En première approximation, les résultats montrent que le déviateur de contrainte de rupture $(\sigma 1 - \sigma 3)_r$ est indépendant de la contrainte de confinement dans la gamme de contrainte testé pour une vitesse de déformation donnée. Les valeurs obtenues à contrainte de confinement nulle, pour différentes vitesses de déformation axiale sont tracées figure 4.

Contrainte de relaxation

Lors des essais, une phase de relaxation a été imposée après un cycle de charge-décharge-recharge jusqu'à la limite de résistance du matériau (pic ou palier), la contrainte obtenue après stabilisation (valeur asymptotique) est appelée : contrainte de relaxation en compression (σ_{pc}). A la suite de cette relaxation, une extension suivie d'une autre relaxation ont été effectuées. La contrainte obtenue est nommée, par opposiition : contrainte de relaxation en extension (σ_{pe}). Les résultats montrent que ces deux contraintes de relaxation sont indépendantes de la température et de la vitesse de sollicitation.

Critère de limite de résistance

A partir des résultats obtenus et de l'hypothèse d'isotropie formulée, il est possible de

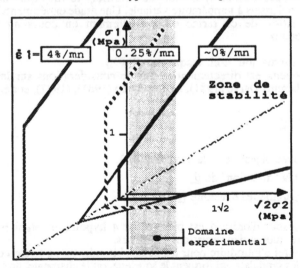

Figure 10 : Courbes limite de résistance à une température de 23°C

tracer dans le plan de contraintes principales : $\sigma_1 - \sqrt{2}\,\sigma_2 = \sqrt{2}\,\sigma_3$, une courbe délimitant la limite de résistance du matériau pour une température et une vitesse données. Nous proposons une courbe confondue avec un critère de Morhs-Coulomb d'angle de frottement : 23°, complétée par des droites parallèles aux axes de coordonnées (Figure 10).

Il apparaît que les contraintes de relaxation correspondent aux valeurs asymptotiques qui seraient obtenues si la vitesse de sollicitation était infiniment lente. Le domaine qui délimite la courbe limite des contraintes de relaxation peut donc être considéré comme un domaine de stabilité du matériau avant endommagement.

Cette observation nous semble importante car elle permet de définir des contraintes que l'enrobé peut reprendre quelque soit la vitesse de chargement.

Nous pensons que ce domaine doit évoluer avec l'endommagement du matériau et poursuivons nos études sur ce point (Néji 1991, Yan 1991).

4 Lois de comportement développée, Di Benedetto (1981), (1985), (1987)

Comme le montrent les résultats expérimentaux, un enrobé bitumineux est un matériau visqueux fortement non linéaire très sensible à la température. Nous proposons d'en modéliser le comportement en exprimant la vitesse de déformation $\underline{D}$ comme la somme de 3 termes :

$$\underline{D} = \underline{D}^{nv} + \underline{D}^{v} + \underline{D}^{\theta}$$

$\underline{D}^{nv}$: vitesse de déformation non visqueuse ou instantanée ;

$\underline{D}^{v}$: vitesse de déformation visqueuse ou différée ;

$\underline{D}^{\theta}$: vitesse de déformation induite par les variations de température, nous ne présentons pas ici ce dernier terme qui nécessite la réalisation d'essais à température variable. Une étude expérimentale visant à la détermination de ce terme est actuellement en cours au Laboratoire Géomatériaux.

a) Vitesse de déformation non visqueuse ($\underline{D}^{nv}$)

Le terme non visqueux est directement déduit de considérations sur les matériaux granulaires non cohérents Darve (1981), Di Benedetto (1981), (1987), et s'exprime :

$$\underline{D}^{nv} = \underline{M}\,(\hat{\sigma},\, h)\,\underline{\dot{\sigma}}$$

où $\underline{\dot{\sigma}}$ est une dérivée objective de la contrainte $(\underline{\sigma})$,

$\hat{\sigma} = \underline{\dot{\sigma}}/\,\|\underline{\dot{\sigma}}\|$ est la "direction" de $\underline{\dot{\sigma}}$

h représente les paramètres d'histoire $(\underline{\sigma},\, Cp,\dots)$

$\underline{M}$ qui est un tenseur d'ordre 4 est précisé par 4 hypothèses établies à partir de considérations théoriques et d'observations physiques.

Il nous paraît naturel d'introduire pour les bétons bitumineux un terme correspondant au comportement des granulats qui forment le squelette, ou la charpente, de ce matériau

b) Vitesse de déformation visqueuse (D^v)

Le terme visqueux $\underline{D}^v$ traduit un comportement purement visqueux et s'exprime par :

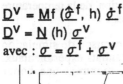

$$\underline{D}^v = \underline{M}f\,(\hat{\underline{\sigma}}^f, h)\,\dot{\underline{\sigma}}^f$$
$$\underline{D}^v = \underline{N}\,(h)\,\underline{\sigma}^v$$

avec : $\underline{\sigma} = \underline{\sigma}^f + \underline{\sigma}^v$

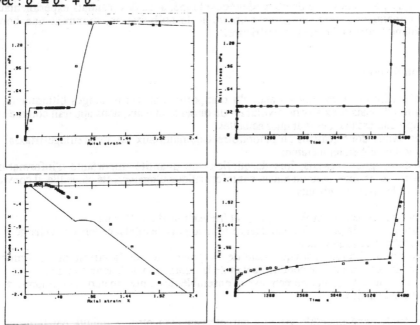

Figure 11 : Comparaison courbe simulée - points expérimentaux pour un essai de fluage
(Cp = 94.2%, $\sigma 3$ = 0 MPa, θ = 23°C)

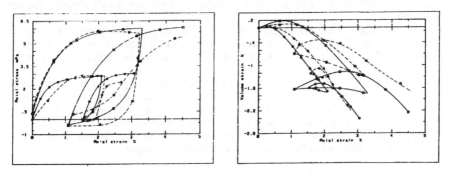

Figure 12 : Comparaison théorie (——) - expérience (---) pour deux essais triaxiaux
(θ = 23°C, Cp = 93.7 / 94.1%, $\sigma 3$ = 0.4 / 0 MPa, $\dot{\varepsilon}$ = 16 / 1 %/mn)

399

Le terme purement visqueux est dû au bitume, sa présence n'est donc pas arbitraire. Il est précisé par 2 hypothèses tirées de la constatation expérimentale suivante

- la forme des courbes est indépendante de la vitesse de sollicitation, de la température et est similaire à celle obtenue pour les autres matériaux granulaires.

Nous présentons figures 11 et 12 quelques exemples de comparaison de simulation numériques, obtenues à l'aide du modèle développé, avec des résultats expérimentaux. Les comparaisons, pour ces divers chemins de sollicitation, montrent une bonne concordance entre la théorie et l'expérience.

5 Conclusion

Afin d'améliorer notre connaissance du comportement des mélanges bitumineux, la réalisation d'essais triaxiaux de révolution de dimension deux, nous apparaît comme une évolution intéressante des pratiques actuelles.

Nous avons donc réalisés au laboratoire Géomatériaux un appareillage triaxial de révolution régulé en température.

Des campagnes d'essais, lors desquelles une quantité importante d'informations facilement accessibles pour un traitement informatique ultérieur a été recueillie, il convient de retenir les aspects suivants :

- le comportement fortement visqueux et irréversible du matériau ;
- l'évolution de la phase de contractance initiale puis de dilatance lors de la charge avec les paramètres de sollicitation ;
- l'aspect qualitativement équivalent de l'augmentation de la vitesse de déformation axiale et de la contrainte de confinement et de la diminution de la température ;
- la similitude de comportement avec l'ensemble des autres matériaux comprenant un squelette granulaire.

Une étude expérimentale visant à caractériser l'anisotropie induite par le mode de compactage des échantillons est, également, proposée. De cette étude, il ressort que les variations relatives des contraintes de rupture selon la direction de compression ou de traction simple sont de l'ordre de 10 à 25 %. Une symétrie de révolution selon un axe perpendiculaire au sens de compactage est également observée. Cette dernière constatation permet de conforter la représentativité des éprouvettes cylindriques obtenues au laboratoire.

La campagne triaxiale et l'évaluation de l'anisotropie ont permis de préciser en première approximation, dans l'espace des contraintes des surfaces limites de rupture en fonction la vitesse de solliciation (figure 10). L'étude de l'évolution de ces surfaces avec la fatigue du matériau nécessite d'être poursuivie.

Bien évidemment, une campagne expérimentale ne peut pas être conçue indépendamment d'une interprétation théorique et des objectifs pratiques. Les résultats obtenus ont servi à la formalisation et à la calibration de notre modèle rhéologique incrémentale. Les comparaisons entre les résultats expérimentaux et les simulations numériques, montrent que ce modèle permet de décrire certaines sollicitations tridimensionnelles qui sont celles existantes au sein des ouvrages.

6 Références Bibliographiques

Addala, F. (1989) Etude du comportement rhéologique des bitumes à moyenne et basse température : mise au point d'un nouvel essai de torsion d'éprouvette creuse, Thèse de

Doctorat, ENTPE - INSA, France.

Arthur, J.R.F. et al (1981) Stress path tests with controlled rotation of principal stress directions, Laboratory Shear Strength of Soils, ASTM (ed. Yung and Townsend).

Assi, M. (1981) Contribution à l'étude du comportement des enrobés bitumineux à la fatigue en cission, Thèse de Docteur-Ingénieur, ENPC, Paris, France.

Bard, E. et al (1989) Anisotropie des enrobés bitumineux en petites déformations, Rapport d'activité du Gréco "Rhéologie des Géomatériaux", Aussois, France.

Bonnot, J. (1983) Essais mécaniques pratiques de formulation et de contrôle des enrobés bitumineux, Rapport général du 3e Colloque International d'essais sur liants et matériaux hydrocarbonés de la RILEM, Beograd.

Bouvard, D. (1982) Rhéologie de milieux pulvérulents : étude expérimentale et identification d'une loi de comportement, Thèse de D. I., Grenoble.

Cambou, B. (1979) Approche du comportement d'un sol considéré comme un milieu non continu, Thèse de D. E., Lyon.

Darve, F. (1978) Une formulation incrémentale des lois rhéologiques. Application aux sols, Thèse d'Etat, Institut de Mécanique de Grenoble.

Di Benedetto, H. (1981) Etude du comportement cyclique des sables en cinématique rotationnelle, Thèse D. I., ENTPE-USMG.

Di Benedetto, H. (1985) Viscous part for incremental non linear constitutive laws, 5th Int. Conf. on Numerical Methods in Geomechanics, Nagoya, Japon.

Di Benedetto, H. (1986) Résultats des essais sur bitume et béton bitumineux, Gréco "Rhéologie des Géomatériaux", Aussois, France.

Di Benedetto, H. (1987) Etude du comportement des géomatériaux, Thèse de D.E., INPG et ENTPE.

Di Benedetto, H et Laradi, N. (1989) Etude expérimentale de l'anisotropie d'un enrobé bitumineux, Rapport d'activité du Gréco "Rhéologie des Géomatériaux", Aussois, France.

Doanh, T. (1983) Contribution à l'étude du comportement de la kaolinite, Thèse D. I., ECP-ENTPE.

Ferry, JD. (1980) Viscoelastic properties of polymers, John Wiley and Sons.

Le matériel LPC (1988) (ed LCPC), Paris, France.

Laradi, N. (1990) Thèse de Doctorat, ENTPE - INSA, France, (à paraître).

Mohkam, M. (1983) Contribution à l'étude expérimentale et théorique du comportement des sables sous chargements cycliques, Thèse de Docteur- Ingénieur, ENTPE - USMG - INPG.

Néji, J. (1991) Etude des remontés de fissures dans les couches de roulement, thèse de doctorat, ENTPE - Ecole Centrale de Paris, France (à paraître).

Torrenti, J.M. (1987) Comportement multiaxial du béton, Thèse de doctorat, ENPC-ENTPE.

Yan, X.L. (1991) Etude du comportement des enrobés aux bitumes polymères, thèse de doctorat, ENTPE - INSA, France (à paraître).

28 RESILIENT MODULUS TESTING OF ASPHALT SPECIMENS IN ACCORDANCE WITH ASTM D4123–82

C.E. FAIRHURST
MTS Systems Corp., Minneapolis, MN, USA
N.P. KOSLA and Y.R. KIM
Department of Civil Engineering, North Carolina State
University, Raleigh, NC, USA

Abstract
Growing demands on the aging U.S. network of interstate,
city, and state roads, and a related urgent need for cost
efficient road construction and repair have resulted in
increased attention to methods of evaluating the materials
used in pavement construction and repair. Several federal
grants, aimed at stimulating research on bituminous materials, have recently been awarded.
MTS Systems Corp., a testing equipment manufacturer,
has recently developed fixturing and unique extensometry
for conducting Resilient Modulus (Mr) testing of bituminous materials in accordance with ASTM D4123-82. Two
software programs have also been developed - one that
controls the indirect tensile strength test, and another
that controls the Mr test.
Mr tests, using a closed loop, servo-hydraulically
controlled loading system, were conducted at the MTS
facilities on laboratory compacted asphalt specimens
inside an environmental chamber at temperatures of 5°C
(41°F), 25°C (77°F), and 40°C (104°F). In addition, tests
were conducted on similar MTS equipment at North Carolina
State University (NCSU) to evaluate the extensometers and
fixturing. Results indicate that Mr decreases with increasing temperature, and increases with increasing load
cycle frequency. ASTM D4123 requires that, after each
test, the specimen be rotated 90° and retested. Lower Mr
values were obtained at the 90° position, indicating that
specimen damage occurs at the initial (0°) position.
Poisson's ratio, if measured, can serve as a good indicator as to whether or not the specimen has been permanently
deformed. Testing procedures occasionally deviated from
ASTM 4123-82. Studies to analyze the time dependent response of asphalt using the Correspondence Principle are
contemplated.
The fixture, extensometers, software, and test results
are described. Modifications to ASTM D4123 are suggested.
Keywords: Resilient Modulus, Bituminous Materials, Mr,
Asphalt, Servo-control, Materials Testing, Closed Loop.

1 Introduction

Over the past 20 to 30 years, interest in the resilient modulus of bituminous materials has been confined primarily to academic research studies (Vinson, 1989). Recently, however, a national concern for improved design and evaluation of pavement materials has led to increased federal funding of several Department of Transportation (D.O.T.) research and development projects on the resilient modulus of materials for use in flexible pavement design. The resilient modulus of both soils and asphalt is of interest for pavement design criteria. These projects are part of a shift towards mechanistic designs, in which the knowledge of the mechanical properties of pavement materials is essential.

Although resilient modulus (Mr) is but one of many parameters used in the design of pavement structures [as defined by the AASHTO (American Association of Safety and Highway Transportation Officials) Guide for Design of Pavement Structures (1986)], it is an important one.

Currently, there is a need for reliable testing equipment and procedures to correctly determine Mr values of bituminous material and, thus, to allow comparison of results between researchers. The goal of the work presented here is to evaluate a recently designed Mr testing system, including the fixturing and transducers (extensometers), by conducting preliminary Mr tests.

2 Explanation of Resilient Modulus (Mr)

The resilient modulus (Mr) of a material is defined as the ratio between an applied stress and the recoverable (rebound) strain that takes place after the applied stress has been removed and a specified time has passed. As an example, suppose a recoverable strain, ϵ, of 0.0002 mm/mm (in/in) is recorded after an applied stress, σ, of 0.2 MPa (30 lb/in^2) is removed. The Mr of the specimen would then be σ/ϵ, or 1000 MPa (150,000 lb/in^2).

The value of Mr for a given specimen will vary to some degree with the magnitude and duration of loading, time allowed for recovery between loading cycles, testing temperature, etc. Thus, clear, precise guidelines must be established so that researchers and laboratories are able to obtain consistent and comparable results.

3 Resilient Modulus Testing

3.1 Equipment
The equipment used to perform Mr testing was designed and built at MTS Systems Corp., Minneapolis, Minnesota, USA, and consists of a load frame, a servo-hydraulically con

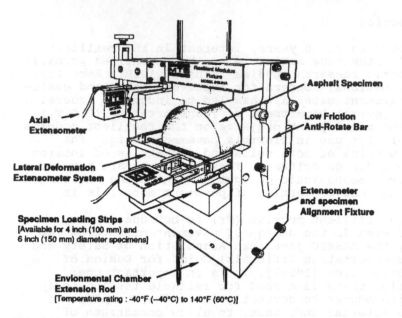

Figure 1 Asphalt Mr Fixture and Extensometers.

trolled actuator, an environmental chamber, a resilient
modulus fixture, control electronics, deformation measurement transducers, a load cell and testing software.

The Mr fixture was designed following the general
guidelines presented in ASTM D4123-82, is shown in
Figure 1. 100 mm (4 in.) and 150 mm (6 in.) diameter
specimens can be accommodated. The fixture can be installed in an environmental chamber which allows testing
temperatures ranging from -40°C (-40°F) to over 100°C
(212°F). A 5000 lb (22 kN) load cell was used for load
measurement, and is protected from overload by the electronic controller interlocks. Vertical (axial) displacements are measured using the MTS Model No. 632.06 extensometer [full scale travel ± 0.15 in. (± 3.80 mm);
range 4 - ± 0.015 in. (± 0.38 mm)]. Horizontal (lateral)
displacements are measured with a newly designed diametral
device [full scale travel ± 0.16 in. (± 4.06 mm);
range 4 - ± 0.016 in. (± 0.406 mm)] shown in Figure 2.
"Ranging" the transducers calibrates the output to a finer
full scale travel and allows higher resolution measurements of "small" deformations.

The diametral device consists of two extensometers with
gauge length extenders, and two specimen adapter brackets.
The brackets are machined to the same radius as the specimen and contact the specimen all along the thickness, t
(see Figure 2), similar to the contact between the loading
strips and the specimen. In this way, each bracket records
the maximum lateral extension (and, upon unloading, contraction) that occurs on each side of the specimen. By

measuring the mean extension (contraction) of the extensometers, errors due to tilting of the brackets are eliminated. The contact area also prevents point loading, as
might occur with an LVDT. The device is located on the
specimen by guiding pins that extend from the side of the
fixture. Before the test begins, the pins are removed and
the extensometer is held on the specimen by springs that
extend between the two brackets. After testing, the
threaded guide pins are used to retract the device and
allow the specimen to be removed. This procedure for
specimen installation and removal, combined with the ease
of attaching the diametral device to the specimen, reduces
the potential for specimen set-up errors, and thus pro-
motes more consistent testing and results.

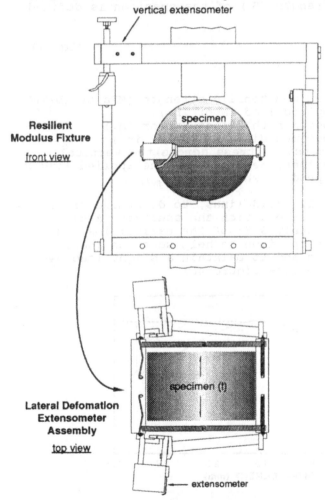

Figure 2   Diametral Measurement Device.

3.2 Software Programs
Two software programs have been developed to assist in conducting the Mr test; one controls the indirect tensile strength test; the other controls pre-conditioning and load cycling of the specimen.

3.3 Testing Procedure
To perform the indirect tensile strength test, the specimen is placed in the fixture and loaded, at a constant axial displacement of 5.04 mm/min (0.2 in/min). The load reaches a peak and then declines as the axial displacement is further increased (see Figure 3). The test is stopped when the applied load has decreased by a pre-determined amount from the peak value, in this case approximately - 40%. The tensile strength (S_t) of the specimen is defined by Eq. 1:

$$S_t = \frac{2P}{\pi tD}$$
(Eq. 1)

where
- S_t = specimen tensile strength (MPa or lb/in^2)
- P = applied load (N or lbs)
- t = specimen thickness (mm or in.)
- D = specimen diameter (mm or in.)

[S_t is the maximum tensile stress across the vertical diameter of the specimen, generated by the applied load P - see Fig. A1 in appendix, where $S_t = (\sigma_{\theta y})_P$].

Following ASTM D4123 guidelines, to determine Mr, it is first necessary to pre-condition the specimen. A small pre-load of approximately 3-5% of the maximum load is first applied to the specimen to help seat the loading strip on the specimen and to maintain a slight load between each load cycle (see Figure 4a).

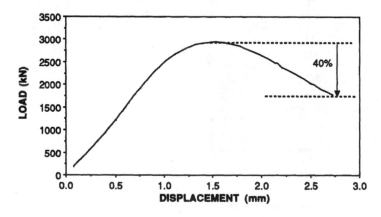

Figure 3   Indirect Tensile Strength Test Curve.

Pre-conditioning of the specimen continues until the rate of deformation of the specimen (i.e. the slope of the line mn in Figure 4c) is essentially constant. When the condition of less than a 2% change in the rate of the vertical and horizontal deformations, over 5 measured cycles, was met, the software will begin taking test data, signaling the end of pre-conditioning. Alternatively, the operator could terminate pre-conditioning (by pressing a defined key) when it is determined that the rates of deformation have become sufficiently constant. When pre--conditioning is terminated by one of the two methods, load and horizontal and vertical deformation data for the next five loading cycles are recorded by the computer.

Each specimen is tested at three temperatures, 5°C (41°F), 25°C (77°F), and 40°C (104°F) and at two specimen rotations, 0° and 90°. At each of the six test states (e.g. 25°C and 90° rotation), the specimen is pre--conditioned and loaded at cycle frequencies of 0.33, 0.5, and 1.0 Hz, corresponding to cycle durations of 3 sec., 2 sec., and 1 sec., respectively. Typical curves of speci-men loading and deformation versus time are shown in Figures 4a, 4b and 4c. The loading pulse used in this example is a 0.1 sec. haversine.

Table 1   Instantaneous and Total Poisson's Ratio and Resilient Modulus Results.

Asphalt Resilient Modulus Test

Test File : ROOMTEMP	Cycle Frequency : 1 Hz
Specimen SN : BETA	Load Duration : 0.1 sec.
Project Number : ALPHA	

Rotation : 0 Degrees

	Instantaneous Values		Total Values	
Test Cycle	Resilient Modulus MPa	Poisson's Ratio	Resilient Modulus MPa	Poisson's Ratio
1	3.728E+03	0.114	2.274E+03	0.234
2	3.841E+03	0.141	2.274E+03	0.237
3	3.784E+03	0.124	2.326E+03	0.247
4	3.812E+03	0.137	2.305E+03	0.252
5	3.900E+03	0.157	2.274E+03	0.240

Rotation : 90 Degrees

	Instantaneous Values		Total Values	
1	3.521E+03	0.123	2.149E+03	0.246
2	3.521E+03	0.104	2.139E+03	0.226
3	3.497E+03	0.113	2.139E+03	0.240
4	3.546E+03	0.111	2.149E+03	0.236

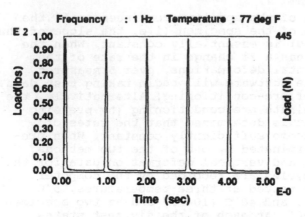

a) Applied Load versus Time.

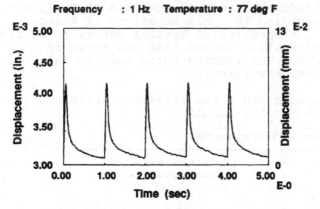

b) Vertical Deformation versus Time.

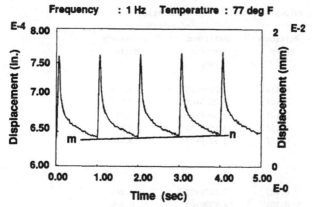

c) Horizontal Deformation versus Time.

Figure 4   Typical Curves of Load (a), Vertical Deformation (b), and Horizontal Deformation (c) versus Time.

For each of the 18 tests, both instantaneous and total values of Poisson's ratio and Mr are calculated. An example is shown in Table 1. Instantaneous values are results calculated from horizontal and vertical deformation measurements taken at the moment the applied load is removed, and before any appreciable specimen relaxation (see ASTM Spec. D4123-82).

3.4 Results and Discussion

The results presented in Figure 5 indicate that Mr decreases and Poisson's ratio increases as the testing temperature is increased. [The applied loads were 1.6 kN (350 lbs) at 5°C (41°F), 1.1 kN (250 lbs) at 25°C (77°F), and 0.2 kN (45 lbs) at 40°C (104°F)]. These results seem reasonable since the ductility of the asphalt binder increases with temperature. It is interesting, however, that an applied load of only 0.2 kN (45 lbs) produces deformations sufficient to induce a Poisson's ratio above 0.5. A Poisson's ratio near 0.5, the theoretical maximum for elastic material, suggests that the applied load may be too high. (An indicated value of Poisson's ratio above 0.5 implies that the specimen is no longer undergoing recoverable deformation only, and that permanent deformation/damage is occurring). ASTM D4123 recommends a load range that would "induce 10% to 50% of the tensile strength" of the specimen. This may be appropriate, but there is no suggestion of the most appropriate range for testing the specimen. It may be possible to determine the correct load range by monitoring Poisson's ratio. If the ratio approaches 0.5, the load is too high and should be reduced. However, it is necessary to maintain a load sufficient to produce measurable horizontal deformations.

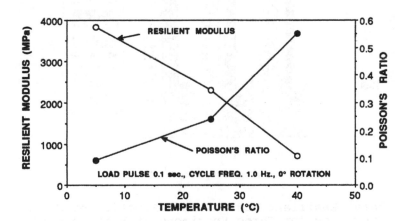

Figure 5   Average Resilient Modulus and Poisson's Ratio as a function of Specimen Temperature.

As mentioned before, the diametral device (and extensometers in general) can be calibrated to finer ranges of the total travel (e.g. a 10% range calibration would amplify the transducer output signal by 10). Matching of the transducer range with expected travel (i.e. "ranging") allows higher resolution measurements of "small" deformations.

Figure 6 shows the change in Mr at different cycle frequencies and specimen rotations. The results indicate that Mr increases slightly with cycle frequency. The tendency for Mr to increase with cycle frequency is also is quite evident in Figure 7, conducted at 40°C (104°F). This is not surprising, since higher cycle frequencies result in shorter rest periods for the specimen between loading pulses, hence less time for rebound strain (ΔV_t), and thus, as Eq. 4 would indicate, higher Mr values.

Figures 6 and 7 also indicate that Mr values at a 0° specimen position are larger than those at 90°. Since the 90° position is always tested after the initial, 0° position, the decrease in the Mr values at the 90° position could be due to internal damage done to the specimen during testing in the initial position. It would seem possible that overloading the specimen in the 0° position would initiate minor (unobserved) fracturing, with a

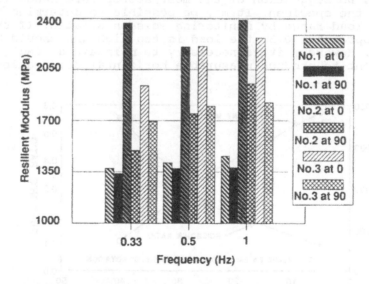

Figure 6   Total Resilient Modulus as a Function of Cycle Frequency for Three Specimens at Rotations of 0° and 90°, and at a Specimen Temperature of 23°C (73°F).

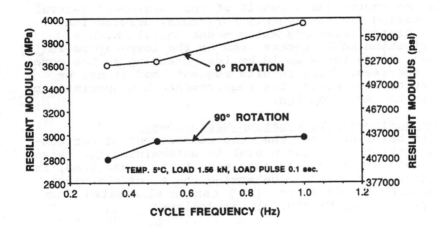

Figure 7   Resilient Modulus as a Function of Cycle Fre-
            quency, at Specimen Rotations of 0° and 90°.

tendency towards vertical splitting, concentrated in the
central region of the specimen, where the tensile stresses
(S_t, see Figure A1 in Appendix) are highest. In the 90°
position, these fractures would be oriented more or less
horizontally - which would cause increased vertical deformation, resulting in a lower Mr, as observed in the results.

It appears that Poisson's Ratio at the 90° rotation is
slightly higher than at 0° in the specimens tested at
lower temperatures (Figure 8). This could possibly be due
to a re-distribution of the applied load into the region

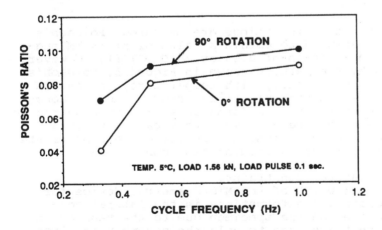

Figure 8   Poisson's Ratio as a function of Specimen Tem-
            perature, at Rotations of 0° and 90°.

outside the center (as a result of the "weakened" central
zone), causing greater overall horizontal deformation
(hence a higher Poisson's Ratio - see Eq. 3). Such a
re-distribution would become less as the temperature is
increases, since there would be less cracking in the more
ductile specimen. These results suggest that it may be
necessary to re-consider the requirement that specimens be
tested in the 90° position.

Need for Horizontal Deformation Measurement.

It should be pointed out that the measurement of horizontal deformation, although useful in determining the value
of Poisson's ratio for the specimen at the given load, is
not necessary in order to determine the Mr of the specimen. Horizontal deformation (ΔH_t) can be eliminated from
the expression for Mr (Eq. 2), as follows;

$$M_r = \frac{P}{\Delta H_t t}(0.27+v) \quad \text{[From D4123]} \tag{Eq. 2}$$

$$v = 3.59\frac{\Delta H_t}{\Delta V_t} - 0.27 \quad \text{[From D4123]} \tag{Eq. 3}$$

Substituting for v (from Eq. 3) in Eq. 2, we obtain;

$$M_r = \frac{P}{\Delta H_t t}(0.27+3.59\frac{\Delta H_t}{\Delta V_t}-0.27)$$

or

$$M_r = \frac{3.59P}{t\Delta V_t} \tag{Eq. 4}$$

If only horizontal deformations are measured, Poisson's
ratio must be assumed to obtain a value of Mr [ASTM suggests a Poisson's ratio of 0.35 at 25°C (77°F)]. However,
as evidenced by the results shown in Figure 5, Poisson's
ratio varies significantly with temperature, so that an
assumed value would probably be in error. Thus, although
measurement of horizontal deformations are not required to
determine the Mr of the specimen, they can serve as a
check on specimen condition, and indicate the appropriate
loading range for the particular test conditions.

Influence of lateral constraint.

One perceived drawback of the MTS diametral device for
measuring horizontal deformations is the lateral constraint generated on the specimen due to the tension in
the springs that are used to hold the assembly in place.
Figure 9 presents the results of tests in which four
different springs were used to evaluate the effect of

412

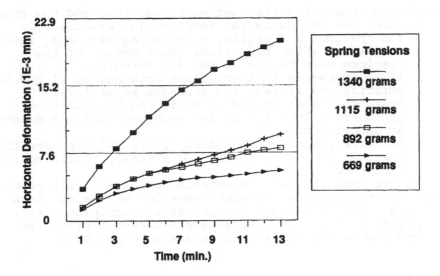

Figure 9   Change in Horizontal Deformation due to Spring
           Tension (no Vertical Load Applied).

spring tension on horizontal extensometer creep. As
Figure 9 suggests, a spring tension of approximately
446 grams pulled the extensometer into the specimen at a
creep rate of less than $4.2E^{-4}$ mm/min. In addition, a
theoretical calculation of the constraint imposed by the
diametral device indicates that the reduction of horizontal expansion, due to the confining influence of the
spring force would be less than 0.1% (see Appendix A). It
is always possible to glue the adapter brackets to the
specimen and eliminate any lateral constraint.

In-situ conditions which vary from location to location
impose varying degrees of lateral constraint when an
asphalt pavement structure is loaded vertically. Laboratory tests involve essentially no lateral constraint, hence
probably yield Mr values lower than those in-situ. It
would be worthwhile to conduct laboratory Mr tests on
specimens with a lateral constraint to determine the
importance of lateral constraint on Mr values. One method
of simulating the constraint could be to install "strong"
springs on the diametral device.

4 Conclusions

The tests conducted with the Mr fixture, while not constituting a complete suite, verified that the system could
produce reliable, consistent results in general agreement
with the results of other researchers (Resilient Modulus
Workshop, 1989). These preliminary tests also raised some

questions concerning the testing procedures, and how these procedures can influence the results. In particular:
- The specimen is fragile; overloading can cause internal damage which may affect the behavior of the specimen when tested at the 90° rotation.
- It is suggested that Mr values obtained at 0° and 90° rotations be listed, and averaged separately.
- Poisson's ratio should be monitored to determine appropriate load levels for each particular specimen or material batch.
- As cycle frequency increases, there is less time for specimen rebound (shorter cycle time), resulting in higher Mr values.

Future work will be concentrated on an analysis of the time and temperature dependent behavior of the specimen material, and on various aspects of the equipment (e.g. temperature effect on extensometers versus LVDT's, improved resolution of lateral extensometers, etc.).

5 References

AASHTO Guide for Design of Pavement Structures (1986). Published by American association of State Highway and Transportation Officials, 444 N. Capitol Street, N.W., Suite 225, Washington D.C. 20001.

ASTM D4123-82 (1987). Standard Test Method for Indirect Tension Test for Resilient Modulus of Bituminous Mixtures. Published by the American Society for Testing and Materials (ASTM), Annual Book of ASTM Standards, Vol. 04.03, pp. 503-505.

Resilient Modulus Workshop Proceedings (1989). **Proc. of the Workshop on Resilient Modulus Testing**, Oregon State University, Corvallis, OR, USA, March 28-30.

Vinson, T.S. (1989) Fundamentals of resilient modulus testing, in **Proc. of the Workshop on Resilient Modulus Testing**, Oregon State University, Corvallis, OR, USA, March 28-30.

6 Appendix

Effect of Horizontal Restraint on Results of Indirect Tension Test for Resilient Modulus of Bituminous Materials (ASTM D4123-82)

The analysis that follows considers a cylinder loaded simultaneously by a radial load P across the vertical diameter, and by a radial load Q along the horizontal diameter. Note: it is assumed that Q is applied over the same radial angle α (but across the horizontal diameters) as the load P (See Figure A1).

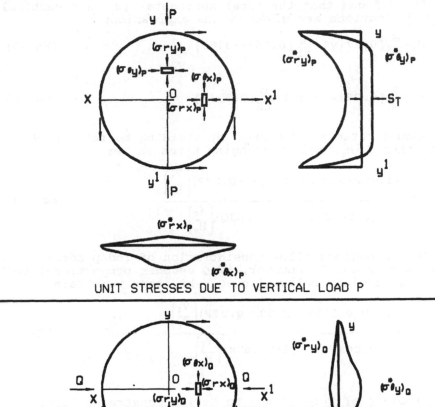

UNIT STRESSES DUE TO VERTICAL LOAD P

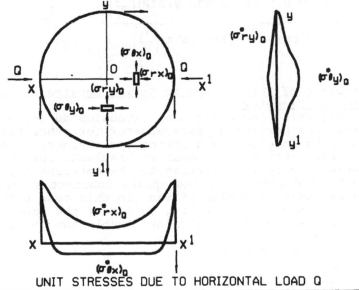

UNIT STRESSES DUE TO HORIZONTAL LOAD Q

Figure A1 Unit Stresses due to Vertical and Horizontal Loads.

It is found that the total horizontal (H_t) and vertical (V_t) deflections are given by the expressions:

$$H_t = \frac{P}{tE}[0.27+\nu] + \frac{Q}{tE}[0.063\nu-3.59] \qquad \text{(Eq. 5)}$$

$$V_t = \frac{P}{tE}[0.063\nu-3.59] + \frac{Q}{tE}[0.27+\nu] \qquad \text{(Eq. 6)}$$

Poisson's ratio may be found by dividing to form (H_t/V_t) [and then V_t/H_t] and rearranging terms to be:

$$\nu = \frac{(3.59P-0.27Q)-(3.59Q-0.27P)\left(\dfrac{V_t}{H_t}\right)}{(0.063P+Q)-(P+0.063Q)\left(\dfrac{V_t}{H_t}\right)} \qquad \text{(Eq. 7)}$$

These expressions allow consideration of independent values of P and Q. However, if Q becomes proportional to P (see discussions below) i.e. Q = kP, then we obtain

$$\nu = \frac{(3.59P-0.27k)-(3.59k-0.27P)\left(\dfrac{V_t}{H_t}\right)}{(0.063P+k)-(1+0.063k)\left(\dfrac{V_t}{H_t}\right)} \qquad \text{(Eq. 8)}$$

Development of Force Q due to Spring Constraint Across Horizontal Diameter.

In measuring the horizontal extension during vertical loading, the horizontal extensometer (or other transducer) may be held in place by a spring (or springs) stretched horizontally between the adapter brackets. The load Q will increase in direct proportion to the horizontal extension (H_t) of the diameter. If the spring constant κ (ie. lbs force per unit extension of the (in this case 6 inch) long spring). [Note: a 3 inch long spring of the same material would have a 'spring constraint' double that of a 6 inch long spring.]

With Q = kP we have

$$H_t = \frac{P}{tE}[0.27+\nu] + k[0.063\nu-3.59] \qquad \text{(Eq. 9)}$$

But $Q = \kappa \cdot H_t$, therefore $k = Q/P = \kappa \cdot H_t/P$, and

$$H_t = \frac{P}{tE}[0.27+v] + \frac{\kappa H_t}{tE}[0.063v-3.59] \qquad \text{(Eq. 10)}$$

$$H_t[1 - \frac{\kappa}{tE}(0.063-3.59) = \frac{P}{tE}(0.27+v) \qquad \text{(Eq. 11)}$$

$$H_t = \frac{\dfrac{P}{tE}(0.27+v)}{1 - \dfrac{\kappa}{tE}(0.063v-3.59)} - \frac{\dfrac{P}{tE}(0.27+v)}{1-\delta} \qquad \text{(Eq. 12)}$$

The numerator is the expression for the horizontal extension when $Q = 0$. Thus, the term

$$\delta = \frac{\kappa}{tE}(0.063v-3.59) \qquad \text{(Eq. 13)}$$

represents the reduction in horizontal expansion due to force Q.

Assuming $v = 0.35$ (ASTM recommendation), we have

$$\delta = \frac{\kappa}{tE}(-3.57) \qquad \text{(Eq. 14)}$$

Although the values for κ and E are not known precisely, we will assume that:
t = 3 inch (ASTM rec.)
E = 0.1 x 10⁵ psi
κ = 2 lbs/inch (for a 6 inch long spring)

Thus $\delta = (2 \times 3.57) / (3 \times 0.1 \times 10^5) = 2 \times 10^{-4}$

So, for these values at least, the reduction in horizontal deformation (extension) due to the lateral "spring-clip" transducer will be

$\delta = 0.02\%$

This suggests that horizontal spring clamps should not pose a problem with the accuracy of the H_t measurement.

Note also that

$$\frac{V_t}{H_t} = \frac{(0.063v-3.59)+\kappa(0.27+v)}{(0.27+v)+\kappa(0.063v-3.59)} \qquad \text{(Eq. 15)}$$

417

is little affected by k. Thus assuming $\nu = 0.35$ (ASTM recommendation) we obtain

$$\frac{V_t}{H_t} = \frac{0.62\kappa - 3.57}{0.62 - 3.57\kappa} \quad [\text{For } \kappa = 0, \frac{V_t}{H_t} = 5.76] \tag{Eq. 16}$$

Not knowing values for E of bituminous material and peak load P, we cannot calculate H_t or k. If we assume $K = 0.01$, then

$$V_t/H_t = (0.06-3.57)/(0.62-0.04) = 3.5/0.5 = 6.05,$$

or 5% above the K = 0 value.

Influence of radial load angle $\alpha_Q < \alpha_p$

If the horizontal device involves a narrower contact arc than for the P loading contacts, the value of $(\sigma^*_{\theta x})_Q$ will be reduced to less than 0.063 (approaching zero as $\alpha_Q \rightarrow 0$). $(\sigma^*_{\theta y})_Q$ must stay always equal to unity. The other two values are likely to remain essentially unchanged.

29 FISSURATION THERMIQUE DE STRUCTURES SEMI-RIGIDES – ESSAIS DE SIMULATION

(Thermal cracking of semi-rigid structures – simulation tests)

L. FRANCKEN
Centre de Recherches Routières, Bruxelles, Belgium

Résumé

Les études entreprises au Centre de Recherches Routières dans le domaine de la fissuration des chaussées ont conduit à développer des méthodes d'analyse et de simulation en laboratoire du phénomène de la fissuration dite "réflective". Dans ce contexte, un appareillage d'essai a été conçu pour permettre l'étude et l'observation de la propagation de fissures ou de discontinuités préexistantes dans la couche de base de structures routières sous l'effet des contraintes d'origine thermique. L'article décrit l'application de cette méthode et donne quelques conclusions pratiques tirées des premiers essais réalisés .

Mots-clés :Essais - Fissuration thermique - Enrobés hydrocarbonés - Géosynthétiques - Interfaces.

1 Introduction : Principes généraux

Le problème de la remontée des fissures affecte de nombreuses chaussées à revêtement hydrocarboné dont la partie inférieure de la structure comporte des discontinuités.
Les cas types sont énumérés au tableau I.
Ce phénomène résulte du fait que les discontinuités préexistantes dans les couches inférieures constituent des zones de concentration de contraintes d'où peuvent repartir de nouvelles fissures dites "réflexion".
Ce phénomène est influencé par deux facteurs extérieurs : le trafic et les contraintes d'origines thermique.
Il comporte trois phases d'évolution :

1) la phase primaire durant laquelle la couche hydrocarbonée de la surface garde son intégrité,
2) l'initiation de la fissure; phase de transition qui correspond au dépassement des critères de performances des couches hydrocarbonées dans la zone de concentration des contraintes.
3) la phase de propagation de la fissure au travers des couches hydrocarbonées.

L'un des moyens proposés pour enrayer ce phénomène, consiste à utiliser des matériaux ou combinaisons de matériaux (constituant un système) comme interface entre la partie

Tableau 1 . Structures routières affectées par la fissuration réflective

Type de structure	Nature des matériaux dans les couches supports	Types de défauts discontinuités initiant la fissuration
Renforcement sur chaussée souple	Couches hydrocar- bonées	Fissure de fatigue faïençage
Chaussée semi-rigide	Grave-ciment Béton maigre	Fissures transversales de retrait hygromé- trique fissures de fatigue
Renforcement sur chaussée rigide	Dalles en béton de ciment	Joints transversaux et longitudinaux fissures de fatigue
Elargissement	Variables	Contiguïté de deux matériaux différents

inférieure de la chaussée fissurée ou présentant des discontinuités, et le revêtement continu de la couche de roulement.

Le tableau II montre que la gamme de produits et de types de systèmes actuellement proposés est extrêmement variée tant par la nature des composants de base que par leur morphologie et la fonction qui leur est attribuée.

Tableau 2 . Caractéristiques et fonctions des matériaux d'interface

Type de produit	NTN	NTT	TIS	GGR	GRM	SAMI
Matériau de base						
Polyéthylène	x	x		x		
Polypropylène	x	x	x	x		
Polyester	x	x	x	x		
Bitume-élastomère						x
Acier					x	
Fonction assurée						
Découplage	x	x	(x)			
Renforcement			x	x	x	
Fluage-glissement	x	x	(x)			x
Déconcentration	x	x				x
Etanchéité	x	x				x
Méthode de pose						
Collage à l'émulsion	x	x	x	x		
Collage au liant pur	x	x	x	x		
Fixation mécanique				x	x	
Coulé en place						x

NTM - Non tissé lié mécaniquement
NTT - Non tissé thermosoudé
TIS - Tissé
GGR - Géogrille
GRM - Grille métallique
SAMI - Membrane de dissipation de contraintes

On recherche en général à prolonger la durée de vie de la structure en accroissant le délai d'initiation de la fissure et le temps nécessaire à sa propagation vers la surface.

Ce résultat est recherché en faisant assurer à l'interface une ou plusieurs des fonctions suivantes :

1) Découplage : rallonger le chemin parcouru par la fissure en privilégiant sa propagation horizontale à l'interface.
2) Renforcement : reprendre les contraintes de traction agissant la base des couches hydrocarbonées.
3) Glissement : Absorber les déplacements horizontaux au doit de la discontinuité par interposition d'un matériau à faible module d'élasticité et/ou déformable plastiquement sous des sollicitations de longue durée.
4) Absorption des contraintes : atténuer la concentration des contraintes à l'extrémité de la fissure.
5) Etanchéité : Maintenir une séparation imperméable entre le revêtement et les couches inférieures.

Pour une interface donnée, la fonction peut varier selon les conditions de sollicitation et le climat. C'est ainsi qu'un produit peut renforcer ou non la structure selon que son module est supérieur ou inférieur à celui du matériau hydrocarboné qui le surmonte, lequel, on le sait, peut varier dans de larges proportions avec la température et la vitesse des sollicitations.

2 Recherches

D'une manière générale les recherches réalisées sur la fissuration se sont attachées plus particulièrement à l'étude de la phase de la propagation d'une fissure (phase 3). Elles font appel aux théories de la mécanique de la fracture et ont pour but principal de faire une évaluation du temps de remontée des fissures vers la surface. Les études entreprises au Centre de Recherches Routières, se sont attachées, dans un premier temps, à recherche les facteurs permettant de prolonger le délai d'initiation de la fissure (phase 1), par une meilleure conception de l'interface. Il apparaît en effet que les chances de succès ainsi que les facteurs sur lesquels on peut avoir prise sont plus importants à ce niveau.Cette recherche fait partie d'un programme comportant les activités complémentaires suivantes:

1) caractérisation des matériaux et systèmes
2) caractérisation des conditions de sollicitation et mise au point d'une méthode d'expertise préalable
3) modélisation sur modèles mathématiques
4) simulation des sollicitations thermiques et mécaniques en laboratoire
5) suivi de la mise en oeuvre
6) suivi du comportement à long terme et en vraie grandeur.

Dans ce qui suit, nous nous intéresserons tout particulièrement à la simulation des sollicitations thermiques en laboratoire pour laquelle nous avons été conduits à mettre au point un appareillage spécial.

3 Simulation des sollicitations thermiques en laboratoire

3.1 Description de l'appareillage
Les contraintes thermiques sont simulées au moyen d'un dispositif dont le schéma est donné sur la figure 1. La caractéristique principale de ce dispositif est de permettre le contrôle précis des variations de largeur des fissures dans des modèles de structures semi-rigides maintenus à une température constante que l'on peut choisir entre 0 et -10°C et, de simuler ainsi le phénomène à une vitesse voisine de celles observées in situ sous l'effet des cycles journaliers de variation de température, c'est-à-dire de

l'ordre de quelques dixième de mm à l'heure. L'éprouvette est constituée d'une dalle en béton fissurée surmontée du produit bitumineux que l'on désire étudier. La déformation est induite par la dilatation ou la contraction du bâti de fixation de l'éprouvette. Ces déformations lentes et parfaitement contrôlables sont obtenues en faisant varier périodiquement la température d'un liquide de thermostatisation circulait dans les barres du bâti de l'appareil d'essai. La température de l'enceinte est contrôlée par un autre circuit indépendant. La couche de base repose sur un lit de billes d'acier permettant le déplacement libre et quasi sans frottement. L'appareillage est équipé de capteurs permettant la mesure des contraintes appliquées, de l'ouverture de la fissure ainsi que du déplacement relatif du revêtement par rapport au support fissuré. L'observation visuelle du phénomène est facilitée grâce à un dispositif d'enregistrement vidéo dont les prises de vue sont commandées par le système de contrôle et d'acquisition des données.

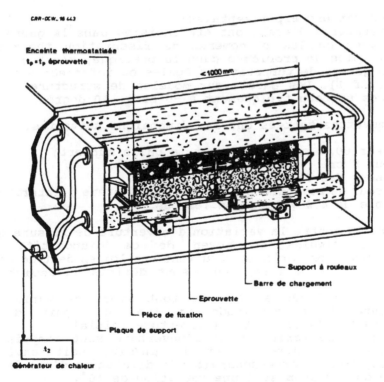

Figure 1 . Schéma de principe de l'appareil de simulation de la fissuration thermique

3.2 Etude du comportement de systèmes d'interfaces

Différents types d'interfaces sont en cours d'expérimentation au CRR au moyen de ce dispositif. Le but poursuivi est de permettre une meilleure connaissance de leur comportement mécanique et de fournir, conjointement avec d'autres essais de caractérisation, les données de base essentielles pour la modélisation des systèmes d'interface et la prévision de leur comportement en vraie grandeur.
On s'est attaché notamment à l'étude de l'influence des paramètres suivants :
. condition physiques extérieures définissant les sollicitations.
. conditions géométriques (épaisseurs des couches, largeur de la discontinuité).
. caractéristiques mécaniques des matériaux constitutifs.

Cette étude est encore insuffisamment avancée pour que des conclusions définitives puissent être avancées, mais les résultats déjà disponibles, dont nous donnerons ci-après quelques exemples, sont riches en enseignements.

3.2.1 Conditions expérimentales

Les températures d'essai ont été choisies dans la gamme de températures où les phénomènes de fissuration thermique posent le plus de problèmes dans la pratique, c'est à dire dans les gammes de températures égales ou inférieures à - 5 degrés (réf 1). Les essais sur modèles de structures avec interfaces se fotn typiquement à -5 et à -10 degrés.

Les éprouvettes testées ont les dimensions suivantes:

. longueur 60,4 cm (2 x 30 pour la dalle support)
. largeur 7 cm
. épaisseur des dalles de béton : 7 cm
. épaisseur de revêtement (y compris le système d'interface)
. variable de 4 à 7 cm selon les cas .

La figure 2a montre la variation d'ouverture de fissure qui peut être simulée au moyen de ce dispositif. En l'occurrence, les variations suivent une loi en dent de scie dont la période et l'amplitude sont définis au départ de l'essai.
Les essais réalisés au CRR se font avec une variation d'ouverture d'une amplitude de 1mm au départ d'une discontinuité d'environ 4mm d'ouverture initiale.
Les vitesses de variation de l'ouverture sont comprises, selon l'essai, entre 15 et 17 µm/min, soit environ 1mm/heure, ce qui correspondrait à la dilatation d'une dalle de béton de 10m soumise à une variation de 10°.

ESSAI DE FISSURATION THERMIQUE

Variation d'ouverture de la fissure en fonction du temps

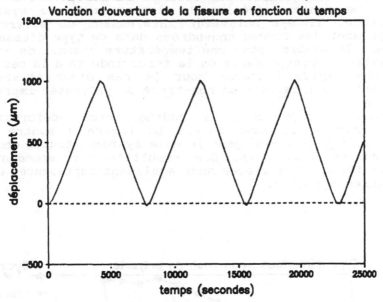

Figure 2a . Variation de l'ouverture (T = - 5°C)

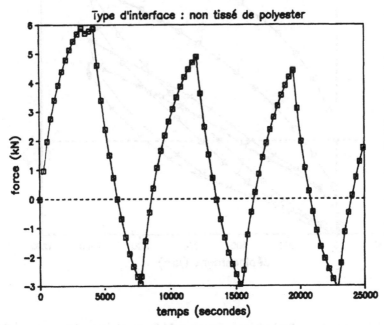

Figure 2b. Variation de la force

3.2.2 Exemples d'applications

Les données enregistrées en cours d'essais permettent de décrire de manière précise le comportement du système d'interface. Il est notamment intéressant de suivre le développement des forces engendrées dans ce type d'essai.
La figure 2b montre, pour une température d'essai de -5°C, la variation correspondante de la force induite à la base de la couche rigide fissurée pour le cas d'une structure comportant une interface en non-tissé de polyester imprégné de bitume.
La figure 3 représente la courbe force -déformation correspondant à ce même essai. La figure 4 montre les résultats obtenus à -10° pour le même système, toutes choses étant égales d'ailleurs. Des résultats correspondant à d'autres types d'interfaces sont également représentés dans les figures 5,6 et 7.

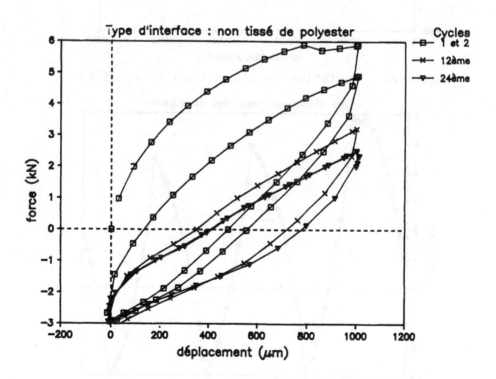

Figure 3 . Relation force-déformation (T = - 5°C)

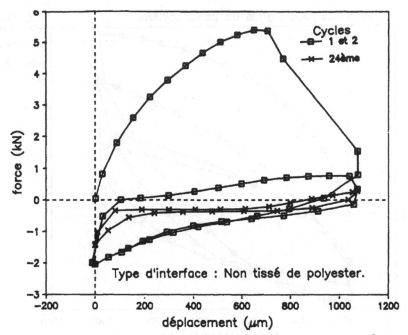

Figure 4 . Courbes force-déformation (T = - 10°C)
avec rupture au premier cycle

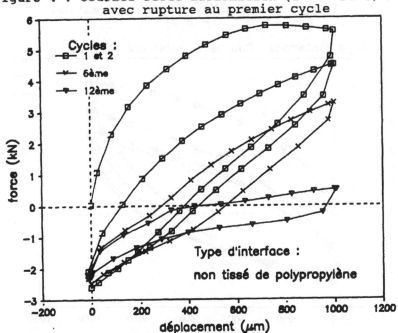

Figure 5 . Courbes force-déformation T = -5)
avec rupture au 12ème cycles

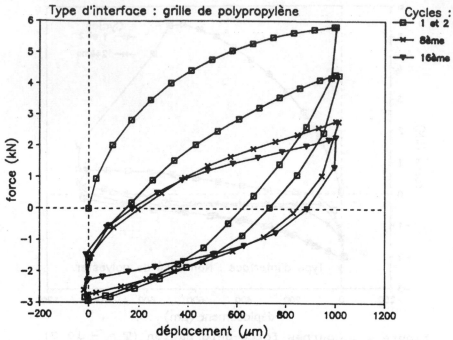

Figure 6 . Courbes force déformation
(T = − 5°C)

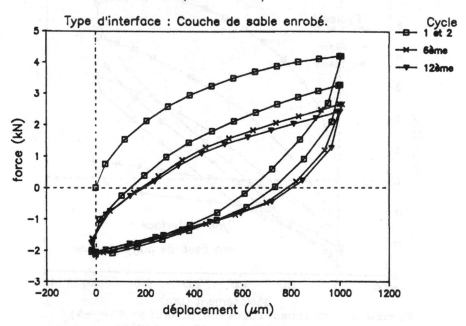

Figure 7 . Courbes force-déformation (T = -5°C)

3.2.3 Résumé de quelques observations

Ces exemples permettent de mettre en évidence les faits et constatations suivantes :

- tous les systèmes étudiés présentent une réponse typiquement non linéaire, caractérisée par une hystérèse très marquée dans les courbes force-déformation,
- le mouvement de la fissure entraîne une déformation permanente qui est plus importante au premier cycle,
- dès le premier cycle, le retour à la largeur initiale de la fissure entraîne l'apparition de contraintes de compression au voisinage de la fissure,
- la force maximale caractérisant la rigidité d'ensemble du système a tendance à décroître de manière continue de cycle en cycle,
- par contre, la force de compression nécessaire à la fermeture de la discontinuité (retour à l'ouverture initiale) ne varie pratiquement pas, et sa valeur maximale ne dépend que de la raideur de la couche de revêtement (déterminée par son épaisseur et son module),
- les performances des systèmes interface-revêtement sont très sensibles à la température, et celles-ci diminuent fortement pour des températures inférieures à -5°(voir figures 3 et 4).

Certaines de ces obervations confirment les résultats déjà obtenus par G. Colombier et all (réf.2) au moyen d'un essai de cisaillement d'interface . Leur étude détaillée doit permettre la modélisation de ces différents types de comportements au moyens de modèles rhéologiques simples incluant les éléments d'élasticité, de viscosité et de frottement en différentes combinaisons.

3.3 Autre possibilité d'utilisation de l'appareillage : l'essai de retrait entravé.

Dans le dispositif qui a et décrit ci-dessus, le contrôle rigoureux de la longueur de l'éprouvette est rendu possible en jouant sur la température du liquide circulant dans les barres du bâti de l'appareil, par ailleurs, la température de l'enceinte est contrôlée par un circuit indépendant.

Il est donc possible de réaliser des variations programmées de la température de l'éprouvette tout en maintenant sa longueur constante à tout moment. Ce type d'essai dit "de retrait entravé" constitue un moyen intéressant pour l'étude des contraintes thermiques induites dans les matériaux routiers.

La figure 8 donne à titre d'exemple, les résultats obtenus au cours d'un de ces essais . On remarque sur ces figures que la contrainte thermique engendrée par une chute rapide de température ne se relaxe que très lentement une fois la température stabilisée à -10°C, et qu'une faible chute supplémentaire de température suffit à faire augmenter à nouveau la contrainte.

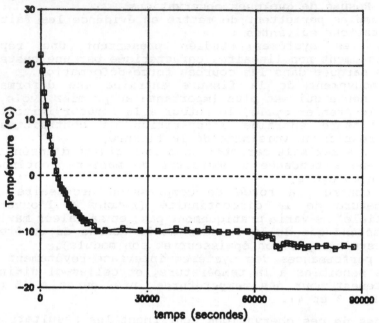

Figure 8 . Essai de retrait thermique entravé
a) variation de la température

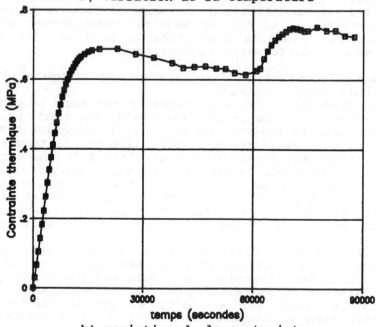

b) variation de la contrainte

L'un des intérêts de cet essai est aussi de permettre l'évaluation du module de rigidité et de l'allongement à la rupture des matériaux hydrocarbonés dans des conditions de sollicitation encore mal connues.

4 Conclusions

Après avoir rappelé les principes généraux et les recherches en cours dans la lutte contre la remontée des fissures dans les structures routières, on présente un appareillage destiné à la simulation de cycles de contraintes thermiques appliquées à des éprouvettes représentant des modèles de structures routières semi-rigides.

Cet appareillage permet le contrôle précis des variations de longueur de fissures ou de discontinuités pré-existantes dans le support, à une vitesse voisine de celles observées in-situ sous l'effet des cycles de variation journaliers de la température.

Il permet entre autre l'étude comparée des performances de différents systèmes utilisés pour entraver ou supprimer la propagation de fissures vers le revêtement bitumineux. Vu sous cet angle, il présente donc un intérêt pratique immédiat.

En outre, il permet d'obtenir des informations complétant utilement les résultats d'études théoriques.

Il devrait notamment aider à l'identification et la caractérisation des propriétés rhéologiques décrivant le comportement des interfaces.

Références

L.Francken : Systèmes d'interface pour le lutte contre la fissuration. Enseignements et expériences. 4e Symposium EUROBITUME Madrid 1989. Vol1 ; Rapport IV.2 pp 757-761.

G. Colombier et al. Fissuration de retrait des chaussées à assises traitées aux liants hydrauliques, Bulletin de Liaison du Laboratoire Central des Ponts et Chaussées, France, n° 156 et 157, juillet et septembre 1988.

30 METHODE D'ESSAI PAR OSCILLATION AXIALE POUR LA DETERMINATION DES CARACTERISTIQUES MECANIQUES ET DU COMPORTEMENT À LA FATIGUE DES ASPHALTES

(Axial vibration test procedure to determine the mechanical properties and fatigue behaviour of asphalt)

R. GUBLER
EMPA, Dübendorf, Switzerland

Sommaire

L'article présente les aspects importants de la technologie d'essais développée pour la détermination des modules complexes par l'essai d'oscillation au moyen de la flexion-cisaillement axiale sur des éprouvettes d'asphalte. En particulier, le concept du déroulement de l'essai depuis l'installation de l'éprouvette jusqu'à l'exécution de l'essai, y compris le planning de l'opération, fera l'objet de cette étude. L'utilisation de l'informatique pour obtenir ces résultats sera exposée. L'influence de la charge dynamique sur les caractéristiques de matériaux bitumineux sera démontrée par un exemple et la méthode d'évaluation pour obtenir la valeur spécifique du matériau non chargé seront décrites. Ensuite sont exposées les méthodes dans un domaine de fréquences et de températures étendu et, d'autre part, pour la détermination des effets de fatigue dans le domaine des températures estivales.

Mots-clés: technologie d'essais, essai d'oscillation axiale, flexion-cisaillement axiale, concept de l'essai, détermination du module, essai à la fatigue, méthode d'essai.

1 Introduction

Les aspects caractéristiques de la technologie d'essai sur de l'asphalte pour la détermination des valeurs mécaniques spécifiques sont présentés dans cet article. Ces valeurs peuvent être obtenues par des essais d'oscillation au moyen de la flexion-cisaillement axiale.

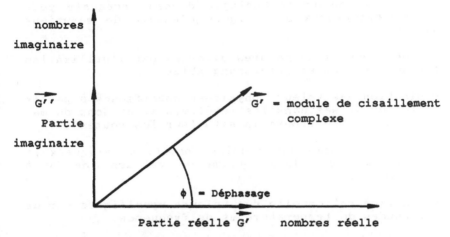

Figure 1. Représentation véktorielle du module complexe
avex G' come component élastiques et G'' come component
visquex et la déphasage

De cette façon, les composantes visqueuse et élastique du
module de cisaillement complexe peuvent être déterminées
séparément. La valeur rhéologique représentée par un vec-
teur peut-être définie par ses composants G'et G". Le
module de cisaillement complexe est alors donné par la
relation.

$G = G' + G"$

La partie réelle du module de cisaillement complexe peut
s'écrire:

$G' = [G] \times \cos \phi$

G' représente une valeur pour l'énergie de déformation
récupérée lors de l'oscillaltion, et donc une valeur pour
l'élasticité du matériau. La partie imaginaire sera définie par l'équation:

$G" = [G] \times \sin \phi$

G" est une valeur pour l'énergie de déformation trans-
formée en chaleur non récupérable pendant l'oscillation.
Ces relations sont montrées sur la figure 1.

Les buts recherchés lors du développement à l'EMPA d'une
installation d'essai appropriée étaient multiples et in-
cluaient les point importants suivants:

- mise au point de la technologie d'essai adéquate pour les caractéristiques spécifiques d'essais de matériaux bitumineux.

- exécution d'essais de natures diverses par l'utilisation d'accessoires d'essai interchangeables

- détermination des valeurs physiques spécifiques à partir des carottes prises dans les revêtements, et donc du matériau comme il se trouve in situ (sur les routes).

- détermination exacte des modules complexes, visqueux et élastiques sur une large gamme de valeurs (de 10 à 10'000 N*mm^{-2}).

- détermination rationnelle des valeurs spécifiques sur de larges bandes de température et de fréquence.

2 Facteurs d'influence sur le déroulement de l'essai

Une installation d'essais adéquate et bien équipée ne suffit pas à elle seule pour obtenir une détermination exacte des valeurs spécifiques recherchées. Le reproductibilité et l'exactitude des valeurs spécifiques dépendent beaucoup plus du concept de déroulement de l'essai dans son ensemble. Cela concerne des facteurs aussi différents que l'utilisation d'amplificateurs de mesurage, le traitement des éprouvettes, l'influence de l'essai sur les éprouvettes et les méthodes d'évaluation. Il devient donc nécessaire de développer des méthodes d'essais qui puissent garantir une qualité élevée et constante de l'essai.

La connaissance des facteurs d'influence possibles sur les résultats est donc nécessaire pour atteindre le but fixé. Diverses sources d'erreur sont à considérer pour la détermination des valeurs spécifiques mécaniques des asphaltes. Les facteurs d'influence sont indiqués dans la table 1, classés selon le moment de leur apparition durant le déroulement de l'essai.

Table 1

Classification chronologique des facteurs d'influence

Type de facteur	Facteurs principaux	Mesures contre les erreurs d'essais
Etat de l'éprouvette	Variation de l'éprouvette, faits antécédents, méthode de fabrication	Recommandation de réception
Paramètres d'essai	Dimensions de l'éprouvette, installation et choix des appareils de mesurage, réglage des amplificateurs de mesurage	Etablissement de la méthode d'ajustage
Influences du déroulement de l'essai	Histoire de l'essai, en particulier de la variation de température. Changement de la structure granulée sous charge mécanique	Projet du déroulement de l'essai
Méthode d'évaluation	Détermination de l'amplitude et des valeurs spécifiques du matériau intact	Evaluation conditionnée

3 Garantie de la qualité de l'essai

Les méthodes développées pour assurer la qualité de l'essai peuvent être classées en quatre catégories dépendant de la méthodique.

Etat de l'éprouvette

L'état des éprouvettes ne peut pas être contrôlé par une amélioration de la méthode d'essai, mais par l'établissement de recommandations relatives à l'extraction ou la production d'éprouvettes. L'essai sur des éprouvettes Marshall livre des résultats erronés car la structure granulée de ces éprouvettes ne correspond pas à celle des carottes provenant d'un revêtement. Les essais sur des éprouvettes fabriquées selon des méthodes de laboratoire ne livrent que des indications comparatives, aussi est-il préférable d'exécuter les essais sur des carottes. Ce problème ne sera pas traité plus avant dans cet article.

Concept du déroulement de l'essai

L'organisation du déroulement de l'essai protège l'éprouvette des influences destructives au cours des essais,
assure l'acquisition des valeurs spécifiques du matériau
intact et garantit une histoire de l'essai clairement définie.

Assistance par l'ordinateur durant l'exécution de l'essai
Cette assistance facilite le réglage de l'installation par
l'assistant de laboratoire et évite les erreurs de manipulation. L'ordinateur contrôle tout le déroulement de
l'essai, qui se passe selon un plan précis à la seconde
près. La température à l'intérieur de l'éprouvette et sa
position sont continuellement réglées par l'ordinateur.
Les phénomènes exceptionnels sont reconnus et documentés,
et toute l'histoire de l'essai est enregistrée dans
l'esprit de la Good Laboratory Practice.

Méthode d'évaluation
Les valeurs spécifiques de l'éprouvette intacte sont déterminées durant l'évaluation et contrôlées quant à des
erreurs éventuelles.

Les méthodes utilisées vont être exposées de façon détaillée.

3.1 Concept du déroulement de l'essai

Réduction au minimum des contraintes mécaniques sur
l'éprouvette. Les carottes prélevées dans le revêtement
sont, ainsi que cela est montré sur la figure 2, collées
dans les cylindres servant de forme extérieure pour
l'éprouvette. Un trou de diamètre intérieur ϕ 50 mm est
foré au centre de l'éprouvette et le cylindre intérieur y
est collé de façon centrée. Puis l'éprouvette sera mise en
place dans l'installation d'essai.

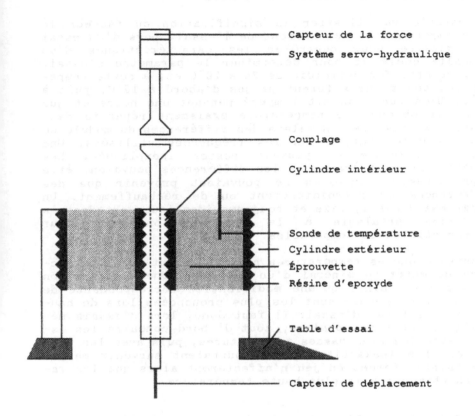

Capteur de la force
Système servo-hydraulique
Couplage
Cylindre intérieur
Sonde de température
Cylindre extérieur
Éprouvette
Résine d'epoxyde
Table d'essai
Capteur de déplacement

Figure 2. Principe du dispositiv d'essai

La fixation utilisée évite une déformation de l'éprouvette, due à la pression exercée par l'encastrement. Ainsi, la structure du matériau minéral reste elle aussi intacte. La position moyenne de l'éprouvette sera assurée
durant toute la durée de l'essai. Les déformations résultant des charges dynamiques d'oscillation sont ainsi symé-
triques par rapport à la position moyenne et ne compren-
nent aucun effet non désiré dû au fluage. L'essai se passe
sous effort contrôlé, il en résulte un meilleur réglage du
comportement et en une moindre dispersion des valeurs
mesurées. Les forces mises en jeu pour la détermination
des valeurs spécifiques mécaniques sont tenues à un bas
niveau afin d'éviter, dans une large mesure, des modifications de la structure du matériau minéral.

Organisation du déroulement de l'essai

Un exemple va illustrer la signification du facteur le
plus important du déroulement de l'essai. Lors d'un essai
préliminaire pour contrôler les caractéristiques d'un
asphalte coulé et pour déterminer le paramètre d'essai,
l'éprouvette fut refroidie de 25 à 20°C et, à cette température, des mesures furent prises d'abord à 15°C, puis à
20°C. Bien que l'on ait tempéré pendant une heure et que
l'on ait obtenu une température également répartie dans
toute l'éprouvette, on releva des différences du module de
8 à 15 pour cent, selon les fréquences utilisées. Une
telle différence ne pouvait rester inexpliquée. Des
répétitions montrèrent que les différences pouvaient être
reproduites et qu'elles ne pouvaient provenir que des
différences du refroidissement ou de réchauffement. Un
programme d'essai, fixe et toujours identique, est donc la
condition préalable à la détermination de valeur
reproductibles et comparables.

L'essai à hautes températures agit fortement sur la structure du matériau minéral d'un asphalte. Une modification
de structure ainsi qu'une modification correspondante de
la valeur du module sont les plus prononcées lors de hautes températures d'essai. Il faut donc, lors d'essais mécaniques sur des asphaltes, tout d'abord conduire les essais avec les plus basses températures, puis avec les plus
hautes. Les inexactitudes qui pourraient survenir malgré
les faibles forces en jeu n'affecteront alors que les valeurs obtenues à la plus haute température.

3.2 Exécution de l'essai avec l'assistance de l'ordinateur

Assistance pendant la mise en place
Les caractéristiques mécaniques des matériaux bitumineux
dépendent largement de la température. Une détermination
optimale des valeurs spécifiques concernées suppose l'existence d'appareillages de mesure adaptés à des forces et
déformations très différentes. Les possibilités de réglage
très variées qui sont ainsi disponibles s'accompagnent
aussi de risques de manipulations erronées. Un logiciel a
été développé pour libérer l'assistant de tout calcul de
conversion et éviter ce risque. Il ne doit plus que re-
lever la valeur des paramètres d'essai, indiquer quelles
valeurs imposées s'appliquent pour l'essai et quel capteur
de déplacement devra être monté sur l'éprouvette. L'ordinateur calcule, à partir de ces données, toutes les mises
au point nécessaires ainsi que les indications de réglage
qui lui seront utiles ultérieurement. Parallèlement, les
contradictions possibles entre les paramètres d'essai dé-
sirés sont recherchées. Les programmes d'essai qui ne peu-

vent être réalisés ou qui endommageraient l'éprouvette seront reconnus à l'avance et ne seront donc pas entamés.
Les gaspillages d'éprouvettes et d'essais sur des éprouvettes chargées et avec des structures de matériau minéral
modifiées de façon inadmissible seront ainsi évités.

Contrôle du déroulement de l'essai

L'ordinateur contrôle de façon continue la température, la
position moyenne de l'éprouvette et aussi pendant l'application de la charge dynamique, la valeur du module. Ce
contrôle constitue la base d'un réglage adéquat. Les températures dans l'éprouvette peuvent être exactement main-
tenues, à $0.2°C$ près. Les phénomènes de fluage sont évités
par le maintien de la position moyenne.

Ainsi l'essai se déroulera automatiquement selon un programme fixe et à la seconde près. Les influences sur les
résultats, provenant de déroulements d'essai différents
dans leur durée, seront éliminées. Les variations du module, de la température dans l'éprouvette et de sa position moyenne, seront enregistrées et documentées. Cette
procédure correspond aux principes de la Good Laboratory
Practice, et permet d'obtenir à tout moment l'évaluation
de la fiabilité des valeurs déterminées.

Evaluation instantanée

Les valeurs des mesures enregistrées sont évaluées aussi-
tôt après leur relevé. Les représentations graphiques instantanées des valeurs mesurées importantes fournissent à
tout moment à l'assistant exécutant l'essai un aperçu de
la situation de l'essai.

La force et la déformation sont enregistrées, à chaque
oscillation, avec 32 mesures individuelles à intervalles
exactement égaux. Cela permet, grâce à une analyse de
Fourier, de déterminer avec la meilleure exactitude possible les coefficients sinus et cosinus de son fondamental
et, ainsi, d'obtenir par calcul les amplitudes de la force
et de la déformation ainsi que les phases intermédiaires.
Cette évaluation est faite plusieurs fois pour chaque
température et chaque fréquence. Par ailleurs, les dévia-
tions des courbes de la force et de la déformation par
rapport à la sinusoïde sont déterminées, et enregistrées
avec la température de l'éprouvette et la fréquence de
charge mesurée.

3.3 Méthode d'evolutions développée

Acquisition des valeurs spécifiques obtenues sur le matériau intact. Des recherches détaillées antérieures ont montré que l'essai pouvait causer des modifications réversibles et irréversibles des matériaux bitumineux. La figure 3 montre, à titre d'exemple particulièrement marquant, les résultats obtenus avec un mélange de bitume et de filler.

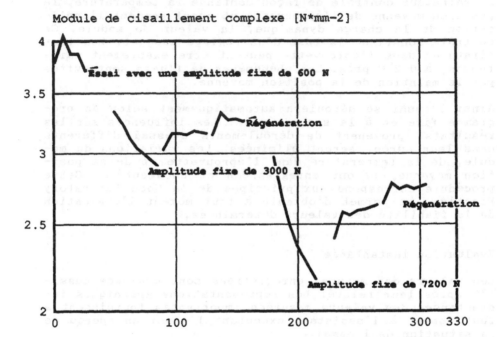

Figure 3. Influence de la charge sur le module complexe à 20 °C d'un mélange de bitume et de filler

Moins nettement, mais fondamentalement de la même façon, des modifications des caractéristiques des éprouvettes ont été relevées sur des matériaux bitumnineux les plus divers. Des modifications irréversibles de l'éprouvette peuvent être largement évitées grâce au concept d'essai dont on vient de parler. Par contre, les modifications réversibles se manifestent, en fonction de la température et de la fréquence des oscillations, déjà avec des amplitudes de la force minimes et nécessaires pour une détermination suffisamment exacte. Elles devront être prises en considération lors de l'évaluation des résultats.

C'est pour ces raisons que, à chaque fois, des essais seront effectués avec les 6 valeurs déterminées et consécutives de l'amplitude de la force, de la déformation et de la phase entre la force et l'amplitude. Cela afin de déterminer une dépendance statistiquement contrôlable avec la durée de la charge dynamique. Si une telle modification des valeurs mesurées est finalement prouvée, la valeur spécifique concernée sera déterminée par une régression linéaire, comme une valeur au moment déterminé où la charge dynamique est appliquée. Les valeurs moyennes seront utilisées lorsqu'aucune dépendance de la durée de charge ne sera statistiquement trouvée.

4 Transposition dans la méthode d'essai

4.1 Détermination des valeurs spécifiques mécaniques

Les méthodes mises au point permettent de déterminer, sur la même carotte et avec une précision suffisante, les valeurs de modules en fonction de la température et de la fréquence dans une bande allant de 10'000 $N*mm^{-2}$ à 100 $N*mm^{-2}$. La température peut varier de -5 à +30°C et la fréquence de 0.1 à 32. Les résultats obtenus peuvent être évalués individuellement à partir de la masse d'informations enregistrées durant l'essai. Les valeurs se plaçant en dehors de la série, par exemple à hautes températures et à basses fréquences, peuvent alors être considérées comme erronées et elles seront éliminées. La figure 4 expose un exemple des valeurs déterminées pour le module complexe en fonction de la température et de la fréquence.

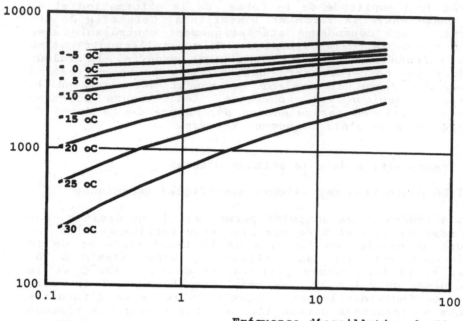

Module complexe de cisaillement [N*mm-2]

Fréquence d'oscillation [s-1]

Températures d'essai indiquées

Figure 4. Module complexe pour diverses températures et fréquences, d'une carotte extraite d'un revêtement AB 16

4.2 Essai de fatigue

La possibilité de suivre continuellement la variation du module, avec un choix approprié du capteur de déplacement et le réglage d'un amplificateur de mesurage, ainsi que de déterminer exactement de très petites valeurs des modules ($10 \ N*mm^{-2}$), permet d'exécuter des essais de fatigue, même avec des températures supérieures à 30°C. Ces essais sont conduits jusqu'à la rupture. Mais même lors de ces essais destructifs, la position moyenne est conservée et la rupture est causée uniquement par la charge dynamique. Selon le type d'essai utilisé, le résultat indique le nombre de cycles de charge (essai à température déterminée), ou la température de rupture (essai à température croissante). La deuxième valeur spécifique, en particulier, permet d'établir une prévision en ce qui concerne la résistance du revêtement testé. La figure 5 montre la relation existant entre le module complexe et la température ainsi que la durée de charge.

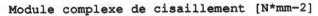

Module complexe de cisaillement [N*mm-2]

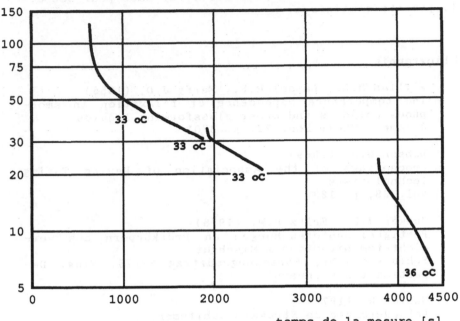

temps de la mesure [s]

Température d'essai comme indiquées

Figure 5. Comportement á la fatigue caractérisé
par la diminuation du module complexe

5 Conclusions

La méthode d'essai mise au point, permet:

- de déterminer le module de cisaillement complexe de 10 à
  10'000 $N*mm^{-2}$, des modules élastiques et visqueux correspondant, avec des dispersions généralement inférieures à 1%, et une reproductibilité inférieure à 5% de la
  valeur.

- selon l'esprit de la Good Laboratory Practice, d'obtenir
  un grand nombre de valeurs caractérisent l'essai.

- d'étudier les détails du comportement mécanique, comme
  la modification de la structure du matériau minéral sous
  charge.

L'article expose les diverses méthodes employées pour atteindre ces buts, et il peut servir de base pour des développements similaires.

Bibliographie

1   William M.L., Landel R.F., Ferry J.D. (1955)
    The temperature dependence of relaxation in amorphous polymers and other glassforming liquids
    J. Amer. Chem. Soc. 77, p. 3701

2   Dobson G.R. (1969)
    Proceedings of the Association of Paving Taring Technologists
    Vol. 38, p. 123

3   Junker J.P., Fritz H.W. (1976)
    Festigkeitsuntersuchungen an Prüfkörpern aus verdichtetem bituminösem Mischgut
    Schlussbericht, Forschungsauftrag 22/75, Eidg. Departement des Innern

4   Gubler R. (1977)
    Anforderungen an Strassenbaubitumen
    Schlussbericht, Forschungsauftrag 8/77, Eidg. Departement des Innern

5   Junker J.P. (1987)
    Entwicklungen zur Bestimmung mechanischer Materialkennwerte an bituminösen Baustoffen, insbesondere an Asphal
    Dissertation, EMPA-Bericht Nr. 215

6   Bernhard A., Gubler R. (1989)
    Filler für bituminöses Mischgut, Forschungsauftrag 3/83 des Eidg. Verkehrs- und Energiewirtschaftsdepartements, Schlussbericht Nr. 172

31 EVALUATION DE MÉLANGES D'ASPHALTE COULÉ AVEC DIVERSES SORTES DE LIANTS AU MOYEN D'ESSAIS PAR OSCILLATION AXIALE ET D'ESSAI À LA FATIGUE

(Evaluation of cast asphalt mixes with various binders by axial vibration and fatigue testing)

R. GUBLER and H.W. FRITZ
EMPA, Dübendorf, Switzerland

Sommaire

L'action de liants sur les caractéristiques de revêtements fut étudiée en déterminant le module complexe de cisaillement et le comportement à la fatigue par des essais d'oscillation. L'établissement de courbes maîtresses pour une température de 20 °C, et leur évaluation, permettent d'obtenir les valeurs spécifiques qui sont instructives pour l'appréciation de mélanges d'asphalte coulé. Les divers asphaltes coulés qui furent testés sont évalués.

Mots-clés: module complexe, essai d'oscillation axiale, essai à la fatigue, asphalte coulé, modification des liants, courbes maîtresses.

Introduction

La mise en place de divers mélanges d'asphalte coulé fut envisagée dans le cadre de travaux de remplacement d'un vieux revêtement en asphalte coulé. Un bitume-polymère ou un asphalte coulé avec addition de Trinidad furent prévus, en plus d'un asphalte coulé au bitume routier normal B 40/50.

Les effets des liants sur les propriétés du revêtement furent étudiés au moyen d'essais par oscillation axiale et d'essais à la fatigue.

Buts

Les essais prévus doivent permettre d'obtenir des informations sur les propriétés du revêtement dans la zone allant des températures d'hiver à celles d'été. La fonction liant le module complexe avec la température et la fréquence, les relations entre ces données et des essais à la fatigue sous des températures estivales constituent la base de ces informations.

Table 1. Matériaux testés

Asphalte coulé avec du bitume routier	2 mélanges
Asphalte coulé avec addition de Trinidad	2 mélanges
Asphalte coulé avec du bitume-polymère	5 mélanges

Etablissement du programme d'essai adéquat

Une gamme aussi large que possible de paramètres d'essai fut donc choisie pour l'exécution des essais dynamiques. La force fut prise comme valeur consignée pour des raisons de technique d'essai. La grandeur de l'amplitude de la force qui restait à choisir fut fixée à 12.5 N*mm^{-1}, soit proportionnelle à la hauteur de l'éprouvette cylindrique de 150 mm de diamètre. La charge sur des éprouvettes de hauteurs différentes reste ainsi identique. Le programme d'essai ci-dessous indique les fréquences et les températures utilisées:

Température   -5 à +35°C, par étapes de 5°C
Fréquence   0.125 s^{-1} à 32 s^{-1}, avec redoublement à
                chaque étape

Ensuite, l'essai de fatigue fut exécuté, afin de déterminer le module à des températures de 45°C. L'amplitude de la force de 20 N*mm^{-1} comme grandeur de référence détermina la hauteur de l'éprouvette. L'essai conduit à une fréquence de 0.25 s^{-1}.

La figure 1 montre l'installation utilisée pour les essais sur les asphaltes coulés.

Le mélange d'asphalte coulé est versé dans le cylindre extérieur. Un trou central est ensuite percé dans l'éprouvette refroidie et le noyau central y est introduit et collé. Cette façon de procéder est nécessaire car le coulage de l'asphalte dans le cylindre extérieur avec un noyau central déjà en place ne donnerait pas d'adhérence suffisante de l'asphalte sur ce noyau, du fait de la trop petite surface extérieure de celui-ci. Les grains glissent pendant l'essai le long du noyau et il en résulte la formation d'une masse bitumineuse sur sa surface qui agit comme une surface de glissement. Un tel phénomène n'a jamais été observé sur le cylindre extérieur. Le coulage est donc une méthode qui se justifie en comparaison avec la fabrication d'une carotte et de son collage.

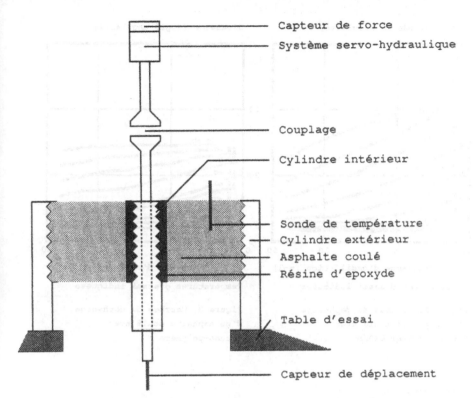

Figure 1. Principe du dispositiv d'essai

Résultat de l'essai

Les figures 2 et 3 montrent les modules complexes ainsi déterminés pour un asphalte coulé avec du bitume routier normal et un asphalte coulé avec du bitume-polymère.

Méthodes d'évaluation

Etablissement des courbes maîtresses de modules.

Six mesures sont prises pour chaque essai déterminé par les paramètres de température et de fréquence. Leur évaluation donne des valeurs dépendantes de la température et de la fréquence pour le module complexe et pour ses composantes élastiques et visqueuses. Dans le but d'obtenir, à partir de ces nombreuses données, les informations recherchées, des courbes maîtresses pour la température 20 °C furent établies en utilisant les principes connus de la recherche rhéologique sur les matières plastiques [1]. La figure 4 montre la superposition des courbes secondaires sur la courbe maîtresse.

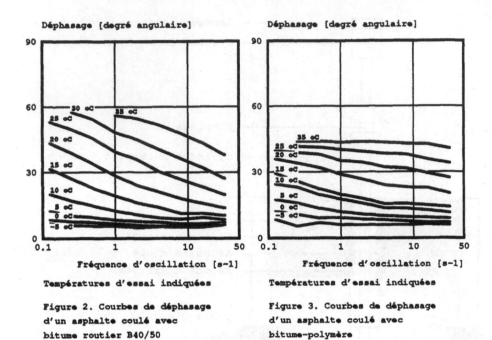

Figure 2. Courbes de déphasage
d'un asphalte coulé avec
bitume routier B40/50

Figure 3. Courbes de déphasage
d'un asphalte coulé avec
bitume-polymère

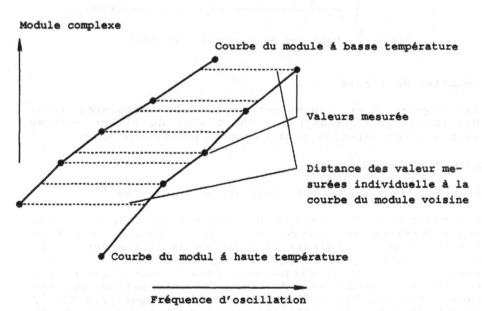

Figure 4. Détermination des distance entre les courbes comme moyenne
des distances individuelles

Les travaux des recherches de l'EMPA cités [3 à 6] et ceux qui ont succédé, ont montré que les courbes des modules complexes à différentes températures peuvent être superposées par une translation horizontale. Les vecteurs de translation nécessaires dépendent de la température. Comme la représentation des modules ainsi que celle des fréquences est à l'échelle logarithmique, les vecteurs de translation correspondent en fait à des facteurs de multiplication avec lesquels les fréquences devront être multipliées pour égaliser l'influence de la température. On peut démontrer de façon théorique que les facteurs de multiplication des fréquences correspondent à un facteur de rapport des composant visqueux d'un modèle rhéologique dont tous les membres présentent la même dépendance de la température.

Evaluation mathématique des courbes maîtresses

Un modèle a été proposé dans la référence [2] pour les courbes maîtresses des bitumes, qui décrit une transition graduelle et continue du comportement élastique au comportement visqueux lors d'une réduction de la fréquence. Ce comportement se rencontre généralement avec les matériaux bitumineux et il vaut aussi pour les asphaltes coulés. Les courbes maîtresses seront dans ce modèle définies par 3 valeur spécifiques:

- la valeur limite d'un comportement purement élastique
  Elle représente la valeur du module complexe pour une fréquence infiniment haute.

  le paramètre b de réponse au cisaillement
  Cette valeur indique la vitesse à laquelle le comportement élastique passe au comportement visqueux. Les valeurs élevées correspondent à une transition rapide, les valeurs basses à une transition lente.

- la valeur limite de la viscosité dynamique avec un comportement uniquement visqueux
  Elle correspond à la mesure relevée pour une fréquence d'oscillation égale à $0s^{-1}$.

Evaluation de l'essai à la fatigue

L'essai de fatigue à hautes températures fut conduit jusqu'au dépassement d'une amplitude d'oscillation de 2 mm. L'asphalte coulé, avec sa haute teneur en liants, ne se rompt pas sous l'effet d'une telle déformation. La mesure du module et l'étude des courbes de déformation indiquent toutefois une destruction importante de la structure du matériau minéral. Le temps déterminé pour ce processus fut converti en un nombre de cycles.

Valeurs spécifiques choisies comme base d'appréciation

Les valeurs spécifiques obtenues furent étudiées en ce qui concerne leur qualification comme base d'une appréciation du mélange examiné. Les valeurs spécifiques choisies permettent de formuler les remarques suivantes:

Valeur limite pour un comportement uniquement élastique.
La valeur représente une extrapolation des mesures aux plus basses températures et elle définit bien les propriétés mécaniques à de très basses températures.

Déphasage minimal
Cette valeur obtenue en prenant la moyenne des déphasages à des températures de -5°C avec des fréquences supérieures à $1\ s^{-1}$, indique le comportement aux basses températures mais elle ne représente plus une extrapolation car elle est basée sur des valeurs mesurées. Une valeur basse correspond à un comportement presque uniquement élastique pour lequel on ne peut plus s'attendre à une relaxation.

Paramètre de réponse au cisaillement b

La valeur indique la vitesse du passage d'un comportement élastique à une comportement visqueux. Une valeur élevée, et donc un passage rapide, signifie qu'entre un comportement purement visqueux et un comportement purement élastique on ne trouve qu'une petite variation de la fréquence. Il s'ensuit, selon le principe de superposition température-fréquence, que pour une fréquence donnée la charge provoque ce changement de comportement déjà avec une modification minime de température. Par conséquent, une valeur élevée indique un petit intervalle de fusion, et inversement une petite valeur résulte en un grand intervalle.

Rapport des modules de courbes maîtresses

Le rapport des modules pour de grandes périodes au module déterminé pour la période 0.01 s montre la tendance existante d'une diminution du module lors des charges de longue durée, et cela permet de faire une constatation au sujet de la durabilité.

Déphasage d'un jour des courbes maîtresses

Le déphasage pour une période de un jour indique la partie restante du comportement élastique pour des charges de longue durée et, selon le principe de superposition température-fréquence, aussi à de hautes températures. Il constitue donc une mesure de la capacité du revêtement à redevenir élastique après un changement.

Nombre de cycles jusqu'à la rupture

L'examen visuel de l'éprouvette après l'essai à la fatigue montre que les plus gros gravillons furent repoussés hors de la masse initiale de l'éprouvette. La valeur indique donc la résistance au déplacement des gravillons ainsi que la résistance à un rattachement dans la structure du matériau minéral. Cette valeur précise donc aussi un aspect de la durabilité.

Résultats des recherches

Les valeurs spécifiques importantes de mélanges d'asphalte coulé, avec des modifications de diverses sortes, sont exposées dans la table 2.

Table 2. Récapitulatif des valeurs spécifiques

Adjuvant/Modification	sans	avec Trinidad	avec polymère
Valeur limite élastique N*mm^{-2}	9'640	11'200	12'300
Paramètre de réponse au cisaillement	0.29	0.25	0.20
Déphasage à -5°C en degrés circulaires	5.1	5.3	6.8
Rapports des modules au module pour 0.01 s			
avec une période de 100 s	63	76	62
avec une période de 1000 s	445	490	327
avec une période de 1 jour	34'000	36'000	14'000
Déphasage de la courbe maîtresse de 1 jour en degrés circulaires	86.5	85.3	74
Nombre de cycles de fatigue	30	28	37

Appréciation des mélanges d'asphalte coulé

Asphalte coulé avec du bitume routier B 40/50

Le mélange d'asphalte coulé, avec du bitume B 40/50, fut utilisé comme matériau de référence. Le comportement des asphaltes coulés de cette sorte est bien connu à partir de l'expérience pratique déjà acquise. Les divergences positives et négatives résultant de l'effet d'adjuvants peuvent, grâce à une comparaison avec ce mélange, être rapportées à l'expérience pratique acquise. Cela permet une meilleure évaluation. Deux mélanges avec du bitume B 40/50 furent testés. Les valeurs spécifiques obtenues concordaient de façon satisfaisante. Cela démontre qu'ils pouvaient être pratiquement utilisés comme matériau de référence.

Asphalte coulé avec addition de Trinidad

Les valeurs spécifiques de ces deux mélanges sont identiques et correspondent à celles des mélanges avec du bitume routier B 40/50.

Asphalte coulé avec du bitume-polymère

Ce groupe de cinq mélanges présente, malgré quelques différence entre les mélanges, un comportement bien reconnaissable et différent de celui des mélanges avec du bitume routier normal. Le déphasage dans la zone des basses températures est nettement plus grand que celui des mélanges avec du bitume routier normal, et laisse ainsi prévoir un comportement favorable aux basses températures. Le rapport du module pour une période de 0.01 s avec le module de la même courbe maîtresse pour une période de 100 s est comparable à celui du mélange de référence. Par contre, le rapport avec de grandes périodes diminue nettement. Cela concorde entièrement avec la plus petite valeur du paramètre de réponse au cisaillement b et laisse prévoir de meilleures propriétés mécaniques aux températures estivales. Le déphasage avec une période de un jour est plus petit, et la part d'élasticité restante plus grande qu'avec le mélange de référence. Un meilleur retour à la forme initiale après chargement à haute température est à prévoir.

Le meilleur comportement lors d'essais de fatigue concorde avec ces observations. Des valeurs individuelles plus élevée du nombre de cycles sont atteintes. La valeur moyenne est nettement plus haute que celle des mélanges avec du bitume routier ou que celle des mélanges avec adjonction de Trinidad. Les différences entre les éprouvettes des différents mélanges sont, en fait, importantes. L'un des mélanges avait le même nombre de cycles que l'asphalte coulé non modifié.

Conclusions

La comparaison des modules à des températures et fréquences données, déterminés sur divers mélanges d'asphalte coulé, permet d'établir des prévisions relatives aux exigences pratiques. Une analyse détaillée de toutes les valeurs d'essais obtenues permet de définir des différences marquantes entre les différents mélanges d'asphalte coulé. Ces résultats et les conclusions qu'on peut en tirer ont été largement commentés dans l'évaluation présentée précédemment.

Les modules à 35°C et 0.125 s⁻¹. pris isolément, n'indiquent aucune tendance. Dans la zone des températures moyennes, les modules d'asphaltes coulés avec bitume-polymère sont plus petits que ceux de l'asphalte coulé non modifié. Le grand déphasage à -5°C indique, par comparaison, un meilleur comportement aux basses températures des asphaltes coulés avec bitume-polymère.

32 A TEST METHOD DESCRIBING THE MECHANICAL BEHAVIOUR OF BASE COURSE MIXES

E.-U. HIERSCHE, K. CHARIF, H. KÖESSL, K. VASSILIOU
Institute for Highway and Railroad Engineering, University of Karlsruhe, Germany

Abstract
Within the framework of a research project, supported by the Joint Study Group of Industrial Research Associations (AIF), comparative investigations were carried out in respect to the mechanical behavior of bituminous base course layers with and without recycled asphalt. A test track was paved, consisting of two subsections, alternatively utilizing base course materials with 30% recycled asphalt and with 100% unused materials. From both sections samples were taken and were prepared as specimen to perform cooling-off tests (executed in TU Braunschweig), dynamic flexural-tensile tests and dynamic creep tests.
The discussion of the test results indicates that the performed tests are suitable for the analysis and the evaluation of bituminous base course materials mixed with recycled asphalt.
Keywords: Behavior of bituminous base course materials with recycled asphalt, cooling-off tests, dynamic flexural-tensile tests, dynamic creep tests

1 Introduction

The use of recycled asphalt in bituminous base course materials is being satisfactorily practiced since several years. The technical as well as the contractual requirements are already formulated for the employment. However, it is up to date not yet adequately clarified, whether the reuse of asphalt materials would influence the homogeneity of the resulting mix and its mechanical behavior. This problem which in the meantime represents one of the main fields of research in the asphalt technology was the objective of a research project, carried out at the Institute for Highway and Railroad Engineering of University of Karlsruhe and supported by the Study Group of Industrial Research Associations (AIF). This paper presents the methodology utilized within the scope of the research project and the results of the laboratory tests, describing the mechanical behavior of bituminous base course materials.

2 Methodology

In order to achieve results as representative as possible to the practical conditions a test track was constructed, consisting of two subsections, in which two different types of mixes were used. In the first subsection 30% recycled asphalt was mixed in the bituminous base course and in the second subsection the bituminous base course was mixed with 100% unused materials, to serve as a reference. The design of both mixes complies with the Specifications for Mix Type CS 0/32 mm according to the ZVTV - StB 86 [1].

Trial mixes were carried out to establish the mix designs. During trial testing, special interest was paid to the comparability of bitumen contents and mineral aggregate gradation between the recycled and reference materials.

The employed recycled asphalt, mixed in as milled materials, was composed mainly of basaltic aggregates. As also planned according to the conception of the research project, these differentiated in color from the unused mineral aggregates (crushed limestone, crushed Rhine gravel chippings, crushed Rhine gravel sand and natural sand). The recycled asphalt was made up of 80% wearing course and 20% binder course materials.

Through the deliberate use of optically, definitely distinguishable mineral aggregates in recycled asphalt versus in unused materials, it was possible to evaluate the achieved degree of homogeneity of resulting mixes by mineralogical methods. The analyses were carried out as well purely optical as also by examining thin sections, taken from the test track, under microscope. The results indicated that a thorough homogeneity, in respect to the mineral aggregates, was not existing.

Both mixes were manufactured in a central plant and were paved in an urban street in Karlsruhe. The ambient conditions during mixing and paving were held constant as far as possible.

The test results of samples, taken from the mixing plant, show that the mix with recycled asphalt tends to indicate a higher bitumen contents and hence a lower air void contents in comparison to the mix without recycled asphalt. The determined differences between the two mixes range within the permissible reproduceability in accordance with DIN 1996 [2].

3 Laboratory tests for describing the mechanical behavior

The investigation of mechanical behavior comprised tests, carried out at low, room and high temperatures. These tests were: cooling-off test, dynamic flexural-tensile test and dynamic compressive test (dynamic creep test).

Samples were cut and drilled from the paved base course layer and after corresponding preparation, in respect with the maximum mineral aggregate size of 32 mm, following specimen were manufactured:

60x60x160 mm specimen (beams) for the investigation of low temperature behavior,

80x60x300 mm specimen (beams) for the determination of flexural-tensile fatigue behavior at room temperature,

150 mm diameter and 75 mm high cores for the evaluation of stability under dynamic compressive loading at high temperature.

3.1 Cooling-off tests

3.1.1 General
The cool-off tests and the investigation of low temperature behavior were carried out at the Institute for Highway Engineering of Technical University of Braunschweig (Prof. Arand). The testing equipment, used in the investigations, was developed there also [3].

Main component of the experimental set up is a processor controlled testing machine. The testing machine is composed of a rigid frame, an interval motor unit, capable of inclusion of the linear strain of the specimen with an accuracy of $5 \cdot 10^{-5}$ mm, and a loading head for the registration of the induced tensile forces as a result of the linear deformations.

3.1.2 Test conditions
In this test the asphalt specimens are cooled-off, starting from an initial temperature, within time dependent definite intervals. During cooling-off the initial length of the beam specimen is held constant. The induced tensile forces, due to the deliberate hindrance of the thermal contraction are recorded time and temperature dependent. The characteristic values, "maximum cryogenic tensile stress", as well as the pertinent "fracture temperature" are registered in case of a fracture of the specimen within the test duration.

The asphalt specimens, to be employed in the test, are conditoned to an initial temperature of +20 °C and then keeping the length constant they will be cooled-off by a temperature rate of -10 °C/h. In the meantime the cryogenic tensile stresses will be temperature dependent recorded until the fracture of the specimen.

3.1.3 Results of cool-off tests
As it is apparent from Figure 1, the 0/32 mm base course materials with 30% recycled asphalt (Mix A) indicate a higher maximum cryogenic tensile stress by approximately 1.3 N/mm² and a lower fracture temperature by 1.8 °C than the 0/32 mm base course materials mixed with unused mineral aggregates (Mix U). Mix A has a higher bitumen contents by 0.3% and is compacted more than the Mix U.

The differences between the arithmetic mean of the ob-

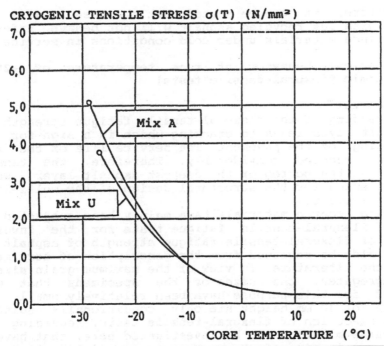

CRYOGENIC TENSILE STRESS σ(T) (N/mm²)

Feature		Mix A	Mix U
Maximum Cryogenic Tensile Stress			
σ_{max}	(N/mm²)	5,024	3,726
Standard Deviation	(N/mm²)	0,292	0,407
Fracture Temperature T_{Frac}.	(°C)	− 28,7	− 26,9
Standard Deviation	(°C)	0,69	1,169
Bulk Density (Beams)	(g/cm³)	2,490	2,416
Standard Deviation	(g/cm³)	0,006	0,021

Figure 1. Core Temperature Dependent Cryogenic Tensile
(TU Braun- Stress and Comparison of Maximum Cryogenic Ten-
schweig) sile Stresses, Fracture Temperatures and Bulk
Densities with Corresponding Standard Devia-
tions Determined on Beam Specimens of Base
Course Materials 0/32 mm with 30% Recycled
Asphalt (Mix A) and with Unused Mineral Aggre-
gates (Mix U)

served characteristics are significant. In respect with
the magnitude of the standard deviations, the mix with 30%
recycled asphalt shows a better homogeneity than the mix,
manufactured with unused mineral aggregates.

Therefore, it can be stated that the base course materials with 30% recycled asphalt will not behave inferior to reference materials under cold conditions in service.

3.2 Mechanical behavior at room temperature of +20 °C (dynamic flexural-tensile tests)

3.2.1 General
The exceeding of the flexural-tensile fatigue strength of an asphalt layer leads to cracking which is a sign for the overloading of the pavement. The service life of the road is hence reduced considerably. Therefore, the tensile strains at the bottom of the deepest asphalt layer serve as a criterion for the structural design of the multi-layer asphalt pavements.

Wearing course materials have mainly been subjected to dynamic flexural-tensile fatigue tests for the investigation of flexural-tensile fatigue strength of asphalt as it can also be reviewed from the description of test methods in the literature. In view of the maximum grain size of the aggregates, the sizes of the specimens that were employed for this purpose have been relatively small. The specimen of 40 mm height are only conditionally suitable for the execution of flexural-tensile tests, regarding the base course materials to be investigated here, that have a maximum grain size of 32 mm. This is especially true, if the homogeneity of asphalt materials has to be judged, using the same specimen.

3.2.2 Properties of the employed beams
The sizes of the specimen to be used in the dynamic flexural-tensile tests were determined to have a width of B=80 mm, a height of H=60 mm and a length of L=300 mm. This was done on the basis of preliminary tests with base course materials having a maximum grain size of 32 mm and considering the fact that, on the first place the specimen would be cut out from the compacted pavement, instead of being manufactured in the laboratory and that, secondly in spite of using large specimen the practicability of the test should still be ensured.

The determined values of bulk density and air void contents of the specimens of both mix alternatives indicated that:

The mean of the air void contents of the specimens with recycled asphalt was 1.4% less than the mean of the specimens with 100% unused mineral aggregates (absolute values: 2.2 and 3.6 Vol.-%, respectively). The calculated mean values for the air voids are significantly different.

There is no significant difference between the standard deviations of air void contents of two mix alternatives (absolute values: 0.6 and 0.4 Vol.-%. respectively).

3.2.3 Testing equipment

The dynamic flexural-tensile tests were carried out using the hydraulic pulsating testing machine of the Institute for Highway and Railroad Engineering of the Karlsruhe University. In the machine the forces are registered by a loading cell which is built in a loading head and the movement of the loading head is recorded by a linear variable differential transducer (LVDT). As well the control of input, as also the data acquisition are managed by a processor, to which additionally a TV-display, a printer and a plotter are installed. After an intermediate storage the data are transferred to a computer and processed by suitable software in respect to the desired final stage, for example, showing the time dependent course of the measured output.

A four-point loading apparatus which was constructed to fit the size of the specimen was used in carrying out the dynamic flexural-tensile tests. It consists of four frames and a base plate. The two outer frames are connected to the base plate and the two inner frames to the loading cell by means of hinge joints. The arrangement of the joints in the frames is made in a way that their axes of rotation coincide with the longitudinal median plane of the specimen. The specimens are fixed in the frames approximately at quarter points of their longitudinal axes. The exact configuration of loading points can be seen in Figure 2. The formation of the loading points as frames whose internal face corresponds to the exterior of the specimen, permits the application of threshold tensile as well as compressive-tensile loadings. Hence, a uniform loading of the specimen over its entire crossection is possible. Furtheron, the longitudinal section of the specimen between the two loading points is subjected to a constant bending moment. The loading is initiated at the upper side of the specimen which corresponds to the top of the paved base course layer.

3.2.4 Testing conditions

The input during the dynamic flexural-tensile testing was the vertical strain at the loading point. The time dependent repetition of this deformation was given by a sine curve, with the frequency of f=5 Hz and an amplitude of 0.12 mm.

On the basis of the preliminary tests, the loading duration was determined to be 30 min, corresponding to 9000 loading cycles, since after this period no correlation could be established between the actual and the input loading path. This was evidently caused by in the meanwhile occurred decrease of stiffness of the specimen which was also recognized at the oscilloscope.

The crack initiation on the bottom part of the specimen towards the end of the loading period was visible to the eye, however, the exact moment of crack appearance could

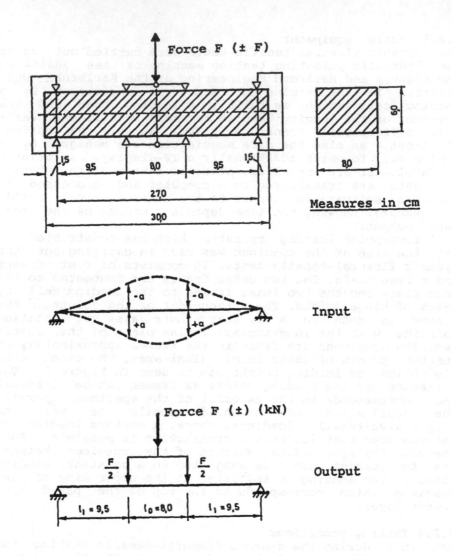

Figure 2. Schematic Representation of the Input and Recorded Output during Strain Controlled Dynamic Flexural-Tensile Test

not be determined during the investigations. With increasing time, the crack appeared on the upper side of the specimen also. With the utilized testing equipment a fracture state cannot occur, even by continuation of the loading, since in the course of time the resistance of the specimen to the prescribed strain converges to a minimum and not to zero. For an optical examination of the homogeneity of the fractured faces, the specimen could be

460

easily broken by hand at the crack location, after dis-
mounting from the testing equipment.

At the determined testing temperature of 20±1 °C, the
reaction of the specimen to the induced deformation is a
force in sine curve form with the same frequency as the
deformation and without a phase angle. The time dependent
force path corresponds to a damped harmonic sine oscillation, i.e. a vibration with decreasing amplitude. The decrease of amplitude with increasing number of loading
cycles can also be seen from the decreasing interval
between the auxiliary curves enveloping the minimum and
maximum peak points of the force path (s. Fig. 3).

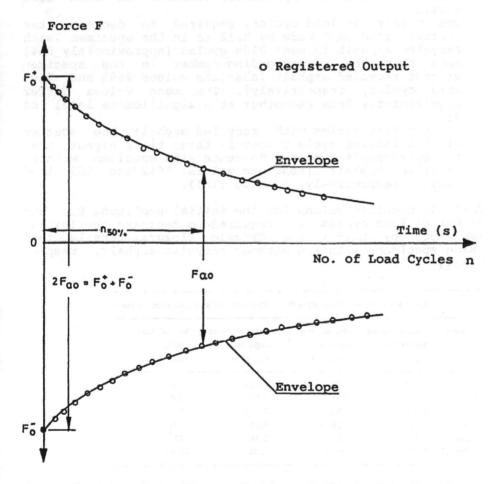

Figure 3. Schematic Representation of the Utilized Values
of Initial Force Amplitude F_{ao} and of $n_{50\%}$ Used
for the Evaluation of Strain Controlled Dynamic
Flexural-Tensile Test

461

3.2.5 Results of flexural-tensile tests

For the evaluation of service behavior of the investigated mix alternatives, the number of loading cycles was used, around which the load amplitude was decreased by half of the maximum value, measured at the start of testing.

On the basis of the results of above investigations, following may be stated:

There is no significant difference between the mean of the initial load amplitude, measured on the specimen with recycled asphalt (2.178 kN) and on the specimen without recycled asphalt (2.233 kN) at a significance level of 95%. Same holds true for the standard deviations of initial amplitudes, measured in both test series.

The number of load cycles, required to decrease the initial load amplitude by half is in the specimen with recycled asphalt in mean 2159 cycles (approximately 86%) more than the corresponding number in the specimen without recycled asphalt (absolute values 4662 and 2503 load cycles, respectively). The mean values differ significantly from eachother at a significance level of 95%.

In the test series with recycled asphalt, the scatter of the loading cycle number is three times higher than the corresponding number in tests with specimen without recycled asphalt (absolute values: 1612 and 563 load cycles, respectively, s. also Tab1).

Table 1. Compiled values for the initial amplitude F_{ao} and number of load cycles $n_{50\%}$ required to decrease the initial amplitude by half in the dynamic flexural-tensile test using specimens with and without recycled asphalt, respectively

Feature	Specimen with recycled asphalt		Specimen without recycled asphalt	
	Initial force amplitude F_{ao} (kN)	No. of load cycles $n_{50\%}$ (-)	Initial force amplitude F_{ao} (kN)	No. of load cycles $n_{50\%}$ (-)
Maximum	2.384	7500	2.429	3525
Minimum	1.854	2250	1.842	1500
Mean	2.178	4662	2.233	2503
Range	0.530	5250	0.587	2025
St.Dev.	0.145	1612	0.163	563
C.Var.(%)	6.66	34.57	7.30	22.49

In the plots of number of load cycles versus air void contents of specimens from both test series a correlation tendency can be recognized which, however, could not be established by means of a regression analysis. However, this

type of testing is often called impuls creep test. The impuls forms may be of sinusoidal, square or triangular waves.

3.3.2 Determination of specimen size, preparation of cores

For the performance of axial creep tests asphalt specimen as slim as possible should be utilized, in order to exclude the influence of the load transmitting plungers.

Although the bituminous base course was paved in a 14 cm lift, it was not possible to comply with the suggested specimen height of 100 ± 5 mm in this research project. During the preparation of the drilled cores, rather large limestone chippings splitted off repeatedly in the bottom of the base course layer. It was only possible to establish the required evenness of the end faces of the specimen at a height of 75 mm.

The friction forces between the load transmitting plungers and the specimen should be minimized, in order to achieve a uniform stress distribution within the total length of the specimen. For this reason, in this investigation, the end faces of all cores were parallelly abraded with a maximum deviation of ± 0.1 mm and before testing treated with a thin film of silicon fat and graphite powder.

3.3.3 Testing conditions

The dynamic creep tests were carried out with an impuls loading. The stress path between the minimum and maximum value was prescribed by a sine function. This stress form has on one side the advantage of simulating the natural loading of the road by a rolling wheel and on the other side the influence of corner discrepancies of the square wave form are eliminated. The pulsating deviator stress was determined to be 0.3 N/mm^2 with a constant axial stress of 0.02 N/mm^2. The testing temperature was held at + 45 °C.

3.3.4 Data acquisition and evaluation procedures

The load cycle dependent distribution of the permanent deformation typical for this configuration of testing is called creep curve (s. Fig. 4) and will be subdivided in three phases: the phase 1 with decreasing deformation increments, the linear phase 2 with constant deformation increments and the phase 3 with increasing deformation increments. The transition between phase 2 and phase 3 is called the inflection point and is a measure for the expected service life of the materials. Since from this point on the increase of deformation rate, resulting from internal structural damage exceeds the decrease of the same, caused by post-compaction. Furtheron, the slope of the linear range is called the rate of deformation and is expressed permill per load cycle.

In the graphical evaluation method, the initial and end points and the slope of the linear range will be determined

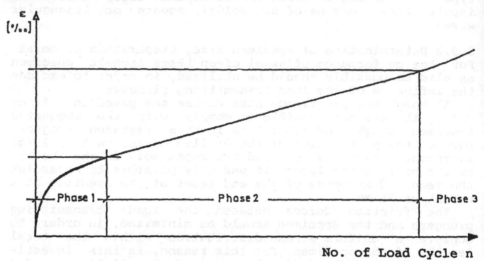

Figure 4. Qualitative Path of the Permanent Axial Strain Versus Number of Load Cycles n in the Dynamic Creep Test (Creep Curve)

with a best fitting straight line. This method is adequately accurate and delivers almost identical values of evaluation even by different persons whose mutual deviations are less than the usual scatters during tests of several specimens.

In a different interpretation of the form of the creep curve, the convex curve at the start of the test develops into a concave curve, without a linear phase and with an inflection point and with overproportionally increase of deformations. Hence, the course of the deformation may be described by a third degree-polynomial. Utilizing the recorded data the four coefficients of the polynomial can be determined by the help of a regression analysis. The inflection point, as well as the deformation rate as the slope of the tangent of the polynomial at inflection point can be calculated.

The evaluation of the investigations was carried out according to both described procedures. However, it was found out that the values of deformation rate in accordance with the regression method were very strongly dependent on the "definition" of test termination. If, for example, the regression analysis was performed once again after neglecting the last test values, differentiating values were determined as deformation rate. Therefore, the results of the graphical procedure were adopted for the calculation of deformation rates.

3.3.5 Results of dynamic creep tests

The mean for the number of load cycles at the inflection point for both mix alternatives is 26% higher according to the graphical procedure than with the regression analysis (s. Tab. 2). The reason for this is the different definition of the inflection point according to both methods of evaluation. While in the graphical method the end point of the linear range is defined as the inflection point, according to the regression analysis, it lies somewhere within the approximately linear range of the regression polynomial. Hence, the corresponding number of load cycles will be less.

Table 2. Comparison of number of load cycles in the inflection point and rate of deformation of core specimens with and without recycled asphalt, respectively, determined according to both evaluation methods

Analysis method	Feature	Cores w. rec. asphalt		Cores w/o rec. asphalt	
		Infl. point $n_{ip}(-)$	Def. rate $\epsilon'(\%_o/n)$	Infl. point $n_{ip}(-)$	Def. rate $\epsilon'(\%_o/n)$
Regression $\epsilon(n)=a_0+...+a_3 n^3$	No. of samples	15	15	14	13
	Mean	1862	$2.884 \cdot 10^{-3}$	2413	$2.870 \cdot 10^{-3}$
	St. Dev.	634	$1.770 \cdot 10^{-3}$	522	$0.945 \cdot 10^{-3}$
	C. Var.	34.0 %	61.4 %	21.6 %	32.9 %
Graphical Lin. range	No. of samples	15	15	14	13
	Mean	2343	$4.113 \cdot 10^{-3}$	3054	$3.211 \cdot 10^{-3}$
	St. Dev.	898	$2.203 \cdot 10^{-3}$	681	$0.984 \cdot 10^{-3}$
	C. Var.	38.3 %	53.3 %	22.3 %	30.6 %

The mean for the number of load cycles at the inflection point of the curves for core specimens with recycled asphalt, determined as well with the regression analysis as according to graphical method of evaluation was 23% less than the corresponding value for the core specimens without recycled asphalt (absolute values - regression analysis: 1826 and 2413 load cycles, respectively, graphical method: 2343 and 3054 load cycles, respectively). This indicates that, asphaltic base course mixes manufactured with addition of recycled asphalt wearing and binder course materials would have a 23% shorter service life under elevated temperatures than the base course mixes, produced of 100% unused mineral aggregates.

The comparison of the standard deviations of load cycle numbers for core specimens with recycled asphalt at the inflection point shows that the scatter of these values are 21% to 32% higher (depending on the procedure of evaluation) than the corresponding values for cores without recycled asphalt (absolute values- regression analysis: 522 and 643 load cycles, respectively, graphical method: 681 and 898 load cycles, respectively). This indicates that, the addition of recycled asphalt adversely influences the homogeneity of the paved base course layers.

The mean for the deformation rate of cores with recycled asphalt is approximately 28% higher than the mean for cores without recycled asphalt (absolute values $4.133 \cdot 10^{-3}$ and $3.211 \cdot 10^{-3}$ permill per load cycle, respectively). The scatter of individual values of cores with recycled asphalt is also higher than the values of cores without recycled asphalt. Referring to the standard deviations, cores with 30% recycled asphalt show a scatter ranging 120% more than the cores without recycled asphalt.

In reviewing above statements, it should be kept in mind that the mix alternative with recycled asphalt was produced with an excess of 0.3% bitumen contents and hence exhibited a corresponding lower air void contents than the mix without recycled asphalt.

4 Conclusions

The influences of cold addition of recycled asphalt wearing and binder course materials to base course mixes had, in the mode it was performed within the framework of this research project, differing results on the behavior of the paved bituminous base course layer:

Behavior at low temperatures:
The base course materials with 30% recycled asphalt from wearing and binder course materials are distinguished by a better homogeneity, as well as by a 7% lower fracture temperature at a 34% higher maximum cryogenic tensile stress against base course materials produced with 100% unused mineral aggregates. This suggests that, in service, the asphalt mix alternative with recycled materials will not behave inferior to the reference mix with 100% unused mineral aggregates.

Behavior at room temperature of +20 °C:
The addition of 30% recycled asphalt prolonged the service life of the paved base course by 86% versus the base course with 100% unused mineral aggregates, although the results exhibit a large scatter.

Behavior at high temperatures:
The addition of 30% recycled asphalt led, in comparison to base course materials with 100% unused mineral aggregates, to a shortening of the expected service life at high temperatures by 23% and to a raise of deformation rate by 28%. In the meanwhile, the homogeneity of the paved layer was considerably adversely influenced at high temperatures.

However, in looking at the above statements, it should be noted that the mix alternative with recycled asphalt was produced with an excess of 0.3% bitumen contents and hence showed a corresponding lower air void contents than the mix without recycled asphalt. In spite of this fact, the investigation results indicate that, at approximately equal air void contents, the mix alternative with recycled asphalt would behave at elevated temperatures inferior to the mix without recycled asphalt.

In the meanwhile, the results of the tests suggest possibilities to avoid the adverse effects of adding recycled asphalt in respect to the mechanical behavior at high temperatures: since the base course with 30% recycled asphalt has adequate potential regarding the mechanical behavior at low and room temperatures, the conception of a stiffer mix during trial mixing would lead to elimination of adverse effects of the behavior at elevated temperatures. This methodology is, however, only valid for the recycled asphalt utilized in this project and can only be conditionally generalized. Since the properties of recycled asphalt materials vary considerably, an utmost importance must be given to the investigations to be carried out for the determination of the properties of the recycled asphalt for utilization within the framework of suitability tests.

The results of this research project illustrate the influences of utilization of recycled asphalt on the mechanical behavior of bituminous base course materials. Still, some questions are unanswered that should help serve as an approach for further attempts to research:

How would the use of recycled asphalt effect the resulting mix, if in both mix alternatives identical bitumen contents and air void contents could be achieved?

In how far can an analogy of asphalt technological and characteristical values be achieved by employing suitability tests which go beyond the scope of conventional methods?

What would the results be, if the maximum grain size of recycled asphalt and the unused mineral aggregates would be identical?

In view of the findings of the present research work, it should be mentioned as a concluding statement that, before demands for use of higher percentage of recycled asphalt or demands for addition of non-preheated recycled asphalt to higher quality mixes can be considered, satisfactory experiences should be available regarding the use of preheated recycled asphalt.

467

5 References

[1] Complementary Technical Specifications and Guidelines for Base Courses in Highway Construction - ZTVT-StB 86. Federal Ministry of Transportation, Department for Highway Construction, Edition 1986.

[2] DIN 1966: Testing of Bituminous Mixes for Highway Construction and Related Fields. Subcommittee Materials Testing/Subcommittee Civil Engineering, Edition August 1988.

[3] Arand, W., Steinhoff, G., Eulitz, J., Milbradt, H.: Behavior of Asphalt mixes at low Temperatures. Part A: Background Information for Evaluation of Rolled Asphalt; Part B: Influence of Mix Characteristics. AIF-Research Project Nr. 5699, 1986.

[4] Recommendations for the Performance of Uniaxial Static Creep Tests with Asphalt Specimens. In: Colloquium 77 - Permanent Deformation of Asphalt Mixes. Institute of Highway, Railroad and Rock Engineering of the Federal Technical University Zurich, Paper Nr. 37, Zurich, November 1977.

[5] Development of Dynamic Tests of Asphalt Mixes - Comparative Dynamic Creep Tests. Joint Research Project of Department for Highway Engineering of Technical University Budapest and of Institute for Highway and Railroad Engineering of University of Karlsruhe, Budapest/Karlsruhe, 1985/86.

33 DEVELOPMENT OF DYNAMIC ASPHALT TESTS – COMPARATIVE DYNAMIC CREEP TESTS

E.-U. HIERSCHE
Institute of Highway and Railroad Engineering, University of Karlsruhe, Germany
E. NEMESDY
Department of Road Engineering, TUB, Hungary

Abstract
Dynamic creep test have been made on asphalt specimens of various dimensions, varying test temperatures, way of loading (sinus, orthogonal) and load intervals. These tests took place simultaneously in Budapest and Karlsruhe by means of hydropulse equipment available at both universities, testifying equivalence of the two testing equipment. Dynamic creep test results were evaluated by two different methods and discussed.
Keywords:
Dynamic testing equipment, creep test, effect of specimen size, testing temperature, load type, load intervals, evaluation method.

1 Antecedents and purpose

Since 1970, the Technical University of Budapest and the University of Karlsruhe (TH) have been related by a cooperation agreement. In frames of this cooperation agreement, the Department of Road Engineering, TUB, and the Institute for Road and Railway Engineering, Karlsruhe, have concluded a special agreement on a research program "Development of Dynamic Asphalt Tests", aimed at creating conditions for developing a single dynamic testing method for bituminous masses. It involved systematic comparison of dynamic testing equipment available at both institutions, applying the same testing parameters, and under variable testing conditions to examine their effects on test results. It has an outstanding technical-scientific significance that it is the first occasion of a systematic comparison between dynamic testing equipment.

2 Description of the testing equipment

2.1 Testing equipment in Budapest
The dynamic asphalt tester is an electronically controlled oil hydraulic press, transformed and complemented at the Department according to individual conceptions. Maximum load of the testing cylinder is 100 kN, maximum adjustable testing frequency is 100 Hz at 0.1 mm amplitude. Both dynamic and path of the equipment may be controlled. Further control is possible by means of accessory signalers (e.g. path recorder, dynamometer).

There are various possibilities to apply dynamic loads (sinusoidal, triangular, etc.), but loading may be also by cyclic load pulses and no-load intervals. In addition to uniaxial compression tests, asphalt prisms may also submitted to bending and plitting tests.

The desk and the testing equipment with asphalt specimens have been incorporated in an air conditioned box of 1000 by 500 by 300 mm capacity. Testing temperature may be kept constant in the range of`-15 to +15°C.

2.2 Testing equipment in Karlsruhe
Also the hydropulse equipment of the Institute for Road and Railway Engineering in Karlsruhe (TH) is an electronically controlled, oil hydraulic testing equipment. Maximum load of the testing cylinder is 63 kN, the max. adjustable testing frequency is 60 Hz, at an amplitude of 0,6 mm. The equipment may be directly computer controlled for force or path.

Different dynamic load applications are possible (sinusoidal, trapezoidal, rectangular, triangular, etc.), but also as periodical or random sequences or given load pulses and intervals.

The desk supporting the hydropulse equipment has a clearance or 700 mm between columns of the loading equipment, and can accomodate specimens up to 700 mm high. Mounting an air conditioning box on this desk (of about 60 by 60 by 60 cm capacity) permits tests in the temperature range of -20 to +80°C.

3 Experimental

3.1 General
Tests were made on cylindrical asphalt specimens of different sizes undergoing timely variable compression-swelling-loads, continuously recording residual strain along the load axis, by means or an inductive plotter.

Specimens were partly tested identically at both research stations, and partly, under individually chosen testing conditions, to increase conclusiveness of the results.

In general, at least four specimens were tested with constant test parameters. In case of rather similar test results on at least three specimens, no further specimens were tested under the given conditions.

3.2 Specimens
Dynamic creep tests were made on asphalt concrete 0/11 mm specimens of three different sizes. Dimensions and essential asphalt technology characteristics of the specimens have been complied in Figs 1 and 2.

Mineral components	Filler (< 0,09 mm) Sand (0.9 ... 2,0 mm) Gravel (> 2 mm) Density	9.6% by wt 35.4% by wt 55.0% by wt 2.842 g/cm^2
Mix	Binder percentage Mineral percentage Density	5.0% by wt 95.0% by wt 2.613 g/cm^2

Fig.1. Data for mineral components and mix

Marshall specimens (Mspec)		
Diameter Height Specific density Porosity	d h Q H	101 mm 62,5 mm 2,547 g/cm^3 2,5% by vol.
Large cylindrical specimen (Lspec)		
Diameter Height Specific density Porosity Compactness (referred to Mspec.)	d h Q H k	120 mm 240-299 mm 2,557 g/cm^3 2,2% by vol. 100,4 %

Fig.2. Characteristics of the applied specimens

Marshall specimens were made in conformity with DIN 1996, part 4, and the big test cylinders according to the "Double pisont principle" first applied by Krass in Karlsruhe, 1971. Here the compacting mould is placed elastically on a plate mobile against the lower compress. Thereby the lower compress contributes to the compaction, hence the mix is also compacted from below, permitting rather uniform compaction of specimens 300 mm high.

While in Karlsruhe cylindrical specimens 300 mm high were

tested, in Budapest the testing equipment coped with max. 240 mm specimen heights. To determine the effect of this height difference, also in Karlsruhe somewhat shorter specimens were tested, too.

To reduce the friction between pressure plate and specimen in dynamic tests, specimen butt ends were ground plan parallel before testing, smeared with silicone grease, and strewn with graphite flakes.

3.3 Test parameters

Test parameters underlyig dynamic creep tests have been compiled in Figs 3 and 4.

Legend: σ_o = maximum stress,
σ_u = minimum stress,
t_o = action time of maximum stress,
t_u = action time of minimum stress.

Three modes of loading prevailed:

Modes I and II of loading comprise load pulses and intervals; maximum, minimum load values and times of maximum stress are equal. Modes of loading I and II differ only by the time of minimum stress (no-load time). Mose of loading III is sinusoidal at 4 Hz frequency.

Specimen type	Specimen height [mm]	Test temperature [°C]	Compressive stress		Load time		Test type Nr
			σ_o	σ_u	t_o	t_u	
Mspec	63,6*	40	0,400	0,012	sinus	4Hz	III
		50					
		50	0,185	0,020	0,2	1,5	I
Lspec	240	40	0,185	0,20		1,5	
		50			0,2		
		50	0,185	0,020		4,5	II

*Mean value

Fig.3. Test parameters in Budapest

Specimen type	Specimen height [mm]	Test temperature [°C]	Compressive stress		Load time		Test type Nr
			σ_o	σ_u	t_o	t_u	
Mspec	62,5*	50	0,185	0,020	0,2	1,5	I
						4,5	II
Lspec	300	40	0,185	0,020	0,2	1,5	I
	240	50					

*Mean value

Fig.4. Test parameters in Karlsruhe

4 Test results

4.1 Evaluation
Quantitative course of dynamic creep curves at the beginning of the tests is seen in Fig.5

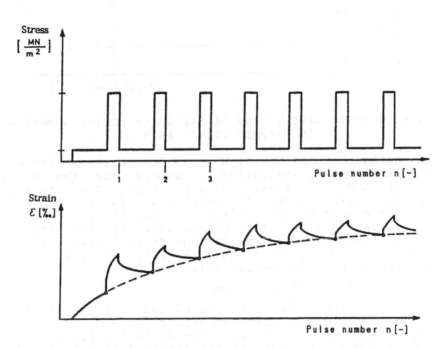

Fig.5. Load pulse-stress, and load pulse-strain diagrams
      in the beginning phase of a dynamic creep test

In the representation of the strain procedure (specimen swelling) spontaneous and time-dependent parts of the strain may be distinguished.

Flotting the strain directly before the next pulse - dash lines in Fig.5. for thousands of load pulses yields the creep diagram in Fig.6, comparising three phases:

Phase 1: Moderation of the initially abrupt deformation
         increase
Phase 2: Deformation increases linearly - in fact, only
         approximately - with load alternations. Later on,
         this section will be considered as straight, and it
         is represented as a straight line in Fig.3.
Phase 3: Deformation grows more than linearly

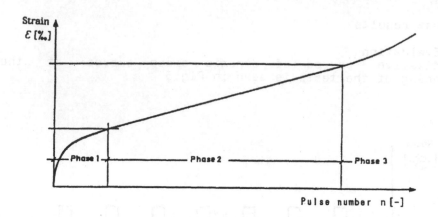

Fig.6. Quantitative course of strain vs. pulse number n
in dynamic creep tests

Transition between phases 2 and 3 is called simply an "inflection point", indicating a measure for the service lives of the specimens, namely here the increased deformation velocity due to structural faults begins to exceed the decrease due to ulterior compaction.

To evaluate the deformation behaviour of the specimen, also the "linear" phase will be considered, determining the number n of load alternations, as well as the specimen height variation. Deformation rate

$$\varepsilon^* = 1000 \cdot dh/h \cdot dn \qquad (1)$$

will be indicated in ‰ per number of load pulses (‰/n)
Here hspec - original specimen height, mm;
    dh - specimen height variation mm in the linear domain;
    dn - number of pulses in the linear domain (phase 2).

Although the ascent has been stated by a possibly accurate fitting of a straight line, this method is exact enough for a proper data recording, and even evaluated by different persons, it yields close data, less acattered than usual in testing several specimens. This has been the evaluation method in Karlsruhe.

According to a different concept of the creep curve course, the right-hand arc passes to the left-hand arc without a linear section about the inflection point, with a more than proportional increase of deformations, a hypothesis underlying the evaluation method in Budapest.

Test results have been processsed by means of a computer program such as:
- First step consisted in computing pulse number n and residual deformation ε_m [‰] from correlated test results

- Second step was to compute a common cubic regression polynomial from correlated measurement results.
Cubic regression polynomial:

$$\varepsilon \, [\text{‰}] = a_0 + a_1 n + a_2 n^2 + a_3 n^3 \qquad (2)$$

yielding - after determination of constants a_0, a_1, a_2, a_3 - the parameters:

n_{ip} = pulse number at the inflection point $\qquad$ [-]
ε_{ip} = residual deformation at the inflection point [%]
$10^{-6} \cdot (d\varepsilon/dn)_{ip}$ = tangent ascent at the inflection point=
$\qquad\qquad\qquad$ = rate of deformation $\qquad\qquad$ [-]

In addition, the correlation coefficient r of the regression polynomial may be determined as rate of accuracy of the obtained statistic correlation. Moreover, the ratio n n_{ip}/ε_{ip} can be determined and applied as material characteristic foor the material behaviour at elevated temperatures.

4.2 Interpretation of results obtained in Budapest
Test results from Budapest have been compiled in Fig.7.

In every test series, the cubic regression polynomial chosen for evaluation the test results from Budapest exhibits a high correlation coefficient, hence a strict statistic correlation.

The underlying test parameters permit direct comparison of test results as follows.

Comparison of Marshall specimens (Symbol of test series A) and large test cylinders (Sts B) at a constant test temperature of $50^{\circ}C$ shows:
- Position of the inflection point does not differ essentially between the two types of specimens.
- Residual deformation at the inflection point is about 30% higher for Marshall-specimens than for large test cylinders, primarily attributed to the increased ulterior compacting of Marshall specimens in the first test phase.
- The ascent of the tangent at the inflection point for Marshall specimens is about 1.8 times that of the large test cylinders. This proportion of deformation rates corresponds to the comparison above of residual deformations, but more obtrusive.
- Ratios n_{ip}/ε_{ip} of Marshall specimens and of large test cylinders hardly differ. Thereby the slight effect of the specimen height on the thermal behaviour stands clear.

Symbol of test series	[-]	A	B	C	D	E	F
Specimen type	[-]	Mspec	Lspec	Lspec	Lspec	Mspec	Mspec
Test temperature	°C	50	50	40	50	40	50
Way of loading (see Fig.3)	Nr	I	I	I	II	III	III
Pulse number at inflection point	n_{ip} [-]	1920	1810	10560	878	17290	3030
Residual deformation at inflection point	$\mathcal{E}_{ip}$ [‰]	13,7	10,9	11,7	12,6	29,2	28,6
Tangent ascent at inflection point	$10^{-6}(\mathcal{E}d/dn)_{ip}$	3650	2089	214	4425	583	3165
Correlation coefficient	r [-]	0,941	0,960	0,948	0,953	0,987	0,928
Ratio	$n_{ip}/\mathcal{E}_{wp}$ [-]	140	166	903	70	592	106

Fig.7. Results from dynamic creep tests in Budapest

Comparison of large test cylinders tested at $T = 50°C$ (Sts B) and $T = 40°C$ (Sts C) shows:
- The inflection points have quite different positions at the two test temperatures. The significant shift of the inflection point between $50°C$ and $40°C$ points to the outstanding effect of test temperature.
- Residual deformation (at the inflection point) does not differ between these test temperatures. The nearly equal deformations at $50°C$ and $40°C$ arise, however, from different pulse numbers.
- The tangent ascent at the inflection point at $50°C$ is about 10 times that at $40°C$. This reduced deformation rate with decreasing test temperature is rather significant.
- The ratio $n_{ip}/\mathcal{E}_{ip}$ at $40°C$ is about 5 times that at $50°C$. In this temperature range thermal behaviour of specimens markedly changes.

Tests on large test cylinders made with different no-load times and otherwise identical test parameters show that increase of no-load times from 1.5 s (Sts B) to 4.5 s (Sts D) significantly affects test results:
- For shorter no-load times, pulse number at the inflection point was about the double of that for longer no-load times.
- For the shorter no-load time, the ascent of the tangent at the inflection point is about half of that for the longer no-load time.
- Residual deformations (at the inflection point) are about the same for both load types.
- For the shorter no-load time, ratio $n_{ip}/\mathcal{E}_{ip}$ is about twice that for the longer no-load time.

Comparison of test results for Marshall specimens exposed to load test III (sinusoidal) at T = 40°C (Sts E) and at T = 50°C (Sts F) leads to the conclusions:
- At the lower test temperature, pulse number at the inflection point was about five times that at the higher test temperature.
- At the lower temperature, tangent ascent at the inflection point is much less (about 1/5) than that at higher temperature.
- Residual deformations are about the same at the inflection point for both test temperatures.
- At the lower temperature, the ratio n_{ip}/ε_{ip} is five times that at the higher temperature.

Comparison beetween load test type I with load pulses and no-load times (Sts A) and sinusoidal load test III (Sts E) shows:
- For sinusoidal loading, pulse number n is higher by about 1,5 times at the inflection point.
- For both load types, the tangent ascent at the inflection point is about constant.
- Residual deformation at the inflection point for sinusoidal loading is about twice that for load test I.
- For sinusoidal loading, ratio n_{ip}/ε_{ip} is about 25% less than for load test I.

Result differences between continuous sinusoidal load test and load pulses with no-load times certainly arise from the higher load peaks in sinusoidal load tests.

4.3 Interpretation of results from Karlsruhe

Test results for Marshall specimens (Mspec) and for both types of large test cylinders (Lspec) of h_{Lspec} = 240 mm and h_{Lspec} = 300 mm have been compiled in Fig.8. and plotted in Figs 9 to 11.

Test series	Specimen type & height [mm]	Test temperature [°C]	Load test type [Nr]	Linear domain n [-]	Δn [-]	Deformation [mm]	Δh[mm]	Δ‰	Rate of deformation single values ε [10^{-4} ‰/n]	mean values
G	Mspec 62,5	50	I	1100...2300	1200	1,27...1,63	0,36	5,76	48,0	39,7
				1100...3600	2500	1,22...1,74	0,52	8,32	33,6	
				1100...4100	3000	1,31...2,01	0,70	11,2	37,6	
H	Mspec 62,5	50	II	1100...3400	2300	1,34...1,80	0,46	7,35	32,0	28,7
				1500...3300	1800	1,26...1,32	0,26	4,16	23,1	
				1500...3400	1900	1,26...1,73	0,37	5,92	31,2	
K	Lspec 240	40	I	3000...10000	7000	1,78...2,11	0,33	1,38	2,0	2,3
				5500...12000	6500	1,92...2,26	0,34	1,40	2,2	
				3500...11500	8000	1,95...2,47	0,52	2,17	2,7	
L	Lspec 300	40	I	5500...13500	8000	2,37...2,80	0,43	1,43	1,8	1,9
				4000...11500	7500	2,52...3,02	0,50	1,65	2,2	
				5000...17500	7500	2,21...2,60	0,39	1,30	1,7	
M	Lspec 240	50	I	1000...2400	1400	1,46...1,90	0,44	1,84	13,1	33,3
				300... 650	350	1,55...2,45	0,50	2,06	38,9	
				500...1200	700	2,10...2,57	0,47	1,96	28,0	

Fig.8. Dynamic creep test results in Karlsruhe

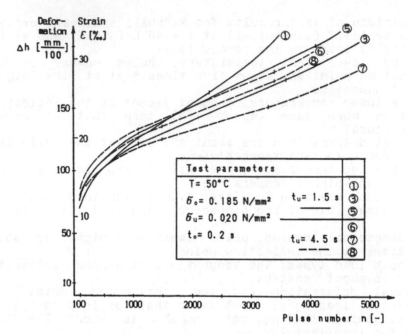

Fig.9. Dynamic tests on Marshall specimens Mspec
(Karlsruhe)

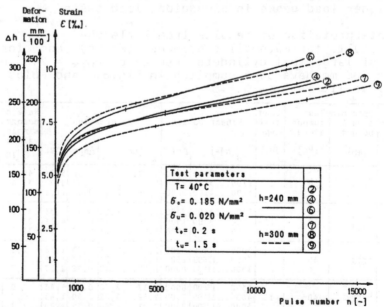

Fig.10. Dynamic tests on large test cylinders Lspec
(Karlsruhe)

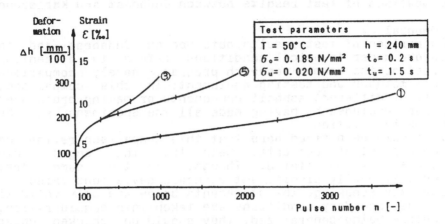

Fig.11. Dynamic tests on large test specimen
(Karlsruhe)

Underlying parameters permit direct comparisons as follows:
- For longer no-load times, for otherwise equal test parameters, Marshall specimens exhibited different deformation rates. Increasing no-load times from 1,5 s to 4,5 s reduced deformation rates from $39,7 \cdot 10^{-4}$ ‰ to $28,7 \cdot 10^{-4}$ ‰ as an average, showing the important effect of recovery times resulting from no-load times on deformational behaviour.
- Large test cylinders of different heights tested at T = 40^{0}C showed about similar linear domains, where relative deformations but slightly differed.
- Thereby there is only a tendential deformation rate difference, for lower specimens increased deformation rates may be expected.
- 40^{0}C, and in particular, 50^{0}C test results for large test cylinders 240 mm high are the most heterogeneous among all the test series. In both test series, linear domains were significantly different both as to length and position, as a function of the rather variable pulse number. Consequently, the deformation rates much differed as a function of test temperature. (Average) deformation rate in test series T = 50^{0}C is about 15 times that for T = 40^{0}C. The important standard deviations of test results point to the inadequacy of three speciemens per test series.

5 Comparison of test results between Budapest and Karlsruhe

5.1 General

Comparison of test results obtained in Budapest and in Karlsruhe under identical conditions refers to a central interest of the common research program, namely comparison of test methods and testing equipment. In this comparison, effects of different asphalt components or making procedures have been excluded by having made all the specimens at the same testing station.

It should be noticed here that this comparison relies on a moderate set of test data. Seen from this angle, only tendences may be stated. Though, for the same test parameters, partly significant standard deviations arise.

Strictly speaking, the statements below are only valid if the actual boundary conditions are taken into consideration, and before being generalized, they should be checked under different conditions.

5.2 Effect of test temperature

The temperature difference between $40^{\circ}C$ and $50^{\circ}C$ primarily affects the service life - pulse number. The lower temperature had about 5 times the pulse number as concomitant. Similarly, ascent of the tangent at the inflection point was 5 to 10 times that at the higher temperature. In the absolute value of compression no effect of temperature between 40° and $50^{\circ}C$ could be observed.

Results concerning the effect of different temperatures under otherwise equal test conditions (see Lspec 240 in Fig.8) show that an increased temperature impairs service life, though, deformations at the inflection point are at about the same height, leading to the conclusion that a given rate of deformation alters the inner texture of the specimen that it may be considered as failed, and a further loosening comes soon about, irrespective of the pulse number and the viscosity (i.e. temperature) of the bitumen.

5.3 Effect of load applications

For sinusoidal loading III service life-pulse number is about twice that for loading I with load pulses and no-load times. In this respect, peak load has to be taken into consideration; it is about 2,2 times higher in sinusoidal load testing.

In continuous fatigue test under sinusoidal loading at $50^{\circ}C$, residual compression was half as much, ascent of the tangent at inflection point about one third of that for load type I with load pulses and no-load times.

This fact may be perhaps attributed that sine waves of continuous loads do not persist for all the effect of a cycle.

Large test specimens under load test I (with shorter no-load times in a cycle) at $T = 50^{\circ}C$ showed surprisingly a service lire 1.5 times longer than for load test II (with longer no-load time in a cycle). This behaviour may be righteously explained by the fact that in both cases, the area under the load/time curve is what counts.

5.4 Effect of no-load times
Tests made in Budapest on large test cylinders show the service life to be shorter for a longer time of load valley action. This surprising behaviour may be explained by that the load valley acts as a load rather than as a no-load time.
Tests in Karlsruhe on Marshall specimens show that longer load valley times tend to improve the deformational behaviour by smaller deformations. Position of the inflection point (service life) does not change.
Correlation between service life and no-load time could not be supported in Karlsruhe by evaluating test data from Budapest. It may be stated instead, that longer no-load times in the tested range from 1,5 s to 4,5 s do not impair the deformational behaviour, and longer no-load times beyond 1,5 sec to shorten test times are unnecessary.

5.5 Effect of specimen dimensions
Behaviour of Marshall specimens and large test cylinders tested at $50^{\circ}C$ by load type I may be compared on hand of measurements.
- Curiously, service life/pulse number values are similar
- For all the Marshall specimens, there is clearly a higher specific compression and a more ascending tangent.
No doubt, the 24 cm height of large test cylinders mechanically provides for better, and more reproducible failure conditions. As against this, Marshall specimens are more popular. Asphalt layer are practically applied in 6 cm layers and successively compacted, truly followed in Marshall specimens.
Large test cylinders of different heights (240 mm and 300 mm) point to the equivalence of both specimen heights under actual boundary conditions and test values. This result means that clearly, specimen deformation is uniform throughout its height.

5.6 Effect of the evaluation method
Evaluation of test data by either the Budapest or the Karlsruhe method show no significant differences between the obtained results, hence, lead to no different statements. Pulse number determined by both evaluation methods - underlying the service life expectation - but slightly differ. Relative deformation values at the inflection point are quite similar. Tangent ascents at the point of inflection evaluated by either method are, however, similar only by order of magnitude.

6 Comparison and prospect

The joint research of the Department of Road Engineering, TUB, and of the Institute for Road and Railway Engineering of the Karlsruhe University (TH) was intended to make comparative dynamic creep tests on specimens, using testing equipment available at the two institutions, keeping standardization in mind. Tests have been concerned with temperature effect, specimen size, loading (sinusoidal, rectangular), no-load time (closer, time of lower load value to act), as well as evaluation methods. Load pulse - strain diagrams have been usually evaluated at the two laboratories by different methods. In Budapest, regression analysis has been applied, determining parameters for a cubic parabola. Parabola inflection point defines service life of the specimen, and ascent of the curve at the inflection point its deformational behaviour. In Karlsruhe, results have been evaluated by a method developed in France and generalized in Germany, where the visually about linear curve section is considered as determinant. Service life is assumed to be that for the pulse number at the end of the linear section, and deformational behaviour is described by the curve ascent in the linear section, the deformation rate. Buth methods have their advantages and disadvantages alike.

The optical-manual method is criticized because of the evaluation subjectiveness in determining the linear section and the service life. This aspect is especially right if data recording is disturbed so that data cannot be recorded at a high exactness.

Objectiveness of the computer method is, however, only apparent, since regression analysis parameters much depend on the number of recorded data. Namely it can be proven that a measured value near the end of the test modifies all the parameters. This is due to the fact that, although a regression analysis optimally approximates the existing data, an extrapolation mostly leads to excessive deviations. Hence, computer evaluation involves also a subjectiveness in deciding when to end a test.

Essentially, test goals have been achieved. Test procedures are about equivalent in both laboratories, and have contributed to the completion of knowledge of asphalt specimen behaviour under dynamic uniaxial loads. As concerns standardization of the testing procedure, however, specification of the specimen type and of the evaluation method relying on knowledge obtained from the experiment above is not yet possible. It requires still further tests, critical considerations, as well as taking experience at other research institutions into consideration.

34 DETERMINATION OF CRACK GROWTH PARAMETERS OF ASPHALT CONCRETE, BASED ON UNIAXIAL DYNAMIC TENSILE TESTS

M. JACOBS
Delft Univeristy of Technology, Road and Railroad Research
Laboratory, The Netherlands

Abstract
This paper describes the preliminary results of a research program
into the applicability of the Absolute Rate Theory (ART) to predict
the cracking of asphalt concrete. This theory is based upon the
assumption that the material is build up with bonds, which will break
or form during the time of loading. For this purpose dynamic tests
with a sinusoidal displacement signal are carried out on dense asphalt
concrete bars at various temperatures. During the fatigue tests the
crack growth in the bars is measured. Based on the results
calculations are carried out with a linear-elastic finite element
method (FEM) program to simulate the uniaxial tensile test. The ART-
parameters to predict the cracking of the dense asphalt concrete slabs
are also determined.
From the results of the measurements the conclusion can be drawn
that there is no relation between the decrease of the force and the
growth of the crack. From the results with the linear-elastic finite
element program it seems that there is no relation between the
measured and calculated crack length-force relation. The description
of the crack process in asphalt concrete with the ART leads to
promising results: the crack process can be described accurately at
each test temperature, whereas the differences between the ART-
parameters at the various test temperatures are only small.
Keywords: Absolute Rate Theory, Dense Asphalt Concrete, Uniaxial
Tensile Test, Crack Growth Measurements, Finite Element Program.

1 Introduction

At the Delft University of Technology research is carried out into
the causes of surface cracking of asphalt concrete structures. The
motive for this research program is the occurence of cracks which
start at the top of the asphalt concrete layer. In this research
program attention is paid to two aspects:
(a) the indication of causes which may result in surface cracking of
   asphalt concrete and
(b) the development of a theory that describes the crack growth proces
   in asphalt concrete. The aim of this part is to get a better
   understanding of the crack mechanism in asphalt concrete so the
   occurence of cracks can be retarded or even prevented.

One of the possible causes of surface cracking in asphalt concrete top layers is the three-dimensional non-uniform stress distribution in the contact area between tire and roadway. Due to this distribution, tensile stresses occur at the top of an asphalt concrete layer which may cause surface cracking.

To get a better understanding in the development of cracks in asphalt concrete, a theory has been developed by Jacobs which is based on the Absolute Rate Theory and the Deformation Kinetics. With this theory the crack initiation and propagation can be determined under a known stress situation.

In this paper first attention will be paid to the deduction of a formula out of the Absolute Rate Theory (ART) and the Deformation Kinetics, which describes the crack growth process in asphalt concrete under a combination of one or more loads. To determine the ART-parameters, uniaxial tensile tests with a sinusoidal signal have been carried out. During these fatigue tests crack growth measurements are carried out in the asphalt concrete bars as a function of the number of load repetitions. In this paper the results of the fatigue tests and the crack growth proces in the asphalt concrete will be given. With these results lineair-elastic finite element method calculations are carried out to investigate the existence of a relation between the crack length and the parameters, which are measured during the fatigue tests (load, displacement, phase lag between load and displacement). Also the ART-parameters of the asphalt concrete are determined, based on the results of the fatigue and crack growth tests.

2 The Absolute Rate Theory (ART)

2.1 Introduction to ART

In this paper attention will be paid to the applicability of the Absolute Rate Theory (ART) to describe the crack process in asphalt concrete. The ART has been chosen because:
(a) The theory has a large field of application. It can be applied to all materials in which bonds can be broken or moved. It can be used to describe the material behaviour of metals, non-cristallic materials, liquids and gasses, on biological and chemical processes. From the large field of application the basic character of the ART is emphasized.
(b) The relations and material parameters can be extended in an easy way from a one-dimensional situation to a two- or three-dimensional situation. The first approach can be uniaxial; afterwards the model can be extended to a biaxial or a triaxial model without change of the material parameters.
(c) In the ART bond-forming and bond-breaking processes are introduced. Essentially it should be possible to describe the healing-process in asphalt concrete by means of this approach.

In this paper only attention is paid to the one-dimensional situation. First of all the theoretical considerations of the Absolute Rate Theory will be given. The crack parameters will be determined by using the results of dynamic uniaxial tensile tests with crack growth measurements.

2.2 Theoretical considerations

In general crack growth processes are investigated at three different
levels: on atomic, microscopic and on macroscopic level. Between these
levels several overlaps exists. Nevertheless conditions at one level
are not used in another level to avoid complexity.

In the ART processes on atomic level are emphasized. It starts from
the principle that cracks at a macroscopic level are produced by a
thermal activated crack process of primary and secundary bonds on a
atomic level. In contradiction to the continuum mechanics the ART
presumes existence of discrete parts which determine the properties
of the material. In this way there exists a relation between fracture,
displacement and deformation on one hand and crack initiation and
propagation on the other.

The ART emphasizes the speed of bond-forming and bond-breaking
processes and the influence of the atomic interactive energy on the
crack growth in a material. Crack initiation and propagation can be
seen as a proces in which the speed of the bond-breaking proces is
larger than the bond-forming proces. If the bond-forming proces is
equal to the bond-breaking proces, the material is external at rest.
Tobolsky and Eyring originally developed the ART for polymers in 1943.
Herrin and Jones and Herrin, Marek and Strauss applied the ART the
first time to describe the shear behaviour of bituminous materials.

The basic mechanism of the ART, according to Krausz and Eyring, is
the forming and breaking of bonds. Bonds will take the position in
which they have the least amount of energy. Bonds with a high energy
level will break and take a position in which the energy level is
smaller. A new bond will be formed with release of the redundant
energy. If temperature is the only energy supplier, as many bonds are
broken as bonds are formed. In case extra energy is supplied to the
system by an external force, the bonds will be activated in the
direction of the force. The material will react elastically if all
bonds will return to their original position after release of the
force. If, however, the force is large enough (this means the energy
level is large enough), more bonds will be broken than new bonds are
formed and the material will crack or will deform plastically. It can
be supposed that during the healing process in asphalt concrete the
redundant energy is so large that new bonds can be formed.

Out of the theory of deformation kinetics, the basic equation for
the ART has been derived by Tobolsky and Eyring in 1943. Since that
time the theory has been applied to describe the material behaviour of
polymers, wood, concrete and asphalt concrete. The basic equation for
the ART can be defined by:

$$\dot{\Delta} = C_1 \exp\left(-\frac{Q_a}{kT}\right) \exp\left(\frac{W}{N\,k\,T}\right) \tag{1}$$

where: $\dot{\Delta}$ = derivation of the displacement with time (mm/s)
   Q_a = free energy of activation, i.e. the energy a bond
      needs to break (J)
   N = number of bonds per volume (1/mm³)
   k = Boltzmann's constant (J/K) = $13.8066*10^{-24}$ (J/K)
   T = absolute temperature (K)
   W = energy exerted by the external force per volume (J/mm³)

Out of (1) several solutions have been derived. For a visco-elastic material Knauss stated that the solution can be described as:

$$\frac{d(A/A_0)}{dt} = -\frac{kT}{h} \exp\left(-\frac{Q}{k}a_T\right) 2\left(\frac{A}{A_0} \cosh\frac{f}{kT} - \sinh\frac{f}{kT}\right)$$ (2)

with: $\dfrac{A}{A_0} = \exp\left(-\dfrac{kT}{h} \displaystyle\int_0^\tau \cosh\dfrac{f}{kT}\, dt\right) -$

$$\int_0^t \frac{kT}{h}\left(\sinh\frac{f}{kT} \int_0^\tau \cosh\frac{f}{kT}\, d\theta\right) d\tau$$ (2a)

and: $\quad dt = \displaystyle\int_0^\infty \phi(H) \exp\left(-\frac{h}{kT}\right)$ (2b)

where:
A = cracked area (mm²)
A_0 = virgin area at t=0 s (mm²)
t = time (s)
h = Planck's constant = $0.662618*10^{-33}$ (Js)
f = elastic energy per bond (J)
$\phi(H)$= probability function

Knauss stated that only elastic energy contributes to the fracture process in a material. This leads to the assumption that:

$$f = \frac{W_{el}}{N}$$ (3a)

Further one can presume in a one-dimensional situation that:

$$\frac{A}{A_0} = \frac{N_0 - N}{N_0}$$ (3b)

with: N_0 = number of bonds per volume in the virgin material.
W_{el} = elastic energy per volume

The solution of Eq.(2) with Eq.(2a), (2b), (3a) and (3b) can be described as:

$$\frac{dN}{dt} = (2N_0-N)\frac{kT}{h} \exp\left(-\frac{Q}{k}a_T\right) \exp\left(-\frac{W_{el}}{N}\frac{1}{kT}\right)$$
$$- N\frac{kT}{h}\exp\left(-\frac{Q}{k}a_T\right)\exp\left(+\frac{W_{el}}{N}\frac{1}{kT}\right)$$ (4)

The following equation can be derived for the elastic energy per volume. Under a dynamic test with a sinusoidal stress or strain signal, the dissipated energy per unit volume per cycle can be described as:

$$W_{diss} = \pi \sigma_0 \epsilon_0 \sin\phi_0$$ (5a)

The total energy per cycle can be described as:

486

$$W_{tot} = \pi \, \sigma_0 \, \epsilon_0 \qquad\qquad (5b)$$

Vectorial substraction of the dissipated energy per cycle W_{diss} from the total energy per cycle W_{tot}, provides a description for the elastic energy per cycle per volume W_{el}:

$$W_{el} = \pi \, \sigma_0 \, \epsilon_0 \, \cos \phi_0 \qquad\qquad (5c)$$

with: σ_0 = stress amplitude
ϵ_0 = strain amplutude
ϕ_0 = phase angle between the stress and strain wave signal

Substituting Eq.(5c) into Eq.(4) the following first-order, non-linear differential equation is found. This equation describes the one-dimensional crack proces in a visco-elastic material under a sinusoidal force or displacement signal:

$$\frac{dN}{dt} = (2N_0 - N)\, \frac{k\,T}{h}\, \exp\,(-\frac{Q}{k}a_T)\, \exp\,(-\frac{t\,\omega\,\sigma_0\,\epsilon_0\,\cos\phi_0}{2\,N\,k\,T})$$

$$- N\, \frac{k\,T}{h}\, \exp\,(-\frac{Q}{k}a_T)\, \exp\,(+\frac{t\,\omega\,\sigma_0\,\epsilon_0\,\cos\phi_0}{2\,N\,k\,T}) \qquad\qquad (6)$$

with: ω = angular velocity (rad/s) = $2^*\pi^*f$
f = loading frequency (Hz)
t = time (s)

The magnitude of t, T, ω, f, σ_0, ϵ_0, and ϕ_0 is dependent on the environment of the material; N_0 and Q_a are material parameters which must be determined from the test results. In the next section further details on the tests and the test material are given.

3 EXPERIMENTAL PROGRAM

In order to obtain the values of the free activation energy Q_a and the number of bonds per volume of the virgin material N_0, an experimental program was carried out. The experimental program consisted of:
(a) dynamic uniaxial tensile fatigue tests on notched beam specimens with a sinusoidal displacement signal.
(b) crack foils are used to follow up the crack growth at the outside of the specimens.
     First of all attention will be paid to the mix composition and the preparation of the test specimens.

3.1 Mix composition, preparation of the test specimens
Only one mix composition is considered because of the fundamental character of the tests to determine the material parameters Q_a and N_0. A dense asphalt concrete has been used; this is a typical Dutch mix (DAB 0/16). Due to a lack of fines in the natural mineral aggregate a factory made filler is added. Table 1 shows the omposition and the properties of the mix.

Table 1. Mix Composition and characteristics of the dense asphalt concrete (DAB 0/16).

Composition	Properties of the dense asphalt concrete		
On sieve C22.4 : 0.0	Bitumen 80/100	: 6.0	($\%$m/m)
C16 : 0.0	Marshall stability	: 9540	(N)
C11.2 : 1.0	Marshall flow	: 2.3	(mm)
C 8 : 17.0	Marshall quotient	: 4250	(N/mm)
C 5.6 : 30.0	Theoretical maximum density:	2417	(kg/m³)
2 mm : 55.0	Density of the mixture	: 2340	(kg/m³)
63 μm : 92.0	Voids in mixture	: 3.2	($\%$V/V)

A direct tensile test was selected for the crack growth experiments. The specimens are sawn out of hand made slabs, which are compacted by means of a roller. The specimens are 0.15 m long and the cross section was about 0.05*0.05 m. At midlenght of two opposite sides, an artificial crack was sawn. The depth of these "cracks" (notch) was varied in relation to the test temperature.

The specimens were glued to the top and bottom loading plate with an epoxy resin. In order to increase the glued area, small PVC-blocks are glued beside the test specimen at the top and bottom loading plate. To ensure a proper alignment of the specimens, hardening of the glue took place under the ram of the dynamic loading system. A picture of the specimen is given in Figure 1.

Use was made of a MTS electro-hydraulic loading system. The shape of the load pulse was a sinusoidal signal without rest periods. The elastic displacement was kept constant for each temperature during the tests (displacement controlled test). The load, the displacement and the phase angle between load and displacement were continuously computer-recorded (HP-A600). The loading frequency is 8 Hz.

Fig.1. Picture of a dense asphalt concrete specimen with crack foil.

3.2 Crack growth measurements

In order to determine the material parameters Q_a and N_0 the growth of the crack during the fatigue test has to be determined accurately. This is possible with crack foils, glued at the outside at the midlength of the specimens (see Fig.1). The thickness of the foil is 5 μm. It is supposed that the size of the crack at the outside of the specimen has the same size as the crack in the cross-section of the specimen.

In case of the uniaxial dynamic tensile test crack foils with a ladder principle has been used. The crack growth at the outside of the specimen will be recorded discontinually. The distance between the rungs of the ladder is 1 mm; the lenght of each rung is 4.1 mm. Each rung of the ladder is connected to a channel of a multi-programmer. The electric circuit with the crack foil is built in such a way that in the uncracked situation the difference in voltage between the top and the bottom of the crack foil will be 0.0 Volt; in the cracked situation the difference is 8.0 Volt.

During the fatigue test every 3 seconds all rungs are scanned. In case a new scan is proportionaly changed with regard to the previous scan, the scan is registered. In Figure 2 a picture of the results of the crack growth measurements are given.

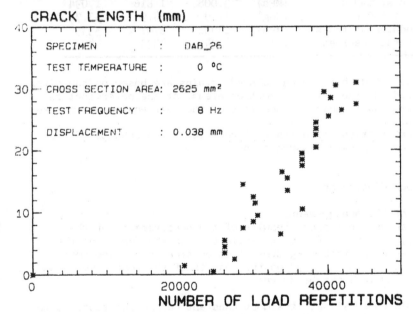

Fig.2. The crack growth proces in dense asphalt concrete
        in a uniaxial tensile fatigue test specimen (DAB_26).

3.3 Test conditions and test program

The crack growth experiments are carried out with a sinusoidal signal with at a frequency of 8.0 Hz at three different temperature levels: 0, 10 and 20 °C. During the tests the displacement is kept constant.

The deformation level is chosen in such a way that the life span of the bars will be about 1.5 hour (40000 load cycles). To get more reliable results at each temperature 3 bars are tested.

With the deflection kept constant during the tests, the changes in force and phase angle and the growth of the crack are recorded continually as a function of the number of load repetitions. In table 2 the characteristics of the tests are given (mean values).

Table 2. Characteristics of the tests and the test specimens.

Test temperature	(°C)	0	10	20
Load frequency	(Hz)	8	8	8
Length of the specimens	(mm)	150.5	150.2	149.7
Height of the reduced area	(mm)	5	5	5
Section area of the specimen	(mm²)	2610	2620	2597
Reduced section area specimen	(mm²)	1571	1621	1998
Displacement amplitude	(mm)	0.019	0.021	0.044
Strain amplitude in the reduced area at t=0 s	(μm/mm)	217	216	382
Force amplitude at t=0 s	(N)	4745	2942	2186
Stress amplitude in the reduced area at t=0 s	(MPa)	3.008	1.816	1.094
Phase angle at t=0 s	(deg)	11.9	24.9	53.7
Stiffness modulus at t=0 s	(MPa)	14962	8707	3335
Life span of the specimen	(s)	4804	4333	5499

It must be noted that the calculated strains are based on the sizes of the reduced and the original section areas. At the determination of the stiffness moduli a part of the mass of the test equipment is taken into account.

In the next section attention will be paid to the results of the fatigue and crack growth measurements.

4 Results and discussion

4.1 Results of the measurements

Figure 3 shows the typical results of the measurements of the displacement, the force, the phase angle between the force and the displacement and the remaining cross section area in the specimen during the fatigue test. Based on the results of the tests the following conclusions can be drawn:

(a) The displacement signal is almost constant during the test. At the point of time the size of the crack increases rapidly, the displacement signal decreases. This is due to the fact that the load signal is not a neat sinusoidal signal any more, but a flattened one.

(b) The decrease of the force cannot directly be related to the growth of the crack or the decrease of the remaining cross section area in the specimen. This is probably caused by the fact that micro cracks

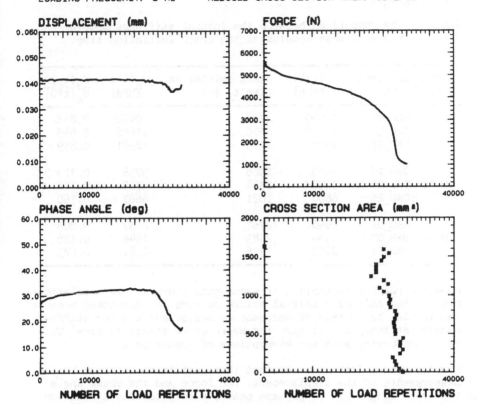

SPECIMEN: DAB_20
TEST TEMPERATURE: 10 °C   CROSS SECTION AREA: 2656 mm²
LOADING FREQUENCY: 8 Hz   REDUCED CROSS SECTION AREA: 1642 mm²

DISPLACEMENT (mm)

FORCE (N)

PHASE ANGLE (deg)

CROSS SECTION AREA (mm²)

NUMBER OF LOAD REPETITIONS

NUMBER OF LOAD REPETITIONS

Fig.3. Typical results of the measurements during the fatigue test.

develop during the fatigue test, which are not registrated by the
crack foil. So the decrease of the force gives an indication of the
macro crack growth in the specimen.

(c) It seems that at the point of time a maximum in the relation
between the phase angle and the number of load repetitions occurs,
the crack starts to grow.

(d) Hordijk and Reinhardt have proved that the uniaxial tensile test
is not a real tensile test but an alternative bending test. This is
due to the fact that in a material small irregularities occur. From
the test results on asphalt concrete this is proved, because in all
specimens the crack grows from one side of the specimen to the
other side and not from two sides to the middle of the specimen.

(e) There is no smooth relation between the number of load repetitions
and the crack length (Fig.2) or the remaining cross section area
(Fig.3). This is probably caused by the occurance of small macro
cracks in the cross section area. These small macro cracks will

491

flow together into one large macrocrack.
(f) In table 3 the initial stiffness modulus $S_m(t=0$ s$)$ is compared with the stiffness modulus at the point of time the crack starts to grow $S_m(t_c)$.

Table 3. The comparison between the initial stiffness modulus and the stiffness modulus at the crack initiation time.

Test temp	specimen no.	$S_m(t=0)$ (MPa)	crack initiation time t_c (s)	$S_m(t_c)$ (MPa)	$\frac{S_m(t_c)}{S_m(t=0)}$
	DAB_22	12500	2786	10432	0.835
0 °C	DAB_26	13800	2575	11658	0.845
	DAB_28	15000	2125	12581	0.839
	DAB_18	7900	1986	5758	0.729
10 °C	DAB_20	7350	2784	5705	0.776
	DAB_25	8800	1821	6687	0.760
	DAB_07	2650	2767	2137	0.806
20 °C	DAB_09	2350	2483	1846	0.786
	DAB_11	2650	1346	2181	0.823

From the results in table 3 it seems that there exists a relation between the corrected initial stiffness modulus (corrected because the initial large rate of decrease is neglected) and the stiffness modulus at the point of time the macro crack starts to grow. This is in conformity with the assumptions of Hopman et al.

From the conclusions it seems that there exists no relation between the measurements of the displacement, the force and the phase angle and the crack length. It is perhaps possible to determine a relation between the decrease of the force and the crack length with a finite element method (FEM) simulation of the uniaxial tensile test. In the next paragraph the results will be given of this simulation.

4.2 Results of the FEM simulation
The simulation of the uniaxial tensile test is carried out with the linear elastic finite element program CRACKTIP, which is developed by Jayawickrama. In this plane stress/plane strain program 2-dimensional 4-CST quadrilaterals and/or constant strain triangles with optional cracktip elements are used.
The purpose of this simulation is the determination of a possible relation between the measured values of the force and the crack length and the calculated values. There exists a relation if the measured force-remaining cross section area relation has the same shape as the calculated one. To compare the measured and the calculated relations, the values of the force and the remaining cross section are restored with respect to their initial value (relative values).
In the FEM program the stiffness modulus is chosen in such a way that the initial value of the measured force (t=0 s) has the same

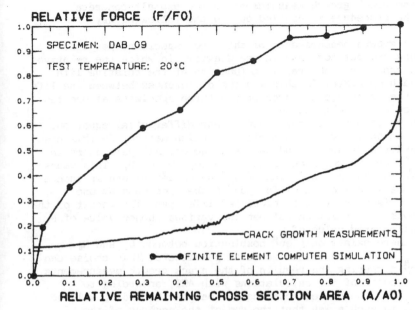

Fig.4. The comparison of the measured and calculated relation between the relative force and the relative remaining cross section area.

value as the calculated force with no crack. The force will be calculated at the crack length of 0 mm and every following 4 mm. In the determination of the measured force-crack length relation the relation between the crack length and the number of load repetitions is supposed to be elliptic. In Figure 4 the results of the measurements and the calculations are given at a test temperature of 20 °C.

It seems that there is no conformity between the measured and calculated relative force- relative remaining cross section area relation. This means that with the assumptions of the FEM calculations (only one macro crack, no micro crack zone, linear elastic material) the crack length in a specimen cannot be based on the decrease of the force which is measured during the fatigue test. In case the calculations are carried out with a visco-elastic finite element programs, with more macro cracks in the specimen or with a micro crack zone (i.e. a zone with a reduced stiffness modulus), the difference between the measured and the calculated relation could be much smaller.

4.3 Results of the calculations with ART

In paragraph 2 a differential equation (formula 6c) is found which possibly can describe the crack growth proces in dense asphalt concrete. There are only two unknown parameters left in this equation: the number of bonds per volume N_0 and the free energy of activation of a bond Q_a.

The numeric values of N_0 and Q_a can be determined by using the

493

results of the crack growth measurement in the asphalt concrete
specimens. The procedure is carried out in the following way:

(a) The crack growth measurements of the three specimens per test
temperature are put together in one diagram, in which the relative
remaining cross section area is a function of the relative life
span of each specimen. In this way the differences between the life
span and the cross section area of the three specimens at one test
temperature can be eliminated.

(b) Equation 6c is a first-order, non-linear differential equation,
which can be solved with a numerical solution method. In this case
the Runga-Kutta-Merson method has been chosen. All parameters in
equation 6c are known (from table 2), except Q_a en N_0. This means
that if N_0 is chosen, the value of Q_a can be determined in such a
way that the mean measured life span of the specimens at one
temperature is equal to the calculated life span. The result of the
calculations is a relation between the various chosen value of N_0
and the calculated value of Q_a.

(c) However there exists only one combination between Q_a and N_0,
because Q_a and N_0 are unique material parameters. This choise can
be made by comparing the results of the crack growth measurements
with the results of the calculations with ART, according to
equation (3b). The correct combination between Q_a and N_0 will be
determined in such a way that the sum of the squares of the
distances between the calculated and the measured values is as
small as possible. In Figure 5 the results of the calculations with
the correct combination of N_0 and Q_a for 0 °C are given.

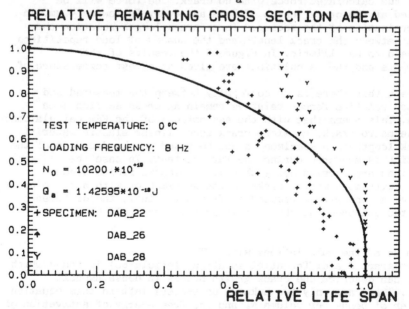

Fig.5. Comparison between the calculated results of the ART
and the crack growth measurements.

It seems that the ART can describe the crack growth proces in dense asphalt concrete accurately. In table 4 the results of the calculations for the various test temperatures are given.

Table 4. The results of the calculations with ART.

Temp	N_0 (1/mm^3)	Q_a (J/bond)
0 °C	$10200.*10^{19}$ ($\pm10\%$)	$1.426*10^{-19}$ ($\pm0.5\%$)
10 °C	$4990.*10^{19}$ ($\pm10\%$)	$1.477*10^{-19}$ ($\pm0.6\%$)
20 °C	$6040.*10^{19}$ ($\pm10\%$)	$1.491*10^{-19}$ ($\pm1.0\%$)

According to Herrin, Marek and Strauss the free activation energy of an asphalt concrete mixture in a shear test at a test temperature of 10 °C is $1.509*10^{-19}$ (J/bond). This means that the results of these preliminary calculations are very promising for further investigations into the applicability of the ART in the crack behaviour of asphalt concrete.

In these further investigations attention has to be paid to the small discrepancies between the calculated values of N_0 and Q_a at the three test temperatures. First of all the value of N_0 must be independent of the test temperature, because the number of bonds per volume is a constant of the material. So for the three test temperatures the value of N_0 should be equal.

Knauss, Francken and Herrin, Marek and Strauss supposed that the value of Qa depends on the test temperature according to equation (7):

$$Q_a = C_1 * (\Delta H - T * \Delta S) \tag{7}$$

with: C_1 = a constant
ΔH = the activation heat
T = test temperature
ΔS = the entropy of the system

This temperature dependence of the free activation energy Q_a is also found in table 4. However it makes no sense to calculate the numerical values of C_1, ΔH and ΔS because the numerical value of the calculated N_0 is not constant at all test temperatures. It can be added that the activation heat ΔH of asphalt concrete (50000 cal/mole) is higher than that of most other materials (tin: 8000 cal/mole and laminated plastic: 16000 cal/mole).

5 Conclusions

From the results of the research program the following conclusions can be drawn:

(1) At the uniaxial tensile fatigue test there is no relation between the decrease of the force and the growth of the macro crack in the specimens.

(2) At the point of time a maximum occurs in the relation between the phase angle and the number of load repetitions, the macro crack starts to grow. The ratio between the initial corrected stiffness modulus and the stiffness modulus at the point of time the macro crack starts to grow depends only on the test temperature.

(3) The crack growth behaviour of (visco-elastic) asphalt concrete can not be simulated with a linear elastic finite element computer program in which only one macro crack and no micro crack zone is assumed.

(4) From the first results of the calculation with the Absolute Rate Theory it can be concluded that the ART can describe the crack growth behaviour of asphalt concrete accurately. Further investigations has to be done to eliminate the small discrepancies between the mutual test temperatures.

Acknowledgement

The investigations were carried out with the support of the Netherlands Technology Foundation (STW).

6 References

Francken, L. (1977) Module complexe des Melanges bitumineux. **Bull. Liaison Labo. P. et Ch., Special V**, 181-198.

Herrin, M Jones, G.E. (1963) The behaviour of bituminous materials from the viewpoint of the absolute rate theory. **Proc. Ass. Asphalt Paving Tech.**, 32, 82-105.

Herrin, M. Marek, C.R. Strauss, R. (1966) The application of the absolute rate theory in explaining the behaviour of bituminous materials. **Proc. Ass. Asphalt Pav. Tech.**, 35, 1-19.

Hordijk, D.A Reinhardt, H.W. (1988) Macro-structural effects in a uniaxial tensile test on concrete. **Proc. 2nd Int. Symp. on Brittle Matrix Composites - BMC 2.**

Hopman, P.C. Kunst, P.A.J.C Pronk, A.C. (1989) A renewed interpretation method for fatigue measurements: Verification of Miner's rule. **Proceedings Eurobitume Symposium 1989**, 556-561.

Jacobs, M.M.J. (1988) The formation of cracks in asphalt concrete top layers (in Dutch). **Delft University of Technology**, Road and Railroad Research Laboratory.

Jayawickrama, P.W. (1985) Methodology for predicting asphalt concrete overlay life against reflection cracking. **MS Thesis, Texas A&M University**, College Station TX.

Knauss, W.G. (1965) The time dependent fracture of visco-elastic materials. **Proc. First Int. Conf. on Fracture (Sendai)**, 1139-1164.

Krausz, A.S. Eyring, H. Deformation Kinetics (1975), **a Wiley-Interscience Publication.**

Tobolsky, A. Eyring, H. Mechanical properties of polymeric materials (1943). **Journal of Chemical Physics**, 11, 125-134.

35 HYPOTHESIS OF CORRELATION BETWEEN THE RHEOLOGICAL BEHAVIOUR OF BITUMENS AND MECHANICAL CHARACTERISTICS OF THE CORRESPONDING BITUMINOUS CONCRETES

G. MANCINI and P. ITALIA
Euron S.p.A., S. Donato M., Italy
F. DEL MANSO
Agip Petroli, Rome, Italy

Abstract
The rheological behaviour of a series of bitumens has been examined, and the sinuisoidally varying stresses and strains of the corresponding bituminous mixes at various temperatures and frequencies have been studied in conditions of linear viscoelasticity. A correlation has been found to exist, in all the cases examined, between the pairs of straight lines in a Heukelom chart for two different bitumens and the pairs of curves which show the variation of the dynamic moduli of mixes containing them, in the range from 0° - 40°C at a frequency of 0.1 Hz. This correlation makes it possible to extract useful information on the practical performance of bituminous mixtures in different conditions of temperature and traffic, starting from simple rheological measurements and using a reference material of known characteristics.
Keywords: Mechanical properties of mixes, Bitumen and mix rheology, Dynamic modulus, Loading time, Viscosity, Modified bitumens.

1 Introduction

The mechanical properties of a bituminous mix are acknowledged to be an important factor in evaluating the performance of a road pavement in service conditions.

It is not however always possible to measure conveniently and sufficiently reliably the necessary parameters (dynamic modulus, creep, etc.) required as a basis for the estimate. In such cases recourse is made to indirect procedures based on graphs or mathematical formulae; but the results are often far from reality on account of the approximations introduced and possibly over-bold extrapolations. We therefore thought it worth-while to study, with the help of practical experiments, the correlation between the rheological characteristics of the bitumen and the mechanical properties of the corresponding mixes, since the former can be easily and quickly measured with ordinary laboratory apparatus.

The experimental verification included measurement of the dynamic modulus of the bituminous mixes at temperatures of 0°,20°, and 40°C at frequencies of 0.1, 1, and 10 Hz; while the corresponding bitumens were measured for penetration at 25°C, for Fraass Breaking Point, for softening point, and for dynamic viscosity at various temperatures.

For the tests pairs of bitumens were chosen having characteristics
appreciably different from one another. Comparison of the results
from each pair showed that the trends of the straight lines which
indicated on a Heukelom Chart the various states of the bitumen
with change of temperature (within a range covering from the Fraass
Point to over 100°C) followed in every case the curves which showed
how the dynamic moduli of the mixes varied with temperature at a
frequency of 0.1 Hz. In each case the point where the curves met,
that is to say the temperature at which the two materials had the
same properties, was the same. The difference between the values
indicating the characteristics of the two materials above and below
that temperature always had the same sign in the case of the bitumens
and of the mixes. This correspondence, taking into account the principle of the superimposition of frequency and temperature, and the
correlation between frequency and the speed of passing vehicles,
can be useful, as will be made clear in the interpretation of the
experimental data, in obtaining a synthetic picture of the practical
performance of a bitumen by comparing its rheological behaviour
with that of a reference material whose mechanical properties are
known. Naturally it is assumed that the hypothetical mix to be compared with that having known characteristics will be prepared following the same procedures as those used for the reference mix.

2 Apparatus used

A Haake Rotovisco RV 20 viscometer with a cone-plate measuring device
PK 100, interfaced with a Rheocontroller RC 20 for controlling and
measuring the test parameters, was used for the bitumens.

Measurements on the bituminous mixes were carried out with an MTS
(Material Testing System) electro-hydraulic apparatus, by means
of which it was possible to apply dynamic loads according to functions
programmable at various frequencies either in tension or in compression, at test temperatures between 0° and 40°C.

This apparatus was used to measure the absolute value of the
complex modulus E*, which defines the mechanical properties of a
material possessing linear viscoelasticity. This value was obtained
by submitting test samples of bituminous mixes, made up as for the
Marshall Test, to sinusoidal dynamic loading. Various granulometric
curves were used for the different materials examined.

3 Results obtained

3.1 First Case: Intersection Point in the range 10°-15°C

The test materials were an 80/100 Bitumen and a 180/200 Bitumen
modified with a radially branched block polymer of butadiene and
styrene. Their properties are given in Table 1 (N°1 and N°2 respectively).

Table 1. Bitumen characteristics

Bitumen number	Penetration at 25°C (dmm)	Softening point (R&B) (°C)	Fraass breaking point (°C)	Viscosity at 90°C (P)	Viscosity at 120°C (P)
1	87	48	-13	93.5	9.6
2	71	106	-18	390	39.2
3	86	49	-13	30	3.3
4	25	95	-14	712	82.7
5	89	47	-12	41.7	5.5
6	140	97	-30	127	28.6

Fig. 1 shows the same values on a Heukelom Chart.

In the case of the modified bitumen the softening point has not been used since it is known that the R & B values of this kind of bitumens do not lie on Heukelom straight lines.

As regards the viscosities it must be remembered that for modified bitumens, which show a non-newtonian behaviour over a wide range of temperature, the results do not depend only on the value of the rate of shear but also on the procedures used to reach that value and the time which elapses between the instant it is reached and the moment at which the viscosity is read. The values shown in Tab. 1 for modified bitumens were obtained by following a standard procedure: At 90°C a shear rate of 8 sec^{-1} was reached in 2 mins and the reading was made after 20 mins. At 120°C, 50 sec^{-1} was reached in 1 min. and the reading made 5 mins later.

Fig. 2 shows, for 0.1 Hz frequency, the variation of the modulus with temperature for both modified bitumen and for the normal one. It can clearly be seen from Figs. 1 and 2 that the physical parameter

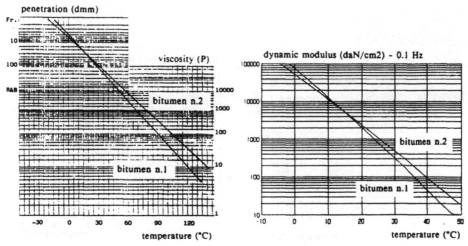

Fig.1 - Rheological behaviour of bitumens No 1 and No 2

Fig.2 - Dynamic modulus of bitumens No 1 and No 2 at 0.1 Hz

which characterises the material increases or decreases when the
polymer is present according to whether the temperature is respectively above or below that at which the two curves intersect. The temperature of the intersection is about the same in the two cases and
is also the temperature at which the bitumens have the same rheological behaviour and the corresponding mixes have the same stress-strain
response.

3.2 Second Case: Intersection Point at -5°C
The materials used were an 80/100 Bitumen and a mixture of 80/100
Bitumen with 20% of Polythene Waste. Their characteristics are shown
in Tab. 1 (Samples $N^°$ 3 and $N^°$ 4 respectively). From a temperature
of -5°C upwards the curve of the modified bitumen, shown by the
graph in Fig. 3, is always above that of the normal bitumen, indicating a greater consistency. On the other hand below -5°C the presence
of the polymer causes the binder to have a lower rigidity.

The mixtures show similar behaviour: when the additive is present
the modulus is lower until a temperature of about -5°C is reached
(see Fig. 4), while beyond that temperature the modified mix always
shows a higher value for the modulus, which means a grater resistance
to deformation, the difference increasing as the temperature increases.

3.3 Third Case: Intersection Point at 35°C
Contrary to what was seen in the previous case, the modified bitumen
has a lower viscosity than the normal bitumen, even at temperatures
above ambient temperature.

The characteristics of the two binders, an 80/100 Bitumen and
a High-Penetration Bitumen modified with 7% of elastomer, are listed
in Tab. 1 (N°5 and N°6).

It can be seen from the graph in Fig. 5 that the values get closer

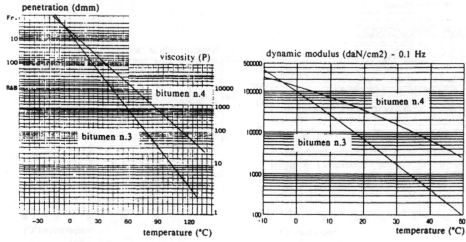

Fig.3 - Rheological behaviour of   Fig.4 - Dynamic modulus of bitumens
bitumens No 3 and No 4   No 3 and No 4 at 0.1 Hz

together as the temperature rises and coincide close to 35°C, while
above this point the difference changes sign.

The dynamic modulus (see Fig. 6) follows curves resembling the
preceding ones, being lower for the modified material until the
temperature reaches a value of about 33°C. At this point there is
an inversion and the unmodified mix begins to show a grater tendency
to deformation.

4 Interpretation of the results

The three cases considered, which exemplify three typical rheological
situations, lend themselves to some considerations which are particularly interesting from the point of view of practical applications.

The first case is the one which will most frequently be met with
in practice: the normal bitumen and the modified one have very similar
rheological characteristics at ambient temperatures, while the corresponding mixtures show very similar responses to dynamic stresses;
but in more severe climatic conditions the improvement in performance
due to additive becomes evident. For example, at 0°C the modulus
is lower, indicating a lower rigidity for the product and thus a
lesser tendency to brittle fracture at low temperatures; while at
20°C and above the material is, on the other hand, decidedly more
resistant to loading and thus less susceptible to deformation phenomena.

In the second example the resistance to deformation of the modified
material is always higher than that of the normal material for temperatures above about - 5°C; while the greater flexibility due to the
presence of the polymer makes itself felt only below that temperature.
This type of mixture can be thought of as a limiting case which
would, hypothetically, be suitable for use as a wearing course in

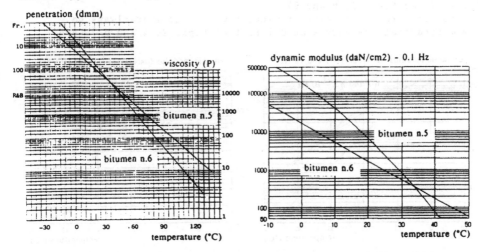

Fig.5 - Rheological behaviour of
bitumens No 5 and No 6

Fig.6 - Dynamic modulus of bitumens
No 5 and No 6 at 0.1 Hz

climates distinguished by exceptionally high temperatures.

An exactly contrary case is represented by the last example, where the modified binder, thanks to the presence of the polymer, is found to be less rigid up to over 30°C, although it proves to be more resistant to deformations at very high service temperatures. The best places for the use of this kind of hypothetical binder would obviously be those having climates with low or very low temperatures.

In the cases examined so far account has been taken of the values of the modulus obtained at 0.1 Hz. In fact it is at this frequency that the modulus curves have an intersection temperature that coincides with that of the rheological straight lines.

The case of dynamic mechanical stresses at low frequencies is however the most interesting from the practical point of view, since in these conditions the stresses transmitted by passing vehicles to the viscoelastic materials which constitute the various layers of the road pavement are more severe. Furthermore, study of the long-term stresses can give useful information regarding the strains caused by thermal contractions (due to rapid changes of temperature), although the time scale involved is appreciably longer.

As regards the behaviour of the mixtures at higher frequencies, and bearing in mind the principle of the superimposition of frequency and temperature, it can be seen that, in each of the three cases examined, going from 0.1 Hz to 1 Hz and on to 10 Hz is equivalent, as far as the modulus is concerned, to a progressive lowering of temperature. In consequence the temperature at which the curves intersect will necessarily be higher at a frequency of 1 Hz than at 0.1 Hz; and similarly will be higher again for a frequency of 10 Hz.

This tendency is clearly shown in the experimental curves for 1 Hz and for 10 Hz referring to the cases previously examined at 0.1 Hz (see Figs. 7, 8 and 9).

From the practical point of view, however, it must always be remembered that the curves at 0.1 Hz (see Figs. 2, 4 and 6), and

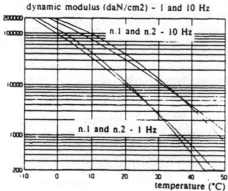

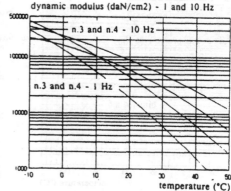

Fig.7 - Dynamic modulus of bitumens No 1 and No 2 at 1 and 10 Hz

Fig.8 - Dynamic modulus of bitumens No 3 and No 4 at 1 and 10 Hz

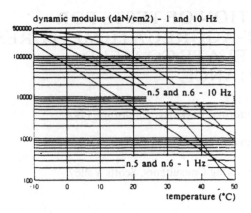

dynamic modulus (daN/cm2) - 1 and 10 Hz

n.5 and n.6 - 10 Hz

n.5 and n.6 - 1 Hz

temperature (°C)

Fig.9 - Dynamic modulus of bitumens No 5 and No 6 at 1 and 10 Hz

therefore also the rheological curves of the bitumens, already supply
qualitative information on the way the moduli behave at other frequencies.

In fact for each temperature the tendencies at higher and lower
frequencies can be derived from those which the curves at 0.1 Hz
show at temperatures which are respectively higher or lower.

Conclusions

The results quoted above are in agreement with the hypothesis that
there is a direct correlation between the rheological behaviour
of bitumens and the viscoelastic response of corresponding bituminous
mixtures to sinusoidal loading.

This correspondence has been discovered operating at low frequencies, which are the conditions in which the viscous component of
the deformation becomes predominant; and a more profound examination
might make it possible to acquire further data which would be useful
for a more complete interpretation of the correlation which has
been disclosed.

References

Collins J.H., Mikols W.J., Block copolymer modification of asphalt
    intended for surface dressing applications. **Proceeding of the
    AAPT**, Vol. 54 (1985) pp. 1-17.
Ferry, J.D. (1970) **Viscoelastic properties of polymers**, 2nd Ed.,
    J. Wiley & Sons Inc, N.Y.
Marvillet J., Verschave A., Duval J., Enrobés au liant bitume-SBS.
    Eurobitume Symposium, Cannes Oct. 1981.

36 EVALUATION OF TESTS FOR ASPHALT-AGGREGATE MIXTURES WHICH RELATE TO FIELD PERFORMANCE

C.L. MONISMITH
University of California, Berkeley, California, USA
R.G. HICKS
Oregon State University, Corvallis, Oregon, USA
F.N. FINN
ARE, Inc., Engineering Consultants, Scotts Valley, California, USA

Abstract
This paper discusses some results forthcoming from the Strategic Highway Research Program (SHRP project), en-titled, Performance Related Testing and Measuring of Asphalt-Aggregate Interactions and Mixtures. The primary purpose of this project is to establish relationships between asphalt binder properties and field performance which includes considerations of fatigue cracking, permanent deformation, thermal cracking, aging, and water sensitivity. A secondary objective is to establish performance related tests which can be used for the design of asphalt-aggregate mixtures or for the prediction of specific specific performance characteristics of mixtures in-service pavements.

Included is a description of the initial methodology used to select the laboratory tests to evaluate mix properties, together with the research hypotheses being tested. A discussion is also included of the methodology to be used to verify that the results of the laboratory tests simulate in-service conditions. It is planned that these simulations will include the use of wheel tracking facilities at the University of Nottingham (permanent deformation), the U. S. Army Cold Regions Research and Engineering Laboratory facilities (thermal cracking), and the Federal Highway Administration Accelerated Load Facility (ALF).

The method to be used to establish performance-based criteria for asphalt-aggregate mixtures is also described, and a preliminary set of criteria related to the laboratory test procedures under consideration are presented.
Keywords: Asphalt-Aggregate Mixtures, Fatigue, Permanent Deformation, Thermal Cracking, Aging, Water Sensitivity, Performance Related Specifications.

1 Introduction

A 50 million dollar research program is currently underway in the United States on asphalt materials as a part of the Strategic Research Program (SHRP) (TRB, 1986). The overall purpose of this program is to improve the in-service performance of asphalt pavements. Specific products resulting from this research effort will include improved specifications for asphalt binders and asphalt aggregate mixtures. The mixture specifications will be supported by accelerated performance related tests that can be used in an asphalt-aggregate mixture analysis system, an example of which is shown in Figure 1.

One of the projects, "Performance Related Testing and Measuring of Asphalt-Aggregate Interactions and Mixtures," (SHRP project A-003A) and with the University of California serving as the prime contractor, has as its objectives:

1. To extend and verify the results obtained by other SHRP asphalt research, the relationship between properties of the asphalt binder and pavement performance.
2. To develop accelerated performance related tests for asphalt-aggregate mixtures that successfully model construction and service conditions.
3. To develop a database derived from laboratory investigations that can be used to verify the asphalt chemical and physical characteristics significant to the performance of asphalt-aggregate mixtures.

To achieve these objectives, the approach shown in Fig. 2 is being followed. It consists of the following steps:

1. Evaluation of various laboratory test systems which measure fundamental properties and which can be used in analyses systems (e.g., like Fig. 1) to provide reasonable measures of performance.
2. Verification of the laboratory test methods by using small or full-scale accelerated tests in test tracks.
3. Once the tests have been shown to provide reasonable measures of performance, it will be possible to verify whether the asphalt properties (defined in another research project) are sensitive to mixture and field performance.

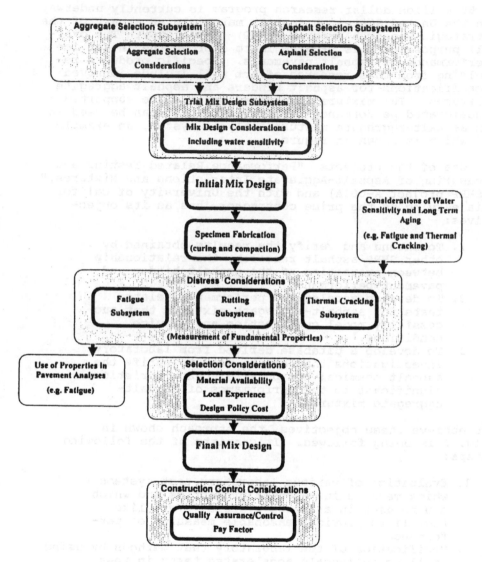

Fig. 1. Framework for an asphalt-aggregate mixture analysis system (AAMAS).

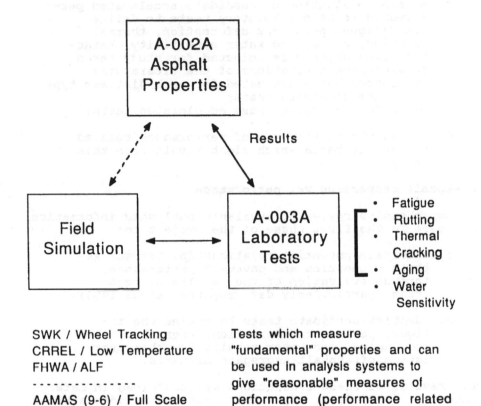

Fig. 2. General approach associated with SHRP project
A-003A.

It is the purpose of this paper to present some of the results obtained to date (March 1990) on this project. Included are:

1. a brief evaluation of candidate accelerated performance related laboratory tests to define the fatigue, permanent deformation, thermal cracking, aging, and water sensitivity characteristics which were selected for study based on extensive evaluations of the literature;
2. hypotheses being evaluated for each distress type using the candidate tests;
3. plans for verifying these accelerated tests; and
4. an example of the type of performance related mixture criteria which might result from this study.

2 Asphalt properties vs. performance

A comprehensive review of available published information was made in the first phase of the project to:

1. ascertain potential relationships between asphalt properties and pavement performance through evaluation of the results of test roads (particularly data reported since 1950); and
2. identify candidate tests to define the fatigue, permanent deformation, thermal cracking, aging, and water sensitivity characteristics of asphalt-aggregate mixtures.

The results of these evaluations are contained in References (1, 4, 5, 6, 7, 10). This section includes a brief summary of item (1) above.

An evaluation of completed pavement full scale experiments by Finn, et al (1990) has confirmed that asphalt properties do influence pavement performance. Though no quantitative and consistent relationships between asphalt properties, asphalt-aggregate mixture properties, and pavement performance were shown from field studies, some qualitative relationships were established between asphalt properties and the five distress modes under investigation. These relationships are summarized in Table 1. Overall, the evaluation indicated that the source of crude oil influences the in-service durability of asphalt and that the method of refining affects asphalt properties.

The study, which included an evaluation of more than 500 references, was important for a number of reasons:

1. to confirm that asphalt cement properties influence the performance of asphalt pavements (the basis for the research approach in the SHRP Asphalt Program);
2. to establish where knowledge "gaps" exist;
3. to provide a basis for ranking asphalt properties from good to poor or from acceptable to unacceptable;
4. to bring together in one document the extensive information developed in various field trials, both in the U.S. and abroad; and
5. to provide a basis for the factors to be studied in the development of performance criteria for the five distress areas.

Table 1. Relationship between asphalt properties and field performanace - 1950 to present

Fatigue and Permanent Deformation	Related to the stiffness of the asphalt -- information is inconclusive because of strong interrelationships between these distress modes and pavement and environmental factors.
Thermal cracking	More strongly related to asphalt properties than is fatigue or rutting; limiting viscosity and stiffness criteria have been proposed.
Aging	Chemical factors influence aging - association more qualitative than quantitative.
Water Sensitivity	Chemical factors influence resistance to the action of water and water vapor.

3   Development of Accelerated Performance Tests

The literature evaluations were conducted, as noted above, to identify candidate tests to define the propensity of asphalt-aggregate mixtures to fatigue, permanent deformation, thermal cracking, aging, and to loss in stiffness or adhesion due to the action of water and water vapor (water sensitivity).

For each of the distress modes, a number of tests have been identified from which the ultimate test procedure will be selected or developed. Hypotheses regarding material behavior have been formulated for each of the distress modes. The test procedure which will be selected

Table 2. Fatigue Test Methods being Evaluated
(From Tangella, et al, 1990)

Test Methods	Interpretation	Advantages	Limitations
Repeated Flexure and Direct Tension	Development of strain vs. cycles to failure	Results can be used for both mix design and pavement analyses	Time consuming, costly
	Determination of dissipated energy	Results can be used for both mix design and pavement analyses	Time consuming, costly
Repeated Diametral	Development of stress vs. cycles to failure	Results can be used for both mix design and pavement analyses	Time consuming
			Complex stress state
		Simpler than flexure	
Direct Tension	Correlation with fatigue test results	Eliminates need for fatigue tests when correlation has been established	Requires considerable testing for particular region to develop correlation
Fracture	Fracture mechanics principles	Eliminates need for fatigue tests	At higher temperatures theory for interpretation more complex

for each distress mode is that which gives the best prediction of behavior and can be used within the framework of Fig. 2.

Results obtained from simulation tests in test tracks or from actual pavement studies will be used to validate the selected test methodologies.

3.1 Fatigue

The candidate methods chosen are shown in Table 2. These include: 1) a number of flexural procedures; 2) a diametral fatigue test; 3) a repeated loading direct tension test; and 4) a direct tension (single load application to failure) test.

Flexural fatigue measurement comprise 4-point beam tests at Berkeley and tests on trapezoidal specimens at SWK/University of Nottingham. Diametral tests, repeated loading indirect tension, are being carried out at North Carolina State University (NCSU) while the repeated loading direct tension test on cylindrical specimens is being done at SWK/University of Nottingham.

The single load application tension tests are being performed at Berkeley as a potential accelerated test to define fatigue response. Similarly, some tests on notched beam specimens using fracture mechanics interpretations are also being performed at Berkeley.

For fatigue, the following hypotheses are being evaluated:

1. Cracking results from a tensile stress or strain (less than the fracture stress or strain-at-break under one load application) at a specific number of stress (or strain) applications, the number of load applications being larger as the magnitude of the stress or strain is smaller, i.e.,

$$N = A(1/\epsilon_t)^b \quad \text{or} \quad N = C(1/\sigma_t)^d$$

   The relationships are dependent on the temperature and mode-of-loading [coefficients (A and B) and (C and D)] and must be established by some form of repetitive load testing; or
2. Cracking results from repetitive stress (strain) applications when either the total energy or the strain energy of distortion reaches some limiting value regardless of the mode of loading to which the specimen is subjected, or
3. A direct correlation can be established between the stiffness and fracture characteristics of a mix and its fatigue response (e.g.,

similar to that established by the LCPC of France); or

4. Results of fracture tests on notched specimens can be used to predict the fatigue response of asphalt concrete mixtures over a range in temperatures.

From experimental evidence such as that illustrated in Fig. 3, the magnitude of tensile stress or strain repeatedly applied appear to be reasonable determinants of the cracking which occurs in asphalt-bound layers subjected to repetitive trafficking. Since actual pavements are subjected to bending stresses, this mode of loading appears most reasonable to define fatigue response using laboratory test equipment.

At the University of California at Berkeley the bending fatigue tests are being conducted at a rate of 100 repetitions per minute, a comparatively slow rate and one in which the influence of rest periods has been shown to be negligible. Since it may be desirable, should the flexure mode of testing be selected, to conduct the test at a faster rate of loading, controlled-stress fatigue tests on pyramidal shaped specimens are being conducted at SWK/Nottingham University at a rate approximately ten times as fast. In addition, since the bending mode of loading requires that specimens be sawed to prismatic or pyramid shapes, it was considered desirable to test specimens not requiring sawing; hence, the direct tension tests at SWK/Nottingham performed on cylindrical specimens.

When testing to define fatigue behavior, the mode-of-loading influences response - with specimens of comparatively low stiffness performing well in the controlled-strain mode and specimens of high stiffness performing well in the controlled-stress mode. Since both controlled-stress and controlled-strain tests are being performed at UCB, consideration is being given to the determination of the total strain energy and the strain energy of distortion in an attempt to eliminate the mode-of-loading variable. Such analyses require a measure of the complex modulus and phase angle for each mixture corresponding to the time of loading and temperature of the fatigue test together with the stress/strain vs. number of load applications to failure. While this approach still requires the conduct of fatigue tests, it may have the potential to sort out mode-of-loading effects.

The Laboratoire Central des Ponts et Chaussees (LCPC) has developed a correlation between the response of a mix in uniaxial loading and its fatigue response. In the LCPC methodology measurement of the stiffness characteristics of the specific mix at different strain levels and temperatures is also required. An evaluation of the LCPC approach will be made to determine its efficacy since the

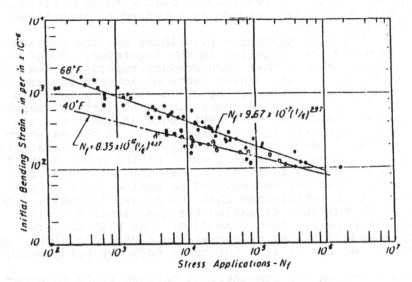

A. Stress vs. applications to failure.

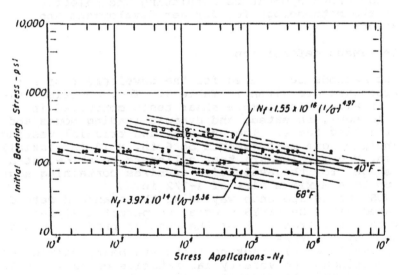

B. Strain vs. applications to failure.

Fig. 3. Typical fatigue results.

use of a direct tension test has the potential to reduce considerably the time required to define fatigue response of asphalt-aggregate mixtures as compared to repetitive load testing.

Use of <u>fracture mechanics</u> principles has the potential to shorten the time required in the laboratory to define fatigue response. Rather than conduct repetitive loading fatigue tests, direct loading tests on notched specimens permit the determination of specific control parameters from which the fatigue response can be estimated. Depending on the size of the nonelastic zone at the crack tip, different interpretations are required. If the majority of the material behaves elastically, the stress intensity factor, (K), governs the response. On the other hand, if the nonelastic zone is large, either the J-integral or the C* line integral may be required to define crack propagation. It is anticipated that the stress intensity factor will be suitable to define fatigue response at low temperatures (e.g., less than 0° C). At high temperatures, however, it may be necessary to consider either the J integral or the C* line integral; both are being evaluated.

By conducting the conventional fatigue tests together with the additional tests described herein, sufficient data will have been obtained to permit the evaluation of all of the above hypotheses permitting the selection of an appropriate methodology for further development and evaluation.

3.2 Permanent deformation

The test methods being used for the development of a rutting procedure are given in Table 3. They comprise uniaxial, triaxial, and simple shear tests conducted in static (creep), repeated, and dynamic loading modes and in wheel tracking tests. The uniaxial and triaxial testing will include some measurements on 6 in. × 12 in. (axial) and 6 in. × 6 in. (shear) test cylinders to allow, at a later stage, for the evaluation of mixes containing aggregates up to a maximum size of 1-1/2 in.

At NCSU tests are underway using the standard method associated with the VESYS program to permit analyses to be made within the framework of the rutting prediction method developed by the FHWA.

Small scale wheel tracking tests are being carried out at SWK/Nottingham University laboratories on 12 in. × 12 in. slabs of asphalt concrete. This may be a satisfactory surrogate for a field simulation test.

Evaluation of available information indicates that permanent deformation (rutting) in an asphalt concrete layer may be caused by a combination of densification (volume change) and shear deformation (plastic deformation with-

Table 3. Rutting test methods being evaluated
(From Sousa, et al, 1990)

Test Method	Advantages	Limitations
Uniaxial Loading		
·Creep	·Easy to implement ·State of stress includes shear stress component	·May not reflect influence of aggregate
·Repeated	·More representative of traffic loading than creep	·Limited stress states in comparison to field loading
·Dynamic	·Permits determination of loss tangent	·Limited stress states in comparison to field loading
Triaxial Loading ·Creep ·Repeated ·Dynamic	·Provides increased range in stress states ·Influence of aggregate better represented	·Limited stress states in comparison to field ·Test system more complex
Simple Shear Triaxial Loading ·Creep ·Repeated ·Dynamic	·All loading conditions have capability to reflect in-situ stress state ·Repeated and dynamic loading reflect traffic loading	·Creep test does not reflect field loading conditions for moving traffic ·Test systems more complex
Test Track	·Simulates field loading conditions ·Good for validation	·Fundamental material characteristics not determined

out volume change). Shear deformations resulting from high shear stresses in the upper portion of the asphalt concrete layer appear to be the primary cause of this mode of distress with the repeated application of these stresses under conditions of comparatively low mix stiffness leading to the accumulation of permanent deformations in the form of ruts at the pavement surface.

Test methods which duplicate the stress states occurring in the upper portion of the asphalt-bound layer have the potential to define the propensity for a particular asphalt concrete to rut under repeated trafficking. Hence, equipment with the capability of directly applying shear stresses is being included in the investigation.

For the tests to be applicable to a wide range in conditions of both loading and environment (particularly temperature), the results of the tests should have the capability of being used in an analysis system which permits the prediction of rutting under repeated traffic loading (Fig. 2). With lateral wander of the traffic in the wheel path there is the possibility that reversal of shear stresses will occur. It may be desirable to include this effect in any test methodology resulting from this investigation. Since none of the existing analytical procedures correctly consider the effects of the shear stresses as described above, development of a procedure to predict rutting under repeated trafficking is considered an essential step in this phase of the study.

A number of organizations make use of the axial creep test in the mix evaluation phase; accordingly, this test has been included in the investigation. It must be recognized, however, that the unconfined creep test does not duplicate representative stress states in the upper portion of the asphalt concrete layer. Nevertheless, it may be a useful test to be considered as a part of a procedure for rutting evaluation.

A direct shear creep test is also included since, as noted above, shear stresses appear to be the primary contributor to plastic deformations. When this test is performed in the unconfined state, however, the test overemphasizes the influence of the binder characteristics. Thus, the test is being conducted in the confined as well as the unconfined state.

There is evidence that materials may respond differently in repeated loading than in creep resulting in a different ordering of rutting propensity (Valkering, et al, 1990). Accordingly, repeated load tests are being conducted in both the axial and shear modes.

While the above tests are relatively straightforward to conduct, they do not exactly duplicate the stress conditions leading to rutting. Some tests are being planned wherein these stress conditions are better simulated and include tests on hollow cylindrical asphalt concrete

specimens subjected to a combination of shear and axial
loads. The hollow cylinder test configuration is considered, at present, to have the potential to provide
stress states in the laboratory which are most representative of in-situ conditions. Also included will be a
direct shear test in which confining pressure and axial
stress can be controlled. While this test is much simpler
than the hollow cylinder test, it may not permit the range
in stress states required to duplicate those occurring in
the upper part of the pavement to be applied. Nevertheless this test has the potential to serve as a useful tool
for mix evaluation.

To insure that the test to define rutting propensity in
shear will provide the proper measures of response for
rutting prediction, a finite element analysis of the test
configurations is being developed using a computer program
which incorporates viscoelastic response characteristics
and permits consideration of "damage" (e.g., excessive deformation).

This same type of analysis is being extended to permit
determination of rutting in an asphalt-bound layer subjected to repetitive trafficking. This is being done to
insure that the results of the laboratory test evaluation
can be used to assess rutting in-situ for a range of traffic and environmental conditions. Wheel tracking tests to
be performed by SWK/Nottingham on the mixes tested in the
laboratory should provide some validation of the methodology.

3.3 Thermal cracking (including thermal fatigue)
The methods selected for evaluation of thermal cracking
are shown in Table 4. A prime candidate is the test in
which rectangular specimens are fully restrained, and the
induced thermal stresses, which occur when the specimens
are subjected to controlled cooling, either once or cyclically, are measured. This test is similar to that already
developed at the Technical University of Braunschweig in
West Germany. The fracture mechanics approach, mentioned
under the fatigue investigations above, also has possibilities.

Other tests, for which some relationship with performance has been gathered, are also involved. These include
tension strength tests, indirect (ASTM D-4123) and direct
(rectangular beam specimens according to Haas, et al,
1973), conducted at different load rates and over a range
of temperatures. Some tests also are being developed to
measure the coefficient of contraction when specimens are
subjected to various cooling rates.

These tests are predicated on the hypotheses that
cracking may result from cold temperatures which occur in
many regions of the United States and elsewhere in the
northern hemisphere. Such cracking may result when the

Table 4. Thermal cracking methodologies being evaluated
(From Vinson, et al, 1989)

Property	Method	Simulates Field Conditions	Application in Existing Performance Models	Existing Experience
Tensile strength and strain at break	Indirect tension or direct tension	No	Indirect application	Extensive prior use
Creep	Direct tension or indirect tension	No	Indirect application	Not widely used
Thermal stress, fracture strength	·Thermal stress - restrained specimen ·Thermal fatigue	Yes	Direct application	Significant prior use
Coefficient of thermal expansion and contraction	Various tests	Yes	Indirect application	Significant prior use

temperature is reduced to a low value, termed <u>low tempera-ture cracking</u>, or may result from thermal cycling, termed <u>thermal fatigue cracking.</u>

Low temperature cracking is illustrated in Fig. 4, which contains the results of tests on a restrained bar of asphalt concrete. As the temperature is reduced, the thermal stress is increased to a point where the thermal stress exceeds the fracture strength of the material, also shown in the same figure as a function of temperature.

Thermal fatigue cracking occurs when temperatures cycle at temperatures higher than those required for low temperature cracking even though the stress in the pavement is typically far below the strength of the mixture at that temperature. Consequently, failure does not occur immediately but develops over a period of time similar to the time required for fatigue cracking associated with traffic load induced strains in the asphalt concrete.

Fig. 5 provides an approximate indication of the temperature regimes where both low temperature and thermal fatigue cracking can occur.

The thermal stress restrained specimen test system provides a measure of thermal stress at break and the associated temperature as the temperature is reduced at some prescribed rate and the asphalt concrete specimen is maintained at constant length. A similar set up is used to study the thermal fatigue problem.

For the fracture mechanics approach which may be used to predict both low temperature and thermal fatigue cracking, fracture tests on specially notched specimens and which can be interpreted using C* line integral concepts are under investigation.

3.4 Aging

Tests for aging comprise procedures to evaluate hardening in the short term (the combined effects of mix preparation and pavement construction) and long-term (five to ten years in service). The methods which have been selected as most promising are given in Table 5.

The short-term aging tests are being conducted on loose mixtures exposed to extended oven heating, extended mixing, and microwave heating methods. Long-term aging tests are being conducted on compacted specimens, and the methods being investigated are extended oven aging and pressure oxidation. At this time the pressure oxidation vessel holds promise for simulating aging in the field.

The effects of aging are being evaluated on the mixtures by such nondestructive tests as dynamic (diametral) modulus, and destructive tests such as the indirect tensile test and the resulting tensile strength at break. Another approach is the measurement of viscosity on asphalt recovered from the mix. For this purpose, only small amounts of the asphalt are required and various asphalt recovery techniques are being investigated.

In general, aging of an asphalt-aggregate mixture results in hardening (stiffening) which results in a change in its performance. The change may be beneficial since a stiffer mixture will have improved load distribution properties and will be more resistant to permanent deformation. However, aging may also result in embrittlement (increased propensity for thermal fracture and fatigue cracking under some circumstances) and loss of durability in terms of wear resistance and moisture susceptibility.

Table 5. Age conditioning procedures being evaluated
(From Bell 1989)

Type	Method	Simulates Field Conditions	Application of Method	Existing Experience
Short term	·Extended heating	Good data from AAMAS	Similar to TFOT	Little on mixes
	·Extended mixing	Simulates plant mixing	Similar to RTFOT	Little on mixes
	·Microwave heating	Not the same	May promote structuring	Very little
Long term (1-10 years)	·Extended heating	Significant aging but at high temperatures	Similar to extended TFOT or FTFOT	Little on mixes
	·Pressure oxygen treatment	Significant aging at low temperatures	Simulates aging due to oxidation	Little on mixes
	·Modified triaxial cell	Difficult to assess	Potential to condition with several "fluids"	None on mixes

The tests which have been selected are based on the hypothesis that two major effects dominate the aging of asphalt-aggregate mixtures.*   These are:

1. Loss of volatile components in the construction phase (short term aging).
2. Progressive oxidation of the in-place mixture in the field (long term aging).

*Other factors may contribute to aging. In particular molecular structuring may occur over a period of time resulting in steric hardening. Actinic light, primarily in the ultraviolet range also has an effect, particularly in desert-like climates.

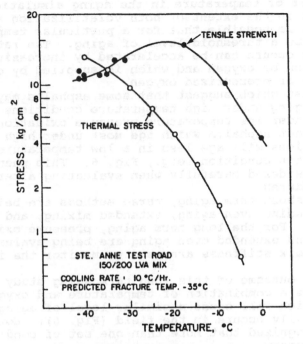

Fig. 4. Prediction of fracture for a restrained strip of asphalt concrete (see
Vinson, et al, 1990).

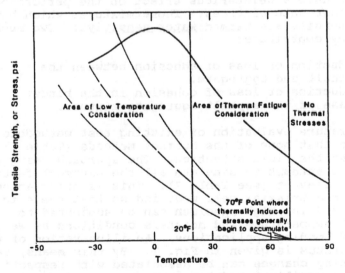

Fig. 5. Temperature ranges associated with different types of thermal cracking (see Vinson, et
al, 1990).

The level of temperature in the aging simulation will play a part in the extent of both volatilization and oxidation. It is possible that for a particular temperature there will be a threshold level of aging. The rate at which aging occurs can be accelerated by increasing the concentration of oxygen and which is supplied by compressed air or pressurized oxygen.

Data exist which suggest that some asphalts may exhibit much more aging under high temperature conditions than they will under low temperature pressure oxidation conditions and that asphalts which age most under high temperature conditions will age less in a low temperature pressure oxidation condition, e.g., Fig. 6. This phenomenon must be considered carefully when evaluating alternative aging procedures.

For the short term aging, three methods are being examined including oven aging, extended mixing, and microwave aging. For the long term aging, pressure oxidation (and air) and extended oven aging are being evaluated. Changes in mix stiffness are used to monitor the influence of aging.

An ideal outcome of this long term aging study will be to determine a combination of temperature and oxygen (air) pressure which will cause a similar level of aging to that which typically occurs in the field (Fig. 6). However, it must be recognized that more than one set of conditions may be necessary.

3.5 Water Sensitivity

Water can have a deleterious effect on the performance of asphalt-aggregate mixtures. Those mixtures which have this propensity are termed <u>water sensitive</u>. Two mechanisms may contribute:

1. Reduction or loss of adhesion between the asphalt and aggregate.
2. Reduction or loss of cohesion in the binder phase of the asphalt-aggregate mixture.

The literature evaluation of existing test methodologies indicated that none of the current methods (Table 6) duplicates the field situation. The approach being pursued is to attempt to simulate all the pavement factors that are relevant (see Table 7). This is being accomplished in a triaxial cell modified so that the 4 in. diameter × 2.5 in. high specimen can be subjected to different temperature and moisture conditions as well as to varied loading applications. An illustration of the test apparatus is given in Fig. 7. By this means, various conditioning changes can be associated with respective changes in resilient modulus.

Table 6. Evaluation of conditioning methods
          (After Terrel, et al, 1989)

Type	Methods	Strengths	Weaknesses
Water	· None · Partially saturated · Saturated	· Easy · More realistic · Easy to achieve	· Not realistic · Difficult to achieve · Too severe
Freeze/ Thaw	· None · 1 cycle · Multiple cycles	· O.K. for mild climate · Reasonable estimate · Realistic for certain areas	· Not applicable for cold climates · Unknown · Time Consuming
Aging	· None · Some	· Easy · More realistic	· Not realistic · Unknown
Traffic loads	· None · Repeated load	· Easy · More realistic	· Not realistic · Unknown

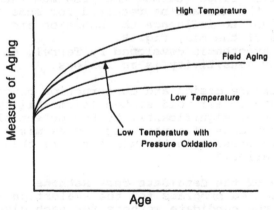

Fig. 6. Possible interactive effects
of temperature and pressure oxidation
on aging.

Table 7. Evaluation of water sensitivity test
          methodologies

Type of Test	Test Method	Simulation of field conditions	Use in Performance Models	Limitations
Loose Mixtures	· Static immersion	Does not	Not applicable	For adhesion only
	· Boiling water	Does not	Not applicable	For adhesion only
Compacted Mixtures	· Immersion	Not known	Not applicable	Do not consider effects of traffic loads or differences in environments
	· Lottman	Fair	Yes	
	· Tunnicliff-Root	Fair	Yes	
	· AASHTO T-283	Fair	Yes	

Damage from both loss of adhesion and cohesion may be
minimized by precluding water from entering the mixture or
by providing for drainage so that water is not retained.
This physical condition could be achieved by constructing
mixtures with void contents (permeabilities) that fall
outside a specific range (termed the pessimum range) where
water can enter the mix and be retained for some time
period, sufficient to influence the adhesion and cohesion
characteristics of the mix, Fig. 8.

The working hypothesis developed by Terrel, et al
(1989) for water sensitivity is as follows:

"Existing mixture design and construction practices
tend to create an air void system (permeability)
that contributes significantly to the potential for
reduced performance of asphalt-aggregate mixtures
when water is present in the mix for more than a
very short period."

3.6 Evaluation of the Candidate Test Methods
A series of detailed programs for the evaluation and
development of the candidate methods for each distress
mode has been prepared. The programs (Hicks, et al, 1989)

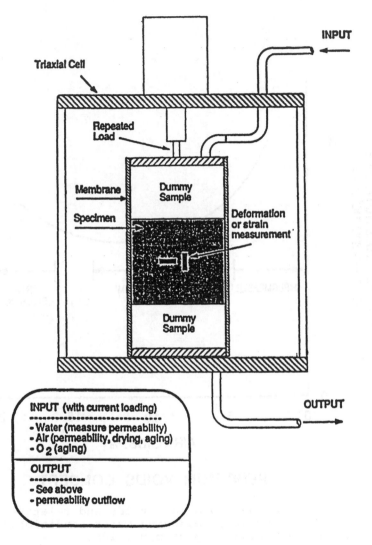

Fig. 7. Test system for evaluating the effect of moisture and temperature conditions while continuously monitoring resilient modulus.

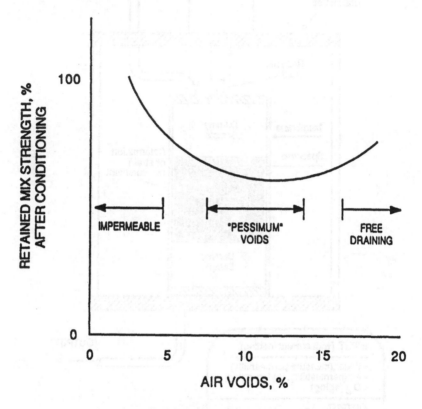

PESSIMUM VOIDS CONCEPT

Fig. 8. Relationship of air voids and relative strength of mixtures following water conditioning showing the region of pessimum voids.

describe the materials to be used, the protocols for the preparation (compaction) of the mix specimens, studies to evaluate the tests, and a cooperative test program which will involve state highway and federal agencies and industry representatives.

The same aggregates and asphalts are being tested in the initial study of all five performance factors. The two aggregates and two asphalts used in this phase of the work were selected from SHRP's Materials Reference Library (MRL). The MRL comprises 13 aggregates and 32 asphalts which have been carefully selected (Cominsky, et al, 1989) to provide the complete range of types that exist in the U.S.A. with respect to their known performance as well as to their chemical and physical properties.

The aggregates are a granite known for its resistance to moisture damage effects (nonstripper) and a moisture susceptible Gulf Coast gravel (stripper). The asphalts, from Boscan and California Valley crudes, have widely different composition and viscosity temperature characteristics. The standard mix is a typical dense graded type with a 3/4 in. maximum stone size. Asphalt contents are based on both Hveem and Marshall designs.

Subsequently, the test procedures selected for each of the performance factors will be used to evaluate mixes prepared using a wide range of the aggregates and asphalts contained in the MRL. The program stipulates that a minimum of four aggregates and eight asphalts will be evaluated.

An initial compaction study has established that three methods of compaction: 1) gyratory; 2) kneading; and 3) rolling wheel, which all provide a shearing action, achieve similar levels of performance. These were evaluated by fatigue (both flexural beam and diametral) and rutting (axial creep and shear creep) measures. A similar result was shown in a recent NCHRP study (Von Quintus, et al, 1988). Accordingly, it was decided that any of these three procedures could be used by the various researchers and the sole criterion for compaction would be the achievement of the required density and air void contents.

Currently (March 1990), the test system evaluation is progressing in all five distress areas. It is expected that the final selection of the tests will be made in July 1990.

Several criteria will be used in the selection of these methods. These include:

1. Sensitivity to mix variables. Particularly to the effect of asphalt type but also to asphalt content and aggregate type and gradation.
2. Simulation of field conditions. Should obviously be simulated as much as possible unless simpler tests yield similar results.

3. Provision of criteria for models. Must be able to provide performance fundamental data that can be used in performance models and design methods.
4. Implementability. Must be reliable, accurate, and reproducible to be suitable for user agencies and industry. (Note that a complex test may be suitable provided it can be controlled by a user friendly computer program).

However, if work were to stop today, the test recommendations would be as follows:

1. Fatigue. Flexural beam test.
2. Permanent deformation. Simple shear test with confinement.
3. Thermal cracking. Thermal stress restrained specimen test.
4. Aging. Use of the pressure oxidation vessel and monitored with a modulus test.

4. Field verification

To insure that the accelerated laboratory tests finally selected reasonably define field performance, verification is planned through the use of controlled field simulation studies, Fig. 2. The verification plan described herein represents our initial planning (March 1990)

4.1 Fatigue
Verification of the laboratory fatigue tests can be accomplished using either test tracks or full-scale pavements. Because of the short time frame associated with this project, test tracks appear to afford the most expeditious way for verification. It is interesting to note that this is the approach which is being followed in France with the LCPC Nantes facility. Table 8 contains a listing of the test tracks under consideration and their suitability to validate fatigue tests. One difficulty which is anticipated in interpreting the test track results is to be able to define the influence of rutting on fatigue response.

At the Nantes facility of the LCPC four test sections will be constructed in April 1990. These encompass the use of three binders including a hard asphalt and a modified asphalt with each mixed one aggregate type and gradation. This particular test program has considerable potential to provide valuable validation data since the fatigue tests being conducted at SWK/Nottingham are similar to those performed by the LCPC.

Table 8. Tests tracks planned for verification of fatigue/ rutting tests

Test Track	Type of Track	Use	
		Fatigue	Rutting
SWK/ Nottingham	Small linear/ single section	x	x
SWK/ Nottingham	Large linear/ multiple section	x	x
LCPC - Nantes	Circular/multiple sections	x	x
LCPC Type (ESSO, Exxon, Elf Asphalt, Nynas)	Small linear - slab specimen	-	x
ALF- FHWA	Linear/single section	x	x

4.2 Permanent deformation

The laboratory tests used to estimate permanent defor- mation can also be verified using either test tracks or full scale pavements. At this time, the use of tests tracks appears to be the most viable way of confirming the laboratory tests (Fig. 9). Table 8 also contains a listing of the test tracks which have the potential to be used in the rutting validation study.

At SWK/Nottingham the small wheel track facility will be used to conduct a full factorial experiment with the two asphalts (California Valley and Boscan) and two aggre- gates (Watsonville granite and Gulf Coast gravel).

The small wheel track device originally developed by ESSO, France, and used by the LCPC is available in the laboratories of Exxon (New Jersey), ESSO (France), Elf Asphalt (Indiana), and Nynas (Belgium). Consideration is being given to the use of this device as well.

4.3 Thermal Cracking

For thermal cracking, validation is needed to confirm both the low temperature and thermal fatigue cracking labora- tory test procedures. Validation will be conducted using both test tracks and full scale pavements (see Table 9)

For the low temperature and fatigue cracking problems the CRREL facility will be used. At this facility the test program will consist of constructing a test section

Table 9. Test facilities to be used for thermal cracking

Test Facility	Type of Facility	Proposed Test Program
Test Tracks		
U.S.A.C.R.R.E.L.	Linear test track w/environmental control	Test sections will be constructed in Fall 1990 or Spring 1991
Field Tests		
Pennsylvania DOT	Field test sections	Constructed in 1976
Alaska DOT	Fairbanks test sections	Construction in 1988
Minnesota DOT	3 field test sections	To be constructed in 1991
Oregon DOT	Bend test sections	Constructed in 1985

and monitonically or cycling cooling the test sections to determine the temperature at which cracking is initiated.

Field test sections also have been identified that may be used to verify the low temperature test procedure:

1. Pennsylvania DOT. These test sections were constructed in 1976 and cracking occurred during the first winter. Six different asphalts were used, and all are available in the MRL.
2. Alaska DOT. Slabs of cracked/uncracked pavements as well as original asphalt/aggregates have been obtained from pavement sections constructed near Fairbanks in 1988.
3. Minnesota DOT. Several test sections will be constructed in 1991 to evaluate the effect of asphalt type on low temperature cracking. Three asphalts will be used, and arrangements are being made to acquire original asphalt/aggregates.

Only one field-test site has been identified (to date). In Oregon ten sections with different asphalt types were

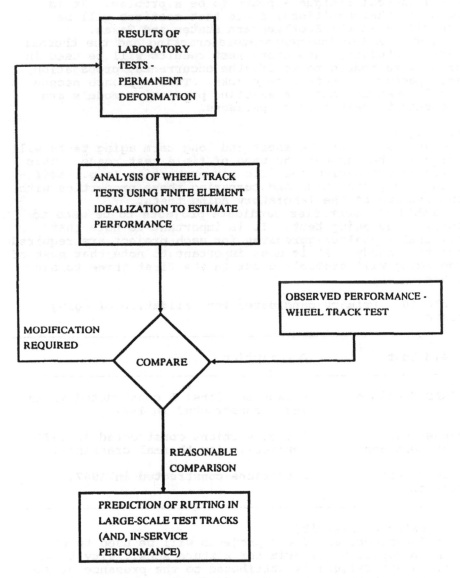

Fig. 9. Proposed approach to validate laboratory test procedure for permanent deformation evaluation.

constructed in 1985, and four additional ones (with polymers) were constructed in 1988. These are in an area
where thermal fatigue appears to be a problem. It is
possible that additional field test sections will be
identified in the Southwestern States and Texas.

For both the low-temperature cracking and the thermal
fatigue studies, laboratory test results will be used in
predictive models to verify the occurrence, propagation,
and spacing of transverse cracks. This approach necessarily assumes that the existing predictive models are
adequate to make such comparisons.

4.4 Aging
The validation of the short and long term aging tests will
primarily be through the used of field test roads. This
involves evaluating specific mix properties (e.g., stiffness, etc.) over time and comparing these properties with
the results of the laboratory aging tests.

Table 10 identifies candidate projects to be used to
confirm the aging test. It is important to note that
original asphalts/aggregates for each project are required
for this study. It is also important to note that most of
the aging will probably occur in the first three to six
years.

Table 10. Primary candidates for validation of aging
tests

Field Test	Description
AAMAS Sections	9 test sections; 5 constructed in 1986 and 4 constructed in 1987.
Pennsylvania Test Sections	7 test sections constructed in 1976 (primarily for thermal cracking).
French Test Sections	3 sections constructed in 1987.

4.5 Water Sensitivity
Both laboratory and field projects will be used to establish appropriate criteria for mixture sensitivity to water. Where failure is attributed to the presence of water, the relative importance of adhesion and cohesion
(loss in stiffness) will need to be assessed by testing
and observation. Fig. 10 illustrates the general approach
to be used and includes several steps.

1. Test asphalt and aggregate materials individually and in loose measures to evaluate adhession and/or stripping potential.
2. Fabricate specimens in the laboratory and subject them to moisture conditioning appropriate for the climate and establish a ratio of unconditional stiffness or strength, S_{wet}/S_{dry}.
3. Test specimens (unconditioned and conditioned) for response to fatigue and rutting to establish the relative sensitivity to these disstress modes.
4. Test the same mixtures in test tracks (laboratory scale); e.g., wheel tracking in both dry and wet conditions.
5. Obtain core samples and original materials from constructed highway pavements and subject them to testing and evaluation similar to that above. In conjunction with other researchers, establish pavement life curves for each project and compare these to laboratory results.

Table 11 is a summary of the test facilities and pavement sections planned for use in this verification effort.

Table 11. Validation of water sensitivity tests

Test Type	Description
Test Track	
ESSO - France	Wheel tracking with water
ELF - Indiana	Wheel tracking with water
Nynas - Belgium	Wheel tracking with water
Field Sections	
Texas	Several test sections with and without additives
Nevada	Several test sections with and without additives
Oregon	Several test sections with and without additives
FHWA	Several test sections with and without additives

5 Performance criteria development

An earlier section of this paper has described the procedures to be used in <u>selecting and/or developing test methods</u> for dense graded asphalt-aggregate mixtures, which can be used to estimate field performance when such mixtures are incorporated as part of the structural section of an asphalt surfaced pavement. A primary consideration in developing specific tests and conditioning procedures will be the ability to evaluate the role of asphalt properties believed to significantly influence pavement performance.

One of the objectives of this investigation is to provide information which can be useful in the development of performance related specifications for asphalt-aggregate mixtures. To achieve this objective it will be necessary to: 1) verify the test methods with regard to their ability to relate to asphalt properties, 2) relate test results to performance, and 3) interpret results as a basis for identifying desired mixture properties.

5.1 Verification of test methods

The verification process has been described in the previous section. Based on completion of this phase, it can be reasonably concluded that a series of test methods will have been successfully developed which can be related to specific types of distress and which reflect the role of asphalt properties.

5.2 Relate test results to performance

Test results will be related to performance by at least two methods: 1) analytical models based on phenomenological test procedures, and 2) analytical models which can be used to predict stress, strain, or deformation needed to incorporate field conditions which influence pavement response to traffic, age, and environment.

To illustrate this process, one can refer to phenomenological test models used to measure fatigue characteristics of asphalt mixtures. Based on laboratory tests (previously verified) a typical analytical model for fatigue relates cycles to failure as a function of the maximum tensile strain in the asphalt concrete. This fatigue test is currently sensitive to the stiffness (modulus) of the asphalt binder and may also prove to be sensitive to such factors as asphaltene content, aging index, and moisture sensitivity. Thus, the influence of asphalt properties can be identified with fatigue characteristics of mixtures.

To extend this information to performance based specifications for asphalt and asphalt-aggregate mixtures, it will be necessary to extend results to field conditions which will include such variables as traffic, environment (rainfall, temperature regimes) and structural design.

Similar phenomenological tests will be developed for permanent deformation; e.g., dynamic compressive creep or dynamic shear creep and low temperature cracking; e.g., montonic or fatigue in each case, an analytical model will be developed which can be used to relate stress, strain, or deformation to distress and which will reflect the properties of the asphalt binder.

5.3 Interpretation of test results

Based on test methods developed in this investigation, which implicity include the steps required for verification, it will be necessary to develop performance prediction models. The prediction models will incorporate information provided by newly developed test methods and will reflect the properties of the asphalt binder as well as the asphalt-aggregate mixture. The prediction models, which will also be verified by comparisons with in-service pavements or full scale test tracks, will provide information required to develop performance based specifications.

An example of performance-related criteria which could be used as basis for performance based specifications for asphalt-aggregate mixtures is shown in Table 12. In this example, provision has been made for the thickness of the asphalt concrete layer (structural design) and for air temperature. An additional requirement to reflect rainfall may be added as necessary; however, in the example, the potential problems related to rainfall are incorporated in the wet/dry stiffness index. It is likely that criteria will also be added to reflect necessary requirements for adhesion of asphalt binder to aggregate. Similar criteria will be developed for a range of layer thicknesses and environments. The specific properties may be expanded to include additional requirements or revised levels for specific criteria.

The information in Table 12 would typically replace current requirements for either the Marshall or Hveem test methods. Additional requirements for construction control will need to be provided as a basis for preparation of guide specifications.

6 Summary

The final products to be developed by the SHRP asphalt research efforts will be performance-based specifications for both asphalt binders and asphalt-aggregate mixtures. In order to achieve these two objectives, it will be necessary to coordinate results from a number of SHRP investigations. This is illustrated in Fig. 13. The A-005 contractor will develop and validate performance prediction relationships based on information provided primarily

Table 12. Example of performance-related criteria

Heavy duty highway - Asphalt concrete layer $\geq$ 8 in.
Environment: MAAT = 50°F; Range - 40°F to 95°F

Mix Characteristic	Distress Mechanism		Thermal Cracking*
	Fatigue	Rutting	
Stiffness, psi	$\geq$ 500,000 psi at t=0.02s, T=68°F	$\geq$ 100,000 psi at t=0.02s T=120°F	$< 2 \times 10^6$ psi at t=30 min, T=40°F
		$\geq$ 30,000 psi at t=100 min T=120°F	
Aging Index S_{aged}/S_o	-	-	2.0
Stiffness Index (Water) S_{wet}/S_{dry}	0.8	0.8	0.8
Min Wet Strength (20°C) (Diametral)	100 psi	100 psi	100 psi
Fracture Strength, psi	-	-	> 700 psi at T=40°F measured at load rate=0.1 min)

*Based on load-induced stresses. Criteria based on
 thermal stresses not yet available.

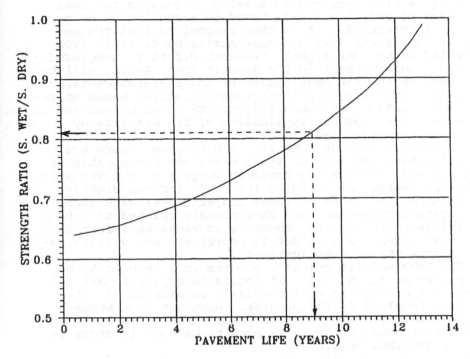

Fig. 10. Laboratory tests and field performance data used to establish water sensitivity.

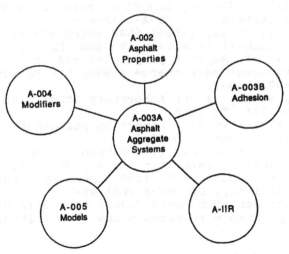

Fig. 11. A-003A interfacing diagram

from A-002, A-003, and A-004 contracts. The A-005 contractor is also required to develop procedures for user agencies to update criteria based on accumulation of pertinent information. It is thus recognized that the performance criteria will not necessarily be final in 1992, when the major SHRP asphalt research effort is concluded.

The final contract in the asphalt area, A-006, will be responsible for the development of the performance-based specifications for asphalt-aggregate mixtures based on results on asphalt and asphalt-aggregate interactions and to provide guidelines for application of the criteria by state highway agencies.

In addition to the SHRP research program, there are complimentary projects underway under the sponsorship of the National Cooperative Research Program of the Transportation Research Board and by the U. S. Federal Highway Administration. Most of these projects deal with asphalt-aggregate mixtures and are thus closely related to A-003A activities. An ongoing procedure to exchange information has been organized in order to coordinate and utilize results from these investigations.

The SHRP asphalt research program is scheduled to be completed by March 1993. It should be apparent that to meet such a schedule will require a commitment of all parties involved in the program. Much has been accomplished in the first two-plus years since the program was initiated; however, there is much to be done in order to meet prescribed deadlines.

7 References

Bell, C.A. (1989) Summary report on aging of asphalt-aggregate systems, SR-OSU-A-003A-89-2.

Cominsky, R.J., et al, (1989) SHRP materials reference library: Asphalt selection process, SHRP-IR-A-80-002.

Finn, F.N., et al, Asphalt properties and their relationship to pavement performance (1990) Literature review SR-ARE-4-003A-89-3.

Hicks, R.G., et al, (1989) Laboratory study plan for SHRP project A-003A, TM 89-8.

Sousa, J.B. (1990) Summary report on permanent deformation in asphalt concrete, TM 89-4.

Tangella, S.S. Rao, et al, (1990) Summary report on fatigue response of asphalt mixtures, TM 89-3.

Terrel, R.L. and Shute, J.W. (1989) Summary report on water sensitivity, SR-OSU-A-003A-89-3.

Transportation Research Board (1986) Strategic highway research program - research plans, Washington, D. C., 1986.

Valkering, C.P., et al (1990) Rutting resistance of asphalt mixes containing non-conventional and polymer modified binders, **Proceedings, Association of Asphalt Paving Technologists, vol. 59**

Vinson, T.S., et al (1989) Summary report on low temperature and thermal crackingj, SR-OSU-A-003a, 89-1.

37 L'ESSAI DE FATIGUE LPC: UN ESSAI VULGARISABLE?
(The LPC fatigue test: a generally applicable test?)

F. MOUTIER
Laboratoire Central des Ponts et Chaussées, Nantes, France

Abstract
The LPC fatigue test, conducted on a trapezoidal cantilever element with strains and temperatures controlled, is one of the fundamental tests used to obtain information about the life of a bituminous mix.

A major campaign of tests aimed at investigating the influence of mix design parameters and compaction energy on the fatigue behaviour of bituminous mixes was the occasion for a special effort to increase the capacity of the fatigue testing systems while at the same time making the test easier and less expensive, and this effort was successful.

Currently, the nationwide capacity of the LPC network is more than adequate to meet the many needs of private and government laboratories (about 60 investigations a year). Concurrently, participants in the road-building process (three contractors and two refineries) have acquired the same resources. All of this gives this test a much less fundamental character (many laboratories are capable of conducting it), and its discriminatory power and good repeatability give it an edge over conventional tests, which are much too coarse.

Although this test is not perfect (it too has its weaknesses, such as the induced thermal effect that interacts with the failure criterion), it has become easy to perfom and is now used almost systematically for major projects (such as motorway construction) and for characterizations of modified binders.

It is a test that could be used generally, especially since the existing data base (more than two hundred results) may make interpretation easier.

The aim of the paper is to give the reader the information he needs to draw his own conclusions concerning this development.

Résumé

L'essai de fatigue LPC, sur console trapézoïdale, à
déformations et températures imposées, est un essai entrant
dans la catégorie des essais fondamentaux destinés à donner
une information sur la durée de vie des enrobés bitumineux.

A l'occasion d'une campagne d'essais importante relative
à l'étude de l'influence des paramètres de composition et
énergie de compactage sur le comportement en fatigue des
enrobés bitumineux, un effort particulier destiné à
augmenter la capacité des chaînes de fatigue tout en
diminuant le coût d'exploitation et la difficulté de
l'essai a atteint son but.

Actuellement la capacité nationale au sein du réseau LPC
est largement suffisante pour satisfaire les besoins
multiples des laboratoires privés et publics (environ 60
études par an). Parallèlement les entreprises routières se
sont données les mêmes moyens (trois entreprises et deux
raffineries en sont équipées). Tout cela donne à cet essai
un caractère beaucoup moins fondamental (de nombreux
laboratoires savent l'exploiter) et son caractère
discriminant lié à une bonne répétabilité l'avantage vis à
vis des essais traditionnels qui sont beaucoup trop
grossiers.

Bien que cet essai ne soit pas parfait (il possède lui
aussi des défauts comme l'effet thermique induit qui
intéragit avec le critère de rupture), son emploi est
devenu facile et presque systématique dans le cadre des
grands chantiers (autoroutiers par exemple) et pour la
caractérisation des liants modifiés.

C'est un essai qui pourrait se généraliser, d'autant que
la base de données acquise actuellement (plus de deux cents
résultats) peut faciliter les interprétations.

L'objet de la communication correspondante est de donner
à chacun les éléments qui lui permettront de tirer les
conséquences d'un tel développement.

1 Réflexion en guise d'introduction

Les laboratoires des Ponts et Chaussées pratiquent depuis
plus de vingt années l'essai de fatigue en flexion à
déformation imposée sur éprouvette console trapézoïdale.

Cette longue expérience concrétisée par l'existence d'un
fichier de plusieurs centaines de caractérisations
complètes de mélanges hydrocarbonés, dont certaines ont
fait l'objet de communications à l'AAPT 88 ou EUROBITUME
89, a permis d'affiner la méthodologie d'essai au point
qu'on peut prétendre maitriser parfaitement l'essai.

Cette maitrise, caractérisée par l'existence d'un
excellent pouvoir discriminant et d'une bonne
reproductibilité permet d'envisager d'alléger l'essai en
diminuant le nombre de répétitions (réduction par 2).

L'allègement consécutif entraine la réduction du volume d'échantillons à prélever ou fabriquer, réduit le volume de sciage, diminue le nombre d'opérations de préparations (mesures physiques et collages) et de montages au point que le coût d'une caractérisation complète devient tout à fait acceptable.

On objectera que cet essai n'est pas un essai simple et que le matériel nécessaire à son exécution est trop spécifique, contrairement à l'usage universel qu'offre une presse.

Néanmoins, pour caractériser le comportement en fatigue des mélanges hydrocarbonés, il n'existe pas d'essais autres que l'essai de fatigue lui même et l'interprétation de la non linéarité au travers de l'essai de traction, tel que les LPC l'ont proposé et qui fait l'objet d'une autre communication.

Si l'on écarte ce dernier essai, qui fait l'objet d'un usage rationnel de presse axiale, il semble qu'il vaut mieux utiliser des matériels spécifiques, comme les matériels développés par les LPC qui, bien conçus et bien caractérisés, simplifient l'usage et garantissent la qualité de l'essai, plutôt qu'une presse non spécifique utilisée en dynamique (donc inutilisable pour autre chose), car le coût de l'unité d'essai devient prohibitif.

L'autre objection pourrait provenir du fait que les éprouvettes employées sont difficiles à élaborer.

De toute façon, l'élaboration d'éprouvettes est nécessaire. Faut-il maintenant employer une géométrie compliquée (trapèze) conduisant à des essais discriminants pas trop sensibles aux effets exothermiques, ou prendre une géométrie simple conduisant à des résultats peu discriminants ? Seules des comparaisons d'essais lèveront le voile !

Enfin le mode d'élaboration des plaques, dans lesquelles les éprouvettes sont extraites, peut contribuer à donner au mélange un comportement semblable, ou dissemblable à celui du même matériau prélevé in situ.

Pour l'instant, le choix du compactage à l'aide de pneumatiques donne entière satisfaction.

Toutes ces affirmations vont être étayées une à une dans le texte qui suit.

NOTA : Le principe de l'essai, la description du matériel LPC ainsi que les résultats récents du plan pluriannuel et de l'étude nature bitume figurent en annexes 1, 2 et 3.

2 L'essai LPC est-il discriminant ?

Quelle qualité un essai doit-il avoir pour être discriminant ?

Pour juger du pouvoir de séparation d'un essai, il faut disposer de deux grandeurs :

- la première, mesurant la répétabilité de l'essai, caractérisée bien souvent par l'écart type de répétabilité σ_r,
- la seconde, mesurant l'ordre de grandeur de la variable mesurée, caractérisée par l'estimation de la moyenne des valeurs apparaissant successivement lors de la répétition de la mesure, cette estimation de la moyenne est souvent notée $\bar{x}$ et a pour définition :

$$\bar{x} = \frac{x_1 + x_2 + \ldots + x_n}{n}$$

Deux cas peuvent se produire pour évaluer une même moyenne m.

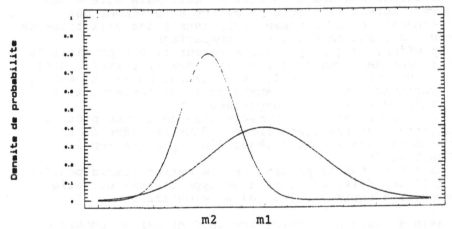

Pour séparer des échantillons de moyennes différentes mais voisines, le cas 2 est nettement supérieur au cas 1. Il nécessitera surtout moins de répétitions.

En clair, plus $\sigma_r/\bar{x}$ appelé coefficient de variation est faible, plus la séparation est facile et donc plus l'essai présente un caractère discriminant.

Les déformations relatives ϵ_6 entrainant une résistance moyenne en fatigue de 10^6 cycles, obtenues avec les chaines de fatigue type LPC s'étalent pour l'essentiel entre 50 et 150. 10^{-6} et la pente des droites correspondantes est statistiquement constante et égale à -0,2 (Log ϵ = f (log N)).

Juger la qualité d'un résultat revient, en appliquant le principe du coefficient de variation à former le quotient de l'estimation de l'écart type correspondant à l'estimation de ϵ_6 ($\sigma_{\epsilon6}$) issus de la régression consécutive au calcul de la droite de fatigue, par ϵ_6 lui-même.

Le pouvoir discriminant (= CV) est donné par

$$CV = \frac{\sigma_{\epsilon_6}}{\hat{\epsilon}_6}$$

Si on prend pour ϵ_6 la valeur de 100.10^6 et $\sigma_{\epsilon6} = 3.10^6$, valeurs moyennes observées pour l'ensemble des études, alors

$$CV = 3.10^{-2}$$

Cette valeur n'a pas d'intérêt en soi, mais elle a son "poids" lorsqu'il s'agit de mettre en évidence des différences de comportement relatives à des variations de teneur en liant en énergie de compactage.

En effet, avec cette faible valeur il est possible de distinguer de façon significative deux ϵ_6 correspondant soit à une variation de 0,3 % de teneur en liant, soit à une variation de 2 % de teneur en vides consécutifs à une variation d'énergie de compactage.

Ces valeurs sont les conséquences de l'interprétation statistique de résultats obtenus dans le cadre du plan pluriannuel qui a fait l'objet d'une publication à EUROBITUME 89.

Sans discuter la pertinence des autres essais possibles en fatigue, faire un choix d'un type d'essai de fatigue reviendra à choisir celui qui a le meilleur CV.

Mais si cela peut permettre de diminuer le nombre de répétitions, un autre aspect tout aussi déterminant pour le choix d'un essai de fatigue doit être abordé, c'est celui relatif à la reproductibilité.

3   L'essai LPC est-il reproductible ?

Pour parler de reproductibilité, il faut avoir plusieurs implantations distinctes de matériels d'essais.

C'est le cas pour le matériel type LPC, puisque 7 laboratoires différents en possèdent.

La maitrise de la reproductibilité nécessite une parfaite connaissance de l'essai, c'est à dire des pièges relatifs aux paramètres influents mais non répertoriés, donc bien souvent non contrôlés.

C'est ainsi que la qualité de l'encastrement des masses sismiques sur lesquelles sont vissées les éprouvettes ont eu une influence certaine dans le passé, ajoutant du désordre dans la structure des résultats et empêchant ainsi des interprétations statistiques convenables.

Ce problème, ainsi qu'un certain nombre d'autres relatifs à la métrologie (calibrage des capteurs et d'enceintes thermiques) réglés, le désaccord entre deux laboratoires distincts est devenue imperceptible.

Pour permettre à chacun de juger du résultat obtenu, voici un exemple récent : celui relatif au transfert des chaines de fatigue du LCPC entreposées à ORLY.

Suite à la fermeture du laboratoire d'ORLY, le laboratoire de NANTES a acceuilli un matériel devenu **naturellement** vétuste.

Après un remontage et révision des enceintes thermiques et du matériel, suivi d'un calibrage systématique des capteurs et d'un étalonnage des enceintes, un essai croisé avec le laboratoire de référence, celui d'ANGERS, a été exécuté.

Des lots d'éprouvettes, aussi identiques que possible (même formulation, même énergie de compactage) ont été distribués entre les deux implantations. Après passage de ceux-ci sur les chaines de fatigue respectives et dépouillement on constate l'existence d'un excellent accord entre les laboratoires. La figure suivante, où on a reporté les deux droites de fatigue sur un même support, le concrétise.

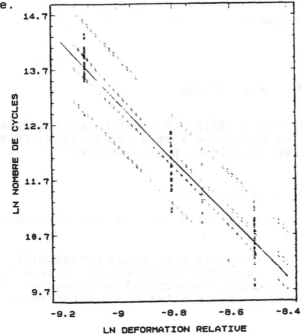

L'analyse statistique confirme l'identité des résultats obtenus comme le montre les éléments de l'analyse (coefficients des variables qualitatives de séparations significativement nuls).

Formule testée F.201 0/14 SG 5,4 % ESSO PORT JEROME

modèle testé :

$$Ln\ Nc = A_0 + A_1.Ln\ Eps + A_2. + A_3.Ln\ Eps.$$

avec   = 0   pour données en provenance de
                    Nantes

      = 1   pour données en provenance
                    d'Angers

   Eps = ε
   Nc = Nombre de cycles

Coefficient	Valeurs	t
A_0	-35.12	-17.5
A_1	- 5.36	-23.6
A_2	0.26	0.08
A_3	0.02	0.08

Conclusion analyse :

$\hat{A}_2 = 0$

$\hat{A}_3 = 0$

$\varepsilon_6 = 1,09.10^{-4}$ p = - 0,186

Cet état de fait a déjà été confirmé antérieurement, ainsi peut on affirmer que l'essai de fatigue LPC est parfaitement reproductible, donc maîtrisé.

4 Difficultés de mise en oeuvre

Trois difficultés majeures sont à vaincre pour exécuter un essai de fatigue :

- **la première, relative à la bonne homogénéité des lots d'éprouvettes,** homogénéité géométrique afin d'éviter les réglages incessants, homogénéité de teneur en vides pour limiter les dispersions inter éprouvettes excessives,

- la seconde, relative à la qualité des opérations
nécessaires au maintien des performances du matériel
(raccordement aux étalons primaires pour les capteurs,
vérification des ancrages d'éprouvettes, entretien des
pièces de liaison, etc ...),
- la troisième, relative à la qualité des réglages des
paramètres imposés (déformation - température), à la
qualité du collage et du montage des éprouvettes, ainsi
qu'à la qualité du pré-conditionnement dont la durée
doit être suffisante.

Par ailleurs, comme le nombre des éprouvettes testées
est important, l'exécution de l'essai nécessite un repérage
et un suivi minutieux des éprouvettes, sachant qu'un temps
contrôlé de murissement après sciage a été jugé jusque là
nécessaire..
Est-ce que ces difficultés en font **un essai plus
difficile que les autres ?**
Tout d'abord, **tout essai bien fait nécessite de résoudre
les mêmes types de difficultés.**
D'autre part, l'essaimage de l'essai nous a permis de
constater, qu'après une formation pratique d'une semaine et
une prise en main d'environ un mois, un technicien était
capable de maitriser parfaitement le matériel, l'essai et
l'exploitation.
**On ne peut donc plus considérer que c'est un essai
d'expert.**

5 Le coût de l'essai

Dans sa version actuelle, l'essai de fatigue nécessite le
passage de 36 éprouvettes partagées en trois groupes
identiques, chaque groupe étant sollicité à un niveau de
déformation donné.
Pour avoir un lot homogène de 36 éprouvettes, la
réalisation de 80 éprouvettes trapézoïdales, à partir de
plaques 600 x 400 x 120, compactées au compacteur de
plaque, est nécessaire.
Cette version lourde de l'essai, retenue pour améliorer
la connaissance de ϵ_6 a nécessité l'emploi de machines
d'essais à têtes multiples, par exemple 4 batis à 2 têtes
ou 2 batis à 4 têtes, enfermées dans 2 enceintes à
température contrôlée au 1/10e de degré pour la valeur
moyenne et à quelque dizième de degré près pour la
régulation.
Avec les matériels annexes, banc de collage, banc de
sciage pour élaborer les éprouvettes, compacteurs de
plaques, malaxeurs et étuves, l'investissement initial peut
sembler très important, ce qui laisse sous entendre, que
seul des laboratoires importants peuvent se permettre
d'exécuter un tel essai.

Or, à l'origine, quand les matériels ont été installés, le rendement dans cette configuration était estimé à 10 études par an.

Après une période de rodage, celui-ci passa à 30 études par an, ce qui rentabilisa d'autant l'investissement.

Le coût d'un essai lourd pu alors être recherché, et à cette date (début 90), amortissement compris, celui-ci est de 40 000 F H.T. dont 5 000 F pour la fabrication des plaques et 5 000 F pour le sciage.

Ce coût a été établi sur la base d'un amortissement portant sur 100 essais lourds.

Il est effectivement dissuasif <u>mais</u> l'information en retour, suite à la réalisation des essais du plan pluriannuel nous a permis d'exécuter avec succès des essais allégés pour certaines études (effet température par exemple).

Les considérations relatives à la reproductibilité et le caractère discriminant affirmé de cet essai permettent maintenant d'envisager la réalisation d'essais allégés. Dans ce cas, le nombre d'essais élémentaires serait réduit de moitié * et le coût total tomberait à 20 000 F pour 100 essais élémentaires.

La capacité d'un ensemble comprenant un bloc à deux têtes, une étuve, un banc de collage, un enregistreur deux voies et divers électroniques serait de 15 études par an pour un investissement initial de 600 000 F H.T. environ (hors moyen d'élaboration des éprouvettes).

Ces investissements sont très acceptables vis à vis de l'achat d'une presse dynamique pour faire de la fatigue en traction compression axiale ou diamétrale par exemple, ce qui n'exclu pas l'achat d'une enceinte thermique - et les moyens de préparation des éprouvettes -

Une expérience de fatigue à déformation imposée est actuellement à l'étude sur une presse de traction compression industrielle de charge nominale 100 kN : pour l'instant, la mise au point du matériel bute sur le problème de l'asservissement à déformation imposée (résolution de la mesure et vibration sont "contradictoires").

Avec les presses industrielles, essentiellement conçues autour de l'étude des produits sidérurgiques, l'adaptation aux essais sur matériaux viscoélastiques en travaillant à déformation imposée pose de nombreux problèmes. C'est peut-être pour cela que l'on préfère faire, à l'aide de ces machines, des sollicitations à force imposée pour laquelle elles sont, de façon évidente, adaptées.

---

* justification en annexe 4 : études de l'incertitude sur ϵ_6 (a = 95 %) lorsque le nombre de répétitions croît.

6 Le choix des modalités d'essai et l'effet exothermique

Tout matériau viscoélastique soumis à un train d'ondes sinusoïdales est exothermique.

L'énergie dégagée au cours d'un cycle par celui-ci est proportionnel aux amplitudes de la force F_0 et de la flèche f_0 ainsi qu'au sin ϕ de l'angle ϕ de déphasage.

$$Q = K.F_0.f_0.\sin \phi.$$

Les énergies mises en jeu évoluent dans le temps suivant l'évolution de F_0, f_0 et ϕ.

De plus, plus une géométrie d'éprouvettes conduit à des charges importantes pour une flèche f_0 donnée, plus l'énergie dissipée est importante.

Donc, le choix d'une forme d'éprouvettes et d'un mode de sollicitation ne sera pas sans conséquences sur le résultat de l'essai et sans interaction avec le critère de rupture.

Par exemple :

<u>à charge imposée</u>

f_0 croit au cours de l'essai, de même que ϕ.

L'énergie dissipée ne fait que croître et celle-ci engendre une diminution du module et une augmentation de ϕ.

Ce type de déséquilibre perturbe sérieusement l'évolution de f_0 en fonction du temps. De plus, l'efficacité de l'enceinte risque d'avoir un rôle à jouer puisque les échanges thermiques seront fonction de la convection en surface.

Donc stabilité ? ou instabilité à charge imposée ? Validité des résultats dans ce cas ?

<u>à déformation imposée</u>

f_0 est constant, l'énergie fait chuter le module, donc F_0 diminue et comme la température θ augmente, ϕ augmente. On montre dans ce cas qu'en l'absence d'endommagement au sein de l'éprouvette celle-ci tend vers un équilibre thermique.

Il faut donc imaginer l'existence d'un processus d'endommagement pour expliquer en partie la forme de F_0 en fonction du temps (Réf. RILEM page 1155, CHAPMAN AND HALL ENDOMMAGEMENT PAR FATIGUE DES ENROBES BITUMINEUX)

Ce qui ne veut pas dire que l'évolution de F_0 en fonction du temps est indépendante de l'effet thermique, mais certainement plus que l'évolution de f_0 dans l'essai à charge imposée.

En clair, pour exécuter un essai de fatigue, il faut choisir la forme d'éprouvette limitant les effets thermiques et choisir le mode de sollicitation qui masque le moins le processus de dégradation (quand cela est possible).

Il semble que les processus de fatigue LPC à déformation imposée répondent relativement bien à ces exigences.

Néanmoins, ceci ne peut constituer un obstacle au développement d'autres moyens ou types d'essais sachant qu'il est difficile actuellement de mesurer la "**pertinence**" de cet essai dans ce domaine.

7 La représentativité des éprouvettes

Un résultat d'essai n'a d'intérêt que si l'échantillon testé est représentatif du matériau qui est mis ou sera mis en oeuvre.

Il sera donc utile de disposer en laboratoire d'un moyen de fabrication et de moulage qui confère au produit final les propriétés du matériau tel qu'il est fabriqué et mis en oeuvre sur chantier.

Les LPC ont développé un moyen de moulage spécifique utilisant le compactage pneumatique.

Ce moyen de moulage est inspiré d'une idée émise par le laboratoire ESSO Mont Saint-Aignan près de Rouen.

Il confère aux mélanges les propriétés recherchées :

- représentativité,
- homogénéité des plaques et des éprouvettes.

De même, le moyen de fabrication, un malaxeur à vis épicycloïdale donne toute satisfaction tant que le matériau in situ est malaxé à l'aide de centrales continues ou discontinues traditionnelles.

8 Conclusions

Sans se poser la question de la pertinence de l'essai de fatigue, un tour d'horizon sur la qualité de l'essai de fatigue LPC vient d'être fait.

On constate que ce dernier présente :

- une bonne reproductibilité,
- un caractère discriminant accentué,
- un coût unitaire modeste compte tenu de l'information apportée, sachant qu'il nécessite la réalisation d'un nombre non négligeable d'essais élémentaires,
- un investissement acceptable,
- un effet exothermique faible,
- une difficulté d'exécution comparable à celle des autres essais de fatigue (ni plus, ni moins).

Cet état de faits, associé à l'expérience étendue de cet essai en France (plusieurs centaines d'études complètes à l'intérieur et à l'extérieur du réseau des LPC) en font un essai vulgarisable.

ANNEXE 1

DESCRIPTION DE L'ESSAI DE FATIGUE LPC

MOYEN D'ESSAI

f : flèche en tête
F : Force en tête

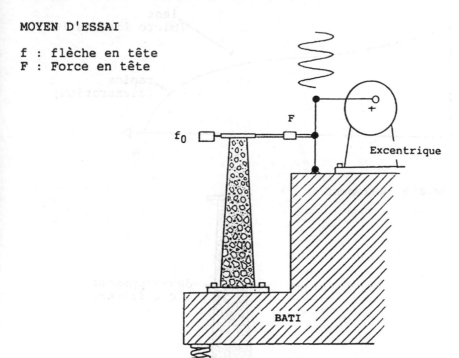

NATURE ESSAI

Essai de flexion à déformation Imposée. Signal sinusoïdal
25 HZ
Eprouvette consolé trapézoïdale (B = 56, b= 25, h = 250, e
= 25)

On impose

$$f = f_0 \cos \omega t$$

On mesure

$$F = F_0 \cos (\omega t + \varphi)$$
$$0 < \varphi < 90$$

On constate : Endommagement

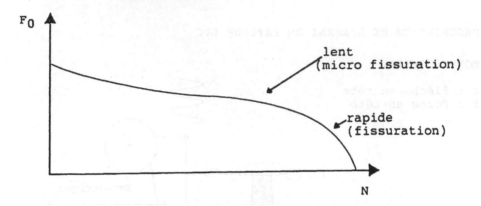

On constate également

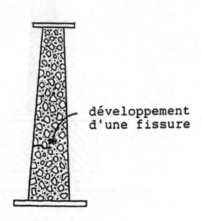

développement
d'une fissure

On est en présence d'un essai de rupture différée
c'est celà un essai de <u>Fatigue</u>

Critère de rupture adopté par LPC

Une éprouvette est "rompue" lorsque sa charge en tête a
diminué de moitié.

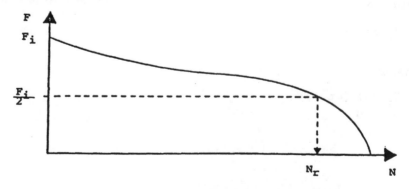

Durée de vie d'une éprouvette

c'est le nombre de cycles N_r pour lequel

$$F = \frac{F_i}{2}$$

DROITE DE FATIGUE

a/ quand on fait n essais sur n éprouvettes différentes
avec une flèche f_o donnée on obtient cette distribution:

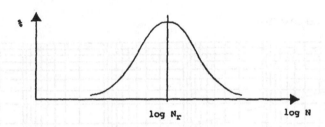

Les durées de vie sont distribuées suivant une loi Log
Normale.

b/ Si on répète, sur d'autres éprouvettes ce type d'essai
avec des flèches f_0, f_1, f_2 il est possible de tracer la
droite de fatigue :

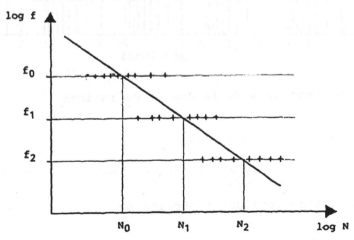

Elle est obtenue par Régression

A chaque flèche correspond un ε_{max} donné par :

$\varepsilon_{max} = K_\varepsilon . f$

K_ε étant un facteur dépendant de la géométrie.

On peut donc substituer f par ε_{max} dans la figure précédente :
On a alors la droite de fatigue LPC.

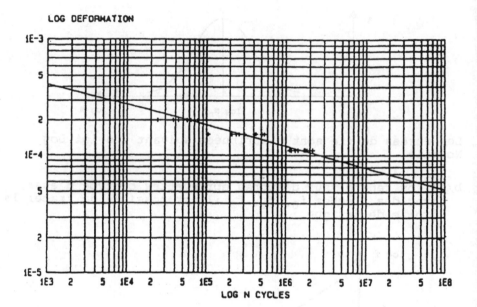

Paramètres importants de la droite de fatigue

<u>pente</u> : voisine de - 0.2 (notée p)

$\underline{\varepsilon_6}$: déformation relative entrainant une durée de vie moyenne de 10^6 cycles.

$\sigma_{x/y}$ = écart type résiduel (fonction de ε_6)

$\Delta\varepsilon_6$ = intervalle de confiance sur ε_6
$$\left[4.10^6 < \Delta\varepsilon_6 < 10.10^{-5} \right]$$

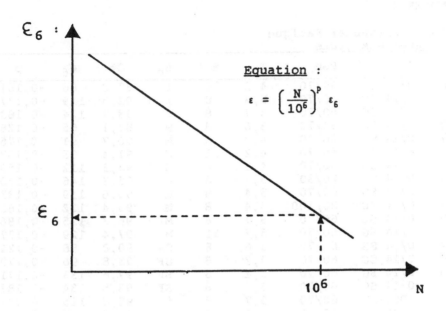

Equation :

$$\varepsilon = \left(\frac{N}{10^6}\right)^p \varepsilon_6$$

Moyen d'essais LPC

ANNEXE 2

Plan pluriannuel Fatigue
1   Résultats bruts

NF	G	Pen	TL	% f	E_C	C%	ϵ_6	p
38	0/14 SG	60/70	4,2	8	f	87,2	86	-0,181
27	0/14 SG	60/70	5,4	8	f	92,8	119	-0,177
30	0/14 SG	60/70	5,7	8	f	93,7	124	-0,183
43	0/14 SG	60/70	3,6	8	M	87,1	66	-0,128
17	0/14 SG	60/70	4,2	8	M	90,7	91	-0,176
65	0/14 SG	60/70	4,2	11	M	91,4	85	-0,167
45	0/14 SG	60/70	4,8	8	M	93,3	112	-0,169
21	0/14 SG	20/30	5,4	8	M	95,3	146	-0,170
20	0/14 SG	60/70	5,4	8	M	95,6	119	-0,183
22	0/14 SG	80/100	5,4	8	M	95,4	132	-0,185
29	0/14 SG	60/70	5,7	8	M	96,3	125	-0,186
64	0/14 SG	60/70	5,7	11	M	97,4	129	-0,187
44	0/14 SG	60/70	3,6	8	SF	90,0	88	-0,124
39	0/14 SG	60/70	4,2	8	SF	93,8	96	-0,179
28	0/14 SG	60/70	5,4	8	SF	97,6	133	-0,171
33	0/14 SG	60/70	5,7	8	SF	98,5	134	-0,187
51	DCT 1	60/70	5,7	8	f	95,8	116	-0,206
87	DCT 1	60/70	3,6	8	M	88,7	57	-0,159
47	DCT 1	60/70	4,8	8	M	94,8	98	-0,173
46	DCT 1	60/70	5,4	8	M	96,7	117	-0,184
48	DCT 1	60/70	5,7	8	M	97,8	126	-0,192
86	DCT 1	60/70	3,6	8	SF	91,5	69	-0,157
52	DCT 1	60/70	4,8	8	SF	97,6	115	-0,193
59	DCT 2	60/70	4,8	11	f	92,1	95	-0,181
54	DCT 2	60/70	5,4	11	f	95,5	114	-0,181
68	DCT 2	60/70	5,4	8	f	93,4	117	-0,193
53	DCT 2	60/70	5,7	11	f	96,7	126	-0,185
60	DCT 2	60/70	5,4	11	M	97,6	122	-0,198
61	DCT 2	60/70	5,7	11	M	97,8	132	-0,181
69	0/14 D	60/70	4,2	4,6	f	83,3	90	-0,164
72	0/14 D	60/70	3,6	4,2	SF	83,6	77	-0,151

NF = Numéro de Formule
G   = Granulométrie
Pen = Classe Pénétration bitume
TL = Teneur en liant (pondérale par rapport à la masse des granulats)
% f = Pourcentage de fines (< 0,08)
E_C = Energie de compactage
    f : Compactage faible
    M : Compactage moyen
    SF : Compactage fort
C% = Compacité (C% = mva/mvr x 100)
ϵ_6 = Déformation relative maximale entrainant une durée de vie moyenne de 10^6 cycles (en 10^{-6})
p = Pente de la droite de fatigue ($\ln \epsilon = f(\ln N)$).

2 Interprétation statistique

2.1 Signification des paramètres

TL : Teneur en bitume pondérale pour des granulats de masse volumique 2,85. Il conviendra de faire une correction si la masse volumique des granulats est différente de 2,85.

ΔC : Ecart de compacité consécutif à une variation d'énergie de compactage par rapport à une énergie de référence dite moyenne. L'écart de compacité mesure la différence entre la compacité obtenue ou espérée par le matériau sur chantier (C) et celle qu'il obtient avec le compacteur de plaques pour le compactage moyen (C_m) ou avec la PCG pour 80 girations (C_{80}).

$$C = C - C_m$$
$$C = C - C_{80}$$

BE:Variable qualitative prenant la valeur 1 pour le BBCM type 1 et 0 pour les autres enrobés.

AL:Variable qualitative prenant la valeur 1 pour le BBCM type 2, 0 pour les autres.

BBCM : Béton bitumineux pour couche mince.

La projection qui suit donne un aperçu de la qualité de la régression (distance "valeur réelle, nappe explicative" en gras) ainsi qu'une idée de la façon dont interviennent les facteurs explicatifs TL et ΔC (effet parabolique de TL important, effet linéaire de ΔC faible).

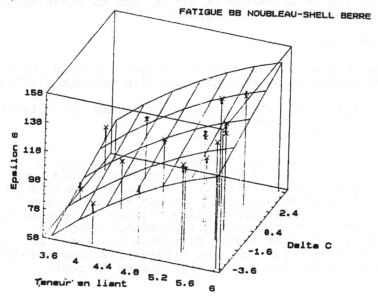

FATIGUE BB NOUBLEAU-SHELL BERRE

2.2 Régressions relatives à la fatigue

La première donne la déformation admissible pour obtenir une durée de vie moyenne de 10^6 cycles (ϵ_6)

$$\hat{\epsilon}_6 = (-125 + 72.TL - 4{,}85.TL^2 + 3{,}3. C - 29.BE + 4{,}4.BE.TL).10^{-6}$$

Intervalle de confiance correspondant :

$$\hat{\epsilon}_6 - 8.10^{-6} < \epsilon_6 < \hat{\epsilon}_6 + 8.10^{-6} \text{ pour } \alpha = 95 \text{ \%}$$

Les suivantes donnent :
- l'estimation de la pente de droites de fatigue

$\hat{p} = -0{,}18$ en coordonnée logarithmique avec ϵ en ordonnée.

$$\hat{p} - 0{,}02 < p < \hat{p} + 0{,}02 \qquad \text{pour } \alpha = 95 \text{ \%}$$

L'estimation de la dispersion résiduelle :

$$\hat{\sigma}_{y/x} = 1{,}576 - 9400 \hat{\epsilon}_6$$

ANNEXE 3

INFLUENCE DE LA NATURE DU BITUME SUR LA RESISTANCE A LA FATIGUE A DEFORMATION IMPOSEE

Historique

Dans un plan pluriannuel de fatigue (3) on a testé avec un même bitume et un même granulat des enrobés différant par :

- leur formulation (teneur en bitume, granularité, courbe granulométrique),
- leur niveau de compactage.

La comparaison de ces résultats avec ceux obtenus lors de caractérisations de matériaux de chaussées quelconques a permis la mise en évidence d'un facteur "NATURE DU BITUME".

A la même époque, quelques-uns de ces bitumes avaient fait l'objet d'études physico-chimiques, essentiellement par Chromatographie sur gel perméable (GPC), et rhéologiques, par mesure du module complexe. On a pu s'apercevoir ainsi que les bitumes les plus structurés étaient aussi les plus résistants en fatigue.

Résultats

Caractéristiques des liants après RTFOT

bitume n°	classe	BA (T°)	p.à 25 °C	IP Pf	FRAASS (T°)	écart -type	Fatigue Eps 6
02	40/50	60	24	-0.5	- 2	4.7	123
03	40/50	53	39	-1.0	- 8	4.6	105
04	60/70	52.5	39	-1.1	- 5	4.1	111
05	60/70	59	37	0.1	- 9	5.0	142
06	60/70	74.5	23	1.7	-10	7.7	150
07	60/70	59	32	-0.2	- 7	4.9	134
08	60/70	53	36	-1.2	- 6	3.8	100
09	60/70	54	35	-1.0	- 8	4.0	103
10	60/70	56	35	-0.6	- 7	4.1	87

Classement par rang des liants après RTFOT

bitume n°	classe	BA (T°)	p.à 25 °C	IP Pf	FRAASS (T°)	écart -type	Fatigue Eps 6
02	40/50	2	2	4	9	4	4
03	40/50	7.5	8.5	6.5	3.5	5	6
04	60/70	9	8.5	8	8	6.5	5
05	60/70	3.5	7	2	2	2	2
06	60/70	1	1	1	1	1	1
07	60/70	3.5	3	3	5.5	3	3
08	60/70	7.5	6	9	7	9	8
09	60/70	6	4.5	6.5	3.5	8	7
10	60/70	5	4.5	5	5.5	6.5	9

La meilleure corrélation entre ϵ_6 et un paramètre caractérisant le bitume est celle utilisant l'écart type du spectre de relaxation (modèle JONGPIER et KUILMAN).

Etude de l'incertitude sur ϵ_6 lorsque le nombre de répétitions croît

Plan pluriannuel   5.7 % de bitume
Formule n° 29

Nombre d'éprouvettes par niveau de déformation	RESULTATS EN FATIGUE				
	origine x/y	pente x/y	Ecart type x/y	ϵ_6 Mdefs	Incertitude α=95 %
2	-39.87	-6.005	.54	131	+/- 13
4	-32.92	-5.204	.48	126	+/- 9
6	-33.56	-5.274	.44	126	+/- 6
8	-33.82	-5.306	.41	126	+/- 5
10	-34,20	-5.346	.39	126	+/- 4
12	-34,59	-5.388	.42	125	+/- 4

Plan pluriannuel   3.6 % de bitume
Formule n° 43

Nombre d'éprouvettes par niveau de déformation	RESULTATS EN FATIGUE				
	origine x/y	pente x/y	Ecart type x/y	ϵ_6 Mdefs	Incertitude α=95 %
2	-55.99	-7.142	.77	57	+/- 9
4	-56.65	-7.263	1.17	61	+/- 8
6	-61.08	-7.754	1.05	64	+/- 5
8	-56.41	-7.270	1.01	64	+/- 4
10	-55.61	-7.190	.94	64	+/- 4
12	-61.37	-7.813	.97	66	+/- 3

Plan pluriannuel   4.8 % de bitume
Formule n° 45

Nombre d'éprouvettes par niveau de déformation	RESULTATS EN FATIGUE				
	origine x/y	pente x/y	Ecart type x/y	ϵ_6 Mdefs	Incertitude α=95 %
2	-44.36	-6.405	.55	114	+/- 15
4	-42.67	-6.218	.50	114	+/- 9
6	-42.07	-6.143	.52	112	+/- 7
8	-39.42	-5.837	.55	110	+/- 7
10	-38.83	-5.782	.55	111	+/- 6
12	-40.16	-5.932	.56	112	+/- 6

Commentaires : pour 6 répétitions, l'incertitude est acceptable

38 THE COMPLEX MECHANICAL INVESTIGATION SYSTEM OF ASPHALTS AT THE TECHNICAL UNIVERSITY, BUDAPEST

E. NEMESDY, K. AMBRUS, I. PALLÓS
and K. TÖRÖK
Department of Road Engineering, TUB, Hungary

Abstract
The conditions of applying a series of asphalt mechanical investigations have been established in the Laboratory of the Department of Road Engineering, Technical University Budapest. The basic idea of the investigation system is the investigation of the asphalt in three important temperature ranges in accordance with real climatic conditions during the endurance of the road with different methods taking the viscoelastic character of the asphalt into consideration. As experiences show this type of mechanical investigation system points out quite well which characteristics are improved and which are deteriorated by the chosen grain size distribution, the quality of the mineral material, the type of the bitumen, the magnitude of its adaptation and the degree of packing.
Keywords:
Complex asphalt investigating system, static and dynamic tests, test methods, important temperature ranges.

1 Introduction

The beginning of the laboratory dealing with bitumen and asphalt at the Department of Road Engineering, Technical University, Budapest dates back to 1950. In the first two decades of the four decade existence till 1990 the laboratory carried out research first of all with bitumen binding material. Then the results tried to help practical road construction by improving the adhesion of cutback bitumen and cationactive bitumen emulsion. The research and practical work in the latest two decades was mainly related to the testing of asphalt materials. The laboratory carried

out significant work and gained wide ranging experiences
designing the asphalt materials of the Hungarian motorways
and main roads and controlling quality. The experiences
obtained in accordance with foreign development have shown
that the Marshall test in itself cannot judge the general
behaviour and quality of asphalt. To realize it a kind of
mechanical test series chosen adequately seemed to be
necessary that in the majority of cases were carried out on
Marshall specimens (rolls) made of asphalt to be tested and
on prism shaped specimens of identical body density. Three
groups of mechanical tests have been developed:

   A. Tests to characterize the hot behaviour of asphalts
      A1. Marshall test at +60°C
      A2. Dynamic creep test at +40°C
      A3. Dynamic wheel tracking test at +45°C
   B. Tests to characterize the fatigue of asphalt at
      medium-cool temperature
      B1. Two directional cambering-bending fatigue test
          at +5°C
      B2. One directional bending-fatigue test at +10°C
   C. Tests to characterize the cold behaviour of asphalt
      C1. Static splitting test series between +5°C
          and -20°C
      C2. Expansion (shrinkage) factor measurement between
          +5°C and -20°C.

The behaviour of the asphalt to be tested can be
described in details by the above investigations in three
characteristic temperature ranges. It is characteristic on
the one hand of the deformation resistance of asphalt in
summer heat, on the other hand of fatigue resistance between
average temperature and melting temperature, at last it is
characteristic of the so-called cold behaviour of asphalt
that detects cracking sensitivity in wintertime.

The complex test series results in several testing
parameters. A certain part of them characterizes all the
advantageous and disadvantageous features of the asphalt
tested well.

In the last one and half decades in the Road Laboratory
of BME the above mentioned mechanical asphalt tests have
been developed not only for routine tests but have been
transformed into a coherent testing system in the listed
composition. In the latest decade almost all the versions of
the Hungarian standard asphalt of characteristic composition
have been tested by this complex mechanical asphalt testing
system. Therefore an experience data basis is at our
disposal where characteristic mechanical parameter groups in
case of asphalt types well known on the basis of practical
behaviour are at our disposal. Thus the data of a testing
parameter group obtained at a new asphalt to be tested have
immediate good, comparative evaluation if they are fit into
the data table collected so far. It becomes immediately
clear how the asphalt to be tested will behave in great
summer heat, what deformation resistance can be expected;

how it will behave in winter in freezing periods, how great cracking sensitivity can be expected at last how great fatigue resistance can be expected to repeated load, all year traffic.

Though the majority of tests is quite well known the complex asphalt testing method with the necessary supplements introduced here, to our best knowledge, has been unknown among the evaluation of asphalts.

The comparison of data obtained, fitting them into the data basis can judge the expected quality, advantageous and disadvantageous qualities of the material, can show whether the technological compromises applied at design are satisfactory, successful, doubtful or perhaps are to be improved. This thorough test is by far more reliable than e.g. the old assessment based only on Marshall test. The main advantage is the possibility of conscious compromise in all the three characteristic temperature ranges.

The application of the complex mechanical asphalt testing system itself has become especially useful and necessary since the widespread practical introduction and application of asphalt made with modified bitumen.

2 Testing the hot behaviour of asphalts

After trying several well-known tests the ones separating the different materials only slightly (e.g. static creep test) or the time-consuming and complicated ones (e.g. triaxial asphalt test) were neglected In this way three characteristic tests remained.

A1. Marshall test at +60°C
This test is generally known. Under restricted congitions it gives comparative data characteristic only of hot behaviour. It has been left in our system as a link with older, classical asphalt test.

A2. Dynamic creep test at +40°C
The essence of the test is in Figure 1 with characteristic details. An electronically controlled hydraulic pulsator of MTS make has been transformed into fatigue-loading equipment and fitted with air-conditioning plant. Making +40°C constant a Marshall specimen is exposed to uniaxial, repeated compressive load with 4 Hz frequency dynamic load of sinus shape. The maximum pressure by periods is 0.4 or 0.6 MN/m^2 (4 or 6 bar). A characteristic permanent compression-repetition number curve $\varepsilon_p\%$ n is obtained as result, it is in Figure 1. As experiences show in the middle of the drawn, fixed strain line the more and more slowly increasing residual compressions after an inflexion point have become more and more quickly increasing compressions. The endurance-repetition number N_k according to our definition is the inflexion point of the curve that can be

stated by calculation without subjective judgement from the third degree polynom describing the testing curve with quite good correction. The place of the inflexion point can be calculated accurately $(\varepsilon_k - N_k)$. Two characteristic parameters are calculated in the inflexion point:

$\dfrac{d\varepsilon}{dn} \cdot 10^6$: Slope of the curve point as the speed characterizing failure

$\dfrac{nk}{\varepsilon_k \%}$: Parameter characterizing hot behaviour; higher value is more favourable from the viewpoint of road construction

On average this hot fatigue testing takes a half day shift it means favourable operation conditions. The data separate the asphalt types characteristically.

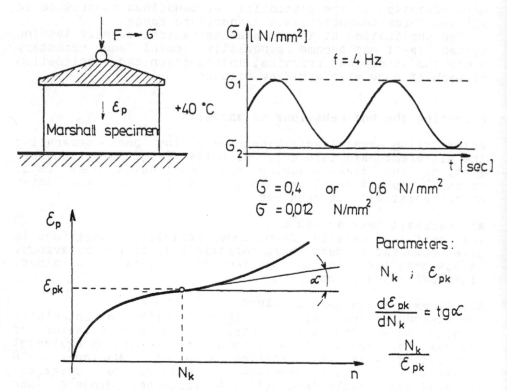

Fig.1. Dynamics creep test at +40°C on Marshall specimen

A3. Dynamic wheel tracking test
The test is carried out by the modification of the English Wheel Tracking Test Instrument according to Figure 2. In the air conditioning case keeping 45°C, on the asphalt board of 300x300x40 mm packed by vibration a solid rubber tyred loading wheel with dimension in the figure with load 0.4

N/mm² rolls to and fro on 200 mm road length at 0.3 Hz frequency in 3 hours. The developing track is scanned by an electronic transducer and it draws longitudinal section six times. After three hours average rut depth mm and the volume of the developing rut mm can be determined; they are two characteristic parameters.

These three tests according to experience characterize the hot behaviour of the asphalt well and separate the effects of different asphalts and binding materials quite well.

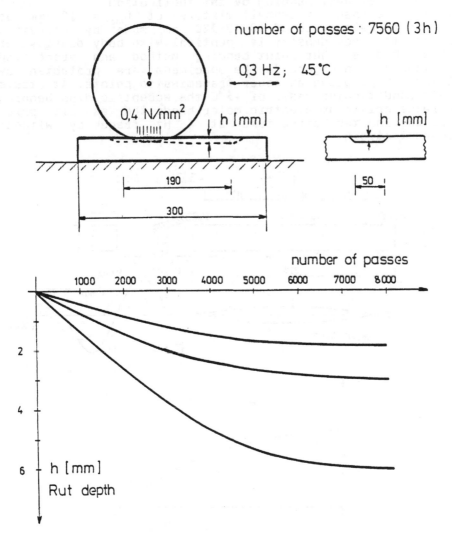

Fig.2. Test of dynamic track formation at +45°C on asphalt sample

3 Characterization of asphalts at medium-cool temperature

Due to their nature fatigue tests are time consuming and show significant standard deviation. It had to be decided that the aim of endurance test at medium temperature is the relative comparison of asphalts from this viewpoint thus data collection for dimensioning to determine tensile elongation and Wöhler curves is possible only by one of the applied tests.

B1. Two directional bending by set inclination
This test is ussed at asphalt mixture of $D_{max} = 12$ mm or finer on asphalt prism of 40x40x320 mm made by vibration packing, the body density is identical with body density of Marshall bodies. Four point bending method and processing relations are in Figure 3. The specimens are protected by metal stirrups glued at power transmission points. At tests in air conditioning cases of $+5^{0}$C the eccentric disc bending apparatus driven by electric motor bends the asphalt prism specimen in two directions at 4 Hz frequency without permanent strain.

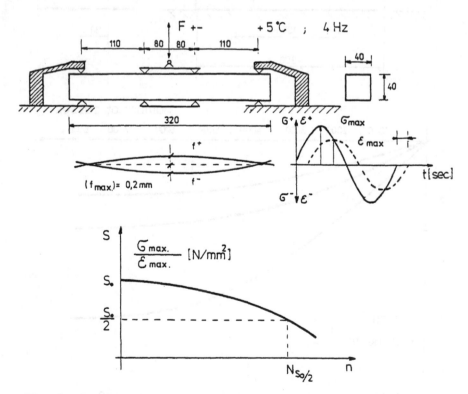

Fig.3. Fatigue by two directional bending with set inclination at asphalt prism

Preset two directional bending limit of s = ±0.20 mm is characteristic of the load. During the load repetition of the fatigue test the rigidity of the specimen gradually decreases and the specimen reaches 0.20 mm set inclination only in case of fracture. Before that however the changing magnitude of loading force on the specimen F [N] and that of f mm inclination together with the repetition numbers $\underline{n}$ belonging to them are recorded in electric way. If the rigidity modulus of the beam specimen continuously decreases from the initial modulus value S_o MN/m^2 = $\sigma_{maxo}/\varepsilon_{maxo}$ to $S_o/2$ during fatigue the international agreement valid for this kind of bending regards repetition number $N_{So/2}$ belonging to $S_o/2$ as fatigue endurance.

B2. One directional inclination with constant stress
The main data of the other bending-fatigue test are in Figure 4. The loading adapter is in the air conditioning case of fatigue apparatus MTS, test is carried out at +10^{0}C. This time the asphalt specimen is a prism of 80x50x150, the body density is identical with that of the Marshall body. Load has effect only in the middle of the 100 mm span in one direction (three pointed bending). The constant bending force F_{max} calculated by formula in Figure 4. acts it may cause 2 N/mm^2 at +10^{0}C (perhaps 8 N/mm^2 at -5^{0}C) bending-tensile force with sine shaped control, 4 Hz frequency.

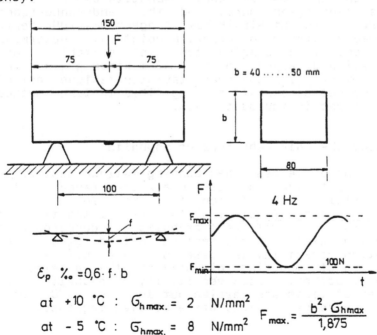

$$\varepsilon_p \text{ ‰} = 0.6 \cdot f \cdot b$$

at +10 °C : $\sigma_{h\,max.} = 2$ N/mm^2

at - 5 °C : $\sigma_{h\,max.} = 8$ N/mm^2

$$F_{max.} = \frac{b^2 \cdot \sigma_{h\,max}}{1.875}$$

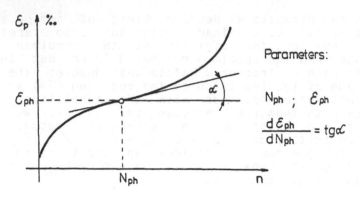

Fig.4. Fatigue by one directional inclination under constant tensile stress

The increasingly greater inclination is measured in the middle in an electronic way at constant loading force. Here plastic strains are accumulating and the permanent elongation ($\mathcal{E}_p$% - n) - repetition number curve is again of the same shape as in case of dynamic creep (fatigue test with uniaxial pressure) in Figure 1. The endurance-repetition number (N_h) here is also determined by the inflexion point of the measured strain curve. The testing curve of the specimen is again substituted by a cubical parabola regression curve and the endurance-repetition number N_h, permanent strain $\mathcal{E}_{ph}$, tangent $d\mathcal{E}_{ph}/dN_h$ derivative (the speed of strain in inflexion point) is determined by calculation. This bending test is suitable for the investigation of asphalt mixture of maximum grain size over D_{max} = 12 mm even for the characteristic fatigue test of core sample cut out of asphalt topping. This testing method has been developed in the past years.

4 Testing the cold behaviour of asphalts

Although the above described bending-fatigue test B2 is also suitable for fatigue in cold temperature range fatigue is not the critical load of asphalt in the cold range, it is the <u>crack causing effect</u> of winter thermal tensile stresses $\sigma_{term} = \alpha E(t_o - t_t)$. Different asphalts are compared here with different parameters. This testing method gives important information especially in case of hard bitumen and the application of modified bitumen on the possible winter behaviour, probable crack sensitivity of asphalts. In fact two types of tests are carried out in the $+5^\circ C$ and $-20^\circ C$ temperature range but they are evaluated together.

C1. Static splitting-tensile tests between +5°C and -20°C

At these tests standard Marshall specimens are applied that are tempered in water bath for 2 hours at +5°C, in freezing box for at least three hours at -5 and -20°C. The main characteristics and evaluating formulae of the splitting test that are known and appear in the Hungarian standard are in Figure 5. The loading adpater used in the 100 kN laboratory cursher is suitable not only for transferring load on the casing but for recording diameter increase perpendicular to load, increasing with pressing load, electronic transducer. In case of investigation the specimens here really split under splitting-tensile stresses caused by compressing force. Splitting-tensile strength σ_{Hh} , splitting modulus E_{Hh} and specific splitting elongation value $\varepsilon_{sz} = \sigma_{Hh}/E_{Hh}$ are determined at every test according to Figure 5 at three temperatures (+5, -5, -20°C).

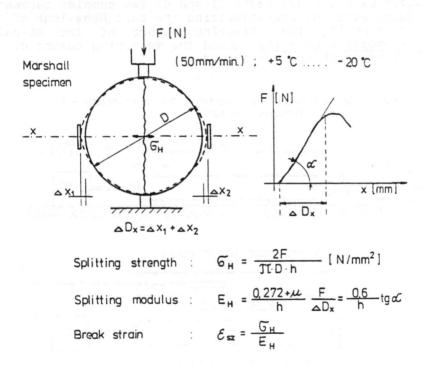

$$\Delta D_x = \Delta x_1 + \Delta x_2$$

Splitting strength : $\quad \sigma_H = \dfrac{2F}{\pi \cdot D \cdot h} \quad [\,N/mm^2\,]$

Splitting modulus : $\quad E_H = \dfrac{0.272 + \mu}{h} \dfrac{F}{\Delta D_x} = \dfrac{0.6}{h} tg\alpha$

Break strain : $\quad \varepsilon_{sz} = \dfrac{\sigma_H}{E_H}$

Fig.5. Static splitting tensile test on Marshall specimens at +5 and -20°C

C2. Measurement of the linear coefficient of thermal expansion of asphalt

To judge the crack sensitivity of asphalt the values of thermal shrinkage coefficients α are to be known in winter temperature range +5 and -20°C. To do this specimens are made in the form of prisms of 40x40x320 mm from the tested asphalt by vibration their body density should be equal to the density of the Marshall body. The two end plates of the prism are limited by glass plates glued on them.

There is constant $-30\pm1^{\circ}C$ in the testing space of the freezing box. Two asphalt specimens of identical material and $+20^{\circ}C$ temperature are put here. In one of them a temperature transducer is included (Figure 6.). The other one is a $\ell = 320$ mm asphalt prism on rollers and longitudinal change (thermal shrinkage) is given by an electronic transducer as a function of time. A regression polinom is put on the curve obtained from the value pairs of temperature differences $\Delta t^{\circ}C$ and shrinkage $\Delta\ell$ mm measured simultaneously and their constants are determined. From it the value series of linear thermal shrinkage coefficient $\alpha = \Delta\ell/\ell\Delta t$ is obtained depending on temperature between +5 and $-20^{\circ}C$.

After carrying out tests C1 and C2 two complex parameters are calculated for characterizing the cold behaviour of the tested asphalt, the measuring number of the so-called <u>cold-elongation capacity</u> $\mathcal{E}_H$ and the measuring number of the fictitious cracking temperature.

Dilatometer in refrigerator box where is the temperature $-30\pm1^{\circ}C$ remains constant

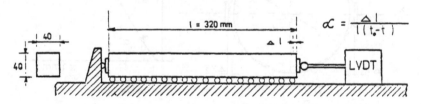

Measurement of temperature

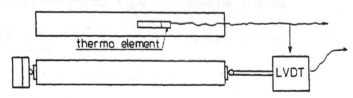

Measurement of dilatation

Fig.6. Determination of the linear expansion coefficient of asphalt prism specimen

The value of $\mathcal{E}_H$ is calculated on the basis of Figure 7 by the formula given there. This complex parameter in fact is the average inclination, incline of the obtained elongation of splitting depending on temperature multiplied by the value of splitting elongation measured at -20°C. The cold behaviour of asphalt is more favourable if the value of complex parameter $\mathcal{E}_H$ is greater for road construction.

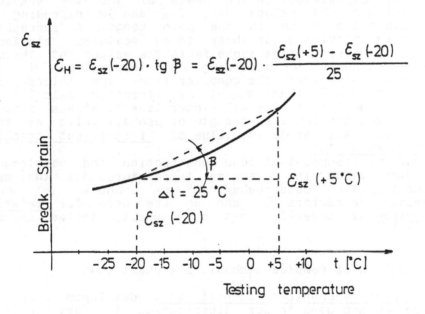

Fig.7. Interpretation of complex quality parameter, of "cold elongation capacity"

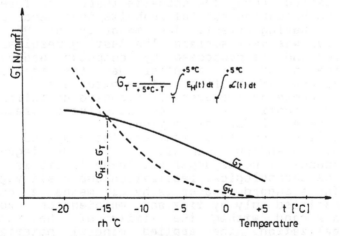

Fig.8. Determination principle of fictitious cracking temperature Rh °C at asphalt prism

Figure 8 shows the meaning and the principle of calculating the fictitious cracking temperature Rh °C. Let's assume that the relaxation of asphalt above +5°C is 100% and under it is 0% that is thermal stresses in cold range are thermally accumulated. On the basis of the two previous tests (Cl and C2) values σ_{Hh}, E'_{Hh} and α depending on temperature are known in the cold range. A regression polinom is calculated for these three measured relations. Also on the basis of the above facts the values of thermal tensile stresses σ_T depending on temperature T are calculated. Afterwards the computer finds the intersection point of curves σ_T-t and σ_{Hh}-t by iteration calculation. Then the magnitude of thermal winter tensile stress reaches then exceeds the tensile strength of asphalt valid at that temperature; thus it is the value of "frictitious cracking temperature Rh °C".

From the viewpoint of road construction and maintenance the lower value the cracking temperature has the more favourable the "cold behaviour" of asphalt is. Cold behaviour parameters ε_H and Rh are used for relative comparison of several types of asphalt, tested asphalt material.

5 Evaluation of complex mechanical asphalt test

The complex mechanical asphalt test developed almost a decade ago and used in our laboratories in fact contains five additional tests besides the usual Marsall test (as B1 and B2 are alternaative cases). In the past quite long period both the standard asphalts used for the construction and reconstruction of the Hungarian motorways and main roads and the materials of thin asphalt layers made from bitumen have been tested. Among the asphalts there were foreign wear surfaces and Hungarian special asphalts that were developed by companies having patent. The majority of 95 asphalts tested so far was wear surface. The testing results are in data carrier and are processed by computer programs from time to time in statistic or other way for the basis of comparative evaluation of the future materials.

Table I. contains the average, standard deviation values of mechanical testing results, characteristic parameters of asphalts type AB-12 for information.

In practice the fact whether the applied standard regulation, standard asphalt is advantageous or disadvantageous with its upper and lower limits is important for both the technological realization and quality guarantee as well. The standard however is by all means a compromise that is justified only by time and maintenance experience. On the other hand within the limits of the standard a certain realization, the applied mineral materials, the quality of the chosen bitumen, grain distribution realized sometimes within broad limits, voids volume of the mineral material, the bitumen replacement 5, the applied bitumen quantity and the free voids volume all significantly influence the expected quality and endurance of the asphalt within the standard limits, too.

Table I.

Test		Characteristics	AB-12 asphalts with standard road bitumen				
			n	mean	variance	min.	max.
		% bit. by wgt. of mix	23	6.26	0.38	5.5	7.2
		Voidless density: g/cm^3	24	2.468	0.038	2.402	2.558
		% air voids	24	3.45	1.20	1.4	6.5
		t_B: %	13	80.6	5.8	69.2	91.3
		Factor of compactibility C	6	30.5	5.89	25	38
		MS: kN	22	12.5	4.0	7.0	22.6
Dynamic creep test	p = 4 bar	N_k	26	18018	14256	2847	43680
		ε_k: ‰	26	25.5	13.8	9.8	75.0
		$10^6 \times d\varepsilon_k/dN_k$	16	573	730	8	2128
Wheel tracking		rut depth: mm	7	2.07	1.26	0.6	4.1
Two directional bending		S_o: N/mm^2	26	13133	1517	9250	15530
		$N_{So/2}$	26	64030	38237	6430	139700
One directional inclination	+10°C	N_h	6	7140	5060	3133	16703
		ε_h: ‰	6	10.8	3.4	8.0	16.6
		$10^6 \times d\varepsilon_h/dN_k$	6	1137	642	309	1730
Static splitting tensile test	+5°C	σ_{Hh}: N/mm^2	26	3.20	0.33	2.56	3.97
		E_H: N/mm^2	26	6835	1810	3646	10600
	−20°C	σ_{Hh}: N/mm^2	17	4.58	0.50	3.78	5.66
		E_H: N/mm^2	17	34605	10628	15360	45900
Expansion		$\propto$ +5°C-on	25	26.0	3.21	20	36
Cold strain		ε_H	17	15.7	7.8	5.0	33.0
Cracking temperature		°C	25	−8.6	3.44	−1.0	−16

The complex mechanical asphalt testing method described here indicates quite clearly the advantageous and disadvantageous qualities of the asphalt, the expected behaviour in the summer period, the expected endurance and sensitivity against winter cracks. In course of comparison highly valuable information and evaluation possibilities are obtained making possible evaluation of asphalts not only among each other. The evaluation of the parameter series characteristic in the three temperature ranges in some cases enabled to judge: how much the applied, perhaps sligthly modified grain distribution, at another time the effect of the modified bitumen, next time the degree of carried out greater or smaller packing improved or deteriorated certain features and which features. The design and application of asphalts have been and will be an economic-technical compromise hopefully a healthy and conscious compromise. It can be reached if the effect of technological decisions, changes can be measured numerically in a reliable way.

At last the complex mechanical asphalt testing method applied at the Department of Road Engineering, Technical University, Budapest has tried to realize it with its experiences and data base increasing and expanding for almost a decade.

6 References

Department of Road Engineering (1978) Technical University, Budapest: Splitting test of asphalt specimens, **Research report**, Budapest

Török,K., Pallós,I., Nemesdy,E. (1985) Development of Asphalt mechanical tests at Department of Road Engineering, Technical University, Budapest. **Research report**, Budapest

Nemesdy,E (1986) Complex laboratory test of factors influencing material quality of asphalts. **Research report**, Budapest

Nemesdy,E., Pallós,I., Török,K. (1989) Special Standard Asphalt Test to Classify Road Building Qualities of Normal and Modified Bitumina. **Eurobitume Symposium**, Madrid.

39 GYRATORY TESTING FOR BITUMINOUS MIX EVALUATION

B.E. RUTH, M. TIA and S. SIGURJONSSON
Department of Civil Engineering , Univeristy of Florida,
Gainesville, Florida, USA

Abstract

A description of the Gyratory Testing Machine (GTM) is presented along with information relating to different testing
procedures. A test procedure using the air-roller equipped
GTM is suggested as a rational method to simulate field
compaction and traffic densification of bituminous mixtures. The results of GTM tests on mixtures that exhibited
both early pavement rutting and resistance to rutting are
presented and discussed in this paper. In consideration of
these and prior test results, it appears that the GTM
parameter, gyratory shear (G_s), provides a reliable measure
of the shear resistance of a mixture which is a function of
aggregate characteristics, bitumen content and degree of
densification.
Keywords: Pavement Rutting, Bituminous Mixtures, Compaction, Gyratory Testing, Mixture Sensitivity.

1 Introduction

The Marshall mix design procedure has been extensively
accepted internationally. The fallacy of this and other
design procedures is their dependency upon the proper
selection and blending of aggregates for the mix. This
selection is often dependent upon local experience and the
use of gradation specifications that have evolved rather
than being based upon a definitive testing procedure. Consequently, the Marshall method only provides a means of
selecting the design asphalt content. Reliance on Marshall
stability criteria to prevent pavement rutting has been
shown to be unreliable. Similarly, Hveem stabilometer
values do not always indicate the resistance of the mixture
to plastic deformation.

Currently, there is limited use of laboratory rolling
wheel testing equipment for evaluation of the resistance of
the design mixture to consolidation rutting and plastic
deformation (shoving). This form of verification testing
is time-consuming and not very suitable for use in a mix
design procedure since it does not provide any test param-

eters that relate to the shear strength. Therefore, the
U.S. Corps of Engineers Gyratory Testing Machine (GTM)
equipped with air-roller is proposed for the simulation of
field compaction and traffic densification to evaluate the
adequacy of mix designs.

The results of the investigation described in subsequent
sections of this paper indicate the desirability of using
the GTM to identify the shear strength and sensitivity of a
mixture to minor changes in asphalt content and aggregate
gradation. Also, details pertaining to GTM procedures and
use of the air-roller to provide variable stress-strain
test conditions are discussed.

2 Gyratory Testing Machine - Operational Characteristics

The Gyratory Testing Machine (GTM) is currently manufactured by the Engineering Developments Company, Inc., of
Vicksburg, Mississippi. This machine was developed based
upon the mechanization of the original hand-operated Texas
gyratory compactor. The current Model 6B-4C GTM has been
designed so that either 6-in. or 4-in. diameter specimens
can be compacted and tested. The original Model 4C was a
smaller machine suitable for only the 4-in. molds. Other
than differences in dimensions, the operational aspects of
both machines are essentially identical.

A schematic sectional view of the GTM is shown in Figure
1 with the major machine elements identified by letter
code. Table 1 lists the name of each element. The machine
may be operated as a fixed shear strain-variable stress
testing device using fixed-roller (D) or hydraulic-roller
(oil-filled upper roller). Procedures for mix design using
the fixed-roller were developed by Kallas (1964). A
standard test method for compaction and shear properties of
bituminous mixtures using the GTM with hydraulic roller can
be found in the ASTM Standards, D-3387-83. The initial
angle of gyration can be set for angles up to 3 degrees or
slightly greater using the lower roller (C) adjustment.
Usually, the angle is set at 1 or 2 degrees using fresh
hot-mix which is placed in the mold (A) with base plate (H)
and inserted into mold chuck (B), secured by bolts, and
then compacted at 690 kPa ram pressure (F) using about ten
revolutions. Then the GTM roller arm is revolved to 180
degree positions to establish horizontal lines on the gyrograph to determine and/or set the angle of gyration (e.g.,
1 degree = 10 divisions). After setting the angle, mixtures may be compacted and tested in the GTM. The shear
resistance of the mixture reduces with excessively high
asphalt content or density level (low air void content)
which results in an increase in the width of the gyrograph.
The hydraulic-roller provides greater sensitivity to
changes in shear resistance than the fixed-roller which
depends totally upon the gyrograph.

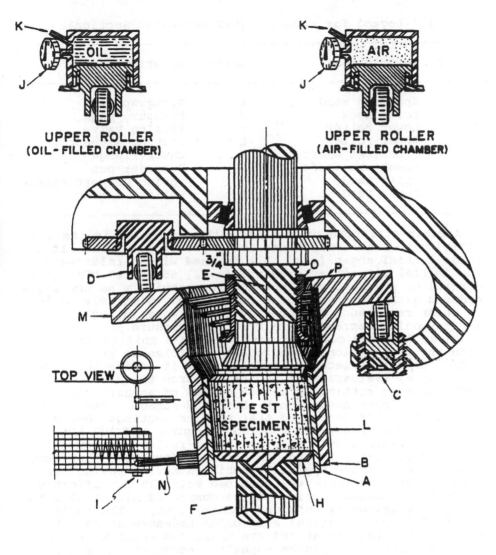

Fig. 1. Schematic Section Through the GTM

The air-roller provides the capability of variable shear strain and stress. Logically, the air-roller equipped GTM provides the best simulation of field compaction and traffic densification. As the shear resistance increases, the angle of gyration (shear strain) is reduced which results in a slower rate of densification. This effect is synonymous with field compaction where density ceases to increase with additional passes of the compaction equipment.

Table 1. Legend for Figure 1 (GTM Schematic Section)

Letter	Description	Letter	Description
A	Specimen mold	I	Gyrograph
B	Mold chuck	J	Pressure gage
C	Lower roller	K	Filling valve
D	Upper roller	L	Heating element
E	Upper ram shaft	M	Chuck flange
F	Lower ram shaft	N	Recorder pen
G	Upper head	O	Spring support sleeve
H	Lower Head	P	Helical spring

Another attribute of the GTM with air-roller is its sensitivity to shear resistance and angle of gyration. If a large initial angle (e.g., 3°) is used with a relatively low initial air-roller pressure (e.g., 55 to 103 kPa), the air-roller pressure can increase substantially as the angle reduces with increased shear resistance. Similarly, a loss in shear resistance can produce an increase in angle and a corresponding decrease in air-roller pressure.

These features combined with the GTM's ability to simulate the orientation of aggregate particles as produced by field compaction methods indicate that it offers the potential for a) evaluation of paving mixtures to assess their resistance to rutting or b) for design of asphalt concrete mixtures. Experimentation by Ruth and Schaub (1966) using the Model 4C GTM indicated that machine settings consisting of a 3-degree angle, 690 kPa ram pressure, and 103 kPa air-roller pressure simulated steel wheel-roller compaction. Twelve revolutions in the GTM produced densities that were comparable to those typically achieved during paving operations. This procedure was combined with GTM densification-traffic simulation testing of the compacted mix at 60 C for use in mix design by Ruth and Schaub (1968). The angle was set at 2 degrees, initial air-roller pressure at 138 kPa, and the ram pressure at 103 kPa using 200 revolutions for densification. The design asphalt content was determined by interpolation of the gyratory shear (G_s) versus number of revolution curves bracketing the design criterion (G_s = 54.0 at 200 revolutions).

Subsequent research was performed by Ruth et al. (1990) using the Model 6B-4G GTM. The differences between the two GTM models were primarily in the air-roller size and distance between upper and lower rollers which affects the moment used in gyratory shear calculations. Analytical conversion indicated that air-roller pressures for compaction and densification using 10-cm diameter molds should be 60.2 and 80.3 kPa, respectively. However, the majority of the tests were performed using 55.2 or 69.0 kPa for compac-

tion and 75.8 or 89.6 kPa for densification, thereby bracketing the analytically determined pressures. Although
density differences were minimal, the best results seemed
to be achieved using the higher air-roller pressures.

3 Evaluation of Rut Resistant Mixtures

Two bituminous concrete mixtures (A and B) which had not
exhibited any significant rutting on heavily traveled
primary and interstate pavements were selected initially
for evaluation using the air-roller equipped Model 6C-4B
GTM with 10-cm diameter molds. Bitumen and aggregates conforming to the job mix formula were heated to about 145 C,
mixed, and compacted for 18 revolutions using a 3-degree
angle, 690 kPa ram pressure, and a 69.0 kPa initial air-
roller pressure. The compacted samples were not extruded
from the molds but allowed to cool, then placed into a 60 C
oven for three hours prior to placement in the GTM for densification testing. The GTM was set at 2 degrees, 690 kPa
ram pressure, and 89.6 kPa air-roller pressure for densification up to 300 revolutions.

The basic GTM data recorded during densification included air-roller pressures and specimen height dial extensometer readings taken initially and after 5, 10, 15, 25,
35, 50, 75, 100, 125, 150, 175, 200, 225, 250, and 300 revolutions. In all cases, the machine was momentarily
stopped to record the data. Upon completion of GTM testing, the specimens were extruded from the molds, tested for
bulk density, and then air void contents and volume of mineral aggregate voids (VMA) were computed using Rice test
data for the maximum theoretical density. The computation
of parameters at specific numbers of revolution were based
upon the measured density and a volumetric relationship
using specimen height as measured by the dial extensometer.

Figures 2 and 3 depict the trends in the test data. The
design asphalt contents for dense-graded mixtures A and B
were 6.3 and 6.5 percent, respectively. These designs
(JMF) were obtained from Marshall testing procedures and
conformed closely to the mix placed on the construction
projects. The mixtures at 0.5 percent greater bitumen
contents than design retained their shear resistance or
gyratory shear (G_s) value. The G_s curves for 1.0 percent
above design (i.e., 7.3 and 7.5 percent) show a reduction
in G_s value with densification. Since the shear resistance
of both mixtures at these bitumen contents fall below a G_s
value of 54.0 prior to 200 revolutions, the mixtures are
not satisfactory at these elevated bitumen contents. However, neither mix A nor B can be considered sensitive to
binder content.

Another example of a good mix is illustrated in Figure
4. Mix D-5 was a dense-graded friction course which is
usually placed 2.5 cm thick and has about 10 to 12 percent

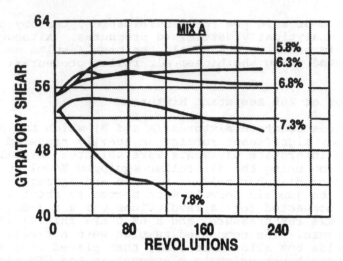

Fig. 2. Mix A: GTM Results

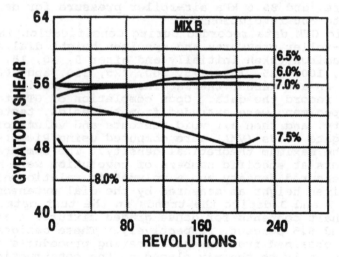

Fig. 3. Mix B: GTM Results

air void content after compaction in the field. The G_s
level is similar to mixture A and B. Since the air void
content and VMA is relatively high, the mix is not sensitive to changes in asphalt content during densification.
However, the low initial G_s value for the 8.5 percent
binder content indicated that the bitumen film thickness
was excessive until it was partially displaced into void
spaces during densification. After 100 or more revolutions, this mix exhibited about the same G_s values as the
low asphalt content mixtures. This observation is typical
of mixtures with high VMA and asphalt content. Conse-

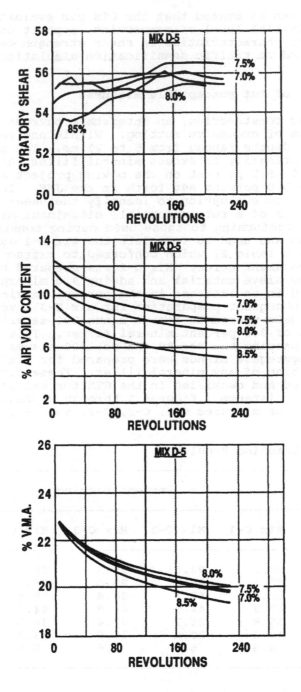

Fig. 4.   Mix D-5:   Bitumen Content Effect

quently, it can be stated that the GTM can evaluate the
effects of initial level of compaction, asphalt content,
and aggregate characteristics on shear strength development
when subjected to traffic densification simulation.

4 Evaluation of Rut Susceptible Mixtures

Shortly after construction, an interstate pavement started
showing signs of excessive rutting. Within one year, the
measured rut depths ranged from 5 to 23 mm. The problem
was related primarily to excess mineral filler which ranged
between 4.7 to 5.5 percent on the paving project as compared to the 3.0 percent set forth in the JMF. This mix
was selected for evaluation to identify the shear strength
characteristics of a rut susceptible bituminous mixture.
   Materials conforming to those used during construction
were obtained and used to formulate the mix C-1 aggregate
gradation (see Table 2) which conformed to extraction test
results of pavement cores. Mix C-2 was prepared by remov-
ing minus 200 sieve material and adding 3.0 mineral filler.
This resulted in an increase in the percent passing the No.
80 sieve. Subsequent preparation for mix C-3 involved the
use of aggregate washed to remove minus 200 material with
the addition of 3.0 percent mineral filler. This reduced
the percent passing the No. 80 and No. 200 sieves. Finally, washed aggregate blends were prepared for mix C-4 without the addition of any mineral filler. These mixtures
were compacted and densified in the GTM for evaluation of
their shear resistance. Figures 5 through 8 present the
test results for mixtures C-1, C-2, C-3, and C-4, respectively.

Table 2. Extraction Results

Sieve Size	Percent Passing				
	Mix C-1	Mix C-2	Mix C-3	Mix C-4	JMF
12.7 mm	98.2	98.5	98.3	98.1	98
9.5 mm	85.5	86.6	85.6	86.1	84
No. 4	59.5	59.1	58.4	59.4	57
No. 10	47.3	44.9	44.9	44.9	44
No. 40	39.8	38.7	37.4	36.8	35
No. 80	18.2	20.6	15.5	13.7	17
No. 200	5.5	5.6	4.5	0.6	3

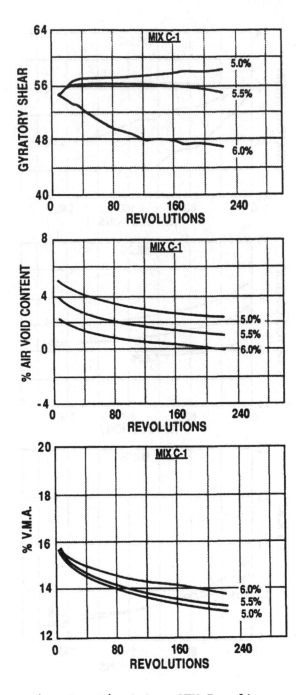

Fig. 5.   Mix C-1:   GTM Results

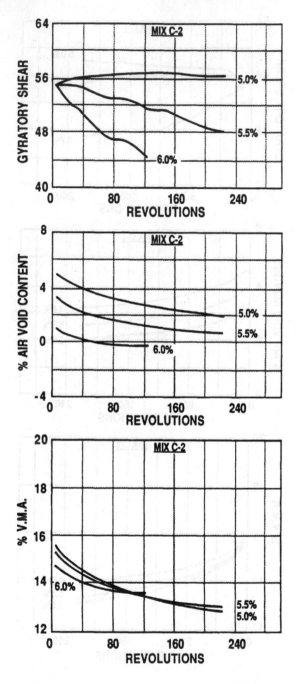

Fig. 6.   Mix C-2:   GTM Results

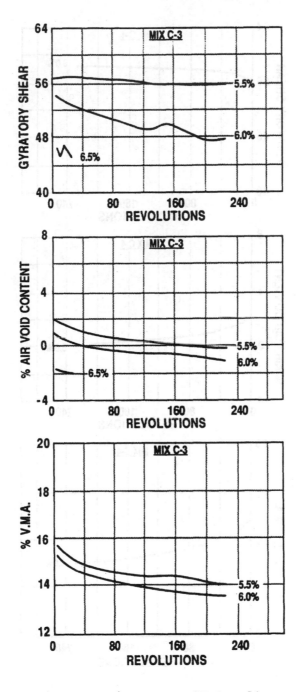

Fig. 7. Mix C-3: GTM Results

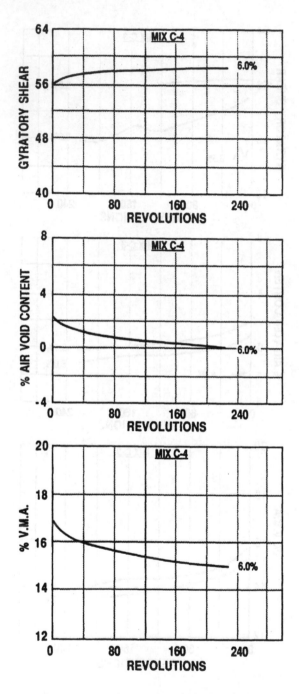

Fig. 8. Mix C-4: GTM Results

The gyratory shear response (G_s) in Figure 5 illustrates that an increase in asphalt content, 0.5 percent above the design value of 5.5 percent, produced a rapid reduction in G_s value, low air void content, and low VMA. Therefore, this paving mix was extremely sensitive to changes in asphalt content. Mix C-2 at the 5.5 percent AC content (Figure 6) exhibited a similar loss in G_s value even though the gradation was identical to mix C-1, except that the percent passing the No. 80 sieve was 2.4 percent greater. The slight reduction in the percent passing the No. 80 and No. 200 sieves did not appreciably alter the G_s response of mix C-3 (Figure 7) as compared to mix C-1, except at the 6.5 percent bitumen content. However, the air void contents for mix C-3 were considerably lower and negative which may have been caused by an error in the Rice density. Removal of almost all mineral filler in mix C-4 permitted the satisfactory use of a 6.0 percent bitumen content.

These results illustrate that the mixture is sensitive to small changes in aggregate gradation and asphalt content. Obviously, aggregate particle shape and surface texture also contribute to the sensitivity or lack of sensitivity to minor changes in mix formulation. Since production of hot-mix can be to some degree variable, it is desirable that mixtures be designed to eliminate any potential for sensitivity as demonstrated by mix C. It appears reasonable to expect that an as-designed mixture and as-designed mixture plus an additional 1.0 percent mineral filler and 0.5 percent bitumen should provide similar shear resistance (G_s values) without exhibiting any drastic reduction.

5 Summary and Conclusions

Achievement of good mixture performance, via high resistance to pavement rutting, is dependent primarily upon the adequacy of the mix design. Current mix design procedures are not generally capable of detecting a sensitive mix. Consequently, minor variations in aggregate characteristics, bitumen content, or level of compaction during construction of bituminous pavements may result in early plastic deformation and excessive rut depths. The detection of sensitive mixtures appears feasible when using the GTM equipped air-roller test procedures outlined in this paper to evaluate mixture designs. Furthermore, it is suggested that a mix design can be evaluated for excess sensitivity if the mix is prepared and tested using both design and 1.0 percent and 0.5 percent higher mineral filler and asphalt contents, respectively. Any major reduction in shear resistance (G_s) will identify a potentially sensitive mix.

Altering a mix to reduce its sensitivity depends upon the factor(s) causing the problem. In general, low VMA, excess natural sand, lack of coarse crushed aggregate, and excessively high mineral filler content can produce a more sensitive mix. The prevailing concept that a good mix design requires a dense-graded mixture with suitable VMA and a four percent air void content, regardless of aggregate characteristics, neglects the potential for different traffic densification rates with subsequent major changes in the shear resistance of mixtures. Therefore, the GTM or similar testing procedure to evaluate the properties of the mixture throughout the range of densification appears extremely desirable.

The conclusions derived from these and prior GTM research investigations are:

1. The combined effects of aggregate particle shape, surface texture, and gradation of the aggregate blend can be evaluated for level of attainable shear strength (G_s value) and for sensitivity to slight changes in mix proportions using the described GTM procedures.
2. Although not addressed here, the GTM air-roller compaction procedure using 12 to 18 revolutions can be used to simulate field compaction.
3. At present, a minimum G_s value of 54.0 should be required for any mixture densified for 200 revolutions. It is probable that an increase in G_s value is necessary as the lift thickness is increased. It is currently estimated that a dense-graded structural mix should have a minimum G_s value of 56.0 when the pavement lift thickness is greater than 50 mm.
4. The GTM densification testing procedure provides information on the shear resistance of the mix regardless of the factors influencing its behavior (e.g., air void content, aggregate characteristics, asphalt content, VMA, or binder properties).
5. The GTM procedure eliminates the need for multi-parameter mix design criteria. Therefore, it can simplify both the design and quality control process.

References

American Society for Testing and Materials. (1989) Annual Book of ASTM Standards, Section 4 Construction, Volume 04.03 Road and Paving Materials; Traveled Surface Characteristics, pp. 373-383.
Kallas, B. F. (1964) Gyratory testing machine procedures for selecting the design asphalt content of paving mixtures. **Proceedings**, The Association of Asphalt Paving Technologists, Vol. 33, pp. 341-362.
Ruth, B. E. and Schaub, J. H. (1966) Gyratory testing machine simulation of field compaction of asphaltic concrete. **Proceedings**, The Association of Asphalt Paving Technologists, Vol. 35, pp. 451-480.

Ruth, B. E. and Schaub, J. H. (1968) A design procedure for asphaltic concrete mixtures. **Proceedings**, The Association of Asphalt Paving Technologists, Vol. 37, pp. 200-227.

Ruth, B. E., Tia, M., Najafi, F. T., and Sigurjonsson, S. (1990) Preliminary Investigation of Testing Procedures for Evaluation of Rutting Resistance. Final report, UF Project No. 4910450425512, Department of Civil Engineering, University of Florida, Gainesville, Florida, USA.

40 RESILIENT MODULUS BY INDIRECT TENSILE TEST*

S.F. SAID
Swedish Road and Traffic Research Institute, Linköping, Sweden

Abstract
With developments in analytical design of pavement structure, there is an increasing demand for knowledge of the mechanical properties of various pavement elements under different conditions encountered in roads, such as tensile strength, stiffness modulus, and repeated-load fatigue resistance of the asphalt concrete mixtures. These parameters provide a good basis for analytical design and performance evaluation of flexible pavements.

In this study, Indirect Tensile Test has been conducted because of its validity and simplicity. However, modifications of the test apparatus and test procedure have been introduced.

Three different types of asphalt concrete mixtures have been examined using indirect tensile method. Two of the mixes are blended with 3% rubber content by total weight. i.e. Rubit (RUB12), which is a dense asphalt concrete used for surfacing, and a porous mix called Rubdrain (RUBD12). The third mix is a conventional dense asphalt concrete (HAB12T) used for surfacing. The maximum size of the aggregate was 12 mm for all types.

Stress-deformation-strain relationships have been established in order to examine the effect of stress and/or strain level on the resilient modulus found by the indirect tensile test and to define an approximate linear viscoelastic zone for bituminous mixtures at different temperatures. Resilient modulus has been measured using the modified indirect tensile method for mixtures studied at four different temperatures.
Keywords: Asphalt Material, Rubberized Asphalt, Indirect Tensile Test, Resilient Modulus, Linear Viscoelastic, Stress-Strain Relation, Temperature.

---

*This job has been conducted by the Dept. of Highway Engineering, Royal Inst. of Technology in cooperation with the Swedish Road and Traffic Research Inst.

1 Introduction

With developments in analytical design of pavement structure, there is an increasing demand for knowledge of the mechanical properties of various pavement elements under different conditions encountered in roads, such as tensile strength, stiffness modulus, and repeated-load fatigue resistance of the asphalt concrete mixtures. These parameters provide a good basis for analytical design and performance evaluation of flexible pavement.

In the last three decades a lot of research has been focused on the response of various layers and materials subjected to possible conditions in the field by using a wide variety of methods and apparatus, which have a provided valuable knowledge platform for further investigation. Indirect tensile test has been conducted in this study because of its validity and simplicity. However, modifications of the test apparatus and test procedure have been introduced.

Three different types of asphalt concrete mixtures have been examined using indirect tensile method. A conventional mix HAB12T, which is a dense asphalt concrete used for surfacing, Rubit (RUB12), which is a rubber granular blended asphalt mix with 3% rubber content, and a porous mix, called Rubdrain (RUBD12) also containing 3% rubber used as a drainage layer, have been used. The maximum size of the aggregate was 12 mm for all types. Rubit and Rubdrain mixtures are marketed by the ABV (NCC) company. The specimens were supplied by the ABV laboratory.

Stress-deformation-strain relationships have been established in order to examine the effect of stress and/or strain level on the resilient modulus found by the indirect tensile test and to define an approximate linear viscoelastic zone for bituminous mixtures at different temperatures.

Resilient modulus has been measured using the modified indirect tensile method for mixtures studied at four different temperatures.

2 Test apparatus

Due to its simplicity and validity the Repeated-load Indirect Tensile Test [1-8], with modification in test apparatus and test procedure, has been used to study the mechanical properties of bituminous mixtures. The principle of this test is that a cylindrical specimen is loaded in the vertical diametral plane, the compressive load is applied through a couple of curved loading strips. The resultant deformation is measured at horizontal diameter.

The expressions given by Schmidt [2] andKennedy et.al. [1] for tensile stress at the center of specimen, resilient modulus, Poisson's ratio and strain across horizontal diameter are the following:

$$\sigma = 2P/\pi tD \tag{1}$$

$$n = 3.4\,[\Delta H/\Delta V] - 0.27 \tag{2}$$

$$M_R = (P/t.\Delta H)[4/\pi + n - 1] = P(n+0.27)/t.\Delta H \tag{3}$$

$$\mathcal{E}_x = (2P/M_R \pi t D)[(4D^4 n - 16D^2 x^2)/(D^2 + 4x^2)^2 + (1-n)] \tag{4}$$

σ = Tensile stress (MPa)
P = Applied load (N)
t = Specimen thickness (mm)
D = Specimen diameter (mm)
n = Poisson's ratio
ΔH = Total resilient horizontal deformation (mm)
ΔV = Total resilient vertical deformation (mm)
M_R = Resilient modulus (MPa)
$\mathcal{E}_x$ = Tensile strain at horizontal diameter

By substituting Equation (1) in Equation (3) and solving for resilient modulus:

$$M_R = (\pi D/2)\,(\sigma/\Delta H)\,(n+0.27) \tag{5}$$

To find the tensile strain at the center of the specimen (x=0), substituting Equation (3) in Equation (4) and solving for strain we get:

$$\mathcal{E}_0 = (2\Delta H/D)[(1+3n)/(4+\pi n+\pi)] \tag{6}$$

If n = 0.35 then

$$\mathcal{E}_0 = 2.1\,(\Delta H/D) \tag{7}$$

where, ΔH and D are defined as before.

The testing machine is capable of applying compressive pulse loads over a range of frequencies and load durations for repeated load testing.

In this investigation the repeated load pulse was applied at a frequency of 0.33 Hz with 0.2 sec loading time and 2.8 sec rest period. The load applied in the vertical diameter plane is sensed by a load cell under the specimen. Deformation across the horizontal diameter of the specimen is sensed by one extensometer instead of two LVDTs, which are normally used. Instron dynamic strain gauge (2620-601, with resolution of $20*10^{-5}$ mm and weight 20 grams), and MTS extensometer (632-11c, resolution $8*10^{-5}$ mm and weight 25 grams) have been shown to be adequate. The strain gauge is fixed to two curved zinc strips (2 mm thick, 10 mm wide and 80 mm long) which are glued at opposite sides of the horizontal diameter. The total weight of extensometer and the two deformation strips is less than 75 gm (See Plate 1 and Figure 1). However, a couple of LVDTs with the frame which carry out the two LVDTs is more than 700 gm which might influence the recorded deformation.

Plate 1. Horizontal extensiometer mounted on the specimen.

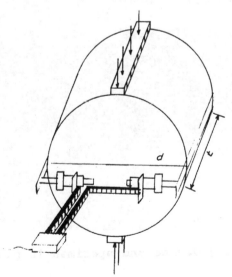

Fig.1. Deformation strips.

The deformation is measured at the same points where the two zinc strips are glued in order to avoid any error in recording deformations due to specimen movement or inadequate contact between extensometer (strain gauge) and specimen. Brown and Cooper [9] have mentioned the shortcomings of indirect tensile test, using LVDTs, in deformation measurement. The recorded deformation was found to be influenced by the strength of the spring used to keep the LVDT core in contact with the specimen, and by the vibration of the frame which carries the deformation transducers. In addition, it is easier to use just one extensometer rather than two LVDTs.

The loading device is shown in Plate 2 and Figure 2. The upper steel loading strip is not fixed on the upper platen of loading device in order to bring a good contact between loading strips and specimen even at low stresses and high stiff asphalt specimens. An alignment bar is fixed to the upper aluminum platen to bring the loading strips into the same vertical plane. Figure 3 shows three different cases which could happen depending on the contact between loading strips and specimen when the loading strips are fixed on the platens. Case 1 shows a good contact, while cases 2 and 3 show insufficient contact which is a result of a rough surface. For more details see reference 22.

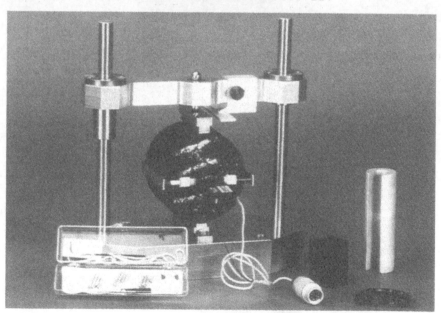

Plate 2. Loading device and specimen in place.

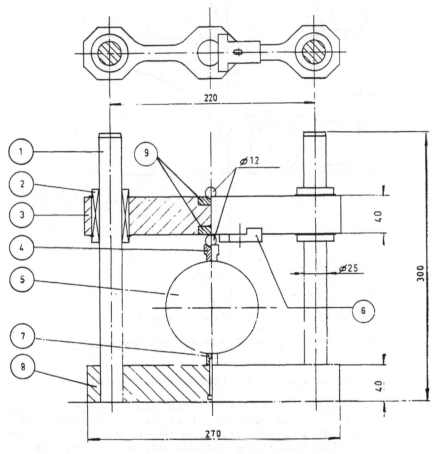

All dimensions in mm

1 Precision Pillar
2 Ball Bushing
3 Upper Plate
4 Float Loading Strip

5 Specimen
6 Alignment Bar
7 Fixed Loading Strip
8 Bottom Plate
9 Hardened Disk

Fig.2. Loading device for indirect tensile test.

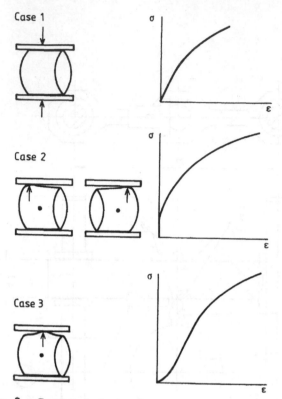

Fig.3. Contact between strips and specimen.

3 Viscoelastic property of bituminous materials

Various investigators have emphasized the effects of stress
or strain levels on the modulus of asphalt concrete mixtures
[2,10-21]. The increasing use of elastic theory in pavement
evaluation and for comparing various asphalt materials in
terms of their rheological characters makes the fundamental
properties in stress-strain relationship an important parameter [16]. With regard to the non-linear viscoelastic behavior of bituminous materials, the stress or strain level has
an essential influence when measuring the moduli. Sayegh
[17] does not exceed a strain value of 4.10^{-5} when measuring
moduli. Monismith et. al. [11] consider the stiffness of
asphalt mixtures independent of applied stress at strains
less than 1.10^{-3} in/in. Bonnaure   et.al. [13] determine
stiffness modulus for strains 5.10^{-6} at -15°C and +9°C;
20.10^{-6} at +30°C. Bonnaure et.al. also report that the phenomenon of non-linearity appears at strains close to
100.10^{-6}. Kennedy and Anagnos [1] and ASTM [3] recommend a
load range to induce 10 to 50 percent of the tensile

strength as determined in the static indirect tensile test, or in lieu of tensile strength data, loads ranging from 4 lb (25 lb according to Kennedy) to 200 lb per inch of specimen thickness can be used. Schmidt [2] uses a stress range from about 1 psi up to 15 psi at the specimen center at 73 °F (24 °C). So a wide variety of limits have been recommended. Therefore, an <u>approximate linear zone</u> (the resilient modulus is independent of applied stress or strain level) should be well defined to make the modulus more reliable.

4 Linearity

The test series consist of testing three specimens from each type of asphalt mixture at temperatures of -10, +5, +25 and +40 degrees Celsius and at a loading frequency of 0.33 Hz. Totally, 36 specimens have been tested with this method. The test procedure is the following:

* The test specimens were stored in a temperature cabinet for 24 hours at testing temperature. (A dummy specimen has shown that it takes at least four hours to bring the specimen from room temperature to -10°C).

* Repeated-load pulse, at a frequency of 0.33 Hz, was applied with 0.2 sec loading time and 2.8 sec rest period.

* The load level was increased gradually up to a level which resulted in a recordable deformation by extensometer.

* A minimum of 50 load repetitions were applied before the first reading.

* The load level was increased gradually to a higher level with about 25 load repetitions at each load level. The maximum load level used was about 4000 N.

* Stresses and strains were calculated at each load level, a curve was fitted by regression analysis for all three specimens, which represent the material response at defined temperature. The zero point has normally been adjusted due to surface irregularities.

The isochronous stress-horizontal deformation-strain diagrams were produced for bituminous mixtures to study the stress and strain effect on resilient modulus using repeated load indirect tensile test. Figure 4 shows isochronous diagram for HAB12T mix. Similar curves have been found for the other two mixtures (See Figure 5 and 6). The stress was calculated at the center of the specimen by Equation 1. These diagrams are represented by stress-deformation-strain

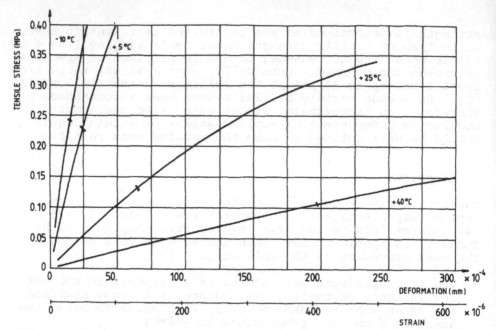

Fig.4. Stress-deformation-strain relationships for HAB12T.

curves and not only by the more familiar stress-strain
curves, hence the deformation distribution on the horizontal
diameter is not uniform [2,4,5,23]. This is done in order
to eliminate the effect of using an assumed value for the
Poissons ratio for strain calculation at the center of the
specimen, which might affect the strain value. However, for
the sake of comparison, the tensile strain at the center of
the specimen is calculated by Equation 7.

The correlation coefficients (R) are higher than 0.94,
except for the porous graded mix tested at 25 degrees
Celsius, which is 0.83. In all the cases, the stress-
deformation relationships show curves which are concave
downwards. This means that strain increases faster than
stress and results in a reduction in resilient modulus. The
nonlinearity relationships are obvious even at low stress
and deformation levels. The non-linear viscoelastic relationships confirm the importance of stress and deformation
levels when measuring the resilient moduli of asphalt
mixtures.

The above discussion shows the dominance of viscous properties at high strain and stress levels. Therefore, and in
order to measure the resilient moduli of bituminous materials with consistent in results, a viscoelastic linear zone
should be defined in which the effect of stress or strain
levels on the resilient modulus is negligible. If resilient
modulus is measured at higher stresses or strains, it could
be justified in some cases in order to simulate field

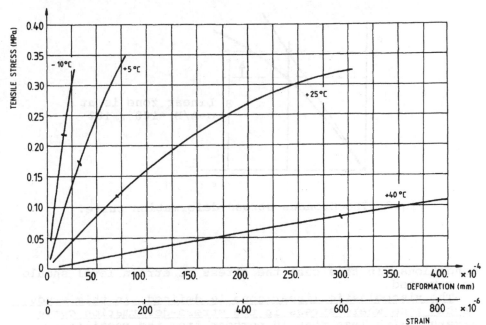

Fig.5. Stress-deformation-strain relationships for RUB12.

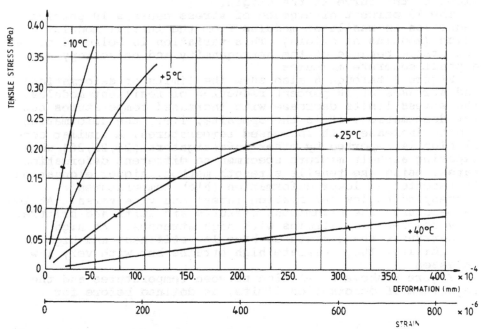

Fig.6. Stress-deformation-strain relationships for RUBD12.

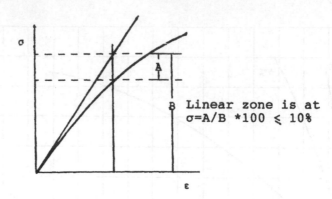

Fig.7. Isochronous stress-deformation curve.

conditions, in this case the stress or strain level should be reported.
The viscoelastic linear zone is defined, in this study, as the zone where stress in the stress-deformation curve diverges with less than 10 percent from the rectilinear stress-deformation relation shown in Figure 7. The rectilinear stress-deformation relation is represented by the slope of the curve at the origin.

The 10 percent divergence of stress cause a 10 percent reduction in resilient modulus from the initial modulus value (modulus at origin). This variation is believed to be acceptable because of the hetrogeneity and unisotropy of asphalt concrete mixtures.

Figure 4 through 6 also show the limits of deformations and stresses at 10 percent reduction of resilient modulus. The stress limits decrease with increased temperatures but for deformations it is the opposite, i.e., the deformation limits increase with increased temperatures. A similar conclusion is reported by other investigators [24,25,26] when exposing asphalt mixture specimen to different deformation rates, using the tensile strength method, higher elongation is reported at lower deformation (high temperatures correspond to low deformation rates) and vice versa. At low temperatures the bituminous mixtures are stiff and brittle. They tolerate low strains with high stresses. At high temperatures these mixtures have low stiffness and high flexibility. They tolerate high strains but with very low stresses.

Figure 8 shows a relation between temperatures and the logarithm of deformation limits, as defined before for

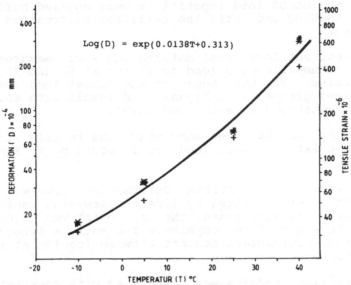

Fig.8. Relationship betweenn the logarithm of deformation
and temperature.

tested mixtures (with 0.2 sec loading time and a frequency
of 20 cycle/min.).The correlation coefficient (R) for all
materials together is 0.986, using regression Equation 8.

$$Log(D) = exp(0.0138T+0.313) \qquad (8)$$

where
D= Horizontal deformation in mm x 10^{-4}
T= Temperature in Celsius.

5 Resilient modulus

In accordance with the discussion in the above section
regarding stress-strain relationships and the effect of
irregularites of specimen surface the following procedure
has been used for resilient modulus measurement. Indirect
tensile test has been used on cylindrical bituminous sample.

* The test specimens were stored in a temperature cabinet
  to the proper temperature.

* Repeated-load pulses at a frequency of 0.33 Hz were
  applied with 0.2 sec loading time.

* A minimum of 50 load repetitions were applied before the first reading and until the resilient deformation became constant.

* The resilient horizontal deformation were measured at the maximum allowable load level and at 50 and 75% of this value. The test began at the lowest load level and increased gradually to higher load levels with about 25 load repetitions at each load level.

* After the test had been completed, the tensile stress was calculated at each load level according to Equation (1).

* Stress-resilient horizontal deformation curve was plotted. Curve fitting by linear regression analysis was used. In such cases, the zero point was adjusted as shown in Figure 9 to compensate for surface irregularities and insufficient contact between loading strips and specimen.

* The resilient modulus was calculated with constant Poisson's ratio according to Table 1 at different temperatures by Equation 5:

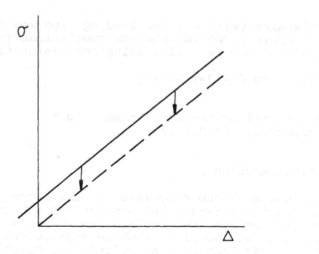

Fig.9. Correction of the stress-strain curve.

Table 1. Constant Poisson's ratio

T°C	n
-10	0.20
+5	0.25
+20	0.35
+40	0.40

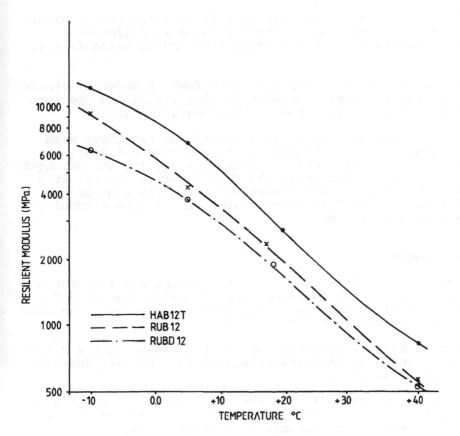

Fig.10. Effect of temperature on resilient modulus.

Figure 10 shows the resilient modulus at different
temperatures. The relationship between temperature and the
log of resilient modulus is not linear. The rubberized mix
(RUB12) shows lower moduli than conventional mix (HAB12T), a
fact which indicates the higher flexibility of the rubberized mix. The porous mix (RUBD12) has shown, as expected,
the lowest moduli. Furthermore, there were no problems in
measuring resilient modulus for the porous mix.

6 Conclusions

1. An approximate linear viscoelastic zone for practical use
   has been defined for asphalt mixtures at different
   temperatures (with 0.2 loading time and 20 cycles/min.).

2. The upper loading strip should be free in order to create
   satisfactory contact between loading strips and specimen
   even at low stresses and stiff asphalt specimens.

3. One extensometer may be used instead of two LVDTs with
   carrying frame. In that case the recorded deformation
   will not be influenced by the following: (1) the strength
   of the spring that keeps the LVDT cores in contact with
   the specimen, (2) the vibration of the frame which carries the deformation transducers, and (3) the extensive
   load on the specimen.

7 References

1. Kennedy T.W and Anagnos J.N, "Procedures for the Static
   and Repeated-Load Indirect Tensile Tests, "Research Report 183-14, Center for Transportation Research, The University of Texas at Austin, 1983.

2. Schmidt R.J, "A Practical Method for Measuring the Resilient Modulus of Asphalt-Treated Mixes", Highway Research
   Record, 404, 1972.

3. ------------, American Society for Testing and Materials,
   ASTM D 4123-82, Method of Indirect Tension Test for Resilient Modulus of Bituminous Mixtures, 1984.

4. Wallace K. and Monismith C.L, "Diametral Modulus Testing
   on Nonlinear Pavement Materials", Proceedings, The Association of Asphalt Paving Technologists, Vol. 49, 1980,
   p. 633.

5 .Kennedy T.W, "Characterization of Asphalt Pavements Mate-

rials Using the Indirect Tensile Test" Proceedings of the Association of Asphalt Paving Technologists, Vol. 46. 1977.

6. Baladi G.Y, Harichandrun R.S and Lyles R.W, "New Relationships Between Structural Properties and Asphalt Mix Parameters", Paper presented at the Transportation Research Board, 67th Annual Meeting, Jan 1988.

7. Hondros G., "The Evaluation of Poisson's Ratio and the Modulus of Materials of a low Tensile Resistance by the Brazilian (Indirect Tensile) Test with Particular Refe rence to Concrete", Australian Journal of Applied Sci ence, Vol. 10, No. 3, 1959.

8. Baladi G.Y., Harichandrun R. and De Foe J.H., "The Indirect Tensile Test, a New Apparatus". Interim Report, College of Engineering, Michigan State University 1987.

9. Brown S.F and Cooper K.E, "The Mechanical Properties of Bituminous Materials for Road Bases and Base courses" Proceedings of the Association of Asphalt Paving Technologists, Vol. 53, 1984.

10.Pell P.S and Taylor I.F, "Asphaltic Road Materials in Fatigue", Proceedings of the Association of Asphalt Paving Technologists, Vol. 38, 1969.

11.Monismith C.L, Epps J.A and Finn F.N, "Improved Asphalt Mix Design" Proceedings, The Association of Asphalt Paving Technologists, Vol. 54, 1985, p. 347.

12.Deacon J.A and Monismith C.L, "Laboratory FlexuralFatigue Testing of Asphalt-Concrete with Emphasis on Compound-Loading Tests", Highway Research Record, No. 158, 1967.

13.Bonnaure F., Gest G., Gravious A. and Uge P., "A new Method of Predicting the Stiffness of Asphalt Paving Mixtures" Proceedings, The Association of Asphalt Paving Technologists, 1977, p. 64.

14.Kallas B.F. and Riley C., "Mechanical Properties of Asphalt Pavement Materials", Proceedings, 2nd International Conference on the Structural Design of Asphalt Pavements, Ann Arbor 1967, p. 931.

15.Kallas B.F and Puzinauskas V.P, "Flexure Fatigue Tests on Asphalt Paving Mixtures", American Society for Testing and Materials, STP 508, 1972, p. 47.

16.Witczak M.W. and Root R.E., "Summary of Complex Modulus Laboratory Test Procedures and Results", American Society for Testing and Materials, STP 561, 1973, p.67.

17. Sayegh G., "Viscoelastic Properties of Bituminous Mixtures", Proceedings, 2nd International Conference on the Structural Design of Asphalt Pavements, Ann Arbor 1967, p.743.

18. Takallou H.B., McQuillen J. and Hicks R.G., "Effect of Mix Ingredients on Performance of Rubber Modified Asphalt Mixtures", FHWA-AK-RD-86-05, Federal Highway Administration, May 1985.

19. Lee M.A. and Emery J.J., "Improved Methods for Characterizing Asphaltic Concrete", Proceedings, Canadian Technical Asphalt Association, Vol. 22, 1977, p.109.

20. Vinson T. "Fundamentals of Resilient Modulus Testing" Workshop on Resilient Modulus Testing, Oregon State University, Corvallis, OR, 1989.

21. Furber B. "Strain and Temperature Effects on Resilient Moduli" Workshop on Resilient Modulus Testing, Oregon State University, Corvallis, OR, 1989.

22. Said S.F, "Tensile and Fatigue properties of Bituminous Mixtures Using Indirect Tensile Method", Ph.D. Dissertation, Department of Highway Engineering Royal Inst. of Technology, 1989.

23. Kennedy T.W and Hudson W.R, " Application of the Indirect Tensile Test to Stabilized Materials", Highway Research Record 235, 1968.

24. Eriksson R. "Dragprov på sandasfalt", Meddelande 85, Statens Väginstitut, Stockholm, 1954.

25. Sugwara T. "Mechanical Response of Bituminous Mixtures under Various Loading Condations", Proceedings, 3rd International Conference on the Structural Design of Asphalt Pavements, London 1972, p.343.

26. Linde S. "Investigations on the Cracking Behavious of Joints in Airfields and Roads" SP Rapport 1988:23, Polymerteknik, Borås 1988.

41 THE METHODS FOR TESTING OF ASPHALT MIXES DEFORMATION PROPERTIES

M. SEKERA
Research Institute of Civil Engineering, Bratislava,
Czechoslovakia
I. GSCHWENDT
Slovak Technical University, Bratislava, Czechoslovakia

Abstract
Standard tests of asphalt mixtures, their stability,
flow and behaviour under static load does not give
the deformation characteristics for calculation of
flexible pavement permanent deformation and rut depths.
In this contribution there are described two new devices
developed at Research Institute of Civil Engineering
in Bratislava for testing the resistency of asphalt
mixtures against the permanent deformations. There are
a principles of method for calculation the permanent
deformation on surfacing of flexible pavement in this
paper and values of material deformation coefficient (K)
as a characteristic of asphalt mixtures for permanent
deformation calculation.

1 Introduction

In the analytical (mechanistic) design method for
flexible pavements used in Czechoslovakia there is
applied mathematical solution of multilayer elastic
half-space for the analysis of stresses and strains
of pavement layers. Subgrade and bound as well as
unbound layers are characterized by elasticity modulus
E and Poisson's ratio μ . These are characteristics of
deformation material properties dependent on loading
particularly its intensity, duration and cycles and on
ambient conditions (mainly temperature). In the analyses
the real values of deformation characteristics need not
be necessarily used, for this aim there can be used the
derived design values. At the evaulation of the
designed pavement construction there are considered the
bending tensile stress in bounded layers and compressive
vertical stress on the subgrade. At present time we have
supplemented the design method for asphalt pavements by
the analysis of permanent deformation that appears in
form of rutting at the pavement surface. The basic

assumptions for analysis of the permanent deformation of
the multilayer asphalt pavement surface deformation is
the possibility of summation of permanent deformation of
separate pavements layers and the subgrade:

$$(1)$$

where   is permanent vertical deformation

- permanent deformation of the layer i,
- permanent deformation of the layer $i+1$.

The permanent deformation of asphalt mixtures (but not
cement bounded mixtures) is considered as a part of the
total deformation. It is derived from the equation for
elastic deformation $/Y_{pr}/$:

$$Y_{perm.} = K \cdot Y_{pr}$$

where $Y_{perm.}$ is material permanent deformation,
      K — material deformation coefficient.

The equation (1) can be supplemented as follows:

$$Y_{perm.} = K_{subgrade} \cdot Y_{subgrade}$$
$$= K_p \cdot y_{perm,p} + \sum_{i=1}^{i=n} K_i / y_{pr,i} - y_{pr,i+1} /$$

where K_p or K_i is the deformation coefficient of the
                 subgrade or of i-layer material.

     The theoretical approach to the problem points at
the deformation coefficient characterizing behaviour
of the visco-elastic asphalt mixtures. On account of its
determination it is necessary to have knowledge of the
elastic and permanent deformations at a certain loading
and certain conditions. The coefficient is to be
understood as resistance characteristics of asphalt
mixtures against the rise of permanent deformation and
"rutting" at the asphalt pavement.

2 Standard tests of asphalt mixtures

The deformation properties of asphalt mixtures are tested
by various methods and procedures. These are divided into
two groups - laboratory and in-situ tests or statical and
dynamic ones. Procedure according to Marshall belongs to
the basic statical tests. The results of the complex or
the extended Marshall test make it possible to evaluate

not only the stability (stiffness) of the mixture but the creep as well. In a very simple procedure there can be estimated the influence of the crushed aggregates or the mixture porosity on its creep at the statical loading. The complex Marshall test was modified in the Czechoslovak specifications (Standard ČSN 73 61 49 Asphalt concrete for road pavements).

The creep test of the asphalt mixtures is in fact the statical test of the mixture resistance against the permanent deformations, so the results are to be considered from the view-point of long-term loading. In the most from among the employed procedures there is used rather relatively low stress e.g. 0,1 MPa. The procedure with varying loading cannot simulate adequately the loading and the effect of the moving vehicle.

Resistance test of the asphalt mixture against the rise of permanent deformations at the wheel loading is used in specialized laboratories. It is connected with a pretending preparatory work of a large-specimen claiming an equipment forcing the wheel to move or to shift the specimen. The testing conditions are similar (only similar) to those on the pavement and the obtained results are supplementing the laboratory results according to Marshall test.

Further, a more perfect testing procedure of the asphalt mixtures can be carried out on pavement constructions in line or circular test tracks. At the testing the real loading of the moving axle is applied and the climatic conditions (influences) can do as well. The testing equipments are complex as for their construction, pretending in price and not standardized. The only circular test track in Czechoslovakia is at the Research Institute of Civil Engineering in Bratislava.

A great number of variable factors and their influence on the deformation characteristics of the mixtures can be applied at the testing of deformation properties on pavements in-situ. However, the results interpretation and properties evaluation of the individual pavement layers are very complex. On the other hand, even without a detailed analysis of the testing methods and procedures there can be stated that in spite of a certain insufficiency the laboratory test methods of deformation properties of asphalt mixtures are advantageous. The resistance of mixtures against the permanent deformations investigated by the specimen test in the course of "rutting" is the most frequent modification and a certain simulation of real conditions.

3 New testing methods and development of testing devices

For testing of deformation properties of asphalt mixtures

at dynamic loadings two types of laboratory devices were
developed at the Research Institute of Civil Engineering.
For testing procedures of mixtures there were designed
two methods. For testing of the asphalt mixtures
resistance against creep and permanent deformation there
was developed "the rutting equipment" called VYKO-VÚIS.
Dimensions of the specimen from asphalt mixture are
600 x 300 mm and its thickness ranges from 60 x 150 mm.
The specimen undergoes repeated loadings due to the
wheel passes (Fig. 1). The standard test is carried out
at the temperature of +50 ^{o}C and at the wheel speed of
130 m . h^{-1}. Contact pressure on the tyre imprint is
0,6 MPa. At the testing there is measured the dependence
of the ruts depth on the number of passes, i.e. on the
number of repeated loadings. The testing device enables
to employ the complete pavement length, the ruts can
reach the length of 600 mm. The resulting indicator of
the asphalt mixture resistance test against the deformations is the absolute value of the formed ruts.
However, it is also necessary to know and to consider the
course of its forming. The relation between the rut
depth y and the number of repeated loadings illustrates
the equation

$$y = a + b \, logN$$

being valid for asphalt mixtures of high quality
From test results of various mixtures obtained from
the device VYKO-VÚIS there can be presented those ones,
which concern the influence of the amount of polymer
admixtures on the permanent deformation magnitude.
In Tab. 1 the test results of three asphalt mixtures
with different binders and the asphalt with polymere
admixtures "SBS" are presented. There is given the
relative deformation referred to the rut depth formed
in specimen 50 mm thick. The values are compared for
number of passes N = 500.

Table 1.

Kind of asphalt binder	A 200	AP 80	AP 65	AP 80 +4 % SBS
Softening point KG (^{o}C)	42.0	45	48	53
Strain,	0.278	0.164	0.136	0.078

The second device, PTD-VÚIS is also designated to the

testing of mixture resistance against the permanent deformations. At the testing there is, however, used a cylindrical specimen like for the Marshall test. The ruts are formed on the surface of the rotating specimen under the steel wheel. The value of load is estimated in a such way that the stress on the contact surface of the test sample corresponds with that in the contact area between the tyre and the pavement, i.e. 0.65 N.mm^{-2}. (The device is shown in Fig. 2). The specimen is tempered to +40 $^{\circ}$C. Its production is simple - the same as for the Marshall test. It provides a whole range of advantages and gives the possibility of comparing the measured values with the stability and deformation according to Marshall test.

For exemplification we present the test results obtained from this device in Tab. 2. The values illustrate the permanent deformation at repeated loading for 3 asphalt mixtures, with the asphalt AP-80 and with the polymeres admixture Bralen PB-25.

Table 2.

Content of the admixture PB-25 in asphalt AP-80 /%/	0	4	8
Deformation at the end of the test	1,41	1,78	1,61

The results obtained from the tests carried out on the above-mentioned devices were compared with the results of other tests, e.g. of the "Creep-test". Generally there is employed the testing procedure according to SCW. In Tab. 3 there are given the test results of the mixtures with different polymer additives, however, with the same content of 4% in the asphalt. Stiffness Sp/60/ ascertained by the "Creep-test" and the permanent deformation measured by the device PTD-VÚIS are presented.

Table 3.

The observed characteristic	Basic mixture	Mixtures with the additives			
		SBS	EVA	PE	PE+SBS
Stiffness modulus of the mixture Sp/60/	21,9	38,6	39,1	39,6	35,8
Permanent deformation /mm/	1,87	1,21	1,08	1,28	0,99

Fig. 1

Fig. 2

There can be estimated by comparison of the values, that the stiffness modulus was as different as the permanent deformation, however the relative changes were not the same.

Another comparison of results obtained from various testing procedures and characteristics according to Marshall test can be done by the help of Tab. 4. Four mixtures of asphalt concrete (coarse grained) with the stability of 9.0 kN, which is higher than it is required by the Czechoslovak Standard, were subjected to the tests. According to the results obtained from the stability test of the mixture against the permanent deformations carried out by the PTD-VÚIS device, the differences in properties of the tested mixture are considerable, however not according to results of Marshall test. New testing methods and the developed devices are more appropriate for evaluation of the resistance of asphalt mixtures against the permanent deformations than the standard methods. The methods with higher distinguishing ability were employed to the comparison of deformation behaviour of mixtures made with modified binder.

4 Application

The permanent deformation of the asphalt mixture depends on many factors, whose significance is evaluated by standard methods but also by the new ones not according to standard. The exemplifying influence of the temperature as well as of the number of the repeated loadings on the rut depth ascertained by the PTD-VÚIS device is given in Fig. 3. (it is concerning the asphalt concrete according the standard). It has been found out, that it is very difficult to design a testing procedure (and to determine conditions for the testing) so that the deformation coefficient K of the asphalt mixture applicable to calculation of the permanent deformation of pavement surface could be directly assessed by the help of the results evaluation. For example, the so-called semi-dynamic creep-test is carried out at the temperature of 40 °C and the pressure of 0.1 MPa. On the other hand the calculation of the elastic deflection of pavements is done at the temperatures of 0, +11, and +27 °C - according to the Czechoslovak design method. According to the method the elasticity moduli of individual materials are determined for the loading time of 0.02 sec. or 60 sec. in the places where some of vehicles stop. With regard to the complexity the problem was solved by derivation of the design values of the deformation coeficient K from the whole set of results obtained from the deformation properties tests of the asphalt mixtures. The values were applied for calculation

Table 4.

Type of mixture	Mixture components in weight %				ratio filler : binder F : A	Physical-mechanical properties of the mixture ascertained by Marshall-test				Results obtained by new devices	
	grains F smaller than 0,09 mm	grains over 4,0 mm	crushed aggregate	asphalt A		stability SM kN	Flow PM 10^{-1} mm	stiffness TM -	VOIDS % %	PTD-výsl. mm	VYKO-výsl. mm
a	8,0	40,0	69,0	6,0	1,33:1	11,6	36	32,5	3,2	6,0	2/
b	9,5	40,6	86,2	5,2	1,82:1	13,9	38	36,6	4,3	3,7	3/
c	7,9	51,4	95,0	5,0	1,58:1	13,8	35	39,4	5,3	2,5	8,1 4/
d	8,2	58,3	95,0	5,0	1,64:1	11,0	39	28,1	3,6	2,4	8,2 4/
accor-ding to ČSN stand.	3-10 1/	38-69 1/	min. 1/ 70	-	-	min. 9,0	23-40	25-40	3-6	-	-

1/ values valid only for aggregate mixture 2/ destruction of the sample due to rutting has already arisen at passes ÷ 2,5 x 10³; 3/ destruction of the sample due to rutting has already arisen at passes of 10⁴; 4/ values obtained at the end of the test / 5 x 10⁴ of passes/

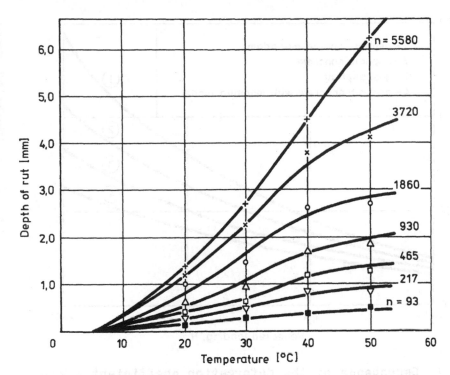

Fig. 3. Rut depth in the specimen from asphalt mixture
at different temperatures in the dependence of
number of loadings

of the pavements permanent deformation, and were compared
then with long - term measurements of the ruttings on
these pavements. The calculated permanent deformations of
asphalt pavements were for $N = 10^5$ up to 10^7 in the range
of 13,5 to 25 mm. The rut depths measured were higher.
The rut depth can reach the value of permanent deformation
multiple by 1,2 up to 1,4. Considering the results
obtained from measurements on pavements there were
proposed following deformation coefficients K:

- for asphalt concrete $\quad\quad\quad\quad\quad 4,5.N^{0,23}$
- for asphalt coated aggregates $\quad\quad 4,5.N^{0,25}$
- for asphalt concrete with modified
  binder $\quad\quad\quad\quad\quad\quad\quad\quad\quad 4,5.N^{0,19}$

The values of the coefficient K according to
mathematical definition are presented in Fig. 4.

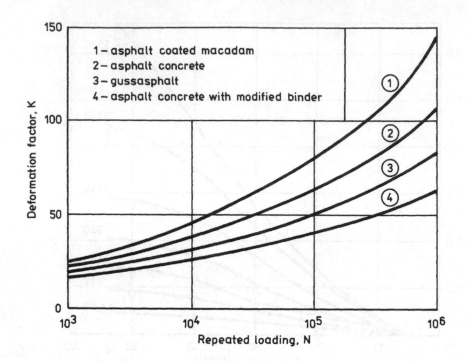

Fig. 4. Dependence of the deformation coefficient K value on number of repeated loadings for asphalt mixtures

The values take into consideration the temperature regime and real way of loading and conditions of pavement construction with surfacing and base course from asphalt mixtures.
These layers share in the total pavement permanent deformation with ratio from 5 to 20%.

42 THE INFLUENCE OF THE QUALITY VARIATION OF CRUSHED STONE PRODUCTS ON THE FATIGUE PARAMETERS OF ASPHALT MIXTURES

I. SUBERT
Institute for Transport Sciences, Budapest, Hungary

Abstract:
    The quality of the crushed stone produts used in the asphalt mixtures in Hungary does not meet some of the requirements. The influence of these quality defects, their effect on the fatigue strength, on the duration can be different.
Keywords: aggregate quality, asphalt mixtures, deterioration process, dynamic split test.

1 INTRODUCTION

The quality of crushed stone one of the most important building material is rather changing in Hungary and it does not satisfy the specifications of MSZ 18291 referring to crushed stone products. Having cleared the relation between pavement defects and stone quality the results of the 1980/81 pavement tests were subjected to defect-cause analyses and 48% of the pavement defects were tracked beck on direct or indirect stone quality. We assessed the demage coused by additional expenses more then onehundred million HUF in 1984.

The fluctuation in the quality of crushed-stone products occurs mainly indirectly (as defect in asphalt mixture) while the effect of characteristics required in the Standard and those of rock tipes of rock-physics on asphalt-mixture specially on its fatigue strength are not fully known.

This study approaches the question of relations between the quality of stone and the design life while exploiting the possibilities of laboratory analyses. The effect of rock type, product characteristics (fluctuation in grading, clay-silt contamination) and those of rock physics on the characteristics of asphalt mixture were examined applying traditional, static and dynamic asphalt mechanical tests. The crushed-stone products were subjected to detailed rock laboratory analyses.

2 TESTS IN ROCK LABORATORY

The laboratory tests were carried out on products made of basalt from Uzsa, andesite from Nógrádkövesd, dolomite from Gánt and limestone from Nagyharsány wich made the mineral composite of asphalt laboratory test carried out later. The sampling and preparation were done by our Institute from the stock of mixing-plants nearby Budapest.

2.1 Stone phisycal tests of crushed-stone products

Results of tests in rock laboratory are summarised in Table 1. and selected according to the place of origin detailed at each product. As according to test results more then half of the products do not satisfy the grading requirements of the standard. The causes of non-compliance were separately marked.

The rock-physical classification (A,B,C,D) of the products from Uzsa "B", those of Nógrádkövesd "C", the dolomite from Gánt "B", while limestone product of Nagyharsány "C". For 12-20mm fractions the wet-deval test results proved to be the critical qualifying tests in two cases, the dry deval test in one case, while in one case at the Los Angeles test when performing classification The grain-strength inhomogeneity of the Nógrádkövesd andesite is the highest among them according to the modified grain-split test.

2.2 The modelling of grading fluctuation

The fraction below 2mm as one component of the asphalt mortar should distinctively be separated because of asphalt technological reasons in the composition formula applied to AB-12 basic mixture. In this case the presence and quality of product 5/12 and its effect on the performance of bitumen mixture was the subject of our investigation. Half of the 5/12 products did not satisfy the Standard specifications. We concluded at our previous investigations that the diversion from the Standard might be higher. The fullfilment of the grading alternatives of asphalt formula detailed later was a further point of wiew.

The following gradings seemed to be optimum: the 5/12 product from Uzsa (U1) is basicly too fine, the volume of particles under specification size is high. So we extracted 11,9 m% from the 2/5 fraction -compared to the total volume- while the the same percentage was added to fraction 8/12 for rendering it standard (U1). The other solution is just the other way round that is a certain percentage was extracted from fraction 8/12 (U2).

Product of 5/12 Nógrádkövesd is actually not satisfactory because its too high parts did not pass the intermediate controlling sieve. The most drastic solution was applied here by extracting 18,9% from fraction 5/8 of the product adding the

Table 1. Results of rock laboratory tests

Mine rock products tests	Uzsa basalt 5/12	Uzsa basalt 12/20	Nógrádkövesd andesite 5/12	Nógrádkövesd andesite 12/20	Gánt dolomite 5/12	Gánt dolomite 12/20	Nagyharsány limestone 5/12	Nagyharsány limestone 12/20	order of goodness U	order of goodness NK	order of goodness G	order of goodness NHS
platiny	29,1	23,7	18,4	16,7	8,0	4,6	19,3	15,4	4	3	1	2
Los A.	13,3	13,3	18,1	17,6	18,8	20,0	24,1	27,1	1	2	3	4
complex ind.	55	55	57	58	57	57	66	63	1	2 3	2 3	4
$MgSO_4$	4,5	8,4	15,5	14,2	4,7	8,6	4,2	9,6	2 1	4 3	3 2	1 4
Na_2SO_4	3,2	4,5	8,5	9,4	2,2	6,6	2,1	8,1	3 1	4 2	2 4	1 3
d.deval	-	1,3	-	1,7	.	2,0	-	1,8	1	3	4	2
w.deval	3,0	5,2	4,6	7,1	2,8	4,0	5,0	7,0	2 2	3 4	1 1	4 3
suval	4,8	5,2	6,2	7,2	6,7	7,4	7,8	9,3	1 1	2 2	3 3	4 4
splitting m5	0	3,0	3,0	8,0	0,7	2,7	2,3	5,3	1 2	4 4	2 1	3 3
dorry-abrasion	-	6,2	-	4,4	-	8,5	-	9,8	2	1	3	4
polishing	0,51	-	0,59	-	0,45	-	0,41	-	3	4	2	1
Δ	0,16	-	0,16	-	0,24	-	0,00	-	2 3	2 3	4	1
adhesion value	100	n.v	100	n.v	50	n.v	60	n.v	1 2	1 2	4	3
adhesion inclinat.	0,35 n.v		0,55 n.v		0,40 n.v		1,25 n.v		1	3	2	4
contaminat inclinat	0,89 n.v		0,90 n.v		0,88 n.v		0,80 n.v		3 4	3 4	2	1
cleaning inclinat.	0,20 n.v		0,18 n.v		0,26 n.b		0,23 n.v		2	1	4	3

same value to fraction 8/12. As matter of fact the volume of parts not passing the sieve became so high that it reaches the extreme value experienced (N3). The opposite of this is N4 sample.

In case of product 5/12 of Gánt we tried to demonstrate the effect of 10% modification of the excess residue that is not passing the intermediate controlling sieve. In case of Nagyharsány we modified the product untill the standard limit. The modified mixture gradings can be seen on Figure 1. Products 5/12 components of test specimens used in asphalt laboratories were delivered in grading as mentioned but washed through on a 0.1 mm sieve in order to ensure the aggregate without contamination.

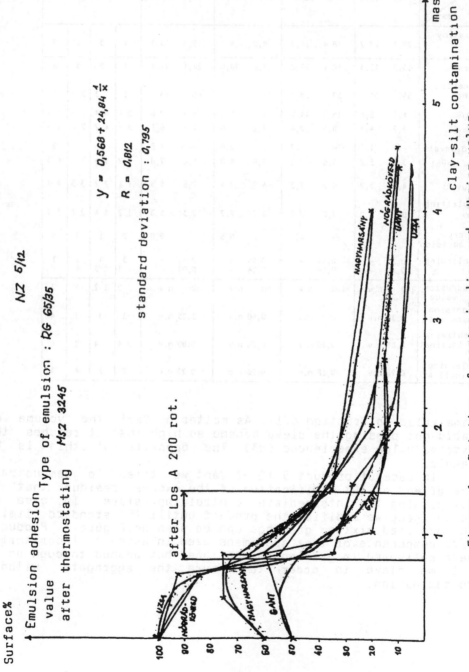

Figure 1. Relation of contamination and adhesion value

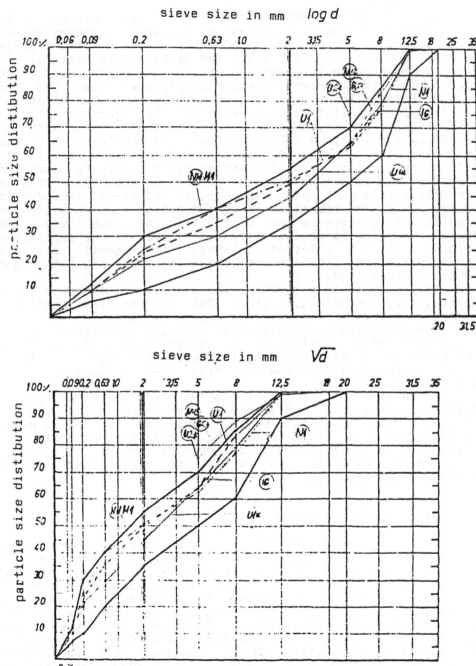

sieve size in mm   *log d*

sieve size in mm $\sqrt{d}$

Figure 2. The grading of mineral frame of mixture variatiens

2.3.Modelling the clay-silt contamination

The clay-silt contamination can be a very serious factor in worthening the quality in case of hidraulic mixtures as well as bituminous mixtures. The possibility that the present series of tests include many rock types and works on the basis of the same composition formula enabled us to determine wether the rock types with different adhesion properties react the same way or differently to receiving or keeping the contamination. The "artifical pollution" was performed by thorough stirring with wet clay and drying the chipping on 180°C. Afterwards the mixture was put into a 200 revolutions Los Angeles drum without balls (simulating a drying drum). As for the test results the self-cleaning property of the contaminated chipping does not depend on the rock type though the mikroroughness, adhesion properties of the chipping surfaces are different. The adhesion values follow the reduction of contamination and do not show significant divergence as a function of rock types. The deviations below 1% contamination is however rather significant (Fig.2.). It is interesting that the small clay-silt contamination of carbonated rock goes together with rather good adhesion properties.

We expressed in percentage the adhesion values experienced at clay-silt contamination after 200 Los Angeles revolution compared to that of the washed chipping (surface%).

The contamination inclination is meant the degree of the actually observed contamination in percentage of artifical pollution (5 mass%). According to the results this ability is almost independent from the rock type and it does not show significant diversion.

At inclination for cleaning means here the percentage the degree of clay-silt contamination remaining after 200 Los Angeles revolution to the maximum received by the type of chipping. We can state from the contamination tests that products 5/12 constituting a considerable part of the aggregate when leaving the drier can be at the most 0.8-1.3 m% of clay-silt contamination.

3 ASPHALT LABORATORY TESTS

For the asphalt laboratory tests we tried to compile the AB-12 asphalt mixture following the Hungarian practice. The bitumen was B-90. When making the mixture our guiding principle was to have the more-or-less same proportion of filler and aggregate and to make the natural and crushed sand proportion be approx 1:1 (Table 2.). According to the essential asphalt laboratory tests significant difference presented itself between the eruptive and carbonated mineral frame in respect of the copaction inclination (so the residual void contant) to such degree that in order to ensure the residual void required we had to apply fine sand instead of the one from

Fehérvárcsurgó which was too rough. We defined the traditional asphalt mechanical properties of the asphalt mixtures (see Table 3.).

Table 2. Composition property of the basic mixtures

composition \ Mark		U1 Uzsa basalt	N1 Nógrádkö- vesd	G 5,7 Gánt	NHM1 Nagyharsá- ny
limestone	m%	6	6	4	3
Fehérvárcsugó	m%	24	24	-	-
Veres sand	m%	-	-	26	26
0/5	m%	28	34	-	33
0/2	m%	-	-	14	-
2/5	m%	-	-	10	-
5/12	m%	42	36	46	38
filler	m%	10,0	10,2	9,6	10,4
crushed stone	m%	54,9	54,6	49,9	49,5
crushed/natural sand		1/1	1/1	22/17	22/18
SZB-90	m%	5,2	5,6	5,7	5,4
residual void cont.		3,7	3,1	3,0	3,0
Grading:					
0,09	m%	10,0	10,2	9,6	10,4
0,2	m%	22,0	22,4	24,2	25,5
0,63	m%	29,2	29,1	36,0	40,3
2,0	m%	45,1	45,4	50,1	51,5
5,0	m%	70,4	65,5	63,6	62,8
8,0	m%	84,0	77,2	84,3	77,0
12,5	m%	99,4	99,0	99,0	100,0
20,0	m%	100,0	100,0	100,0	-
density:		2,792	2,720	2,793	2,704

Properties of the used bitumen:

penetration on 25°C	72
softening point	47
density g/cm^3	1,0215

Table 3. Asphalt mechanical test results

Type of ashalt mixture (mineral frame)		Characters	B%	tb%	Static split +5°C S_aM (g/cm³)	σ_{hH} (N/mm²)	Static splits -10°C S_aM (g/cm³)	σ_{Hh2} (N/mm²)	E_{Hh} (N/mm²)	Dynamic split +20°C n_H	ε_H %o	tg10⁻³	Dynamic split +5°C n_H	ε_H %o $\cdot$ tg10⁻³		Dynamic elastic mod $\cdot10^3$ (N/mm²)
Uzsa basalt	U1	basic mixture	5,2	31,4	2,49	3,27	2,49	4,10	-	2641	3,25	0,221	15595	1,08	0,046	24,75
	U1x	grading fluct	5,2	78,0	2,48	4,03	n.v	n.v	n.v	5781	4,14	0,345	13354	0,75	0,020	38,16
	U2x	grading fluct	5,2	79,8	2,47	3,58	n.v	n.v	n.v	3251	2,83	0,532	14876	1,03	-0,002x	47,87
	Usz	contaminated	5,2	83,9	2,49	3,7	2,48	4,49	-	3890	2,65	-0,074	15053	1,28	-0,004x	28,33
Nógrádköved andesite	N1	basic mixture	5,6	81,4	2,40	3,62	2,40	4,88	36452	1471	2,50	-1,272	13474	1,60	0,094	30,84
	N3	grading fluct	5,6	80,0	2,40	3,62	n.v	n.v	n.v	1972	4,14	0,480	7177	0,98	-0,051x	35,80
	N4	grading fluct	5,6	77,9	2,40	3,62	n.v	n.v	n.v	3081	4,07	0,011	9235	0,98	-0,010x	34,43
	Nsz	contaminated	5,6	79,7	2,41	3,83	2,41	4,72	(73000)	2787	2,61	-0,437	8782	1,99	0,048	27,00
Gánt dolomite	G5,3	basic mixture	5,3	77,5	2,48	3,18	-	-	-	1693	2,66	-0,424	5596	1,52	-0,066x	14,90
	G5,7	basic mixture	5,7	82,6	2,48	3,01	2,49	3,80	24048	646	3,60	1,287	3627	1,84	0,280	29,74
	IG5,7	grading fluct	5,7	87,7	2,49	3,00	2,50	3,93	34825	1359	5,01	1,760	7221	3,71	0,320	17,98
	szG5,7	contaminated	5,7	90,5	2,51	3,05	2,49	4,09	(20600)	3619	6,55	0,893	5356	1,01	0,220	26,03
Nagyharsány limestone	NH1	basic mixture	5,3	77,4	2,41	2,78	-	-	-	1408	2,10	-0,267	13012	2,19	-0,075x	20,25
	NHM1	basic mixture	5,4	81,5	2,42	3,06	2,42	3,63	27885	2334	3,38	0,072	8986	1,22	0,098	30,90
	NH3	grading fluct	5,4	84,6	2,43	3,19	2,41	3,70	26906	2420	2,75	-0,016	5529	1,45	0,112	27,91
	NHsz	contaminated	5,4	80,1	2,43	3,00	2,43	4,17	(120100)	1824	3,18	-0,130	8602	1,42	0,028	20,03

3.1 Asphalt mechanical tests

From among the asphalt mechanical tests -considering the laboratory conditions- we decided in favour of static split and dynamic-fatigue splitting tests on +20 and +5 C adopted by the Road Construction Department of the Budapest Technical University. Details of the test results are summed up in Table 3.

In the dynamic tests we applied the method of reading the value displayed in regular intervalls (300, 600 repetition number)it widened the method with the analyses of section that can be considered linear as follows. It considers the inflection point marking the beginning of the deterioration the standard mean value of the linear sections and calculates the initial specific deformation and furder it determines the rise (steepness) of the linear section also (direction tangent).

4 THE EFFECT OF THE CHANGES IN PROPERTIES OF CRUSHED STONE ON THE PERFORMANCE OF ASPHALT MIXTURE

Summarising the results of asphalt laboratory tests of AB-12 type asphalt mixtures made of products of several rock types and of several places of origin and the results of asphalt mechanical fatigue tests we searched the effects of the fluctuation in grading, contamination and rock type knowing the effect of product properties (contamination, grading).

4.1 The effect of fluctuation in grading

The effect of fluctuation in grading of a crushed stone product cannot be detected from traditional asphalt laboratory and static asphalt mechanic features (see Table 3.). The following can be stated from the results of dynamic fatigue tests:
- The fatigue strength decreased because of the modification in the aggregate up to 2mm.
- The fluctuation of grading of product applied is a significant effect on the fatigue strength only if it (the fluctuation) changes the continuity of the aggregate of the asphalt mixture especially if it increases the volume of fine particles.
The results according to the approaching method of the third degree and those of the linear approaching method of valuation were examined.

According to the results of both processing methods significant diversion was detectable in the effect of fluctuation in grading. The fatigue strength of the basic mixture of dolomite from Gánt is worse than those of the samples contaminated or altered in grain size distribution (grading). The reson is possibly be connected with the present asphalt

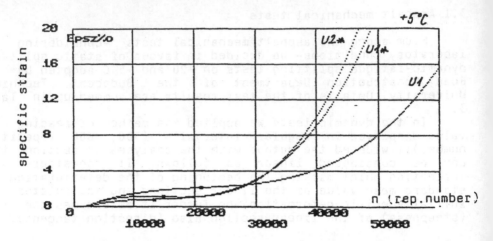

Figure 3. Impact of grading fluctuation

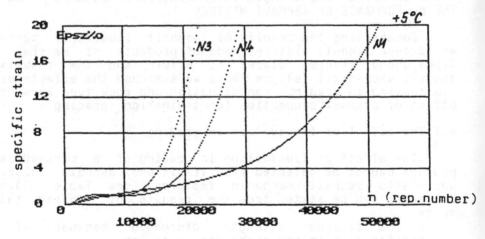

Figure 4. Impact of grading fluctuation

mixture design method. The basic mixture that is satisfactory a
to the traditional asphalt mechanical properties is however   no
satisfactory according to the fatigue strength tests making sur
that the present Marshall-method applied to asphalt mixtur
design made by the use of carbonated rocks is   someho
inadequate.

4.2 The effect of clay-silt pollution

The effect of clay-silt pollution could not be   reveale
from traditional asphalt laboratory and static asphalt mechan
properties (Table 3.). According to the dynamic - fatigue tests

- the effect of contamination according to dinamic-fatigue test results carried out on +5°C more unfavourable with the exception of the product of Gánt
- the results of dinamic split tests carried out on several tests temperatures were in some cases contradictory.

According to the linear approaching processing method at the tests carried out on 5°C unfavourable effects can be revealed in each case with the exception of clay-silt contamination of the Gánt dolomite (Fig.5-6.). It can be generally stated that the effect of clay-silt contamination proved to be the same order as the changes in fatige strength caused by the fluctuation in grading but it was no means up to our expectation. The effect of clay-silt cotamination and the wearing course exposed to weather and traffic can only be compared only a certain conditions. Thus the test series were completed with a series of test specimens where the effects of 5/12 Uzsabánya basalt fraction contaminated was considered after 28 days of soakig.

The having tests carried aut and the new dynamic split tests after soaking did not prove the destructive effects of remaining clay-silt contamination so destructive effects of clay-silt pollution cannot possibly be tracked back on the adhesion between chipping grains and the asphalt mortar but rather on the destructive concentration of contamination in exhauster dust.

4.3 The effects of rock types

The effects of rock types have definetly showed up even at the traditional asphalt laboratory tests making difficulties in the composition design formula. Significant differences showed up between the performance of aggregate consisting of eruptive and carbonated crushed stone products even at the basic test wich presented itself mainly in the significant divergence of composition ability and the Marshall residual void content. There is no definite relation between the divergence of the residual void content and the dynamic fatigue test results. Further asphalt laboratory research works have to clear whether the mixture design parameters and methods of conventional and standard asphalt parameters -like bitumen content, bitumen concentration in the residual void- in case of close relation between the results of the dinamic fatigue tests can be maintained or other more characteristic properties have to be specified.

The different fatigue performances of carbonated rocks can be revealed on the basis of dynamic fatigue test results (Table 3, Fig.7-8) this, however, cannot be definitely attributed to the divergence of rock types just because of the different performances revealed at asphalt laboratory tests. So in other approch the dinamic fatigue features of asphalt mixture made of carbonated rocks are behind those revealed at eruptive rocks in

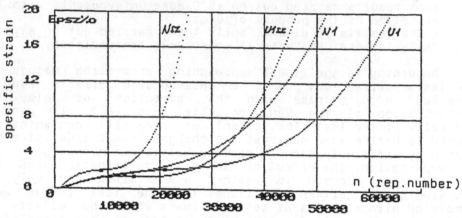

Figure 5. Impact of contamination

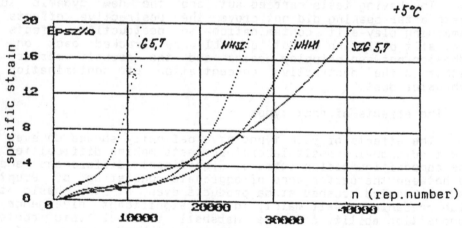

Figure 6. Impact of contamination

case the mixture is designed according to exactly to the same properties as those of the present design methods of asphalt mixtures (grading, saturation, residual void content). The fact that these mixtures showed in cases favourable dynamic-fatigue properties effected by the changing of grading (even that of the contamination) that can approach or even reach those of the eruptive rocks.

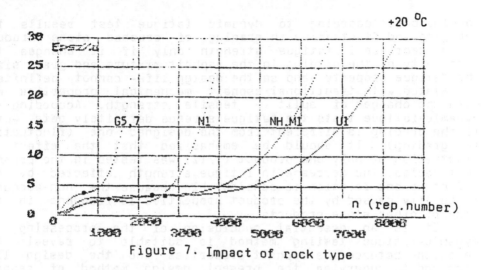

Figure 7. Impact of rock type

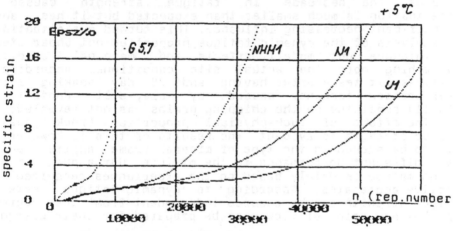

U   basalt from Uzsa
N   andesite from Nógrádkövesd
G   dolomite from Gánt
NHM limestone from Nagyharsány

Figure 8. Impact of rock type

5 SUMMARY

The mineral structure essentially effects the performances of bitumen mixtures. The asphalt laboratory test results did not show significant divergencies in the case of the traditional or the static asphalt mechanical tests that is suitable to draw

conclusions. According to dynamic fatigue test results th
changing and fluctuation in grading of crushed stone produc
couses decrease in fatigue strength only if it changes th
continuity of the grading in the asphalt mixture and grain size
The fatigue property -and so the design life- cannot definitel
be related with traditional asphalt mechanical properties no
with the changes of split - tensile strength. According t
dynamic-fatigue tests the fatigue strengs definitely gets wors
if the grading is different from the designed one (fluctuatio
in grading). It should be emphasized that the effect o
fluctuation of only one product (5/12) was tested in the presen
test series. The decrease in fatigue strength effected by th
defect of one fraction is considerable because the non-adequac
in quality caused by the product properties is frequent in th
case of other stone products.

It is an essential conclusion of the processing o
dynamic-fatigue testing method is suitable to reveal th
relations between the asphalt properties and the design lif
time and to supervise the present design method of asphal
mixtures. The decrease in fatigue strength caused b
contamination is much smaller than excpected but it has, anywa
a significant decreasing influance. This turned our attention t
the analysis of the dynamic fatigue properties not under steril
circumstances but modell effects and marginal condition
approaching more the actual site conditions. According t
dynamic-split tests after having and 28 day soaking furthe
deterioration tendency caused by clay-silt contaminatic
actually remaining on the chipping grains was not revealed.

The effect of rock-physical properties (rock-types) c
fatigue strength is evident. The results of this is however nc
only to be seeked in the type of mineral frame applied but i
their different performance in the mixture. Our asphalt mixtur
design method is mainly based on the experiences recquired wit
eruptive aggregates. According to dynamic-fatigue tests th
fatigue strength can be ensured independent from the aggregat
type but new point of view must be prepared for their design.

6 References

1. WEISE, H.: Grundlagen für die Beurteilung der Ermüdur
   bituminöser Bindemittel - Gesteins-Gemische aus Material
   technischer Sicht
   Die Strasse, 17.k. 3.sz. 1977. p.: 105-113 á:5, t:1, b:27
2. UDRON - BONNAVRE - GRAVOIS: Új eljárás az aszfaltkeveréke
   kifáradási élettartamának előzetes meghatározásározására
   Revue Générale des Routes et des Aerodromes 56.k. 583.sz. 1982
   p.: 55-64 á:7, t:11, b:14

3. BONNOT, I.: Relations entre caracteristiques des granulats propietés des mélanges et comportement des chaussées
Revue Générale des Routes et des Aerodromes 59.k. 621.sz. 1985. jul/aug. p.: 14-19 á:7 5249. KZ 85-12

4. WEIS, V.: Lebensdauer und Gebrauschwest von bituminösen Fahrbahndecken - Erfahsungen und Vergleiche an Hand von Versuchsstrecken
Gestrate Journal, 1983. dec. p.: 6-11 á:4

9. GRAGGER: A kőzetfajta hatása a bitumenes rétegek tartósságára és ellenállóképességére
Strassen und Tiefbau 1980. 1.sz. p.: 6-14 á:8, t:3

5.. DR. NEMESDY E. - PALLÓS I: - DR. TÖRÖK K.: Aszfaltmechanikai vizsgálatok kifejlesztése
Kutatási jelentés a BME Útépítési Tanszék II. rész Bpest., 1985. En. 229014/81. p.: 229

6. PALLÓS I.: Gyengébb minőségű ásványi anyagok felhasználásának lehetőségei és korlátai az aszfaltanyagok gyártásához
V. Budapesti Útügyi Konferencia Budapest 1982. okt. 26-28. 2. kötet p.: 171-177

7. DR. GÁLOS M.: A kőzetfizikai vizsgálatok szerepe az útépítésben felhasznált anyagok minősítésénél
V. Budapesti Utügyi Konferencia Budapest 1982. okt. 26-28. 3. kötet p.: 106-112

8. SUBERT I.: Forgalmi igénybevétel hatására bekövetkező zúzottkő aprozódás becslése, szemcsehasító vizsgálattal
Mélyépítéstudományi Szemle XXXVI.évf. 1986. 3.sz. p.: 124-132

10. Útépítésben felhasznált zúzottkő minőségének hatása a burkolat élettartamára
KTI Döntéselőkészítő tanulmány Témaszám: 4002.22241.4 Témafelelős: Subert István

PART FOUR
ADDITIONAL PAPER

43 DEVELOPMENT OF A METHOD OF STATIC COMPACTION FOR THE PREPARATION OF BITUMINOUS CONCRETE SAMPLES FOR DYNAMIC CREEP TESTING

A. MARCHIONNA and G. ROSSI
Department of Hydraulics, Transportation and Roads,
Rome University, Italy

Introduction

The main mechanical properties of a bituminous mixture
which characterize it for the purpose of use in road
pavement, are:
- workability;
- compactability;
- resistance to permanent deformation;
- resistance to fatigue cracking;
- resistance to cracking at low temperatures;
- durability (resistance to ageing, stripping and erosion).
The principal difficulties facing a formulation study
consist in the identification of a satisfactory compromise
between the contrasting requirements which are needed to
satisfy the properties listed above; for example,
resistance to permanent deformation on the one hand and
resistance to cracking on the other.
The formulation study normally entails one or more
forms of mechanical testing being carried out on the
different mixture compositions; these tests are later
backed up by theoretical calculations.
The most recent trend, which is also employed in Italy,
involves more than one test; thus, one can examine more
than one property of the mixture and so obtain an optimal
formulation for the mixture's particular use (i.e., road
base, wearing course, bituminous thin layers).
The formulation involves two steps:
firstly, establishment of a volumetric composition for the
mixture on the basis of theoretical studies;
secondly, examination of the mechanical properties of the
mixture using conventional tests, such as the Marshall
Test, and those which are aimed at determining a particular
mechanical characteristic, such as resistance to indirect
stress and resistance to permanent deformation.
For assessing the latter property, static or dynamic creep
tests are carried out. These can be carried out using The
equipment needed for these tests is not excessively
sophisticated.

The present study lies follows the trend of the general practice and, in particular, has as its purpose the identification of a method of preparation for dynamic creep test samples which dispenses with the need for special equipment and is at the same time reliable.

A previous study of the formulation of bituminous mixtures for use in semi-rigid pavements, in particular in the layer underneath the wearing course resting on the layers linked with cement, used the creep dynamic test to examine the resistance to permanent deformation of such mixtures.

Taking as an initial point of reference the problems which manifested themselves in this case, a method of static compaction for the preparation of the representative test samples with aggregates of different natures (crushed and rounded) based on the A.S.T.N. norms, different from that initially adopted, was finalized.
The present paper presents the first results of a comparative study carried out to assess the reliability of the two methods.

Preparation of test samples via static compaction in a single layer and subsequent extrusion

In order to carry out creep tests with the equipment available, it is first necessary to prepare test samples with a height of 140 mm and a tollerance of +/- 2 mm (given the dimensional constraints of the testing equipment).
The procedure which was initially adopted drew on the methodology described in A.S.T.M. norm D3496 which foresees the static compaction of the bituminous mixture arranged in a single layer within a cylindrical punch with a diameter of 100 mm.
The punch is closed by ends which, under the action of the load, can move inside the cylinder until they reach the sample height required.
In order to be able to predetermine the density of the test samples, the procedure envisaged by the A.S.T.M. norm is modified.
The mass of the mixture to be compacted and the maximum static load are determined by producing some pilot test samples.
In this way, it is possible to obtain packing levels in the order of 96-98% of the densities obtainable on samples compacted according to the Marshall procedure (h=63.5% mm, d=102 mm, 75 blows per face).
The operative procedure envisages the possibility of applying a maximum compaction load of between 15 and 18 kN and is carried out as follows:
1 - filling the punch with the mixture of inert materials

and bitumen;
2 - a compaction cycle which reaches a load of 5 kN in 1
minute and immediate release of the load;
3 - reaching the maximum load in one minute followed by
immediate release of the load;
4 - reaching the maximum load in one minute, maintaining
such load for 5 minutes and then releasing;
5 - cooling to room temperature and extrusion of the test
sample.
It was noticed, however, that when using mixtures of
aggregates which were not particularly tough, a certain
percentage of the grains in contact with load heads and the
lateral surfaces of the die fractured during the compaction
phase.

Preparation of test samples via static compaction in three layers and subsequent core boring

In order to contain this effect which, obviously, can give
rise to considerable complications which are difficult to
evaluate in the chain "test sample preparation - assessment
of the dynamic creep test results", an alternative method
of compaction was developed which foresees:
1 - the use of punches larger than the test sample (d=152,
h=238 mm),
2 - the compaction of the mixture placed in three layers;
in order to avoid the formation of a discontinuous surface
in the test sample during the passage from one layer to
another, after the application of the static load, the
interfacing surface between the three layers should be re-
arranged,
3 - the interposition of disks of deformable materials (for
example wood shavings) between the rigid steel surfaces
(bottom and top loading heads of the punch) and the
bituminous mixture so as to avoid the breaking of the inert
materials in contact with the load head,
4 - the maximum load applied to each layer and the overall
mass of the mixture are to be calibrated each time, in
relation to the compositional characteristics of the
mixture and in function of the compaction level which it is
intended to obtain; for this purpose it is necessary to
proceed with a pilot experiment on each type of mixture it
is intended to study,
5 - core boring of the test samples of the desired diameter
and reduction to the desired height.
This procedure allows for the load to be dosed so as to
avoid the fracturing of the aggregate's grains.
As an example, we will describe the pilot experiment
carried out on a mixture prepared from aggregates of a
calcareous nature to determine the maximum static load to

be applied to the various layers so as to obtain the desired grade of packing whilst avoiding grain fracturing.

The packing obtained was controlled layer by layer, the portions being divided into 3 cylindrical elements with a thickness equal to a third of the overall height. Below are shown the values of the grades of packing obtained with the loads initially adopted on each layer (which led, however, to grain breaking) and those obtained by applying the maximum load compatible with the conservation of the grains' integrity to each layer:

Test Sample no.1

Layer	1	2	3
Load (kN)	20	25	30
G.A. (%)	97	97	100

Test Sample no.2

Layer	1	2	3
Load (kN)	10	15	20
G.A. (%)	94	96	99

If one wishes to avoid fracturing of the grains, it is necessary to accept lower compaction levels which, in this case, are around 96%.
Observation of such data leads to the conclusion that, a greater unformity in the distribution of the density along the height of the test samples can be obtained by reducing the loads applied to the second and third layers.

Dynamic creep test

The dynamic creep test subjects the test samples to tri-axial stress, the vertical stress varies with time according to a sinusoidal law. This test (more than the static creep test) subjects the materials to closer stresses than those induced by traffic in the superstructures.

These tests were carried out using F.D.E.S. equipment (Esso dynamic creep) at the Road Materials Laboratory of the Department of Hydraulics, Transportation and Roads of the "La Sapienza" Rome University.
This equipment is a simplified version of the electrical hydraulic machinery and allows for the application of vertical sinusoidal stress controlled at a frequency of 10 Hz.
This stress, the extent of which can be varied within certain limits, is exercised by a jack and rotating mass. The vertical thrust exercised on the samples is indicated in Fig. 1. This thrust varies between a minimum value of σH and a maximum value of σmax as seen in the following expression:

$$\sigma max = \sigma H + \sigma S + \sigma O * sen \, wt \qquad (1)$$

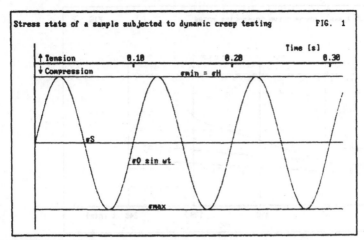

Stress state of a sample subjected to dynamic creep testing   FIG. 1

σH represents the isotropic thrust;
σS is the vertical thrust of the jack;
σ0 is the extent of the sinusoidal thrust exercised on the
rotating mass.
The isotropic thrust and the temperature inside the
triaxial cell are fixed at the start of the test.
Their values can be varied between 0 and 0.4 MPa and
between 20 and 60 °C.
The typical trend of the permanent deformation variation
curves in function of the number of load cycles is shown in
Fig. 2.
These curves, which are called creep curves, have three
distinct parts:
- the first represents the initial deformation of the
samples due to the σS stress;
- the second is characterized by a deformation which varies
linearly in time due to the effect of the dynamic load;
- the third is characterized by a rapid increase in the
permanent deformation and finishes with the breaking of the
sample.
        This final part is linked to the particular test
conditions and does not reflect the material's behaviour
under operational conditions.
The characteristic parameter which the test is used to
obtain is the deformation rate (TD) which is supplied by
the following expression:

$$TD = d\epsilon \ / \ dN \qquad\qquad (2)$$

where ϵ represents the permanent unit vertical deformation
of the test sample; and
N the number of load cycles corresponding to the
deformation E.

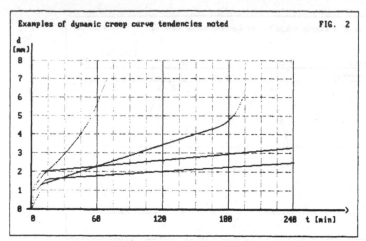

Examples of dynamic creep curve tendencies noted   FIG. 2

The value of TD represents the slope of the second part of
the creep curve. This value represents the permanent
deformation under a load cycle.
The values of TD depend on the characteristics of the
mixtures' formulation and the test conditions (temperature
and thrust).
Under equal test conditions, the least deformed mixtures
are characterized by low TD values.
It has been noted that under equal test conditions, the
value of TD varies in relation to the compaction method
utilized for the preparation of the test samples.

Characteristics of the mixtures studied

With the aim of assessing the extent to which the two
methods of compaction described above ensure homogeneity
and uniformity of the test samples, a plan for testing was
drawn up which envisaged the preparation of test samples
composed of mixtures of differing compositions.
The plan for the mixtures consisted of two phases:
the first phase involved comparing the methods of
preparation of the test samples;
the second phase assessed the repeatability of the test
method as a whole (preparation of test samples - dynamic
creep testing).
In order to control the two methods of preparation, four
batches each of 10 samples were analyzed.
The mixture composition of each batch differed from that of
the others.
      In particular, two batches were prepared using the
single-layer static compaction procedure; the remaining two
batches with triple-layer compaction.

For one mixture only, a further twenty test samples were
prepared which allowed for assessment of the repeatability
of the test method for the deformation rate level alone.

Test samples prepared with single-layer static compaction

The first mixture studied (Mixture n° 1) is characterized
thus:
- granulometric assortment 0/25 mm
- soft rounded aggregates of alluvial origin (56%)
- aggregates obtaining from crushing leucitic rocks (36%)
- 7% cement as welding filler
- 4.5% bitumen B 80/100.
Figures 3 and 4 show the granulometric assortment of the
mixtures being studied and the assortment of the aggregates
utilized respectively.

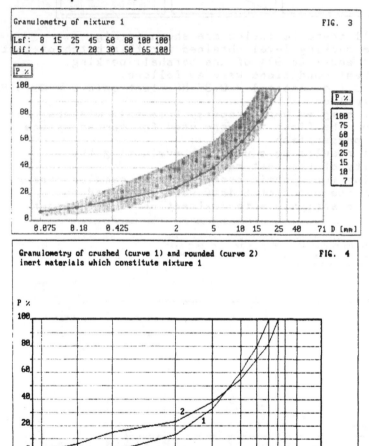

Granulometry of mixture 1 — FIG. 3

Granulometry of crushed (curve 1) and rounded (curve 2)
inert materials which constitute mixture 1 — FIG. 4

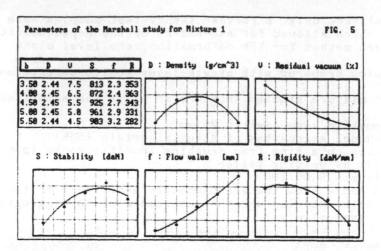

Parameters of the Marshall study for Mixture 1   FIG. 5

b	D	U	S	f	R
3.50	2.44	7.5	813	2.3	353
4.00	2.45	6.5	872	2.4	363
4.50	2.45	5.5	925	2.7	343
5.00	2.45	5.0	961	2.9	331
5.50	2.44	4.5	903	3.2	282

D : Density [g/cm^3]

U : Residual vacuum [%]

S : Stability [daN]

f : Flow value [mm]

R : Rigidity [daN/mm]

The Marshall characteristics are shown in Fig. 5.
The average packing level obtained for this batch of test
samples was equal to 97% of the Marshall packing.
The creep test conditions were as follows:
T = 25 deg.C   σ vertical = 0.35 MP   isotropic σ = 0.10 MPa
σ 0 = vertical σ
Thirty test samples were prepared from this mixture, the
results from the first ten were used for comparing the
methods of preparation.
The overall results were used for evaluating the
repeatability of the test method.
After testing for dynamic creep, 10 test samples chosen at
random from the batch of thirty were subjected to
granulometric analysis after extraction of the binder.

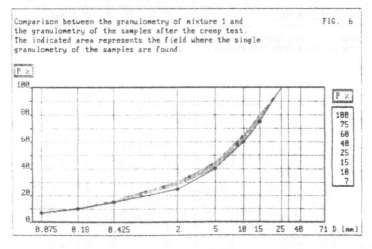

Comparison between the granulometry of mixture 1 and   FIG. 6
the granulometry of the samples after the creep test.
The indicated area represents the field where the single
granulometry of the samples are found.

Examination of Fig. 6, which compares the initial
granulometry of the mixtures studied and the smelting of
the 10 granulometries from the samples,reveals considerable
differences in the granular fraction 0.425/25 mm.

A second experimental phase entailed the preparation of
a batch of 10 test samples with inert materials
characterized by a greater toughness.

The mixture studied (Mixture n° 2) has the following
characteristics:
- granulometric assortment 0/15 mm
- aggregates originating totally from the crushing of
leucitic rocks
- 7% cement as welding filler
- 6% of bitumen 180/200.

Figs. 7 and 8 show the granulometric assortment and the
Marshall experiment parameters respectively.

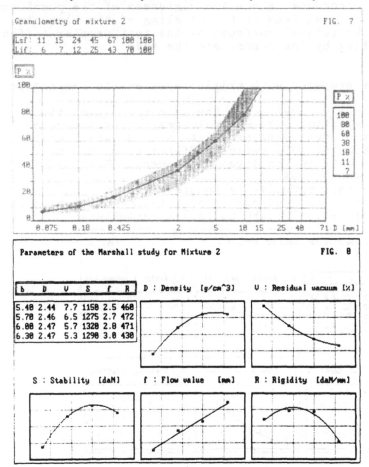

This batch of test samples, for which an average packing level of 96.5% was obtained, was tested in the following conditions:

T = 30 °C   vertical σ = 0.40 MP   isotropic σ = 0.10 MPa

The analyses carried out after creep testing on 5 samples chosen at random did not reveal any modifications to the granulometry as a result of the compaction procedure.

From this point of view, the results obtained with the samples, prepared with Mixtures 1 and 2 indicate that the single-layer compaction procedure in the preparation of bituminous mixtures presents:
- a high level of toughness;
- good form characteristics;
- maximum aggregate dimension <=1/3 of the punch diameter.

On the negative side, when the inert materials are less tough, this procedure, given the rigidness of the punch and the load heads, can lead to the breaking of the elements, both along the lateral surfaces of the test samples (effect of the circling by the press) and the ends.

Test samples prepared with triple-layer static compaction

For the purpose of evaluating the extent to which this method of compaction reduces grain fracturing phenomena, a third series of test samples (Mixture n° 3) was prepared. The composition of this series differed from that of Mixture n° 1 only in the level of binders, which in this case was fixed at 5.0%.

The experimental operative procedure envisaged the preparation of a reduced number of pilot samples which led to the determination of the following parameters:

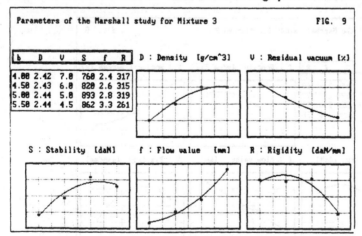

Parameters of the Marshall study for Mixture 3   FIG. 9

b	D	V	S	f	R
4.00	2.42	7.8	760	2.4	317
4.50	2.43	6.8	820	2.6	315
5.00	2.44	5.8	893	2.8	319
5.50	2.44	4.5	862	3.3	261

D : Density [g/cm^3]   V : Residual vacuum [%]

S : Stability [daN]   f : Flow value [mm]   R : Rigidity [daN/mm]

- mass of the bituminous mixture,
- maximum static compaction load in relation to the layer to be compacted,
so as to obtain a level of packing in the order of 98% of the packing evaluated according to the Marshall procedure (average ascertained packing level = 97.5%).
 Fig. 9 shows the tendencies for the characteristic parameters of the Marshall study.
It could be seen that the grain breakage was negligible. This mixture was subjected to dynamic creep testing with the following stress conditions:
T = 30 °C   vertical σ = 0.35 MP   isotropic σ = 0.10 MPa
The fourth mixture prepared (Mixture n° 4) had the following characteristics:
- granulometric assortment 0/25 mm
- rounded aggregate of alluvial origin
- 7% cement as welding filler
- 5% bitumen 80/100.
Fig. 10 shows the trends of the sizes noted during the course of the Marshall test.
This mixture was subjected to testing under the following conditions:
T = 30 °C   vertical σ = 0.25 MP   isotropic σ = 0.00 MPa.
During the course of compaction, even though a level of packing equal to 98.5% of the Marshall packing was obtained, no appreciable grain breaking took place, as can be seen from the macroscopic observation of the test samples at the conclusion of the core boring operations. This is confirmed by the granulometric analyses carried out after the extraction of the binder on a group of 5 test samples chosen at random after having carried out the creep test (see Fig. 11).

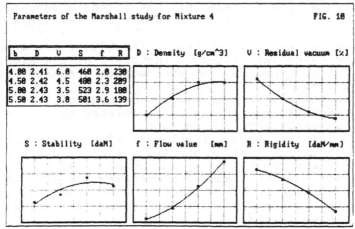

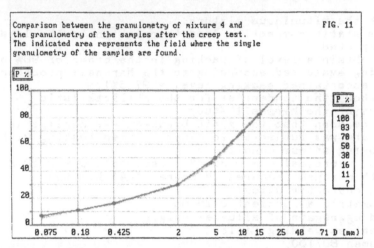

Comparison between the granulometry of mixture 4 and
the granulometry of the samples after the creep test.
The indicated area represents the field where the single
granulometry of the samples are found.

FIG. 11

It can be observed that in this case the modification of
the granulometry after the core boring operations, is very
limited if compared with that found for Mixture n° 1 and
shown in Fig. 6.

Statistical interpretation of the results

In order to evaluate which of the two compaction procedures
is more reliable, the experimental results of the batches
of 10 test samples were compared.
The Shapiro-Wilk test was used to verify the normality of
the experimental distributions beforehand; these appeared
normal within the 5% reliability level chosen.
Table 1 shows the deformation rate, volumic mass and % of
residual vacuums, the average [M], the standard deviation
[S.D.], the variation coefficient [C.V.] and the Range
[Xmax-Xmin].
Table 2 shows the reliability levels for 95% of the average
and rejection types calculated on the basis of the averages
and the rejection types of the test samples prepared with
the four mixtures.
An opinion on the validity of the two methods of
preparation of test samples can be formulated by comparing
the size dispersions shown in Table 1.
     Since such sizes have differing levels, reference is
made to the variation coefficient since such parameter is
independent of the "level" of the measured sizes.
As concerns the volumic masses, it emerged that both the
procedures allow comparable values to be obtained, but that
the experimental dispersions are less common for those
prepared in three layers.

Table 1

```
-------------------------------------------------------------
Mixture 1                   M      S.D      C.V.    Xmax-Xmin
T.D.%/Mcycles           :  8.24   3.381    41.031     11.8
M.V.A. (g/cm^3)         :  2.40   0.015     0.625      0.02
Residual vacuum  (%):      7.44   0.258     3.468      0.81
-------------------------------------------------------------
Mixture 2                   M      S.D      C.V.    Xmax-Xmin
T.D.%/Mcycles           :  3.26   0.986    30.254     2.54
M.V.A. (g/cm^3)         :  2.38   0.013     0.546      0.04
Residual vacuum  (%):      8.65   0.500     5.780      1.58
-------------------------------------------------------------
Mixture 3                   M      S.D      C.V.    Xmax-Xmin
T.D.%/Mcycles           : 31.30   5.468    17.470    16.30
M.V.A. (g/cm^3)         :  2.40   0.011     0.458      0.03
Residual vacuum  (%):      6.83   0.180     2.635      0.37
-------------------------------------------------------------
Mixture 4                   M      S.D      C.V.    Xmax-Xmin
T.D.%/Mcycles           : 72.47  19.719    27.210    66.33
M.V.A. (g/cm^3)         :  2.39   0.005     0.209      0.01
Residual vacuum  (%):      5.13   0.140     2.729      0.29
-------------------------------------------------------------
```

A similar consideration holds for the residual vacuums, the
most moderate dispersions are obtained with this method of
compaction.

Table 2

```
-------------------------------------------------------------
Mixture 1                      Range M              Range σ
T.D.%/Mcycles           :  5.822 - 10.658      2.326 - 6.173
M.V.A. (g/cm^3)         :  2.389 -  2.411      0.010 - 0.027
Residual vacuum  (%) :     7.255 -  7.625      0.177 - 0.471
-------------------------------------------------------------
Mixture 2                      Range M              Range σ
T.D.%/Mcycles           :  2.555 -  3.965      0.678 - 1.800
M.V.A. (g/cm^3)         :  2.371 -  2.389      0.009 - 0.024
Residual vacuum  (%) :     8.292 -  9.008      0.344 - 0.913
-------------------------------------------------------------
Mixture 3                      Range M              Range σ
T.D.%/Mcycles           : 27.389 - 35.211      3.761 - 9.983
M.V.A. (g/cm^3)         :  2.392 -  2.408      0.008 - 0.020
Residual vacuum  (%) :     6.701 -  6.959      0.124 - 0.329
-------------------------------------------------------------
Mixture 4                      Range M              Range σ
T.D.% / Mcycles         : 58.365 - 86.575     13.563 -36.002
M.V.A. (g/cm^3)         :  2.386 -  2.394      0.003 - 0.009
Residual vacuum  (%) :     5.030 -  5.230      0.096 - 0.256
-------------------------------------------------------------
```

Examination of the dispersions of the deformation
rates allows for a global evaluation of the validity of the
tests in relation to the operative chain "test sample
preparation - dynamic creep testing".
Such a comparison is also favourable to the three-layer
method of preparation; in fact, the data for the batches of
test samples prepared in a single layer are more dispersed,
with C.V.% values equal respectively 41% and 30% against
the values of 17% and 27% for the batches prepared in three
layers.
When the comparison between the methods of preparation of
tests samples is finished, the evaluation of the
repeatability of the test method as a whole begins.
This phase envisages the preparation of two batches each of
30 test samples for each of the four mixtures used in the
previous phase.
For the same mixture, one batch of test samples is prepared
with the single-layer and one batch with three-layer
compaction method.
This phase of the study is still in progress; when
concluded, it will be possible to have an overall picture
of the repeatability of the two testing methods as a whole
for various layers of size and deformation rate.
Table 3 shows the data for the first batch of 30 test
samples with a composition similar to Mixture n° 1 and
prepared according to the single-layer procedure.

Table 3

	M	S.D	C.V.	Range
T.D.% / Mcycles :	7.82	3.118	39.869	12.32
M.V.A. (g/cm^3) :	2.40	0.012	0.500	0.02
Residual vacuum (%):	7.62	0.232	3.045	0.88

After having verified the normality of the distribution of
the three sizes measured, the reliability intervals are
calculated at 95% of the average and rejection types which
are shown in Table 4.

Table 4

	Range M	Range σ
T.D. % / Mcycles :	6.655 - 8.985	2.476 - 4.207
M.V.A. (g/cm^3) :	2.396 - 2.404	0.010 - 0.016
Residual vacuum (%):	7.533 - 7.707	0.184 - 0.313

Since the analyzed sample is sufficiently large, it can be
considered that the sample rejection type provides a
sufficiently reliable guide to the rejection type of the

whole; this hypothesis makes it possible to determine the repeatability, r, of the test which is given by the following expression:

$$r = S.D. * 1.96 * \sqrt{2} = 8.64 \qquad (3)$$

(maximum admissible rejection between two results obtained in similar test conditions).

CONCLUSIONS

Resistance to permanent deformation is one of the properties of bituminous mixtures which should be kept under control, since dynamic creep testing is very suitable for its assessment.
The reliability of these tests is nonetheless conditioned by the representability of the test samples.
Amongst the possible methods of test sample preparation, attention has been focused on those based on static compaction, since it requires relatively simple equipment.
The study we have carried out allows for the finalization of a preparation method which offers an alternative to that which envisages static compaction in a single layer in a punch of dimensions equal to the sample's final dimensions. The method which has been finalized foresees the static compaction of the mixtures in a punch with dimensions greater than those of the actual test sample. The compaction is carried out in three layers chosing the appropriate load to be applied to each layer.
The test sample is then reduced to the desired dimensions by core boring.
The advantage of this method of preparation over that previously used is that it avoids the shattering of the inert materials near the sample faces, especially when the inert materials used are not particularly resistant.
    Statistical analysis of the data for the tests carried out shows that using such a method allows for test samples with a lesser dispersion in terms of volumic mass and residual vacuums with respect to those prepared in a single layer to be obtained.
A global assessment of the reliability of the tests, i.e., of the entire range of test sample preparation and creep test procedures can be obtained by analysing the dispersion of the fluage rate values.
This analysis, even though conducted on a reduced number of samples, shows that even from this point of view the method of preparation proposed appears satisfactory.
In order to investigate this aspect in greater detail, a plan of testing, which is still under way, was drawn up for determining the repeatability of the test method.

We would like to thank the laboratory technicians Mr. F.
Amoroso and Mr. A. Di Curzio for their precious and
fruitful collaboration.

Bibliography
F.Giannini, L.Domenichini, A.Marchionna, G.Rossi
   Etude de l'influence du pourcentage de granulat concassè
sur
   la déformation permanente des mélanges bitumineux au
moyen des
   essais de fluage dynamique.
   Eurobitume Symposium, Cannes Oct. 1981
B.Célard, ESSO Road Design Technology, 4th Inter. Conf.
   Structural design of Asphalt Pavements
   Proceddings, Ann Arbor, 1977
Jacques Bonnot, Essais mécaniques pratiques de formulation
et de
   controle des enrobés bitumineux
   RILEM Belgrade - Sept. 1983
P. Ferrari - F. Giannini, Ingegneria stradale vol.II
   ISEDI
Fidélité d'une méthode d'essai, **AFNOR NF X 06-041 1970**

INDEX

This index is based on the keywords assigned to their papers by the individual authors. The numbers are the page numbers of the first page of the paper in which the relevant term is used.

Milton Keynes UK
Ingram Content Group UK Ltd.
UKHW021934071024
449327UK00022B/1799

9 780367 863463